Treatise on Heavy-Ion Science

Volume 5

High-Energy Atomic Physics

Treatise on Heavy-Ion Science

EDITED BY D. ALLAN BROMLEY

Volume 1: Elastic and Quasi-Elastic Phenomena

Volume 2: Fusion and Quasi-Fusion Phenomena

Volume 3: Compound System Phenomena

Volume 4: Extreme Nuclear States

Volume 5: High-Energy Atomic Physics

Volume 6: Astrophysics, Chemistry, and Condensed Matter

Volume 7: Instrumentation and Techniques

Treatise on Heavy-Ion Science

Volume 5
High-Energy Atomic Physics

EDITED BY

D. ALLAN BROMLEY
Henry Ford II Professor of Physics
Yale University
New Haven, Connecticut

PLENUM PRESS • NEW YORK AND LONDON

Library of Congress Cataloging in Publication Data

Main entry under title:

Treatise on heavy ion science.

Includes bibliographies and indexes.
Contents: v. 1. Elastic and quasi-elastic phenomena— —v. 5. High-energy atomic physics.
1. Heavy ions—Collected works. I. Bromley, D. Allan (David Allan), 1926- .
QC702.7.H42T74 1984 539.7 84-8384
ISBN 0-306-41575-5 (v. 5)

A Division of Plenum Publishing Corporation
233 Spring Street, New York, N.Y. 10013

Printed in the United States of America

For Pat, Lynn, and David

Preface to Treatise on Heavy-Ion Science

After a long gestation period, heavy-ion physics and chemistry is now, worldwide, the most rapidly growing area of nuclear science, and the concepts, techniques, and instrumentation of this heavy-ion work are finding ever-widening application in other areas of science and technology. Although there remain broad regions at higher energies, with heavier projectiles, and at higher excitations and angular momenta where heavy ions still provide gateways into the totally unknown, intensive studies over the past two decades have provided a sound framework of understanding of many of the salient features of interactions induced by these new heavier projectiles and a basis for coherent planning of future studies.

It thus seemed appropriate, at this point in the history of the field, to pull together in one place and in as coherent a fashion as possible, an overview of what has been accomplished and some enlightened speculation about where we go next. It is my hope that these volumes will provide a definitive reference for those working in this and neighboring fields—both students and professional scientists; beyond that I would also hope that they will make accessible to a much wider audience in other sciences and technologies some of the richness of heavy-ion research, and perhaps help to stimulate the transfer of techniques and concepts that I have already mentioned.

I have been singularly fortunate in planning these volumes in being able to persuade internationally recognized authorities to write in their areas of special expertise and interest, and also fortunate that I have not had to restrict them to any artificial or externally imposed procrustean restrictions on the scope or length of their chapters. I have asked each author to include sufficient background to make the chapter accessible to students and to the nonspecialist, to provide a broad selection of illustrations, and to feel free to extrapolate and to speculate about future directions.

In inviting contributions to these volumes I have made arbitrary decisions concerning both topics and contributors, and I make no claim to completeness. Indeed, a few chapters that I would have liked to include do not appear because of illness or other reasons which prevented their authors from completing them.

I should like to take this opportunity to thank all the authors represented for taking time in already full schedules to share with a wider audience their special experience and expertise in heavy-ion science. As was inevitable from the outset in a multiple-author venture of this scope—with over 65 different authors— my original scheduling and deadlines proved unrealistic. To all those authors who responded to them and produced manuscripts on or before the original deadline—in many cases, I am aware, at substantial personal cost—my most sincere thanks and appreciation. To them, too, however, go my apologies for the delay in bringing their work into print. I have delayed publication for over a year so that I might include a number of very important chapters without which the work would have been obviously incomplete.

Volumes 1–4 of the *Treatise on Heavy-Ion Science* are devoted to aspects of heavy-ion nuclear science, beginning with an overview of the historical development of the science and some of its simpler interactions—elastic and quasi-elastic, fusion and quasi-fusion phenomena —and moving from them to compound system phenomena and to much more complex and less well understood phenomena involving very heavy nuclear species and very high energies. Volume 5 is devoted to high-energy atomic physics, an entirely new field of science first made accessible by the availability of a broad range of heavy-ion beams, and one still very much in its infancy. Volume 6 considers the impact of heavy-ion studies on other sciences including astrophysics, chemistry, surface physics, condensed matter physics, materials science, and heavy-ion-induced fusion power. Volume 7, the concluding volume of this treatise, is devoted to some of the instrumentation peculiar to heavy-ion science and its applications.

Special thanks go to Ellis Rosenberg and Bennett K. Ragnauth of Plenum Press with whom it has been a pleasure to work on these volumes, and to Mary Anne Schulz for all her help in producing them. And I would also acknowledge my indebtedness to the Alexander von Humboldt Stiftung for a Humboldt Award that I was privileged to hold during part of the time these volumes were in preparation.

New Haven, Connecticut D. Allan Bromley

Contributors to Volume 5

Jack S. Greenberg, A. W. Wright Nuclear Structure Laboratory, Yale University, New Haven, Connecticut 06511

Walter Greiner, Institut für Theoretische Physik, Johann Wolfgang Goethe Universität, Robert-Mayer-Strasse 8–10, 6000 Frankfurt am Main, West Germany

Indrek Martinson, Department of Physics, University of Lund, S-223 62 Lund, Sweden

Joachim Reinhardt, Institut für Theoretische Physik, Johann Wolfgang Goethe Universität, Robert-Mayer-Strasse 8–10, 6000 Frankfurt am Main, West Germany

Paul Vincent, Department of Physics, Brookhaven National Laboratory, Upton, New York 11973

Preface to Volume 5

It is rare, in modern science, when an entirely new field opens up; high-energy atomic physics is such a field. With the availability of precisely controlled beams of heavy species, spanning the periodic table, it has become possible to study the atomic system under entirely new conditions. Vacancy production in heavy-ion collisions is orders of magnitude greater than anticipated; study of one- and two-electron systems with atomic numbers ranging all the way to that of uranium and above becomes possible, testing quantum electrodynamics and relativistic atomic phenomena in new ways; and in very heavy systems the Coulomb field becomes supercritical ($Z > 170$) leading to entirely new physics and chemistry. In this volume, Reinhardt and Greiner set the stage with a comprehensive treatment of the theory of the atomic system emphasizing the high field aspects. Greenberg and Vincent have provided an equally comprehensive overview of the experimental situation focussing on collision systems, while Martinson reviews the current situation in beam-foil spectroscopy where passage of rapidly moving ions through thin target foils produces atomic excitation states that are otherwise extremely difficult, if not impossible, to produce and study.

New Haven, Connecticut D. Allan Bromley

Contents

1. Heavy-Ion Atomic Physics—Theoretical

Joachim Reinhardt and Walter Greiner

2. High-Energy Atomic Physics—Experimental

Jack S. Greenberg and Paul Vincent

3. Beam-Foil Spectroscopy

Indrek Martinson

Contents of Volumes 1–4, 6, 7

Volume 3: Compound System Phenomena

Volume 4: Extreme Nuclear States

Volume 6: Astrophysics, Chemistry, and Condensed Matter

Volume 7: Instrumentation and Techniques

1

JOACHIM REINHARDT (top) holds a postdoctoral appointment in the Institute for Theoretical Physics at the University of Frankfurt where he received both his undergraduate and graduate education and the Ph. D. degree in 1980. During his graduate program he spent periods as a Visiting Fellow at the Lawrence Berkeley Laboratory, at Vanderbilt University, and at the Wright Nuclear Structure Laboratory at Yale.

WALTER GREINER (bottom) is Professor and Director of the Institute for Theoretical Physics at the University of Frankfurt. He received his undergraduate education at this university and his graduate training at the University of Freiburg, graduating with the Ph. D. in 1961. During the 1962–1964 period he held Assistant Professorships at the University of Maryland and Florida State University followed by a brief research associateship at the University of Freiburg before moving to the Theoretical Physics Chair at Frankfurt in 1964. In 1974 he was awarded the Max Born Prize and Medal jointly by the Institute of Physics in London and the German Physical Society. In 1982 he received the Otto-Hahn-Prize of Germany and an honorary degree from the University of Witwatersrand in Johannesburg, South Africa.

Heavy-Ion Atomic Physics—Theoretical

JOACHIM REINHARDT AND WALTER GREINER

1. General Introduction

The physics of ion–atom collisions embodies a large number of phenomena and spans a long historical development. The interesting range of energies goes from the meV to GeV regions, the incident particles may be protons as well as uranium ions. Here we will concentrate on the description of *inner shell processes* in collisions of heavy ions. Progress in this field has been largely stimulated by the advance of accelerator technology. This has been especially true in recent years through the advent of heavy-ion accelerators, particularly the GSI facility at Darmstadt, which are capable of accelerating even the heaviest stable atoms to energies at the nuclear Coulomb barrier. As a result, ion–atom collisions can be studied where the atomic electrons (or, more precisely, the electron–positron field) are subject to quite extreme conditions in terms of field strength and violence of the collison. The interesting new physics ecountered in close collisons of very heavy ions will stand in the center of this article.

From a theoretical viewpoint two basic motivations for studying heavy-ion atomic collisions can be discerned:

i. As an application of complex scattering theory the mechanisms of electron excitation may be studied in detail for a wide range of conditions and parameters.

ii. New phenomena may become observable which transcend the

JOACHIM REINHARDT AND WALTER GREINER • Institut für Theoretische Physik, Johann Wolfgang Goethe Universität, Robert-Mayer-Strasse 8–10, 6000 Frankfurt am Main, West Germany.

limits of traditional atomic physics. In particular, the quantum mechanics of relativistic binding and quantum electrodynamics (QED) of strong fields may be tested.

While the second aspect may be of greater fundamental interest, a thorough understanding of the reaction mechanism is prerequisite if conclusions are to be drawn. Let us start, however, with a brief summary of the new phenomena expected to occur in very heavy atoms (cf. Chapter 2). Much of the related theory has been developed in the years since 1968 by the school in Frankfurt and the group in Moscow.

The binding of electrons in the Coulomb field of a nucleus is explained by the theory of QED. The description is based on the solution of the single-particle Dirac equation. Effects of field fluctuations are very weak and can be calculated with extreme precision in perturbation theory, owing to the smallness of the fine-structure constant $\alpha \simeq 1/137$. The interaction with the charge of the nucleus, however, is characterized by the coupling strength $Z\alpha$, which can exceed unity in superheavy (quasi-) atoms. To study the effects of strong binding let us look at the electronic energy terms as a function of nuclear charge Z, displayed in Figure 2.1. With growing Z the levels rapidly gain in binding energy. Under the (unphysical) assumption of a pointlike central nucleus the eigenenergies $E_n(Z)$ for $s_{1/2}$ and $p_{1/2}$ states even obtain an infinite slope. The solutions break down at the value $Z \simeq 137$ which is associated with a collapse of the wave functions.

Introducing a finite nuclear radius the bound states can be followed smoothly into the region $Z\alpha > 1$. The wave functions, however, now become highly contracted and sensitive to the nuclear charge. The binding energies increase steadily until the value of $2mc^2 = 1022$ keV is reached. The required "critical" charge is, for the 1s state, $Z_{cr} \simeq 173$. Energy shifts due to radiative corrections (vacuum polarization and self energy) are not expected to change this value significantly (Section 2.4).

The physical significance of this border and the consequences of a further increase of Z are best vizualized in Dirac's hole picture which accounts for the many-body character of QED. It can be translated into a proper quantum field theoretical description afterwards. To ensure the stability of matter, all states of the negative energy continuum, $E < -mc^2$, of the Dirac equation are assumed to be occupied with (unobservable) electrons; holes in this sea are identified with positions. If Z exceeds Z_{cr} the bound state continues to gain binding energy and enters the continuum as an additional state. At this point the system becomes unstable: The supercritical level will be filled by two (due to spin degeneracy) electrons while two free positrons will be emitted to infinity. This process may be expected since at $Z = Z_{cr}$ the threshold for pair creation vanishes, provided that the bound state is not occupied already.

The effect of spontaneous positron production is of fundamental interest since it marks a change of the QED ground state upon variation of the parameter Z. In the presence of a supercritical nucleus the usual particle-free vacuum state is no longer stable—it decays into a charged vacuum. This new ground state is stabilized against further decay by the Pauli principle. It is impossible to create an empty space free of real particles in the vicinity of a supercritial nucleus. Even if one succeeds in removing all the electrons, the K shell will be refilled after about 10^{-19} s by spontaneous pair creation.

In principle, such an effect should not be confined to highly charged atoms. Pairs can be created in all sufficiently extended potential wells with a depth greater than $2mc^2$. The process is related to the famous Klein paradox and can be interpreted as the tunneling of a particle through the gap of the Dirac equation. Therefore spontaneous pair creation is exponentially suppressed for potential steps wider than the electronic Compton wavelength. It cannot be observed at macroscopic potential barriers.

The only chance to observe both the effects of relativistic binding and the process of spontaneous positron production therefore rests in heavy-ion collisions where a supercritical charge can be assembled at least for a short interval of time. To test the implications of strong fields the reaction mechanism has to be understood. Although the underlying theory, i.e., quantum mechanical scattering theory based on the electromagnetic interaction, in principle is well understood, actual calculations may become very arduous. Even the most simple prototype case of proton-hydrogen collisions still attracts much attention. In heavier systems additional complications are introduced by the presence of many electrons, relativistic effects, etc.

Depending on the process to be described and on the parameters of the collision several approximation methods have been developed. In view of the intended application to heavy systems, the discussion here will concentrate on the semiclassical form of the theory. The colliding nuclei are assumed to follow well-defined classical trajectories while the electrons are described by quantum mechanics. The dynamics is determined by the time-dependent Schrödinger (or Dirac) equation. As described in Section 3.2, the usual method of solution relies on an expansion of the wave function in some complete basis. This results in a set of coupled differential equations in time which are solved approximately in perturbation theory by numerical methods. Special care has to be taken that the boundary conditions at $t \to \pm\infty$ are met.

Depending on the impact energy the description can be based on different pictures. If the ion velocity is sufficiently high or the charge ratio of projectile and target nuclei is small, an atomic description is justified. The projectile may be considered to cause a weak perturbation of the target electron wave function. In the opposite limit, i.e., for slow, near symmetric collisions the electrons will be able to adjust adiabatically to the combined charge of both nuclei. The time dependence will then be adequately described in terms of molecular orbital (MO) or "quasimolecular" basis states. If the condition for adiabaticity is well satisfied, transition rates in general will be low. Excitations can be caused, however, at those localized regions of internuclear distance where two molecular levels become (nearly) degenerate in energy. Since such processes have clear experimental signatures their observation has inspired the development of the quasimolecular model. During the last decade several of these specific molecular couplings have been intensively investigated (Section 4.3.3).

Prerequisite for any such calculations is the solution of the stationary molecular problem at all values of the internuclear distance. For nonrelativistic systems (with respect to the binding energy) the solutions have been obtained long ago by the pioneers of quantum chemistry. The corresponding relativistic problem, i.e., the two-center Dirac equation, has been solved only recently. The exact inclusion of relativistic effects would be of no great relevance for ordinary molecular physics. Since energetic beams of very heavy ions have become available, however, it is possible to simulate molecules with charge $Z = Z_1 + Z_2 > 1/\alpha$ where relativistic binding becomes essential.

The innermost bound electrons are moving fast enough to adjust to the positions of the nuclei, which in close collisions have a relative velocity $v \sim 0.1c$. The idea of superheavy quasimolecules was developed independently by the two groups in Frankfurt and Moscow mentioned above. In the limit $R \to 0$ "quasiatoms" with charges up to $Z \sim 190$ can be formed transiently in heavy-ion collisions.

Heavy-ion collisions, therefore, should be a unique—though not ideal—tool to test the effects of strong binding described above. In Section 4.4 the peculiar characteristics of inner shell excitation in superheavy collisions are discussed. Quite in contrast to the situation in light-ion collisions the drastic shrinkage of the wave functions with decreasing internuclear distance now, e.g., leads to large rates for vacancy creation and to the emission of δ electrons with high kinetic energy. These predictions have been fully confirmed by recent experiments, supporting the validity of the underlying physical picture. Attempts to perform a spectroscopy of superheavy quasimolecules, i.e., to deduce the molecular binding energies from observed excitation rates, are discussed in Section 4.4.3.

An independent channel of observation is given by the emission of noncharacteristic x-ray radiation (Section 5). Molecular orbital x-rays (MOX) which are emitted during the course of the collision carry information on the transition energies and strengths of the intermediate quasimolecular system. The theoretical framework for the calculation of radiation spectra and angular distributions is given in Section 5.

The question of greatest fundamental interest is discussed separately in Section 6. In close collisions of supercritical systems, $Z_1 + Z_2 > Z_{cr}$, the adiabatic $1s\sigma$ level is predicted to join the lower continuum. Holes which are present in this level—lacking the availability of fully stripped ions they have to be created dynamically during the same collision—can be emitted without expenditure of energy as positrons. The process of spontaneous positron creation, however, is quite slow compared to the collision time. Furthermore, pair creation also can be induced by the Fourier spectrum of the nuclear motion, a process which itself is strongly enhanced in high-Z systems. In a given collision there is no physical way to distinguish between the different contributions. A fully dynamical theory accounting for the various excitation and emission processes has been worked out. The predicted very steep increase of positron production rates with nuclear charge $Z_1 + Z_2$ as well as the variation with kinematic parameters and the shapes of positron spectra meanwhile have been established in several experiments. No definite conclusion about the "diving process" could be drawn up to now.

A fruitful area of research has opened in the border region between atomic and nuclear physics. At bombarding energies above the Coulomb barrier atomic excitations can be affected by a nuclear reaction. Various interference effects have been proposed which might lead to the construction of an "atomic clock" for nuclear reaction times. A success of this program clearly would have strong bearing on the understanding of, e.g., deeply inelastic nuclear reactions. On the other hand, the presence of sufficiently long reaction times in superheavy collison systems would offer a unique way to study the properties of superheavy quasiatoms, e.g., by the observation of structures in the MOX spectrum. In supercritical systems, moreover, the process of spontaneous positron creation could be strongly enhanced. This route may ultimately lead to the confirmation of the concept of the charged vacuum. Experiments presently under investigation seem to show the predicted effects and might indicate the formation of rather long-lived superheavy nuclear systems.

2. Quantum Electrodynamics of Strong Fields

2.1. Introduction

Despite the long tradition and high sophistication of present-day atomic physics, one fundamental topic had not received much attention for a long time and was fully understood only in recent years: The effects of strong electronic binding due to the Coulomb field of a highly charged nucleus with effective coupling constant $Z\alpha > 1$. In the absence of stable or long-lived superheavy nuclei this problem might seem academic if it were not for the possibility of high-Z heavy-ion collisions.

In such experiments for a short time up to $Z \simeq 190$ charge units can be assembled in a localized region of space. Thus a highly charged nucleus is simulated, or rather an intermediate superheavy quasimolecule. Theory predicts remarkable modifications of the atomic wave functions due to the relativistic motion of electrons in the vicinity of a highly charged nucleus. Furthermore, if the charge exceeds a critical value $Z_{\rm cr}$ a new fundamental effect is expected: an empty atomic K shell becomes unstable and decays by positron emission, resulting in the transition to a new, charged vacuum state.

Although it will turn out that the short time scale will interfere with attempts to draw conclusions on the properties of the united atom system (cf. Sections 4–6), the underlying questions have inspired much experimental and theoretical work on the atomic physics of heavy-ion collisions. Due to the principal interest the present review will give highest priority to the collisions of very heavy ions. We will devote this chapter to a short survey of the theory of electronic binding in high-Z atoms. More detailed expositions are found in (Re 77, Ra 78a, Br 78b).

2.2. The Dirac Equation with an External Potential

In relativistic quantum theory spin-1/2 particles are described by the Dirac equation*

$$[\gamma^\mu (p_\mu - eA_\mu) - m]\, \psi(x) = 0 \tag{2.1}$$

where γ^μ, $\mu = 1,\ldots, 4$, are the four Dirac matrices and ψ is a bispinor wave function. An electromagnetic four-component vector potential $A_\mu(x)$ has been introduced in (2.1) by minimal coupling. For stationary states in a static potential $eA_0(\mathbf{r}) = V(\mathbf{r})$, Equation (2.1) can be written, using $\gamma_0 = \beta$, $\gamma_i = \beta\alpha_i$,

$$H_D \psi_n(\mathbf{r}) \equiv [\boldsymbol{\alpha} \cdot \mathbf{p} + \beta m + V(\mathbf{r})]\, \psi_n(R) = E_n \psi_n(\mathbf{r}) \tag{2.2}$$

In the standard representation the 4×4 matrices $\boldsymbol{\alpha}, \beta$ are given by

$$\alpha = \begin{pmatrix} 0 & \sigma \\ \sigma & 0 \end{pmatrix}, \qquad \beta = \begin{pmatrix} I & 0 \\ 0 & -I \end{pmatrix} \tag{2.3}$$

In most of the following presentation we will assume that $V(r)$ is a classically prescribed external potential. Usually in atomic physics this is a very good approximation, due to the high mass ratio of nuclei and electrons.

* Throughout this chapter we mostly use natural units, $\hbar = c = 1$ and sometimes $m_e = 1$.

Assuming, for the present, spherical symmetry the angular and radial dependence in (2.2) can be separated. The spin orbit angular momentum operator

$$K = \beta(\boldsymbol{\sigma} \cdot \mathbf{L} + 1) \tag{2.4}$$

commutes with the Hamiltonian. Its eigenvalues are $\kappa = l$ for $j = l - 1/2$ and $\kappa = -(l+1)$ for $j = l + 1/2$, where l and j are the values of orbital and total angular momentum. Equation (2.2) is separated by the ansatz

$$\psi_{\kappa\mu} = \frac{1}{r}\begin{pmatrix} u_1(r)\chi_\kappa \\ iu_2(r)\chi_{-\kappa} \end{pmatrix} = \begin{pmatrix} g(r)\chi_\kappa \\ if(r)\chi_{-\kappa} \end{pmatrix} \tag{2.5}$$

with the spherical spinors

$$\chi_\kappa^\mu = \sum_{m=\pm 1/2} (l\,\tfrac{1}{2}\,j\,|\,\mu - m\; m\; \mu)\; Y_{1\;\mu-m}\;\chi_m$$

μ is the magnetic quantum number, i.e., the projection of j on the quantization axis.

The Dirac equation (2.2) then leads to a set of two radial coupled first-order differential equations

$$\begin{aligned} \frac{d}{dr}u_1 &= -\frac{\kappa}{r}u_1 + [E - V(r) + m]u_2 \\ \frac{d}{dr}u_2 &= -[E - V(r) - m]u_1 + \frac{\kappa}{r}u_2 \end{aligned} \tag{2.6}$$

In the idealized case of a singular Coulomb potential $V(r) = -Z\alpha/r$, corresponding to the assumption of a pointlike infinitely massive central charge, the Dirac equation can be solved analytically [for details cf. (Ak 65)]. The resulting energy spectrum is given by the well-known Sommerfeld formula

$$E_n = m\left[1 + \left(\frac{Z\alpha}{n - |\kappa| + [\kappa^2 - (Z\alpha)^2]^{1/2}}\right)^2\right]^{-1/2} \tag{2.7}$$

where $n = 1, 2, \ldots$ is the main quantum number.

For the deepest bound state Equation (2.7) gives $E_{1s} = m[1 - (Z\alpha)^2]^{1/2}$. It is obvious that the solution (2.7) breaks down at a charge $Z\alpha = |\kappa|$, i.e., $Z \simeq 137$ for all s and $p_{1/2}$ states. The energy levels as a function of $Z\alpha$ exhibit a root singularity with infinite slope at the critical point. The breakdown is best reflected in the behavior of the radial wave functions. Independently of the main quantum number the dependence at small r is given by

$$g(r) \propto r^{[\kappa^2 - (Z\alpha)^2]^{1/2} - 1} \tag{2.8}$$

Thus even in light atoms the relativistic (point nucleus) s and $p_{1/2}$ wave functions exhibit a mild, integrable divergence at the origin. For $Z\alpha > |\kappa|$, however, the wave functions start to oscillate like $\cos([(Z\alpha)^2 - \kappa^2]^{1/2} \ln r + \delta)$ and lose their physical meaning. This would correspond to a "collapse to the center." Mathematically speaking, the Dirac Hamiltonian loses its self-adjointness as $Z\alpha$ becomes larger than $|\kappa|$. Several authors have applied formal methods to construct self-adjoint extensions to the Hamiltonian (e.g., Ca 50, Bu 81).

In realistic physical situations, however, this problem does not arise: Since the electric field is generated by a source having finite extension the potential is practically cut off at the nuclear charge radius R. We can write

$$V(r) = \begin{cases} -Z\alpha/r & \text{for} \quad r > R \\ -f(r)\, Z\alpha/R & \text{for} \quad r < R \end{cases} \tag{2.9}$$

where simple models for the cutoff function are $f(r) = 1$ (corresponding to a charged spherical shell) or $f(r) = \frac{1}{2}[3 - (r/R)^2]$ (homogeneously charged sphere). To determine the bound states of eq. (2.2), solutions of the radial differential equations in the regions $r < R$ and $r > R$ have to be found, which are regular at $r \to 0$ or $r \to \infty$, respectively. The logarithmic derivatives of these solutions have to be matched on the sphere $r = R$. In this way a unique boundary condition is specified for the exterior wave function and there is no difficulty in constructing solutions for $Z\alpha > |\kappa|$.

Partly motivated by speculations on the existence of superheavy nuclei, several authors have studied deeply bound levels in square well (Sch 40) and extended nucleus Coulomb potentials (Po 45, We 58, Vo 61, Re 69). The first systematic calculation of eigenenergies of the Dirac equation in superheavy elements was performed by Pieper and Greiner (Pi 69). The resulting level spectrum as a function of Z is shown if Figure 2.1. Contrary to the case of a point nucleus (dashed lines) the energies decrease smoothly with growing nuclear charge. They can be traced down to the border of the lower continuum, $E = -m$. For the most strongly bound state E_{1s} becomes negative at $Z > 150$ (i.e., an atom with a filled K shell would lose some of its weight!) and reaches $E = -m$ at the critical charge* $Z_{\text{cr}} \simeq 173$. The same thing happens for the $2p_{1/2}$ state at $Z_{\text{cr}} \simeq 185$, while the higher states require much larger charges to reach $E = -m$.

The level diagram Figure 2.1 shows drastic deviations from the nonrelativistic values. Large energy splittings between states which are normally near degenerate are observed. For example, the "fine structure"

* The energies used for the determination of Z_{cr} and shown in Figure 2.1 include the effect of electronic screening, which has been calculated within the relativistic Hartree–Fock approach (So 74, Fr 77).

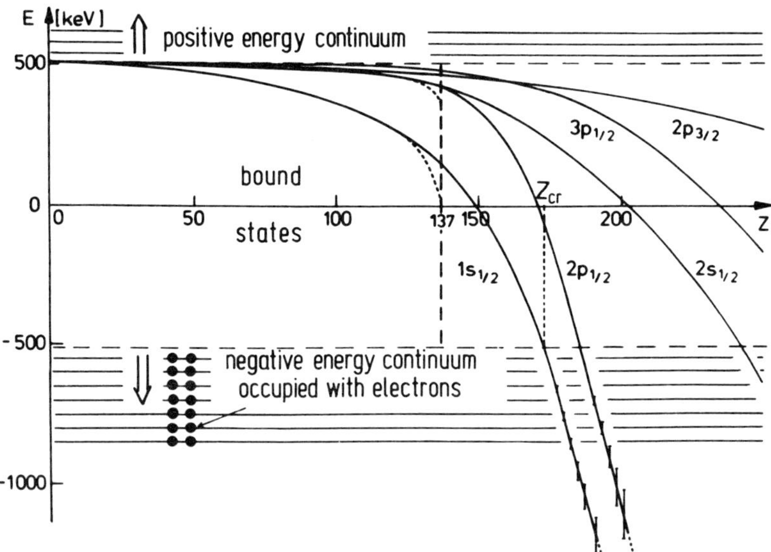

Figure 2.1. Energies of the most deeply bound atomic states as a function of the nuclear charge Z. Whereas the solutions for point nuclei break down at $Z \simeq 137$ (dashed lines), all bound states can be traced down to the lower continuum of the Dirac equation if the finite nuclear extension is taken into account. Beyond the critical charge Z_{cr} a change of the electron vacuum state is predicted.

splitting between the $2p_{1/2}$ and $2p_{3/2}$ states becomes as large as 800 keV. Also the degeneracy of states which differ only by the sign of κ like $2p_{1/2}$ and $2s_{1/2}$, valid for an exact $1/r$ Coulomb potential [cf. eq. (2.7)], is severely broken.

Of course, strong relativistic effects will also be visible in the wave functions. Figure 2.2 shows the radial density $r^2\,|\psi(r)|^2$ of the $1s$ electron in the field of an extended nucleus. The Schrödinger hydrogenic $1s$ wave function is given by $\psi_{1s}(r) = 2(Z/a_B)^{3/2}\,e^{-rZ/a_B}$, where $a_B = \lambda_e/\alpha = 52918$ fm is the Bohr radius. The axes in Figure 2.2 have been scaled appropriately in order to make the shape of the nonrelativistic density independent of the charge Z. Deviations therefore are solely due to relativistic effects. While the shape of $r^2\,|\psi(r)|^2$ does not alter much in the region of the ordinary Periodic Table, the wave function becomes severely contracted as $Z\alpha$ approaches or exceeds the value 1. Although a complete "collapse of the wave function" is averted, the scaled Bohr radius a_B/Z loses its importance if $Z\alpha > 1$ and the only relevant length scale is determined by the nuclear radius R. The inset in Figure 2.2 gives the electron density at the origin, $|\psi(0)|^2$, as a function of Z. Apart from the Z^3 dependence known from the nonrelativistic solution, the density increases by additional three orders of magnitude going to $Z = 180$.

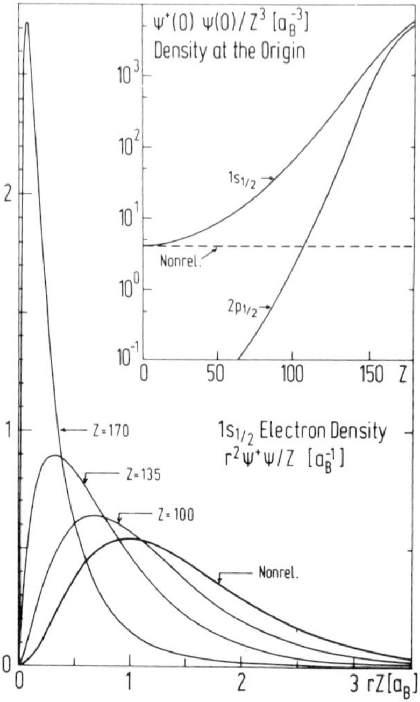

Figure 2.2. Radial electron density of the $1s_{1/2}$ state is superheavy atoms with nuclear charges $Z = 100$, 135, and 170. For comparison the nonrelativistic Schrödinger wave function is drawn. The axes are scaled with powers of Z in order to make the nonrelativistic density independent of the charge. The inset shows the drastic increase of the electron density at the origin over its nonrelativistic value. The wave functions of the $1s_{1/2}$ and $2p_{1/2}$ states become very similar in the region $Z\alpha > 1$.

The figure also displays the electron density at the origin for the $2p_{1/2}$ wave function. For a nonrelativistic p wave, characterized by the sharp angular momentum $l = 1$, this value is exactly zero. The $2p_{1/2}$ Dirac bispinor, however, carries a mixture of $l = 0$ (upper component) and $l = 1$ (lower component) orbital angular momentum, cf. Equation (2.5). In atoms with $Z\alpha > 1$ the "small" and "large" components of the wave function become comparable in magnitude. As a consequence the distinction between $s_{1/2}$ and $p_{1/2}$ states is diminished. This explains the steep rise of the $2p_{1/2}$ density as well as the strong increase of its binding energy as a function of Z in Figure 2.1.

Judging from Figure 2.2 it is quite obvious that superheavy atoms (or quasimolecules, for that matter) will have characteristic properties quite foreign to ordinary atomic physics. As discussed in the following chapters, heavy-ion collisions can serve to study some of these properties.

2.3. Supercritical Electron States

In addition to the shrinking of the relativistic wave functions in high-Z atoms a qualitative change of the spectrum is predicted if the nuclear charge exceeds the critical value Z_{cr}: A bound state vanishes from the gap and becomes degenerate with the negative energy continuum. If the Dirac equation (2.6) is solved for $Z > Z_{cr}$, a narrow resonance is found at an energy $E_{res} < -m$ which smoothly continues the position of the $1s$ state $E_{1s}(Z)$ as a function of Z.

An approximate treatment may be based on Fano's theory of autoionization (Fa 61). This problem is formally similar (Mü 72a, b) to the supercritical atom since it is connected with a bound state embedded in a continuum (although here the electron–electron interaction is responsible for the coupling). We are interested in the eigenstates of the Dirac Hamiltonian $H_D(Z)$ for a supercritical charge $Z > Z_{cr}$. $H_D(Z)$ can be split into two parts

$$H_D(Z) = H_D(Z_{cr}) + V'(Z, r) \simeq H_D(Z_{cr}) + Z'U(r) \tag{2.10}$$

where the extra potential $V'(Z, r) = V(Z, r) - V(Z_{cr}, r)$ has been assumed to be simply proportional to the extra charge $Z' = Z - Z_{cr}$, neglecting the weak Z dependence of the nuclear charge distribution. The supercritical Hamiltonian (2.10) is to be diagonalized with respect to the known solutions of $H(Z_{cr})$,

$$\begin{aligned} H(Z_{cr})\,|\phi\rangle &= E_0\,|\phi\rangle = -m\,|\phi\rangle \\ H(Z_{cr})\,|\psi_E\rangle &= E\,|\psi_E\rangle \qquad \text{for} \quad E < -m \end{aligned} \tag{2.11}$$

Here $|\phi\rangle$ is the $1s$ bound state eigenfunction at $Z = Z_{cr}$ and $|\psi_E\rangle$ is the s continuum. To understand the main effect we have restricted the basis to one discrete state plus one continuum, since higher states are widely separated in energy. The matrix elements of $H_D(Z_{cr} + Z')$ in this truncated basis are

$$\begin{aligned} \langle\phi|\, H_D(Z_{cr} + Z')\,|\phi\rangle &= E_0 + Z'\,\langle\phi|\,U\,|\phi\rangle \equiv E_0 + \Delta E \\ \langle\psi_E|\, H_D(Z_{cr} + Z')\,|\phi\rangle &= Z'\,\langle\psi_E|\,U\,|\phi\rangle \equiv V_E \\ \langle\psi_{E'}|\, H_D(Z_{cr} + Z')\,|\phi\rangle &= E\,\delta(E' - E) + Z'U_{E,E'} \end{aligned} \tag{2.12}$$

The continuum solutions of the supercritical Dirac Hamiltonian

$$H_D(Z_{cr} + Z')\,|\Psi_E\rangle = E\,|\Psi_E\rangle \tag{2.13}$$

can be expanded as

$$|\Psi_E\rangle = a(E)\,|\phi\rangle + \int_{E' < -m} b_{E'}(E)\,|\psi_{E'}\rangle\, dE' \tag{2.14}$$

with the normalization condition

$$\langle\Psi_{E'}|\,\Psi_E\rangle = \delta(E' - E) \tag{2.15}$$

Insertion of (2.14) into (2.13) leads to a coupled system of integral equations for the expansion coefficients. An explicit solution can be found, if the matrix element for continuum rearrangement, $U_{E',E'}$ is neglected. This will be justified if the value of Z' is small; otherwise the continuum $|\psi_E\rangle$ could be prediagonalized to eliminate $U_{E',E}$. Under this condition the solution of (2.13) (2.15) takes the form

$$a(E) = \frac{V_E^*}{E - E_0 - \Delta E - F(E) + i\pi\,|V_E|^2}$$
$$b_{E'}(E) = \delta(E - E') + \frac{a(E)\,V_{E'}}{E - E' - i\varepsilon} \tag{2.16}$$

where the energy shift $F(E)$ is given by the principal value integral

$$F(E) = P\int_{E'<-m} dE'\,\frac{|V_{E'}|^2}{E - E'} \tag{2.17}$$

The amplitude $a(E)$ describes the admixture of the bound state $|\phi\rangle$ to the supercritical continuum. The square $|a(E)|$ therefore gives the probability that a 1s electron bound in $|\phi\rangle$ is moved to the continuum at energy E as the interaction $Z'U(r)$ is switched on,

$$|a(E)|^2 = \frac{|V_E|^2}{[E - (E_0 + \Delta E) - F(E)]^2 + \pi^2\,|V_E|^4} \tag{2.18}$$

This expression obviously shows resonance behavior. If the variation of V_E with energy is neglected we have $F(E) \simeq 0$ and (2.18) takes the form of a symmetric Breit–Wigner resonance curve

$$|a(E)|^2 = \frac{1}{2\pi}\,\frac{\Gamma}{[E - (E_0 + \Delta E)]^2 + \Gamma^2/4} \tag{2.19}$$

centered around $E_{\text{res}} \simeq E_0 + \Delta E = \langle\phi|\,H_D(Z)\,|\phi\rangle \simeq -m - Z'\delta$ with a width

$$\Gamma = 2\pi\,|V_{E_{\text{res}}}|^2 \simeq Z'^2\gamma \tag{2.20}$$

Numerical calculations (Mü 72b) have given values of $\delta \simeq 29$ keV for the "diving velocity" and $\gamma \simeq 0.04$ keV for the reduced width of the supercricital 1s state. Note thet eq. (2.20) is not valid close to the threshold, $Z' \lesssim 3$, since V_E is strongly suppressed due to the repulsion of the wave function $|\psi_E\rangle$.

The resonance character of the supercritical state, summarized in Equations (2.18)–(2.20), is confirmed by an inspection of the exact solutions of the Dirac equation. As described in detail by B. Müller *et. al.* (Mü 73b), the continuum wave functions for a $1/r$ potential can be expressed analytically in terms of Whittaker functions. These have to be matched at the nuclear surface to the interior solutions which are regular at the origin. This procedure results in radial wave functions with asymptotic behavior (for $E < -m$)

$$u_1 \to [(E+m)/\pi p]^{1/2} \cos(pr + \Delta)$$
$$u_2 \to [(-E+m)/\pi p]^{1/2} \sin(pr + \Delta) \tag{2.21}$$

The relevant physical phase shift δ is obained from Δ by subtraction of a logarithmic Coulomb phase,

$$\delta = \Delta - \delta_{\log} = \Delta - y[\ln(2pr/|y|) + 1] - \pi/2 \tag{2.22}$$

where $y = Z\alpha E/p$ and $p = (E^2 - m^2)^{1/2}$. In supercritical atoms the phase shift δ is found to go through a resonance at the energy anticipated above. An example for the sharp variation of δ with energy is shown in Figure 6.3. The result of the phase-shift analysis can be linked to the approximate solution obtained in the autoionization model by the relation

$$\tan \delta = -\frac{\Gamma/2}{E - (E_0 + \Delta E) - F(E)} \tag{2.23}$$

which may be used to extract exact values for the position and width of the resonance.

Some further insight into the nature of the Dirac solutions in strong fields may be gained if the system (2.6) is reduced to a single second-order differential equation. If the small component u_2 is eliminated and u_1 is transformed according to $\chi(r) = u_1(r)\,|E - V + m|^{-1/2}$, the Dirac equation can be reduced to the form of a Schrödinger equation

$$\chi'' + p^2(r)\,\chi = 0 \tag{2.24}$$

with the local effective momentum

$$p^2(r) = (E - V)^2 - m^2 - \frac{\kappa(\kappa+1)}{r^2} + \frac{\kappa}{r}\frac{V'}{E - V + m} - \frac{3}{4}\left(\frac{V'}{E - V + m}\right)^2 - \frac{1}{2}\frac{V''}{E - V + m} \tag{2.25}$$

Equation (2.25) also applies to spin-0 particles (Klein–Gordon equation) if the terms involving derivatives of the potential are removed. From (2.24) an effective potential may be deduced which exhibits a barrier at intermediate values of r. Details on the structure of the Dirac solutions and on the effective momentum description have been discussed in (Ze 72).

The classically allowed region of (2.24) (oscillating solutions) is characterized by $p^2(r) > 0$. If the depth of the potential exceeds $2m$, this condition can be met in two different spatial regions, $r < r_-$ and $r > r_+$, for energies in the lower continuum, $E < -m$. The limits are essentially defined by the borders of the gap of the Dirac equation, i.e., $E - V(r_\pm) \pm m = 0$. This explains the emergence of resonances in supercritical potentials: A particle localized in the bound state region $(r > r_-)$ can escape to the continuum region $(r > r_+)$ by tunneling through the gap of the Dirac equation. Second quantization or Dirac's hole picture tell us that the "particle" must be a positron.

Let us look at the physical interpretation of the "diving process" described above: With increasing nuclear charge Z the atomic K shell gains binding energy while its spatial extension shrinks. As Z grows beyond the critical charge Z_{cr} the $1s$ bound state ceases to exist. According to the usual quantization rule all states at $E < -m$ must be assumed occupied. Consequently, since $\int_{-\infty}^{-m} dE\,|a(E)|^2 \simeq 1$, the charge density of the supercritical continuum $\int_{-\infty}^{-m} dE\, \Psi_E^+(r)\, \Psi_E(r)$ contains two additional units of charge compared with the subcritical case. The spatial distribution of this excess charge resembles just a normal atomic K shell. This phenomenon has been termed real vacuum polarization (Mü 72b, Ra 74) in contrast to the usual virtual vacuum polarization which essentially results from a displacement of the vacuum charge distribution generated by the states at $E > m$ and $E < -m$.

According to this discussion nothing remarkable will happen if the nuclear charge is slowly increased beyond Z_{cr} in a Gedankenexperiment, as long as the K shell is occupied. The spreading of the $1s$ state over the continuum, however, has a profound effect if it was prepared empty initially. The hole state will decay spontaneously, leading to the emission of a positron (Vo 61, Pi 69, Ge 69) which is allowed to escape to infinity. Thus, in sharp contrast to the situation usually encountered in atomic physics, the normal vacuum $|0\rangle$ of the electron–positron field, defined by the absence of real particles, becomes unstable under the influence of a supercritical potential. Instead, since the binding energy exceeds the threshold $\Delta E = 2mc^2$, pair creation will lead to a new charged vacuum state $|0'\rangle$ (Mü 72b, Fu 73, Ra 74). The charge of the new vacuum is limited by the Pauli principle to twice (due to spin degeneracy) the number of "dived" electron levels.

The same conclusion is reached if one starts from the charged vacuum (which is the state of lowest energy if $Z > 150$) and identifies a hole in the K shell with the presence of a bound positron, i.e., a one-particle state built upon $|0'\rangle$. As Z grows beyond Z_{cr} this particle becomes unbound and is emitted, leaving the charged vacuum as the stable ground state.

In principle the process of spontaneous pair creation is not limited to supercritical atoms; it can occur in any potential well which is deeper than $2mc^2$. In fact, an effect closely related was first unwittingly observed

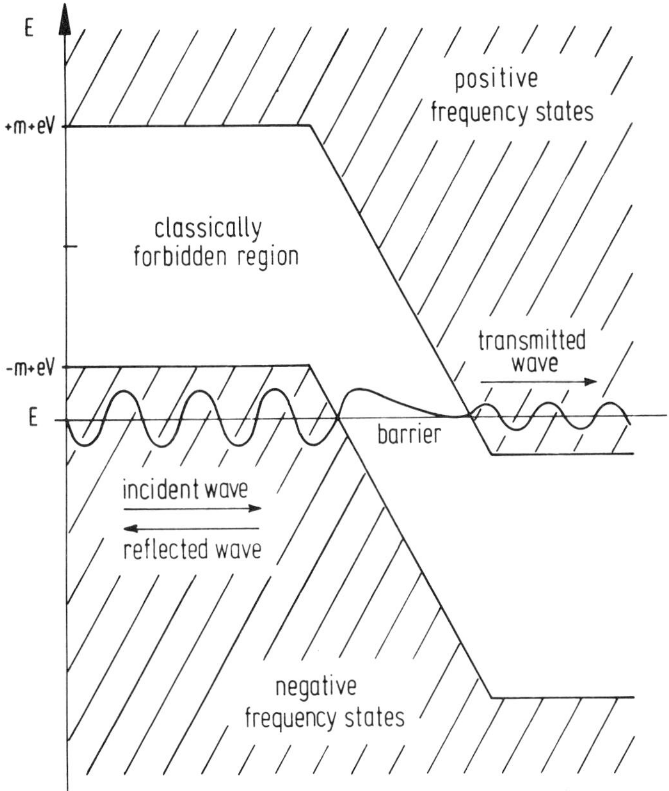

Figure 2.3. A potential barrier higher than $2mc^2$ exhibits Klein's paradox, i.e., an anomalous behavior of the reflection coefficient for electrons or positrons, caused by a tunneling through the gap of the Dirac equation. This is related to the occurrence of spontaneous pair creation.

immediately after the emergence of Dirac's theory and has become known as Klein's paradox (Kl 29): The reflection of an electron incident on a repulsive potential step of height $V > m$ shows irregular behavior for incident energies $m < E < V - m$; cf. Figure 2.3. The reflection coefficient can exceed unity, while the "transmission coefficient" becomes negative, which is clearly paradoxical in a single-particle theory. This transition into the forbidden region also is found for smoothly rising potential steps (Sa 31). The penetrability, however, decreases exponentially if the extension of the barrier becomes larger than a few Compton wavelengths. Therefore the effect cannot be observed on a laboratory scale with macroscopic potentials.

If the Dirac theory is properly interpreted in a many-particle framework, the negative current is immediately understood to be formed by outgoing positrons. The potential barrier thus will create electron–positron pairs. This process will be possible also, if no incident particle is present. Going to the limit of a barrier with infinite extension and height we arrive at the well-known result that a homogeneous electric field E continuously produces pairs. The creation rate was found to be (Sch 51)

$$w = 2\frac{(eE)^2}{(2\pi)^3}\sum_n \frac{1}{n^2} e^{-n\pi m^2/eE} \tag{2.26}$$

If the extension of the electric field is bounded, on the other hand, only a finite number of particles can be localized in the potential well (assuming Fermi statistics). Pair creation then will cease if all the accessible states in the well, i.e., those with energy $E < -m$, are filled (Hu 40). This is just the case in a weakly supercritical atom which may be seen as a narrow potential well supporting only one electron state below $E = -m$. Otherwise the "filling up" must be treated self-consistently since it will reduce the depth of the potential. For a further discussion of Klein's paradox see (Be 63, Do 71, Ha 81).

Finally we point out that the discussion in this section has been based rather naively on the hole picture. The conclusions can be justified, however, in the full field-theoretical formulation which treats particles and antiparticles on equal footing and is the basis of quantum electrodynamics. This is demonstrated in detail in (Mü 81) and in the reviews quoted in the Introduction.

2.3.1. The Point Nucleus Problem

Although having no direct bearing on heavy-ion collisions, it is interesting to observe how the concept of the charged vacuum contributes to the resolution of a long-standing problem: What happens to the electrons in the field of a point source if the charge exceeds $Z = 137$? As noted in Section 2.2, the Dirac solutions for states with $\kappa = \pm 1$ ($s_{1/2}$ and $p_{1/2}$) in a potential $V = -Z\alpha/r$ collapse for $Z\alpha > 1$. A physically reasonable procedure to investigate this phenomenon may introduce a finite cutoff radius R for the potential and study the limit $R \to 0$. This has been done in (Gä 80, Gä 81).

Starting with an extended "nucleus" with charge $137 < Z < 274$ it is observed that all $j = 1/2$ states gain in binding energy as R decreases and eventually become supercritical. This is demonstrated in Figure 2.4a for a nucleus $Z = 150$ with radius $R = R_0(2.5Z)^{1/3}$. The shrinking radius R here plays the same role as the increasing charge Z in Figure 2.1. According to the previous discussion the states reaching $E_n = -m$ will be filled by spontaneous pair creation, forming a charged vacuum. This obviously cannot go on forever as R approaches zero: The growing vacuum charge will screen the Coulomb potential, eventually preventing higher bound states from

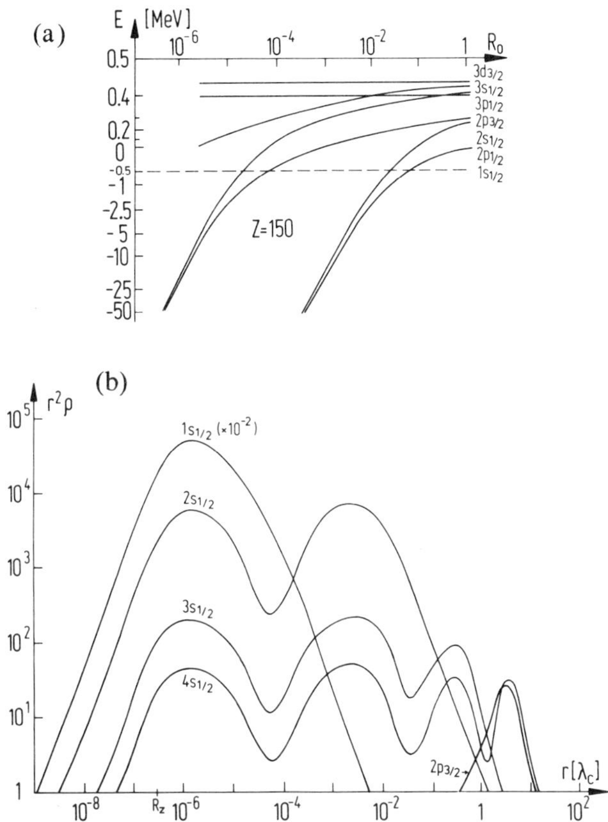

Figure 2.4. (a) Bound state energies of the single-particle Dirac equation in the field of a shrinking nucleus with charge $Z = 150$ and radius $R = R_0(2.5Z)^{1/3}$ as a function of the radius parameter R_0. (b) Radial electron densities (in arbitrary units) for several resonances for a nucleus with $Z = 150$ and $R_0 = 10^{-5}$ fm. From (Gä 80).

becoming supercritical. To obtain a consistent picture the electron–electron interaction cannot be neglected.

The electron densities in the case $Z\alpha > |\kappa|$ for small cutoff radius R have an interesting regular structure. Figure 2.4b shows the radial single-particle density $r^2\rho$ for several $ns_{1/2}$ states calculated with $Z = 150$, $R_0 = 10^{-5}$ fm. In the supercritical case bound states have been constructed out of the continuum wave functions at resonance energy by disregarding the oscillatory tail at distances beyond the outer turning point; cf. Section 6.2. The periodic structure in this logarithmic plot can be understood from an analytical expression for the density. In the limit $pr \ll 1$ and $E \ll -m$ the exact solution of the Dirac equation reduces to

$$r^2\rho = \text{const} \times \left[1 - 2\,|\kappa|\,\frac{Z\alpha + \gamma}{\kappa^2 + (Z\alpha + \gamma)^2}\,\sin\left(2\gamma \ln \frac{r}{R}\,\phi\right)\right] \tag{2.27}$$

where $\gamma = [Z\alpha)^2 - \kappa^2]^{1/2}$ and ϕ denotes a phase which is independent of E and the sign of κ. Evidently the density of all $j = 1/2$ states is periodic in the logarithmic variable $\gamma \ln(r/R)$. The consecutive $ns_{1/2}$ and $(n+1)p_{1/2}$ states have an increasing number of maxima. Due to the logarithmic dependence in each case, nearly the whole charge is assembled in the outermost maximum. The vacuum charge therefore consists of a series of concentric shells, each of them containing four charge units. According to Equation (2.27) the extension of this structure is solely determined by the potential radius R and shrinks to a point together with the central charge.

To obtain a consistent description of the screening caused by the vacuum charge cloud, Gärtner (Gä 81) employed the WKB solutions of the squared Dirac equation (2.24). Assuming $|E| \ll |V|$ and $Z\alpha \sim 1$ the charge density becomes approximately

$$r^2\rho = 4 \sum_{E_n < -m} C_n \sin^2\left[\int_R^r dr'\,\gamma(r')/r' + \pi/4\right] \tag{2.28}$$

where a screened effective charge has been introduced through $V(r) = -Z(r)\,\alpha/r$, leading to $\gamma(r) = \{[Z(r)\,\alpha]^2 - \kappa^2\}^{1/2}$. A self-consistent solution for the vacuum charge distribution is obtained from the Poisson equation

$$d^2Z/dr^2 = 4\pi r\rho(r) \tag{2.29}$$

The form of (2.28) indicates that the "vacuum shell structure" is not destroyed by screening. It can persist up to the distance r_0, where $\gamma(r_0)$ vanishes, i.e., where $Z(r_0) \simeq 137$. This is confirmed by an explicit solution of Equations (2.28), (2.29). Figure 2.5 demonstrates how the effective charge is reduced successively with shrinking nuclear radius R as additional bound states join the charge vacuum. This process continues until an effective charge $Z \sim 137$ is reached.

For an outside observer, e.g., a bound electron, therefore the systems "nucleus + vacuum shell" looks like an ordinary nucleus with $Z\alpha \sim 1$ if the cutoff radius is sufficiently small. In the limit $R \to 0$ this leads to the conjecture that the effective coupling strength $Z\alpha$ for massive pointlike particles cannot exceed the value 1. If pointlike Bosons could be produced (Ba 81d) this limit would be further reduced to the value $Z\alpha \sim 1/2$. The corresponding problem for the case of equal masses (supercritical positronium) has not yet been solved.

Results somewhat related to the problem discussed above have been obtained in (Mü 75a) for the case of highly charged abnormal nuclei of

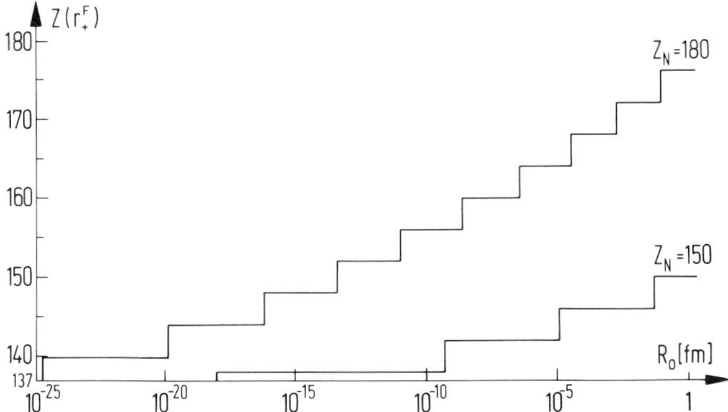

Figure 2.5. Effective charge as a function of the radius parameter R_0 seen by an outside observer. The nuclei with $Z = 150$ and 180 are screened by the vacuum charge. From (Gä 81).

ordinary density. Calculations based on the relativistic Thomas–Fermi model predict that the nuclear charge Z is strongy screened due to spontaneous pair creation and that the remaining effective charge increases with the logarithm of Z only. Moreover, the resulting Coulomb potential cannot exceed a limiting value of ~250 MeV since the proton charge is largely neutralized by the vacuum electrons in the interior of the nucleus. More refined calculations of this effect must take into account also charged pions and muons (Ra 78a).

2.4. Radiative Corrections

The theory of quantum electrodynamics applied to atomic systems is characterized by two different parameters defining the coupling strength: $Z\alpha$ and α. In high Z atoms the effective strength $Z\alpha$ can approach or even exceed unity. For light atoms the lowest-order terms of an expansion in $Z\alpha$ are sufficient to determine the level shifts with high accuracy. In the superheavy regime, on the contrary, it is essential to treat the electron–nucleus interaction exactly, i.e., to all orders in $Z\alpha$. Fortunately, however, the electron–electron coupling constant α remains small in all cases and can be safely treated by perturbative methods.

From this argument, no dramatic modifications of the radiative corrections would be expected in superheavy atoms. In the limit $Z \to Z_{cr}$, however, the situation is not so clear, since the energy gap between the lowest bound state and the neighboring negative energy continuum disappears. Reliable calculations for this case have been performed only quite recently. They confirm that the level diving will not be affected significantly by radiative corrections. Such a result already had been conjectured earlier (So 73) on the basis of limiting field theories (Ra 71).

Any calculations of QED effects in strong fields have been based on Furry's bound state interaction picture (Fu 51). They rely on an expansion of the second quantized field operators in terms of the complete set of solutions of the Dirac equation with external potential. In this way an expansion in $Z\alpha$ is avoided.

Energy shifts due to vacuum polarization, i.e., the effects of virtual electron–positron pair creation, can be calculated in several ways.* Basically the charge density induced by the nucleus (Di 34) is given by the ground-state expectation value of the 0-component of the current operator, formulated in second quantization

$$\hat{j}_\mu(x) = -\frac{|e|}{2}\,[\hat{\bar{\psi}}(x), \gamma_\mu \hat{\psi}(x)] \tag{2.30}$$

where $\hat{\psi}$ is the field operator of the electron–positron field in the Heisenberg picture. It can be expanded in a complete set of basis wave functions

$$\psi(x) = \sum_{q>F} \hat{b}_q \phi_q(x) + \sum_{q<F} \hat{d}_q^+ \phi_q(x) \tag{2.31}$$

where the "Fermi surface" F has been introduced which separates occupied from empty states. The operators $\hat{b}_q$ annihilate an electron in the single-particle state $\phi_{q'}$ while $\hat{d}_q^+$ creates a positron in the state ϕ_q. $\hat{\psi}^+(x')$ and $\hat{\psi}(x)$ and consequently the creation and annihilation operators satisfy the usual Fermion anticommutation relations. Using these properties we have the charge density

$$\begin{aligned}\rho_{\mathrm{VP}}(x) = \langle 0|\hat{j}_0|0\rangle &= \frac{e}{2}\left(\sum_{q<F} \phi_q^+\phi_q - \sum_{q>F} \phi_q^+\phi_q\right)\\ &:= e\,\tilde{\sum_q}\, \phi_q^+\phi_q\end{aligned} \tag{2.32}$$

Here the "tilde sum" is defined by

$$\begin{aligned}\tilde{\sum_q} = \frac{1}{2}\left(\sum_{q<F} - \sum_{q>F}\right) &= \sum_{-m<q<F} + \frac{1}{2}\left(\sum_{q<-m} - \sum_{q>-m}\right)\\ &:= \sum_q^{\mathrm{HF}} + \sum_q^{\mathrm{Uh}}\end{aligned} \tag{2.33}$$

It is equal to the Hartree–Fock sum $\sum^{\mathrm{HF}}$ over all occupied bound states in the gap region, $-m < E_q < +m$, plus the "Uehling sum" which accounts for the effects of vacuum polarization.

* The discussion will be restricted to time-independent electrostatic potentials.

To avoid the unpleasant summation over all bound and continuum states required in (2.32), elegant methods have been worked out which make use of the electron propagator (Wi 54). From the definition of the Feynman propagator S_F as the expectation value of a time-ordered product of field operators (Bj 65)

$$iS_F(x, y) = \langle 0 | T(\hat{\psi}(x), \hat{\bar{\psi}}(y)) | 0 \rangle$$
$$= \begin{cases} \sum_{q>F} \phi_q(x) \bar{\phi}_q(y) \exp[i(t_y - t_x) E_q] & \text{for} \quad t_x > t_y \\ -\sum_{q>F} \phi_q(x) \bar{\phi}_q(y) \exp[i(t_y - t_x) E_q] & \text{for} \quad t_x < t_y \end{cases} \tag{2.34}$$

it is readily shown that (2.32) can be written as

$$\rho_{\mathrm{VP}} = ie \,\mathrm{Tr}[S_F(x, x')\, \gamma_0]_{x' \to x} \tag{2.35}$$

It is advantageous to go to a mixed representation, eliminating the time variable $t - t'$ by a Fourier integration

$$iS_F(\mathbf{x}, \mathbf{x}', t - t')\, \gamma_0 = \frac{1}{2\pi i} \int_C dz\, e^{-i(t-t')z}\, G(\mathbf{x}, \mathbf{x}', z) \tag{2.36}$$

$G(\mathbf{x}, \mathbf{x}', z)$ is the Green's function of the stationary Dirac equation at the complex energy value z. It has poles at the position of the bound state energies and two branch cuts, $E < -m$ and $E > +m$, on the real axis. The choice of the integration contour C in the complex energy plane defines the boundary conditions for S_F at $t \to \pm\infty$. C passes below (above) the real axis near states which are occupied (empty), thus defining the position of the Fermi level. For practical calculations the contour C can be deformed to the imaginary axis and a spherical-wave decomposition of $G(\mathbf{x}, \mathbf{x}', z)$ can be introduced.

As it turns out, the dominant contribution to vacuum polarization is contained in the lowest-order term proportional to $\alpha(Z\alpha)$. An explicit expression for the potential of the vacuum polarization charge distribution induced by a point nucleus can be derived (Ue 35):

$$V_{\mathrm{Uh}}(r) = -\frac{2\alpha(Z\alpha)}{3\pi r} \int_m^\infty dt\, e^{-2tr} \left(1 + \frac{m^2}{2t^2}\right) (t - m)^{1/2}\, t^{-2} \tag{2.37}$$

which is easily generalized to the case of an extended source distribution. Equation (2.37) can be obtained from the one-loop correction to the photon propagator.

The Uehling potential falls off exponentially like e^{-2mr}. Therefore vacuum polarization has an extremely small influence on wave functions which have an extension large compared with the electron Compton wavelength. For example, in hydrogen the $1s$ energy shift amounts to only 9×10^{-7} eV, i.e., $\Delta E/E^b_{1s} \sim 7 \times 10^{-8}$. Although the effect increases strongly with nuclear charge Z, the energy shift remains rather small. In the heaviest atoms of the Periodic Table we have $\Delta E/E^b_{1s} \sim 10^{-3}$. At even higher Z the Uehling energy shift was calculated by (Pi 69). It doubles approximately as Z is increased by 10 charge units. Near the critical charge $Z_{\rm cr}$ the energy shift reaches about $\Delta E_{\rm Uh} \sim -12$ keV, i.e., $\Delta E_{\rm Uh}/E^b_{1s} \sim 10^{-2}$. According to $\Delta E_{1s}/\Delta Z \simeq -30$ keV at $Z = Z_{\rm cr}$ (Mü 72b) the critical charge would be lowered by less than one unit.

Gyulassy (Gy 74, Gy 75) investigated the effects of vacuum polarization in all orders $\alpha(Z\alpha)^n$, $n > 3$ (only odd powers are present due to Furry's theorem). The corresponding Feynman graphs are depicted in Figure 2.6a (the double line stands for the full electron propagator in the external field). Since $Z\alpha > 1$ it was essential to include the finite extension of the nucleus. Just below $Z = Z_{\rm cr}$ he obtained $\Delta E^{(n \geqslant 3)} = +1.15$ keV, half of this value originating from the $\alpha(Z\alpha)^3$ term. This result demonstrates that the Uehling potential accounts for the main effects of vacuum polarization and that higher-order terms will not in any way modify the behavior of the diving bound state. No irregularity appears as the $1s$ level gets supercritical. In an independent approach to vacuum polarization in strong electric fields Rinker and Wilets (Ri 75) explicitly performed the summation (2.32) using exact electron wave functions. Their calculation supports the result just quoted.

The electronic self-energy, caused by the interaction of the electron with the virtual photon field, is even more difficult to evaluate. The relevant

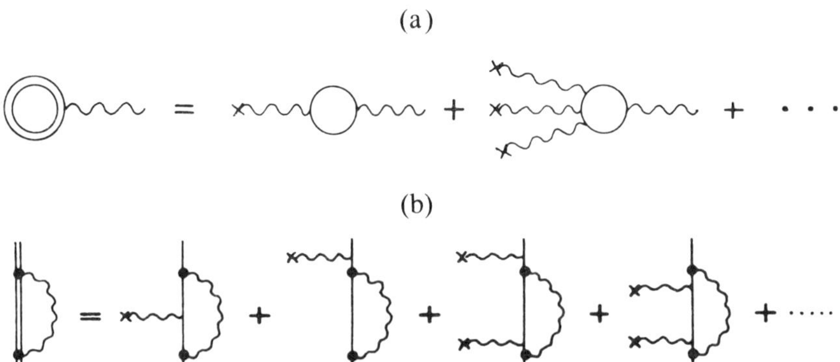

Figure 2.6. Feynman graphs for (a) vacuum polarization and (b) electron self-energy in an external field.

Feynman graphs are shown in Figure 2.6b. The (unrenormalized) energy shift for a bound state ϕ_n can be written

$$\Delta E = 4\pi i\alpha \int d(t_2 - t_1)\, d^3x_1\, d^3x_2\, \bar{\phi}_n(x_2)\, \gamma^\mu S_F(x_3, x_1) \times \gamma_\mu \phi_n(x_1)\, D(x_2 - x_1) \tag{2.38}$$

where $D(x)$ is the photon propagator.

In high-Z atoms (2.38) has to be solved using the full electron propagator S_F. Nonperturbative calculations have been performed by Desiderio and Johnson (De 71). For Z not too large the result is reproduced by an analytical approximation derived by Erickson (Er 71). Mohr (Mo 74) has studied extensively the self-energy for the case of point nuclei up to $Z = 137$. The result has a singularity at $Z\alpha \to 1$. Cheng and Johnson (Ch 76) extended the calculation of (De 71), including extended nuclei and electronic screening) up to $Z = 160$. They could not decide, however, whether ΔE_{SE} remains finite in the limit $Z \to Z_{\text{cr}}$ since here the numerical error increased very rapidly.

Very recently Soff *et al.* (So 82) were able to extend the calculations up to the critical charge. At $Z = 170$ they obtained $\Delta E_{\text{SE}} \simeq +10.9$ keV, which demonstrates that the self-energy remains finite and small. In fact the two types of radiative corrections (vacuum polarization and self-energy) have opposite sign and nearly cancel.

The calculated values of the 1s self-energy in various heavy atoms are collected in Figure 2.7. The results of Mohr (Mo 75), dots; Cheng and Johnson (Ch 76), circles; and Soff *et al.* (So 82), crosses, are included in the figure.

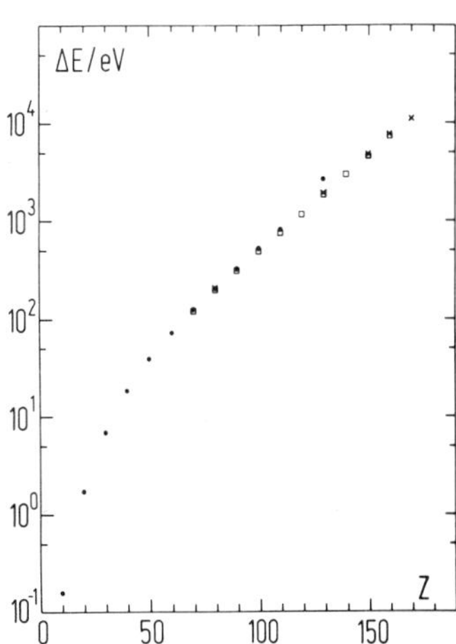

Figure 2.7. The 1s self-energy in heavy and superheavy atoms as a function of nuclear charge Z. From (So 82).

3. Theory of Electronic Excitations

3.1. Introduction

Out of the vast variety of theoretical approaches that have been developed to describe electron excitations in atomic collisions (see, e.g., Mo 65, Mo 77) only a few aspects will be discussed here. In view of the intended application to the collision of very heavy ions, the whole discussion will be based from the outset on the semiclassical approximation. The assumption of classical motion for the nuclei is justified if the Bohr–Sommerfeld parameter η is large,

$$\eta = \frac{Z_1 Z_2 e^2}{hv_1} \gg 1 \tag{3.1}$$

For a given scattering angle this condition guarantees that a localized trajectory with a definite impact parameter can be ascribed to the nuclear motion. The parameter η may be obtained from a comparison of the minimum internuclear distance with the deBroglie wavelength of the relative motion.

Furthermore the nuclear motion should not be affected by the energy ΔE transferred to the electron shell, i.e.,

$$\Delta E / E_1 \ll 1 \tag{3.2}$$

Both conditions usually are very well satisfied in heavy-ion collisions. The full quantum mechanical theory and its semiclassical limit are treated, e.g., in (De 81, He 81).

This chapter deals with the time-dependent Schrödinger (or Dirac) equation describing the dynamics of electronic motion in a collision. The general discussion is followed by the inspection of a few special approaches. In the next chapters the theory, formulated in terms of a basis expansion technique, will be applied to calculate specific processes, all related to inner shell excitation.

3.2. The Semiclassical Theory

3.2.1. The Coupled-Channel Equations

The description of electronic excitations in an atomic collision takes a particularly simple form if one starts from a classical picture for the nuclear motion. The nuclei then are just moving charges acting as sources of an electromagnetic field $A_\mu = (\mathbf{A}, V)$. Let us trace the fate of a single electron that is bound initially to one of the colliding nuclei. Its motion is treated quantum mechanically by the time-dependent Schrödinger (or, in the relativistic case, Dirac) equation

$$(i\partial/\partial t)\, \Phi_i(\mathbf{r}, t) = H_{\mathrm{el}}(\mathbf{r}, t)\, \Phi_i(\mathbf{r}, t) \tag{3.3}$$

where H_{el} is the electronic Hamiltonian

$$H_{\mathrm{el}}(\mathbf{r}, t) = (1/2m)[\mathbf{p} - e\mathbf{A}(\mathbf{r}, t)]^2 + eV(\mathbf{r}, t) \tag{3.4}$$

in the nonrelativistic case and

$$H_{\mathrm{el}}(\mathbf{r}, t) = \alpha \cdot [\mathbf{p} - e\mathbf{A}(\mathbf{r}, t)] + eV(\mathbf{r}, t) + \beta m \tag{3.5}$$

for the Dirac equation $\mathbf{p}$ is the canonically conjugate momentum operator to the electron coordinate $\mathbf{r}$. Equation (3.3) is valid in any inertial system. It is natural to use the center-of-mass system of the two nuclei (CMN) which is inertial up to corrections of the order m/μ, i.e., the ratio of the electronic and the nuclear reduced mass. This ratio is a very small number so that recoil effects in the electron–nucleus system can be safely neglected for most purposes.

The electromagnetic potential A_μ takes the role of a given classical external field. In the absence of incident photons it can be calculated from the current generated by the moving nuclei

$$A_\mu(\mathbf{r}, t) = \int D^R(\mathbf{r}, t; \mathbf{r}', t')\, j_\mu(\mathbf{r}', t)\, d^3r'\, dt' \tag{3.6}$$

where D^R is the retarded Green's function of the wave equation. In the Coulomb gauge the timelike component reduces to the instantaneous interaction

$$V(\mathbf{r}, t) = -\frac{Z_1 e}{|\mathbf{r} - \mathbf{R}_1(t)|} - \frac{Z_2 e}{|\mathbf{r} - \mathbf{R}_2(t)|} \tag{3.7}$$

assuming pointlike sources. The expression is easily generalized to the case of finite nuclei which will become important for heavy collision partners. The magnetic and retardation effects contained in the vector potential $\mathbf{A}(\mathbf{r}, t)$ usually are neglected since they are of higher order in v/c. We will come back to this question in Section 4.4.

Equation (3.3) has to be solved with the initial condition $(t \to -\infty)$

$$\Phi_i \to \phi_i(r)\, e^{-iE_i t} F_i(\mathbf{r}, t) \tag{3.8}$$

where ϕ_i is an atomic eigenstate belonging to one of the separated atoms and E_i is its energy. The phase factor $F_i(\mathbf{r}, t)$ serves to give a translational boost to the electron so that it stays attached to the moving nucleus. The role of the factor $F_i(\mathbf{r}, t)$ will be discussed in detail in Section 3.2.4.

The probability for excitation of an electron from state ϕ_i (before the collision) to state ϕ_j (after the collision) is obtained by projecting the time-dependent wave function Φ_i on the final state and taking the square

$$P_{i\to j}(b) = |\lim_{t\to\infty} \langle \phi_j(\mathbf{r}, t) \mid \Phi_i(\mathbf{r}, t)\rangle|^2 \tag{3.9}$$

Here we have indicated that the transition probability depends on the impact parameter b. Since the nuclear motion is assumed to be independent of the electron dynamics, the differential excitation cross section is given by a simple product

$$d\sigma_{i\to j}/d\Omega = (d\sigma/d\Omega)_{\text{scatt}} P_{i\to j}(b(\theta)) \tag{3.10}$$

with the cross section $(d\sigma/d\Omega)_{\text{scatt}}$ for scattering the projectile under the angle θ. In the case of identical projectile and target atoms this expression has to be symmetrized. The total cross section is obtained by integration over all impact parameters

$$\sigma_{i\to j} = 2\pi \int_0^\infty bP_{i\to j}(b)\, db \tag{3.11}$$

To calculate any of these observables one has to solve Equation (3.3) which is a second-order partial differential equation in four dimensions (in the relativistic case there are four coupled differential equations of first order). Due to the complexity of this problem up to now, only very few attempts have been made at a direct numerical solution of Equation (3.3) (see Section 3.4.3). The standard method relies on an expansion of the time-dependent wave function Φ_i in some suitable basis set ϕ_j*:

$$\Phi_i(\mathbf{r}, t) = \sum_j a_{ij}(t)\, \phi_j(\mathbf{r}, t) \tag{3.12}$$

The square of the amplitude $a_{ij}(t)$ gives the probability that an electron starting from level i occupies the basis state ϕ_j at a given time t. Insertion of (3.3) into (3.12) and projection on $\langle \phi_k|$ immediately leads to

$$i \sum_j \langle \phi_k \mid \phi_j\rangle \dot{a}_{ij} = \sum_j \langle \phi_k | H - i\partial/\partial t \,|\phi_j\rangle\, a_{ij} \tag{3.13}$$

This can be abbreviated using matrix notation

$$\mathsf{i}\mathsf{N}\dot{\mathsf{a}}^T = \mathsf{M}\mathsf{a}^T \tag{3.14}$$

* Here and in the following the sum will stand for summation over the discrete and integration over the continuous spectrum, if required.

with the overlap matrix

$$N_{kj} = \langle \phi_k \mid \phi_j \rangle \tag{3.15}$$

and the coupling matrix

$$M_{kj} = \langle \phi_k | \, H - i\partial/\partial t \, | \phi_j \rangle \tag{3.16}$$

a^T is the transposed matrix of amplitudes where each column belongs to a different initial condition.

Equations (3.13) or (3.14) constitute a set of coupled ordinary differential equations in time which is fully equivalent to (3.3) as long as a complete set of basis states is employed. This assertive statement, however, is of limited value for practical calculations since one is forced to work with a finite, truncated basis set. Therefore the outcome of a calculation will critically depend on two conditions: The choice of the type of basis states and the number of states which are to be included. The answer to these questions depends on the process under consideration and on the impact velocity. Under favorable circumstances only few states will be strongly coupled or the excitation amplitudes will be so small that perturbation theory is valid. Very often, however, an extensive numerical treatment of the coupled channel equations cannot be avoided.

Frequently the expansion (3.12) is slightly modified to include an energy-dependent phase factor

$$\Phi_i(\mathbf{r}, t) = \sum_j a_{ij}(t) \, \phi_j(\mathbf{r}, t) \, e^{-i\chi_j(t)} \tag{3.17}$$

where

$$\chi_j = \int_{t_0}^{t} dt' \langle \phi_j | \, H \, | \phi_j \rangle \tag{3.18}$$

and t_0 is an arbitrary reference time. In this way the diagonal elements of the coupling matrix M are eliminated and the coupled-channel equations read (assuming an orthogonal basis)

$$i\dot{a}_{ij} = \sum_{j \neq k} \langle \phi_k | \, H - i\partial/\partial t \, | \phi_j \rangle \, a_{ij} e^{-i(\chi_j - \chi_k)} \tag{3.19}$$

3.2.2. Coupling Operators

The full coupling matrix M_{jk} exhibits two different terms: $\langle \phi_k | \, H \, | \phi_j \rangle$ due to the electronic Hamiltonian and $\langle \phi_k | \, \partial/\partial t \, | \phi_j \rangle$ containing the time derivative operator. The two parts are often termed "potential coupling" and "dynamic coupling." The latter is present if the basis states ϕ_j are time dependent. Transitions then arise due to the fact that the electron wants to "stay behind" and does not completely follow the motion implied by the development of the basis state. Usually the time dependence of ϕ_j will be a parametric one, via the internuclear separation coordinate which follows the classical trajectory $\mathbf{R}(t)$.

It is often advantageous to use basis states which are quantized along the internuclear axis (the axis of cylindrical symmetry of H) and therefore are rotating with time. To be specific, let us assume that the scattering takes place in the x–z plane and let $\mathbf{r}' = (x', y', z')$ denote the electronic coordinates in a rotating system with the z' axis connecting the scattering nuclei. We have to remember that (3.3) is valid only in an inertial frame, $\mathbf{r} = (x, y, z)$, and that the time derivative has to be performed accordingly for fixed electronic coordinates. If the rotating coordinates are to be used, the chain rule can be applied:

$$\partial/\partial t|_{\mathbf{r}} = \partial/\partial t|_{\mathbf{r}'} + \partial\mathbf{r}'/\partial t \cdot \mathbf{\nabla}_{r'}$$

Using the rotation matrix $x' = x\cos\theta + z\sin\theta$, $y' = y$, $z' = -x\sin\theta + z\cos\theta$, this immediately yields

$$\partial/\partial t|_{\mathbf{r}} = \dot{R}\,\partial/\partial R + \dot{\theta}(z'\,\partial/\partial x' - x'\,\partial/\partial z') \tag{3.20}$$

The bracket corresponds to the generator of rotations, i.e., the angular momentum operator L_y. Generalizing to the case of spinor wave functions the dynamic coupling matrix in the rotating frame is

$$\langle\phi_j(\mathbf{r}', \mathbf{R})|\,\partial/\partial t\,|\phi_k(\mathbf{r}', \mathbf{R})\rangle = \dot{R}\langle\phi_j|\,\partial/\partial R\,|\phi_k\rangle - i\omega\langle\phi_j|\,j_y\,|\phi_k\rangle \tag{3.21}$$

Here $\omega = \dot{\theta}$ is the angular velocity of the rotation and j_y is the operator of total angular momentum of the electron. The same result is obtained using the rotation operator, $|\phi'\rangle = \exp[-i\boldsymbol{\theta}(t)\cdot\mathbf{j}]\,|\phi\rangle$. An explicit transformation of the Dirac equation to a rotating system is given in (Mü 76b).

Equation (3.21) exhibits two kinds of dynamical couplings: the radial coupling resulting from the change of the internuclear distance and the rotational coupling which is due to the varying orientation of the coordinate system. The latter is also called Coriolis coupling and can be related to the fictitious forces arising in a noninertial frame of reference.

It is very useful to split the coupling matrix into various parts, as shown in (3.16) and (3.21). One should bear in mind, however, that this division does not have an absolute meaning since the ratio of the various contributions is dependent on the choice of the basis ϕ_j. In some situations there exists a preferred choice and then an observable physical process may be attributed to a particular kind of coupling.

3.2.3. Properties of the Excitation Amplitudes

The amplitudes $a_{ij}(t)$ describing the transition of an electron from state i to state j in the course of the collision satisfy several useful symmetry relations. These may be used for the reduction of computational effort and for checks of the numerical accuracy. The given relations are exact if the solutions of the coupled differential equations are inserted, irrespective of the choice and size of the basis.

We start from the observation that the coupling matrix M has to obey the condition (Gr 65)

$$i\dot{\mathsf{N}} = \mathsf{M}^+ - \mathsf{M} \tag{3.22}$$

where M^+ denotes the Hermitian conjugate of M. This equation follows from the requirement of probability conservation for the time-dependent wave function (3.12), $d/dt \langle \Phi_i | \Phi_j \rangle = 0$. The coupling matrix is Hermitian if the overlap matrix is constant in time. This is true in particular for an orthonormal basis, satisfying $N_{ij} = \delta_{ij}$. For the rest of this section we will make this assumption.

(i) The orthonormality condition

$$\sum_k a_{jk}^*(t)\, a_{ik}(t) = \delta_{ij} \quad \text{or} \quad \mathsf{a}\mathsf{a}^+ = 1 \tag{3.23}$$

can be verified by differentiation with respect to time and use of (3.14) and (3.22). This condition implies the orthogonality of the wave functions Φ_i at every instant of time

$$\langle \Phi_i(t) | \Phi_j(t) \rangle = \delta_{ij} \tag{3.24}$$

(ii) The identity

$$\sum_k a_{kj}^*(t)\, a_{ki}(t) = \delta_{ij} \quad \text{or} \quad \mathsf{a}^+\mathsf{a} = 1 \tag{3.25}$$

can be deduced in a similar way. It guarantees the completeness of the set Φ_k at any time if one starts from a complete basis ϕ_k.

(iii) In a time-symmetric collision one has also

$$a_{ij}(\infty) = a_{ji}(\infty) \quad \text{or} \quad \mathsf{a}(\infty) = \mathsf{a}^T(\infty) \tag{3.26}$$

In contrast to (3.23), (3.25) this identity is valid only in the limit $t \to \infty$. It reflects the principle of detailed balance which equates the transition rates in both directions of a given reaction if the interaction is invariant under time reflection. Thus (3.26) is valid only for collisons with a symmetric nuclear trajectory, i.e., $R(t) = R(-t)$ and $\omega(t) = \omega(-t)$ which holds for Coulomb scattering neglecting loss of energy and angular momentum. In this case the coupling matrix element satisfies

$$M_{ij}(t) = M_{ij}^*(-t) \tag{3.27}$$

which leads to (3.26) if inserted in the complex conjugate of the differential equation (3.19). In general (3.27) is correct only up to a phase factor and (3.26) holds only for the absolute values (Sh 73).

(iv)

$$a_{ij}(\infty) = \sum_k a_{ik}(0)\, a_{jk}(0) \quad \text{or} \quad \mathsf{a}(\infty) = \mathsf{a}(0)\, \mathsf{a}^T(0) \tag{3.28}$$

This equation holds under the same conditions as (3.26). It can be derived using (3.27) and the unitarity property of the time development operator. The identity shows that it is sufficient to calculate the excitation amplitudes $a_{ij}(0)$ for the incoming branch of the trajectory only. (3.28) is immediately generalized for time-asymmetric collisions: the sets $a_{ik}(0)$ and $a_{jk}(0)$ then have to be calculated for different kinematical conditions.

3.2.4. *Electron Translation Effects*

The theory formulated so far, as expressed in Equations (3.14), (3.16), and (3.9) may lead to serious difficulties if one tries to apply it carelessly. This has inspired a large number of investigations. The subject is reviewed extensively in (Th 78, De 81). We also refer to the work of Heinz *et al.* (He 81) for details.

In any practical calculation the basis employed will consist of wave functions ϕ_j which are formulated either in terms of the atomic coordinates r_n, $n = 1$ or 2, or of the molecular coordinates $\mathbf{r}$ with respect to the CMN. At any rate, for large internuclear distances the states will be centered at one of the two nuclei (1 or 2).* The basis, however, will *not* depend on the state of motion of the nuclei. While the wave functions depend on r_n only (for $R \to \infty$), the dynamic coupling operator has to be evaluated keeping r fixed. Differentiation by the chain rule

$$\partial/\partial t|_{\mathbf{r}} = \partial \mathbf{r}_n/\partial t|_{\mathbf{r}} \cdot \nabla_{\mathbf{r}_n} + \partial/\partial t|_{\mathbf{r}_n} \tag{3.29}$$

then leads to the asymptotic coupling matrix element

$$M_{ji} = \langle \phi_j | \, \partial/\partial t|_r \, | \phi_i \rangle \to -\mathbf{R}_n \langle \phi_j(\mathbf{r}_{n'}) \, | \, \nabla_{\mathbf{r}_n} | \phi_i(\mathbf{r}_n) \rangle \tag{3.30}$$

* This is not quite true in the case of asymptotic degeneracy in symmetric systems, where suitable linear combinations of gerade and ungerade states have to be taken. Furthermore, the continuum requires special consideration, since these states are not localized in space.

If the initial and final state of the transition belong to the same atom ($n'_j = n_i$) this coupling does not go to zero at large internuclear distance. Rather, it becomes a constant proportional to the matrix element of the momentum operator and to the nuclear velocity.

The transition amplitudes $a_{ij}(t)$ long after the collision thus will not approach constant values but will oscillate in time. Furthermore one can show that the solutions depend on the choice of the origin of the coordinate system. The physical reasons for these unpleasant features are quite obvious: The basis states satisfy the wrong boundary conditions. While in the actual collision the electron travels together with its nucleus, the basis states are centered at a fixed position in space. These states are not asymptotic solutions to the time-dependent Schrödinger equation (3.3), hence the nonvanishing coupling (3.30).

Bates and McCarroll (Ba 58) noted that this deficiency can be cured if the basis states are multiplied by a plane-wave phase factor $F_i(\mathbf{r}, t)$ which represents the translational momentum and energy of the atomic electron being attached to its moving nucleus. The new states, appropriately named "traveling orbitals," are

$$\tilde{\phi}_i(\mathbf{r}, \mathbf{R}) = F_i(\mathbf{r}, t)\, \phi_i(\mathbf{r}_n, \mathbf{R}) \tag{3.31}$$

In the asymptotic region, where the nuclei move with uniform velocity v_n on straight lines, the electron translation factor F_i has the form*

$$F_i(\mathbf{r}, t) = \exp[im(\mathbf{v}_n \cdot \mathbf{r} - \tfrac{1}{2} v_n^2 t)] \tag{3.32}$$

where $n = n(i) = 1, 2$. If the origin of the system is fixed at the nuclear center of mass (CMN) we have

$$\mathbf{R}_n = \pm M_n/(M_1 + M_2)\, \mathbf{R} \equiv \tfrac{1}{2}(\lambda \pm 1)\, \mathbf{R} \tag{3.33}$$

with the mass asymmetry parameter

$$\lambda = (M_1 - M_2)/(M_1 + M_2) \tag{3.34}$$

The validity of (3.32) is easily checked if one applies the operator $H - i\partial/\partial t|_{\mathbf{r}}$ to the product wave function (3.31). The kinetic energy operator $-\Delta/2m$ and the time derivative $\partial/\partial t|_{\mathbf{r}}$ both produce extra terms which cancel each other, so that

$$(H - i\partial/\partial t|_{\mathbf{r}})\, F\phi_i = F(H - i\partial/\partial t|_{\mathbf{r}_n})\, \phi_i \tag{3.35}$$

* For finite R deviations from a straight line trajectory can be included if the second phase factor in (3.32) is replaced by $\frac{1}{2}\int^t v_n^2(t')\, dt'$.

Therefore $\tilde{\phi}_i$ satisfies the correct asymptotic condition if ϕ_i correlates to an atomic eigenstate of H_n.

Actually (3.31), (3.32) just describe a Galilei transformation of a Schrödinger wave function between two frames moving with constant relative velocity $\dot{\mathbf{R}}_n$; see (Le 63), (Nu 76). The corresponding transformation to a noninertial, accelerated frame which is valid everywhere on the nuclear trajectory also has been worked out (Sch 77).

According to (3.35) the problems associated with (3.19) are eliminated if the total wave function Φ_i is expanded in the corrected basis (3.31), i.e.,

$$\Phi_i(r,t) = \sum_j a_{ij}(t)\, F_j(\mathbf{r},t)\, \phi_j(\mathbf{r},t) \tag{3.36}$$

This leads again to the coupled-channel equations (3.19) with modified matrices M and N,

$$N_{kj} = \langle \phi_k |\, F_k^* F_j \,| \phi_j \rangle \tag{3.37}$$

$$M_{kj} = \langle \phi_k |\, F_k^* (H - i\partial/\partial t)\, F_j \,| \phi_j \rangle \tag{3.38}$$

The spurious asymptotic couplings are eliminated, as can be seen from (3.35). The overlap and coupling matrices connecting states which are attached to different nuclei (exchange type) are modified by the plane wave phase factor $\exp(\pm im\mathbf{v}\cdot\mathbf{r})$, otherwise for $n(k) = n(j)$ (direct type) $F_k^* F_j = 1$. In practical calculations the matrix elements (3.37) and (3.38) are difficult to handle due to their velocity dependence. Repeated calculation of M and N is required for each new value of the impact parameter or energy. This can be avoided by using a low-velocity approximation, retaining only terms linear in v.

The modification of the theory expressed in (3.31), (3.32) is adequate if an atomic basis set is used, consisting of states each centered on one of the moving nuclei. In sufficiently slow collisions, however, the electronic motion will become adiabatic, which calls for an expansion in terms of molecular states. During the interaction the electron feels the combined attraction of both nuclei. Therefore it is inconsistent with the spirit of the molecular model to use the full atomic translation factor $F_i(\mathbf{r},t)$ throughout the collision. Somehow the translational correction has to be smoothly adjusted as a function of the internuclear distance R. Schneiderman and Russek (Sch 69) and Thorson and Levy (Th 69, Le 69) were the first to introduce a modified translation factor of the form

$$F_i(r,t) = \exp[imf_i(\mathbf{r},\mathbf{R})\,\mathbf{v}\cdot\mathbf{r}] \tag{3.39}$$

$f_i(\mathbf{r},\mathbf{R})$ is a "switching function" which should behave smoothly as a function of r and R but otherwise is quite arbitrary. Only the asymptotic form is known for sure:

$$\lim f_i(\mathbf{r}, \mathbf{R})\,\mathbf{v} = \begin{cases} \mathbf{v}_1 & \text{if} \quad r_1 \ll r_2 \\ \mathbf{v}_2 & \text{if} \quad r_2 \ll r_1 \end{cases} \tag{3.40}$$

Furthermore it is natural to require

$$\lim f_i(\mathbf{r}, \mathbf{R}) = 0 \qquad \text{if} \quad (R/r) \to 0 \tag{3.41}$$

since in this limit the electron sees just one nonmoving (in the CMN frame) center. Since the operators $-\Delta/2m$ and $\partial/\partial t$ in (3.38) now also act on the switching function $f_i(\mathbf{r}, \mathbf{R})$ the explicit form of the coupling matrix becomes rather complicated (Ta 75). In the low-velocity approximation, neglecting terms of order v^2 or higher, derivatives of f_i, and effects of the acceleration $d\mathbf{v}/dt$, the expression simplifies greatly and we obtain the coupling matrix

$$\mathsf{M} = \mathsf{H} + \mathbf{v} \cdot (\boldsymbol{P} + \boldsymbol{A}) \tag{3.42}$$

The terms $\mathsf{H} + \mathbf{v} \cdot \boldsymbol{P}$ constitute the "potential" and "dynamical" couplings $M_{kj} = \langle \phi_k | H | \phi_j \rangle - i\mathbf{v} \cdot \langle \phi_k | \nabla_\mathbf{R} | \phi_j \rangle$ already discussed, and

$$\begin{aligned} \mathbf{A}_{kj} &= -i\langle \phi_k | (\nabla_\mathbf{r} f r)\, \nabla_\mathbf{r} + \tfrac{1}{2}(\nabla_\mathbf{r}^2 f \mathbf{r}) | \phi_j \rangle \\ &= im\langle \phi_k | [H, f\mathbf{r}] | \phi_j \rangle \\ &= im(E_j - E_k)\, \langle \phi_k | f\mathbf{r} | \phi_j \rangle \end{aligned} \tag{3.43}$$

The last equality is valid if the basis consists of eigenstates of the molecular Hamiltonian H. The correction (3.43) is easy to evaluate since only the matrix elements of the dipole operator $f(\mathbf{r}, \mathbf{R})\mathbf{r}$ are required. $\boldsymbol{A}$ has the effect of canceling that part of the dynamic coupling matrix $\boldsymbol{P}$ that only represents translations.

The construction of an optimal switching function has been the subject of numerous investigations since the introduction of the idea in 1969. Thorson and collaborators (Le 69, Le 71, Se 73, Ra 79) studied in detail transitions to the continuum in the $p + \mathrm{H}$ molecular system. They introduced various Ansätze for the switching function, e.g., $f(\mathbf{r}, \mathbf{R}) = \frac{1}{2} \tanh[\beta(R)(r_1 - r_2)]$, and minimized the coupling strength $\boldsymbol{P} + \boldsymbol{A}$. Drastic reductions were found for certain values of the parameter β. The so-defined "optimal switching function" was found to depend on the particular states involved (Ra 79), which might be seen as a shortcoming of this approach.

Recently Vaaben and Taulbjerg (Va 81) introduced a simple parameter-free switching function motivated by physical reasoning. They argued that the local translation velocity of the electron at each point in space is determined by the direction of the Coulomb force it feels. If the force vector

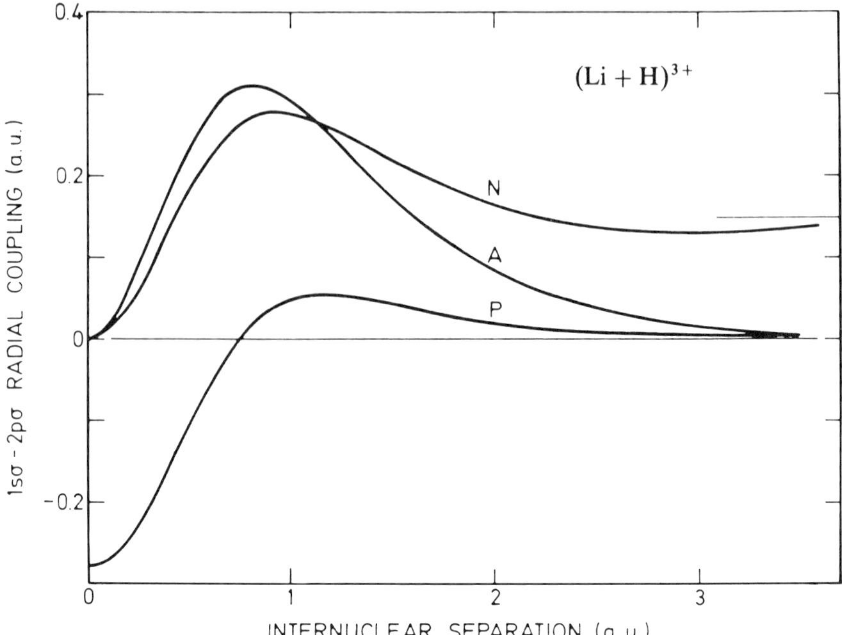

Figure 3.1. Radial coupling matrix element between the states $1s\sigma$ and $2p\sigma$ of the Li + H quasimolecule calculated without translation factors (N), with plane wave translation factors (P) of type (3.34), and with the adiabatic translation factors of Equations (3.39) and (3.44). From (Va 81).

points at one of the nuclei or at the nuclear center of charge, the velocities $\mathbf{v}_1$, $\mathbf{v}_2$, or 0 (if the center of mass and the center of charge coincide) are used. $f(\mathbf{r}, \mathbf{R})$ effects a smooth transition between these limits. This idea leads to the universal form

$$f(r, R) = \tfrac{1}{2}[(Z_2 r_1^3 Z_1 r_2^3)/(Z_2 r_1^3 + Z_1 r_2^3) + (Z_1 Z_2)/(Z_1 + Z_2)] \quad (3.44)$$

Figure 3.1 shows an example for the effect of translation factors (Va 81). The radial coupling matrix elements between the $1s\sigma$ and $2p\sigma$ states of the molecular system $(\mathrm{Li} + \mathrm{H})^{3+}$ are given for different choices of the translation factor. N is calculated neglecting any corrections, P includes plane wave translation factors of the form (3.32), and A is the result obtained with the adiabatic switching function (3.44). Clearly both A and P get rid of the spurious long-range coupling displayed by the uncorrected matrix element N. The plane wave result P, however, leads to large couplings at small internuclear distance due to the high translational velocity assigned to the electron. This is corrected by the use of a switching function.

The situation is quite different in relativistic systems which have been

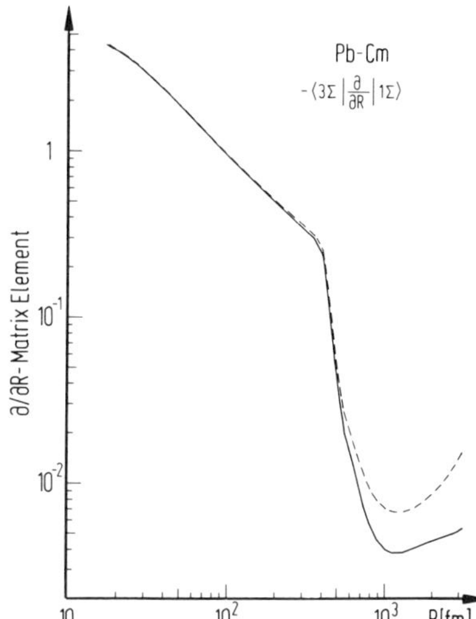

Figure 3.2. Radial coupling matrix element between the 1Σ and 3Σ states (adiabatic notation) in the superheavy quasimolecule Pb + Cm calculated with (solid line) and without (dashed line) translation factors. From (He 81).

studied by Heinz *et al.* (He 81). Here the radial coupling becomes large at close nuclear separations due to the relativistic contraction of the wave functions. Figure 3.2 shows the radial coupling matrix element between the states 1Σ and 3Σ in the Pb–Cm quasimolecule (one-electron approximation) calculated with a switching function similar to (3.44). Only at large values of R does the translational factor have an appreciable effect (dashed line: uncorrected; solid line: corrected matrix element). The break at 400 fm is caused by an avoided crossing between the $2s\sigma$ and $2p_{3/2}\sigma$ (diabatic notation) states.

In closing this section we should remark that "in principle" the expansion (3.12) is possible for any complete basis set. It is only a matter of convenience to introduce translational factors, as long as one projects on the proper moving orbitals (3.8), (3.9) at $t \to \pm\infty$. Indeed coupled states calculations using a molecular basis without translation factors (see, e.g., Pi 77) or with plane-wave factors (e.g., Wi 80) have been performed successfully.

However, one is necessarily forced to work with an incomplete truncated basis. Therefore it is important to span that part of the Hilbert space that is required to describe the excitation process under investigation with as few states as possible. Translation factors help to account for some "trivial" aspects of the electronic motion.

3.3. Excitations of the Many-Electron System

The discussion up to now was concerned with single-electron excitations only. This is clearly insufficient since usually many atomic electrons are present at the beginning of a heavy-ion collison. Furthermore, if pair creation is to be described, the antiparticle continuum must be treated properly.

It turns out, however, that the presence of many electrons does not lead to any additional problems if the electron–electron interaction is neglected or approximated by an effective single-particle potential (Thomas–Fermi, Hartree–Fock, etc.). Under this approximation the electrons can influence each other through the Pauli exclusion principle only. But since the time-dependent wave functions $\Phi_i^{(+)}$ of two electrons initially in different states remain orthogonal throughout the collision, cf. Equation (3.24), the Pauli principle has no effect on the excitation rates. If one is not interested in correlations between several particles or holes, the excitation rates are given by an incoherent summation of single-particle transition probabilities (Re 73, Re 79b). This remains true also if the electron–electron interaction is approximated by a mean screening potential common to all electrons.

The many-particle aspects are most conveniently described in the language of second quantization. We will start with a brief look at the relativistic, quantum field theoretical description of interacting electrons and photons. The notation is that of (Bj 65), in particular $x^\mu = (t, \mathbf{x}), \boldsymbol{\gamma} = \beta\boldsymbol{\alpha}, \gamma_0 = \beta$. We will work in the Heisenberg picture and introduce a time-independent state vector $|F\rangle$ defined by the preparation of the collision system. The electron–positron and the photon field now are described by field operators $\hat{\Psi}(x)$ and $\hat{A}_\mu(x)$. They have to fulfill a set of coupled nonlinear operator equations. For $\hat{\Psi}$ we have

$$[i\partial/\partial t - H(x)]\, \hat{\Psi}(x) = (e/2)\, \gamma^0 \{\gamma_\mu\, \hat{\Psi}(x), \hat{A}^\mu(x)\} \tag{3.45}$$

The left-hand side is just the familiar Dirac operator of (3.3), incorporating the interaction with the external (nonquantized) potential generated by the nuclei, (3.32). The inhomogeneous source term results from the interaction with the quantized electromagnetic field $\hat{A}_\mu$. This in turn is generated by the current of the electron–positron field, i.e.,

$$\Box \hat{A}_\mu(x) = \hat{j}_\mu(x) \tag{3.46}$$

or

$$\hat{A}_\mu(x) = \int d^4y\, D^R(x-y)\, \hat{j}_\mu(y) \tag{3.47}$$

where

$$\hat{j}_\mu(x) = -(e/2)[\hat{\bar{\Psi}}(x), \gamma_\mu\, \hat{\Psi}(x)] \tag{3.48}$$

The analog (3.6) of (3.47) already was used for the external field. Free photons and also vacuum fluctuations of the electromagnetic field are not included. The anticommutator in (3.45) and the commutator in (3.48) have been introduced to guarantee the charge symmetry of the theory. For a discussion of (3.45)–(3.48) see (Kä 72).

The field operator $\hat{\Psi}(x)$ now is expanded in a Fock decomposition

$$\hat{\Psi}(x) = \sum_{i<F} \hat{d}_i^+ \Phi_i^{(+)} + \sum_{i>F} \hat{b}_i \Phi_i^{(+)} \tag{3.49}$$

where $\Phi_i^{(+)}$ denotes a complete set of classical wave functions. Later we will require that each $\Phi_i^{(+)}(t \to -\infty)$ correlates to a definite atomic state ϕ_i in the sense of (3.8). $\hat{d}_i^+$ and $\hat{b}_i$ are creation operators for holes and annihilation operators for particles, respectively. They are defined with respect to the "Fermi level" F, the boundary between initially occupied and empty states.

The operators $\hat{\Psi}(x)$, $\hat{\Psi}^+(x)$ describe a fermionic field and satisfy the usual equal time anticommutation relations in coordinate space. From these follow the anticommutation relations for the Fock operators, i.e.,

$$\{\hat{b}_j^+, \hat{b}_i\} = \{\hat{d}_j^+, \hat{d}_i\} = \delta_{ij} \tag{3.50}$$

The anticommutators for all further combinations of Fock operators vanish. The ground state $|F\rangle$ is defined such that it is annihilated by the operators $\hat{d}_i$ and $\hat{b}_i$,

$$d_i |F\rangle = 0 \quad \text{for} \quad i < F, \qquad b_i |F\rangle = 0 \quad \text{for} \quad i > F \tag{3.51}$$

(Of course this notation does not mean that the initial occupation of states necessarily is determined by the ordering of energies.) (3.51) is valid in the limit $t \to -\infty$. Insertion of (3.49) into (3.45) leads to a set of first-order differential equations in time for the Fock operators (Ki 80a). In general they will have a very complicated time dependence due to the nonlinear interaction term in (3.45). If electron–electron correlations can be neglected or approximated by a one-body operator, however, the description becomes nearly trivial.

Rihan *et al.* (Ri 78) have argued that this is a reasonable approximation for inner shells in heavy-ion collisions. They defined a relaxation time for transitions caused by the nondiagonal part of the electron–electron interaction. This turned out to be larger than the typical collision time by about two orders of magnitude in the system U + U.

Approximating the right-hand side of (3.45) by zero, the $\hat{b}_i$, $\hat{d}_i$ become constant if the classical wave functions $\Phi_i^{(+)}$ in the Fock expansion of the field operator are identified with the set of solutions of the time-dependent single-particle Dirac equation (3.3). Since the basis $\Phi_i^{(+)}$ satisfies the boundary condition at $t = -\infty$ the number operators $\hat{d}_i^+ \hat{d}_i$, $\hat{b}_i^+ \hat{b}_i$ do not describe physically observable particles after the collision. Instead, one has to employ a set which correlates to a definite final state of the separated system after the collision, $t \to +\infty$. Using $\Phi_i^{(-)}$, the dynamic solutions of (3.3) satisfying outgoing boundary condition, one has the alternative expansion

$$\Psi = \sum_{i<F} \hat{\hat{d}}_i^+ \Phi_i^{(-)} + \sum_{i>F} \hat{\hat{b}}_i \Phi_i^{(-)} \tag{3.52}$$

Equating (3.49) and (3.52) yields a canonical transformation between the two sets of particle and hole operators, where the expansion coefficients $\langle \Phi_j^{(-)} \mid \Phi_i^{(+)} \rangle$ are just the amplitudes $a_{ij}(\infty)$ discussed above

$$\left. \begin{matrix} \hat{\hat{d}}_i^+ \\ \hat{\hat{b}}_i \end{matrix} \right\} = \sum_{k<F} \hat{d}_k^+ a_{ki} + \sum_{k>F} \hat{b}_k a_{ki} \qquad \text{for} \quad \left\{ \begin{matrix} i<F \\ i>F \end{matrix} \right. \tag{3.53}$$

The number of particles created in a state above the Fermi level, $i > F$, is

$$N_i = \langle F | \, \hat{\hat{b}}_i^+ \hat{\hat{b}}_i \, | F \rangle = \sum_{k<F} |a_{ki}|^2 \tag{3.54}$$

and the number of holes in a state below the Fermi level, $i < F$, is

$$N_i = \langle F | \, \hat{\hat{d}}_i^+ \hat{\hat{d}}_i \, | F \rangle = \sum_{k>F} |a_{ki}|^2 \tag{3.55}$$

These simple results contain a summation over all possible many-electron configurations with a particle (hole) in the level i. If one is interested in more detailed information on the final state, additional coherent terms arise (Re 79b, Re 80a). The number of correlated particle–hole pairs is given by the expectation value of the product of number operators

$$N_{ij} = \langle F | \, \hat{\hat{b}}_i^+ \hat{\hat{b}}_i \hat{\hat{d}}_j^+ \hat{\hat{d}}_j \, | F \rangle = N_i N_j + \left| \sum_{k<F} a_{ki}^* a_{kj} \right|^2 \tag{3.56}$$

The same formula holds also in the case of particle–particle and hole–hole correlations if the plus sign is replaced by a minus sign. The first term of (3.56) describes statistical coincidences while the sum contains coherent correlation effects. When the excitation rates are low, the second term becomes dominant, since the probability for multiple excitations decreases rapidly. To analyze experiments which do not distinguish between several states (e.g., spin degeneracy) additional incoherent terms have to be added (So 80). The correlations of electron–positron pairs are discussed in (Re 81a).

3.4. Special Approaches

3.4.1. Variational Calculations

The formulation introduced in Section 3.2 starts from the time-dependent Schrödinger (or Dirac) equation, which may be solved approximately by expansion of the wave function in a truncated basis set and integration of the coupled channel equations. To find alternative methods of calculating electronic excitations, the TDSE may be replaced by the variational principle

$$\delta I = \delta \int_{-\infty}^{\infty} dt \, \langle \phi | \, H - i\partial/\partial t \, | \phi \rangle = 0 \tag{3.57}$$

where for simplicity $|\phi\rangle$ has been assumed to be normalized, $\langle \phi \mid \phi \rangle = 1$. For a discussion of (3.57) applied to atomic collisions, cf. (Ma 74b, Kl 78). This formulation has the advantage that $|\phi\rangle$ may be a trial wave function, depending on a set of time-dependent variational parameters $\lambda_i, i = 1 \cdots N$. The variational principle (3.57) then leads to equations of motion for the parameters λ_i. Their solution determines a wave function $|\phi\rangle$ which is "optimal" within the limits set by the chosen parametrization.

Since the parameters $\lambda_i(t)$ which determine the character of the trial wave function are controlled by the collision dynamics, this approach should be well suited for the intermediate range of collision velocities, where the electronic motion follows neither the atomic nor the molecular basis. The success of the variational treatment, of course, will depend critically on the choice of the trial wave function. Variational calculations have been performed in (Ch 68) where a time-dependent effective nuclear charge $Z(t)$ has been used as a parameter. During recent years the variational method has been extended and applied to ion–atom collisions by Kleber and collaborators (Kl 78, Kl 79). We will follow their presentation.

The Euler–Lagrange equations that can be derived from (3.57) by variation with respect to λ_i^* take the form

$$\begin{aligned}(\partial/\partial\lambda_i^*) \langle \phi | \, H \, | \phi \rangle = i \sum_{j=1} \dot{\lambda}_j (\langle (\partial/\partial\lambda_i) \, \phi \mid (\partial/\partial\lambda_j) \, \phi \rangle \\ - \langle (\partial/\partial\lambda_j^*) \, \phi \mid (\partial/\partial\lambda_i^*) \, \phi \rangle)\end{aligned} \tag{3.58}$$

where it has been assumed that $|\phi\rangle$ depends on the complex conjugate parameters λ_i^* solely through the normalization constant and that it has no explicit time dependence. For constant H and λ (3.58) reduces to the Rayleigh–Ritz variational equation.

To describe excitations it is important to allow the parameters to become complex. The real part can be viewed as a generalized coordinate;

the imaginary part then turns out to be the corresponding canonical momentum. The use of (3.58) is best seen when it is applied to a definite problem. In close ion–atom collisions the projectile penetrates the shell of the target electrons and causes them to feel an enhanced attraction. To some approximation this effect may be described by a time-dependent effective charge $Z(t)$ which will vary between the values Z_t and $Z_p + Z_t$. To describe excitations of the K shell of the target atom it is thus intuitively reasonable to choose as a trial wave function a scaled hydrogenic 1*s* state, centered at the target atom,

$$\phi(r) = \pi^{-1/2}[(\lambda + \lambda^*)\, e^2/2]^{3/2} \exp(-\lambda e^2 r_t) \tag{3.59}$$

The real part of $\lambda = \lambda_1 + i\lambda_2$ is proportional to the inverse of the K-shell Bohr radius, $\lambda_1 = \langle\phi|\, 1/r\, |\phi\rangle$, while the imaginary part corresponds to an average radial velocity of the electron, $\lambda_2 = (-1/e^2)\langle v_r\rangle = (i/e^2)(\langle\phi|\, \partial\phi/\partial r\rangle - \langle\partial\phi/\partial r\, |\phi\rangle)$. Equation (3.59) leads to two coupled differential equations (Kl 78)

$$\begin{aligned} \dot{\lambda}_1 &= \tfrac{2}{3}(\lambda_1 e^2)^2\, \lambda_2 \\ \dot{\lambda}_2 &= \tfrac{2}{3}(\lambda_1 e^2)^2 [Z(t) - \lambda_1] \end{aligned} \tag{3.60}$$

with

$$Z(t) = Z_t + Z_p[1 + 2\lambda_1 e^2 R(t)] \exp[-2\lambda_1 e^2 R(t)]$$

which have to be solved by numerical integration with initial conditions $\lambda_1(-\infty) = Z_t$ and $\lambda_2(-\infty) = 0$. In a completely adiabatic collision the parameter λ_1 would follow the charge $Z(t)$. In a real collision, however, the electronic motion deviates from the so-defined path of minimum energy and a radial motion is induced through λ_2. Radial vibrations will persist even after the collision, which means that the electron wave function contains an admixture of excited states. Transition probabilities can be deduced from a Fourier analysis of these "breathing mode" monopole vibrations.

The variational equations have been applied successfully to the problem of Coulomb ionization in highly asymmetric collisions (Kl 79). Agreement with experiment is achieved if attention is paid to the recoil of the target atom and if dipole polarization is included through a second variational parameter in the trial wave function (3.59).

The variational approach can be extended to the relativistic problem without basic difficulties. Instead of (3.59) the corresponding Dirac spinor wave function has to be used and the equations of motion become more involved. A first result (Kr 81) is shown in Figure 3.3. Here the mean electronic radius and the mean radial velocity, $\langle 1/r\rangle^{-1}$ and $\langle v_r\rangle$, respectively, are shown as a function of collision time for a 4.7-MeV/u collision of Pb + Cm. The K electron of the Cm atom is seen to be accelerated to the nuclear center

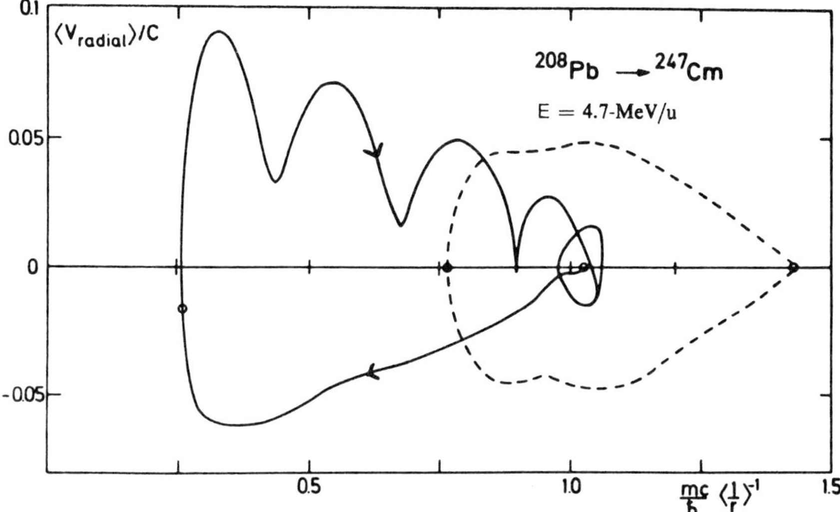

Figure 3.3. Trajectory of the $1s\sigma$ electron in the phase space defined by the shell radius $\langle 1/r\rangle^{-1}$ and the radial velocity $\langle v_{\text{radial}}\rangle$ for a 4.7-MeV/u Pb + Cm collision, calculated with the variational method. Solid line: relativistic calculation leading to strong excitation. Dashed line: nonrelativistic calculation with negligible excitation. From (Kr 81).

until a highly contracted (UA) wave function is formed. Strong monopole excitations are induced in the outgoing branch of the collision, quite analogous to the result of the coupled-channel analysis in Section 4.3. The trajectory in Figure 3.3 does not return to the starting point but moves on a closed curve in phase space, signaling that the final state is excited. For a comparison the figure also shows the result of a nonrelativistic calculation which fails to produce significant excitation.

A detailed study of the variational method applied to relativistic systems has not yet been performed. This approach may supplement the view of the excitation process that has been obtained from the coupled-channel analysis.

3.4.2. *The "Optimal" Basis*

The solution of the TDSE is to be expanded in some set of basis states. When solving the resulting coupled-channel problem this set has to be truncated. Therefore it can happen that the true time-dependent wave function $\Phi_i(t)$ leaves the subspace which is spanned by the basis functions $\phi_n(t)$. Therefore it is desirable to determine a basis set which in some way follows the development of $\Phi_i(t)$ or to have at least a quantitative measure for the quality of the basis. One such criterion was given by Theis *et al.* (Th 79, Th 81).

They consider an expansion of the type (3.17) in a basis of eigenstates of a model Hamiltonian H'

$$H'(\mathbf{r}, \xi)\, \lambda_m(\mathbf{r}, \xi) = \omega(\xi)\, \lambda_m(\mathbf{r}, \xi) \tag{3.61}$$

$H'(\mathbf{r}, \xi)$ depends on a (real) time-dependent parameter ξ which may be defined in various ways. The resulting coupled differential equations have the form (3.19), where in general both types of coupling, potential (H) and dynamic ($-i\partial/\partial t$), are acting. The discussion may be carried further, if it is assumed that $H'(\mathbf{r}, \xi)$ has the same functional dependence on ξ as the true Hamiltonian $H(\mathbf{r}, \rho)$ has on the time-dependent parameter ρ. In the case of an atomic collision, ρ generally is represented by the internuclear distance $R(t)$. Assuming further that the potential coupling operator $H(\rho) - H'(\xi)$ may be replaced by the first-order Taylor expansion $(\rho - \xi)\, \partial H'/\partial\xi$ the coupled-channel equations become

$$\dot{a}_l = -\sum_{m \neq l} a_m e^{-i(\chi_m - \chi_n)} \langle \lambda_l |\, \partial H'/\partial\xi \,| \lambda_m \rangle [\dot{\xi}/(\omega_m - \omega_l) + i(\rho - \xi)] \tag{3.62}$$

To determine an "optimal" trajectory for the parameter $\xi(t)$ Theis (Th 79) proposed a variational principle

$$\delta \int_{-\infty}^{\infty} dt \sum_{m \neq l} |\langle \lambda_l |\, \partial H'/\partial\xi \,| \lambda_m \rangle|^2 (\dot{\xi}^2/(\omega_m - \omega_l)^2 + (\rho - \xi)^2) = 0 \tag{3.63}$$

The functional in (3.63) is not derived in a rigorous or unambiguous way from (3.62) but it should serve the purpose to minimize the strength of excitation, integrated over the full collision.

A somewhat related discussion was given by D. Jakubassa (Ja 79a), who optimized a parameter which splits the potential due to the projectile nucleus in two parts: one included in the model Hamiltonian which determines the basis, the other part acting as a perturbation. Contrary to (3.63) such an approach does not allow for a varying degree of adiabaticity during the course of the collision.

Variation of (3.63) with respect to ξ yields the Euler–Lagrange equation

$$(\xi - \rho)_l = \frac{(\ddot{\xi} + 1/2\dot{\xi}\partial/\partial\xi)\, |\langle \lambda_l |\, \partial^2/\partial\xi^2 \,| \lambda_l \rangle|}{\langle \lambda_l |\, (\partial H'/\partial\xi)^2 \,| \lambda_l \rangle - |\langle \lambda_l |\, \partial H'/\partial\xi \,| \lambda_l \rangle|^2} \tag{3.64}$$

The deviation $(\xi - \rho)_l$ describes how the "optimum" basis states lag behind the adiabatic basis, and thus it is a measure for the violation of adiabaticity.

In the case of rotational coupling $[\rho \equiv \theta(t)]$, the deviation can become very large: The electronic wave function does not follow the swift rotation of the internuclear axis. Equation (3.64) also has been evaluated for radial

$[\rho \equiv R(t)]$ inner shell excitations in the collision of very heavy ions. For example in the system Pb + Pb even at energies well above the nuclear Coulomb barrier the "best" trajectory $\xi(t)$ to describe the $1s\sigma$ electron was found to deviate from the nuclear Rutherford orbit by no more than 1 fm (at the point of closest approach). This demonstrates that the inner shells in superheavy collision systems have to be described in the quasimolecular framework, even at rather high collision energies. Of course this argument does not exclude induced transitions, but the virtual excitations during the collision will be stronger with respect to any other basis of the type considered here. The adiabaticity of the $1s\sigma$ state is also demonstrated by the fact that the calculated "true" energy of the $1s\sigma$ electron in the course of the collision (So 80), defined by the expectation value $\langle \Phi_{1s}^{(+)} | H | \Phi_{1s}^{(+)} \rangle$, differs from the adiabatic value by less than 10%.

3.4.3. *Direct Integration Methods*

Nearly all theoretical studies of electronic excitation processes in ion–atom collisions rely on a basis expansion of the time-dependent wave function followed by coupled-channel calculations or perturbation theory. The most straightforward approach, i.e., the direct numerical integration of the time-dependent Schrödinger equation (TDSE), has been generally neglected. In nuclear physics, in contrast, the corresponding but more complex task to integrate the time-dependent Hartree–Fock (TDHF) equations has received much attention (for a review cf. Chapter 1 by Davies, Devi, Koonin, and Strayer in Volume 3 of this series).

Partly motivated by this development, V. Maruhn-Rezwani *et al.* recently studied the "basic problem" of atomic collision theory, the system $p + \mathrm{H}$ (Ma 79b). The one-electron TDSE was solved numerically with finite difference integration algorithm on a space grid with about 6000 mesh points. For simplicity the calculation was restricted to two space dimensions (cylinder symmetry), which is valid for head-on collisions, $b = 0$. Compared to the situation in nuclear physics, problems are introduced by the singular, at $r = 0$, and long-range nature of the Coulomb force.

The integration starts with the electron placed in a $1s$ orbital of the target nucleus. Probabilities for particular charge exchange or excitation channels can be obtained by projecting the final wave function on the corresponding travelling orbitals. More simply, the charge density can be integrated over the half space of the projectile to yield the total probability of charge transfer.

The calculations of (Ma 79b) show a rapid tunneling of the electron density back and forth between the two nuclei. This leads to marked oscillations of the final transfer probability as a function of impact energy. Such an effect has been known from multistate coupled-channel calculations of charge exchange in symmetric, resonant collisions. It is ascribed to the interference of two indistinguishable channels, where the electron follows the $1s\sigma$ or $2p\sigma$ molecular orbital. For a recent reference cf. (Lü 81).

Despite the high numerical expense the direct integration method may have a promising future for describing collisions in the intermediate velocity region, where neither adiabatic nor sudden pictures can be relied upon. The treatment has to be extended to noncentral collisions (rotational coupling). Ionization can be described by integrating the electron flux through the boundary of the collision region. Such calculations are under study (Hü 81). To describe collisions of very heavy ions correctly, relativistic effects are essential, i.e., a numerical integration of the time-dependent Dirac equation would be required.

In the final step also the electron–electron interaction has to be included which leads to the atomic TDHF problem (nonrelativistic or relativistic). The $p + \text{He}$ system has been investigated recently in this framework (Sa 81). The demanding task to integrate the atomic TDHF equations for collisions of very heavy ions has not yet been attempted.

An approximative but highly successful treatment of the many-electron problem is available in the Thomas–Fermi approach. Therefore it seems natural to describe a heavy-ion collision in the framework of the time-dependent Thomas Fermi (TDTF) theory. The TDTF equations date back to F. Bloch and H. Jensen. They can be written in a fluid dynamical form. Neglecting inhomogeneity and exchange corrections the equations read

$$
\begin{aligned}
&\rho(\mathbf{r}, t) + \nabla[\rho(\mathbf{r}, t)\, \nabla\chi(\mathbf{r}, t)] = 0 \\
&\chi(\mathbf{r}, t) + \tfrac{1}{2}[\nabla\chi(\mathbf{r}, t)]^2 + \tfrac{5}{3}c_1\rho(\mathbf{r}, t)^{2/3} + V(\mathbf{r}, t) \\
&\qquad + \int \frac{\rho(\mathbf{r}', t)}{|\mathbf{r} - \mathbf{r}'|}\, d^3r' = 0
\end{aligned}
\tag{3.65}
$$

where $\rho(\mathbf{r}, t)$ is the electron density, $\chi(\mathbf{r}, t)$ is a velocity potential, $V(\mathbf{r}, t)$ denotes the nuclear Coulomb potential, and $c_1 = 3/10\ (3\pi^2)^{2/3}$.

Horbatsch and Dreizler (Ho 81) have solved (3.65) numerically and studied the time development of the electron density and flux in a collision. Since shell effects are important, the results will not give an accurate description of excitation processes. The electron density obtained in the TDTF approach, however, may be used to generate a realistic screening potential, depending on collision energy and impact parameter. Using this potential the solution of the single-particle TDSE can be performed in the usual way. Such a dynamical treatment of electron screening is preferable to the usual static prescription (Ho 81).

3.4.4. Classical Electron Trajectories

Most of the theoretical models used in atomic scattering theory are attempts to find an approximate solution of the time-dependent Schrödinger equation. However, one may try a completely different approach and treat both electrons and nuclei as classical particles. The scattering problem then, in principle, is solved quite easily by numerical integration of the set of Hamilton's equations $q_i = \partial H_{\text{class}}/\partial p_i$, $p_i = -\partial H_{\text{class}}/\partial q_i$.

To account for the nondeterministic nature of the quantum problem, a large number of integrations has to be performed with random initial conditions, leading to a statistical description of the final state. The initial conditions for the electronic motion are defined by a microcanonical distribution: The probability density is distributed homogeneously on a five-dimensional hypersurface in the electron phase space, defined by the requirement of constant energy $H_{\text{class}}(q_i, p_i) = E$. E is identified with the quantum mechanical binding energy in the initial state.

The classical trajectory method was introduced by Abrines and Percical (Ab 66) and applied to the system $p + \text{H}$. Since that time charge transfer and ionization in collisions of highly stripped ions have been calculated quite successfully in many cases. The agreement with experiment is best at intermediate collision velocities $v_e \simeq v_N$. One reason for the accomplishments of the classical model is founded in the fact that the classical microcanonical distribution in momentum space agrees with the quantum mechanical result (the agreement is exact for nonrelativistic hydrogenic states). This does not hold true, however, in coordinate space. The model can be improved if the microcanonical distribution is replaced by a Wigner function (Ei 81).

The classical trajectory model has been extended by Teubner *et al.* (Te 80) to the case where both the electronic and nuclear motion are relativistic; see Section 4.6. The classical many-body problem with inclusion of the Pauli principle is studied in (Ki 80c).

4. Theory of Inner Shell Excitation

4.1. Introduction

Since the beginning quantum mechanics a wealth of theoretical models and calculations has been developed pertaining to the physics of inner shell excitation in atomic collisions. The present discussion does not strive for completeness, only some of the basic ideas and results will be treated. The main emphasis will lie on methods applicable to very heavy collision systems, where important progress has been made in recent years. More detailed expositions of various aspects, in particular for light systems, can be found in the reviews (Br 76, Me 77, Br 78a, Mo 78, Bu 79, De 81) and in the older works of (Mo 65, Ma 74b).

The discussion will rely on the semiclassical approximation for the nuclear motion and on the independent electron model. Basis expansions of type (3.12) will be used, since this method has proven most successful and flexible up to now. The existing models then can be classified according to the type of basis employed.

(1) If the projectile acts only as a weak perturbation, a time-independent atomic basis will suffice. This is a good approximation in highly asymmetric systems, i.e. (to be definite we will always assume that the projectile is the lighter collision partner)

$$Z_1/Z_2 \ll 1 \tag{4.1}$$

(2) For "slow" collisions the electrons will adjust to the field of both nuclear centers so that a molecular basis has to be used. A rough criterion whether a molecular state can be formed is obtained comparing the velocities of electronic and nuclear motion. The collision will be adiabatic if

$$v_{\text{nuc}}/v_{\text{el}} \ll 1 \tag{4.2}$$

(3) In intermediate cases where neither the limits (1) nor (2) give a good description of the electronic motion, an expansion in a basis of "pseudostates" may offer calculational advantages. Any convenient complete set can be chosen; it will not necessarily consist of eigenstates of a Hamiltonian.

Depending on the violence of the collision, either the excitation process

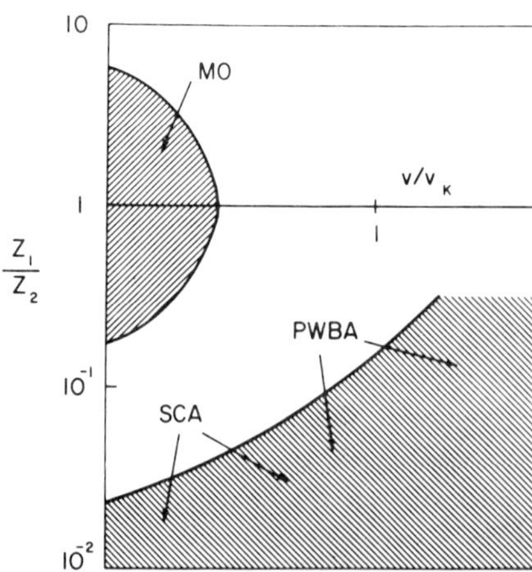

Figure 4.1. Schematic graph indicating the region of validity for various approximation methods in dependence of the charge ratio Z_1/Z_2 and the nuclear velocity v (in units of the K-shell velocity v_K). The plane wave Born (PWBA) or first-order semiclassical (SCA) approximation is valid for asymmetric or fast collisions, while slow near-symmetric collisions call for a quasimolecular (MO) description. From (Ma 75a).

can be treated in perturbation theory or the coupled differential equations (3.19) have to be solved with a sufficient number of channels. The regions where the assumptions of the atomic or molecular model are fulfilled are illustrated schematically in Figure 4.1, taken from Madison and Merzbacher (Ma 75a).

4.2. Atomic Models

Inner shell excitation through charged particle impact can be treated in a relatively simple way, if the target atom is disturbed only weakly by the projectile. This is guaranteed, if condition (4.1) is satisfied, i.e., in highly asymmetric collision systems. Then it is adequate to describe the wave function of the active electron in terms of unperturbed states of the target atom. Since furthermore first-order perturbation theory very often will be sufficient, the transition probability $i \to f$ is

$$P_{i \to f}(b) = Z_1^2 e^4 \left| \int_{-\infty}^{\infty} dt\, e^{i\Delta E t} \langle \phi_f | \frac{1}{|\mathbf{r} - \mathbf{R}(t)|} | \phi_i \rangle \right|^2 \tag{4.3}$$

This approximation was first discussed in detail by Bang and Hansteen (Ba 59). It has since become known under the (somewhat too general) name SCA model (semiclassical approximation). The problem presented by (4.3) is very similar to the semiclassical description of the extensively studied process of nuclear Coulomb excitation. The evaluation, however, is complicated by the fact that the projectile deeply penetrates the electron shell to be excited. Only comparatively recently has the full solution of (4.3) been obtained, mainly through the efforts of groups in Bergen (e.g., Aa 76, Am 78) and Basel (Pa 78a, b). The subject has been reviewed in (Ha 75, Ko 80).

The solution of the four-dimensional integral (4.3) is facilitated by separating the electronic and nuclear coordinates. Either the familiar multipole expansion of $1/|\mathbf{r} - \mathbf{R}|$ or the Fourier representation of this function can be employed. Both approaches—one formulated in coordinate space, the other in momentum space—lead to single integrals over a trajectory factor, which depends on the nuclear motion only, multiplied by an electronic form factor.

If the repulsion between the colliding nuclei is neglected, the projectile moves on a straight line trajectory with constant velocity, $\mathbf{R}(t) = \mathbf{b} + \mathbf{v}t$, which greatly simplifies the calculation (Ba 59). As it turns out the straight line SCA is equivalent to the plane wave Born approximation (PWBA) reviewed in (Me 58).

In the PWBA approach the total transition cross section is

$$\begin{aligned}\sigma_{i\to f}^{\mathrm{PWBA}} &= \frac{1}{2}\left(\frac{Ze^2}{\pi v_1}\right)^2 \int_{q_0}^{q_{\max}} q\,dq \left|\int\int d^3r\,d^3R\,e^{i\mathbf{q}\cdot\mathbf{r}}\phi_f^*\frac{1}{|\mathbf{r}-\mathbf{R}|}\phi_i\right|^2 \\ &= 8\pi\left(\frac{Ze^2}{v_1}\right)^2\int_{q_0}^{q_{\max}} dq\,F_{fi}(q)/q^3\end{aligned} \tag{4.4}$$

with the transition form factor

$$F_{fi}(\mathbf{q}) = \int d^3r\,e^{i\mathbf{q}\cdot\mathbf{r}}\phi_f^*(\mathbf{r})\,\phi_i(\mathbf{r}) \tag{4.5}$$

The integration covers the range of permissible values of the momentum transfer to the electron, where $\mathbf{q}=\mathbf{k}_i-\mathbf{k}_f$ is the difference between initial and final nuclear momentum. Both the integrated PWBA cross section (4.4) and the "impact parameter dependence" (defined by a partial wave expansion) agree with the straight line SCA result (Ta 77), although the condition for classical motion (3.1) might seem to conflict with the region of validity of the Born approximation.

To calculate total cross sections the straight line (or plane wave) approximation is often sufficient since here distant collisions are dominant. In the impact parameter dependence $P(b)$, however, close collisions can be singled out where the Coulomb deflection of the projectile may not be neglected. Therefore the SCA integral has to be solved using Rutherford hyperbolas for the nuclear motion. The trajectory can be expressed conveniently in the parametrized from

$$\begin{aligned}R(w) &= a(1+\varepsilon\cosh w) \\ t(w) &= (a/v_\infty)(w+\varepsilon\sinh w)\end{aligned} \tag{4.6}$$

where $a = Z_1Z_1e^2/2E_{\mathrm{cm}}$ is half the distance of closest approach in central collisions, $\varepsilon=[1+(b/a)^2]^{1/2}=1/\sin(\theta_{\mathrm{cm}}/2)$ is the eccentricity of the orbit and v_∞ the asymptotic relative velocity. Small corrections can arise from the energy loss of the projectile and from the influence of electron screening on the scattering trajectory. As a consequence of Coulomb repulsion the excitation rates can be reduced substantially, since the distance of closest approach is increased and the velocity $\dot{\mathbf{R}}(t)$ is slowed down compared with the straight line approximation. The effect of Coulomb deflection can be approximated by a simple multiplicative factor (Ba 59)

$$P(b)|_{\mathrm{Coul}} \simeq e^{-(\Delta E/v)a}P(b)|_{\mathrm{str.\,line}} \tag{4.7}$$

The full (hyperbolic) SCA theory is obtained as the classical limit of the distorted wave Born approximation (DWBA) (Pa 78a) which is much more difficult to evaluate. Quantum corrections are usually small, even if the condition (3.1) is not too well satisfied, i.e., $\eta \simeq 1$.

At large scattering angles the ionization probability (An 76a) shows a strong variation of the type $P(\theta) \propto A(1 + B \cos \theta)$. This can be explained as a geometrical effect caused by the interference of excitation taking place on the two half-trajectories, which have a relative orientation angle θ ("broken line effect").

Several additional corrections are necessary to obtain full agreement with experiment. Atomic screening can be included through the use of reduced effective charges and the replacement of hydrogenic binding energies by empirical values. More rigorously, the wave functions and energies should be calculated from a screened atomic potential, e.g., in the Hartree–Fock–Slater approach.

For all but the lightest target atoms relativistic effects are important, which calls for the use of Dirac wave functions (Aa 76, Pa 78b). This might seem surprising since binding energies are not much affected by relativistic corrections. The ionization process, however, is very sensitive to the electron density at small values of r. As discussed in Chapter 2., Dirac wave functions are enhanced at the origin by a factor $r^{\gamma-1}$, where $\gamma = [1 - (Z_2\alpha)^2]^{1/2}$. The resulting increase of the potential coupling matrix elements at small distances [or equivalently the enhancement of high momentum components in the form factor $F(q)$] raises the ionization rates. In heavy atoms the difference between relativistic and nonrelativistic results can exceed an order of magnitude even if the projectile is a light particle.

In close collisions (large scattering angles of the projectile) an additional effect, not included in the original SCA, comes into play. The atomic basis used is centered at the target nucleus which has a finite mass and thus does not represent an inertial frame, due to nuclear recoil. Therefore the potential has to be supplemented by an additional coupling operator (Ci 65, Am 78). In the simplest form this correction is proportional to the target acceleration

$$H_{\text{recoil}} = -\ddot{\mathbf{R}}_2 \cdot \mathbf{r} \tag{4.8}$$

This operator modifies the dipole transition amplitudes. The shape of $P(b)$ at small impact parameters can be significantly altered by the recoil effect.

One further correction has to be discussed, which marks the limits of the atomic SCA model. Even in quite asymmetric collisions, cf. (4.1), the wave function of the target electron is distorted by the attractive force exerted by the projectile nucleus. This could, of course, be described correctly going to higher orders in the perturbation series. In a more physical picture one can try to retain the first-order expression and, instead, modify the energies E_k and wave functions ϕ_k entering (4.3). The dominant effect will be an increase of binding energies due to the presence of the projectile nucleus. This in turn leads to a reduction of ionization rates, since now a larger energy transfer is required. Corrections due to the "binding" and "polarization" effects have been discussed by Brandt and co-workers (Ba 73). In lowest order an improved approximation of the binding energies is given by the expectation values of the two-center Hamiltonian in the atomic basis ("distortion approximation"). This implies the replacement of the phase factor $(E_f - E_i)t$ by

$$\int_{t_0}^{t} dt' \, (\langle \phi_f | \, H(t) \, | \phi_f \rangle - \langle \phi_i | \, H(t) \, | \phi_i \rangle)$$

As a simpler prescription, constant energies can be substituted, e.g., those of the united atom limit. In general the basis should be made time dependent. This approach, however, leads to the molecular model discussed in the next sections. A dynamical description of binding and polarization effects has been attempted in the framework of Section 3.4.1 (Kl 79).

Recently Reading and collaborators (Fo 77, Be 80a) have developed efficient numerical methods to treat Coulomb ionization and charge transfer in asymmetric systems in all orders using an atomic pseudostate expansion.

Briggs (Br 75) suggested a simple extension of the SCA model to the case of slow near symmetric collisions. If the excitation occurs mainly at close distances, the molecular wave functions and energies may be approximated by their united atom limit, ϕ_k^0, E_k^0. The transition amplitude

$$a_{fi} = i \int_{-\infty}^{\infty} dt \, \langle \phi_f^0 | \, \frac{Ze^2}{|\mathbf{r} - \mathbf{R}_1(t)|} + \frac{Ze^2}{|\mathbf{r} - \mathbf{R}_2(t)|} \, | \phi_i^0 \rangle \, e^{i(E_f - E_i)t} \tag{4.9}$$

is a coherent sum of two terms which have exactly the same form as the usual SCA expression (4.3).

The predictions of the SCA (and PWBA) model are governed by the parameter $q_0 = \Delta E / v_\infty$ which coincides with the minimum momentum transfer in (4.4) if $\Delta E / E_1 \ll 1$. The reciprocal value is the adiabatic radius $r_{\text{ad}} = v_\infty / \Delta E$. This quantity can be identified with the typical period of oscillation of the exponential factor in the Fourier integral (4.3). Therefore it is quite plausible that r_{ad}, and not the radius of the electron shell, determines the characteristic length scale in the collision.

It turns out, for example, that the weighted impact parameter dependent excitation probability $bP(b)$ has its maximum at $b = r_{\text{ad}}$ (Ba 59, Ba 79b). In slow collisions this value can be much smaller than the electron shell radius (to be specific, we have $v_{\text{el}} / v_\infty \simeq r_{\text{shell}} / r_{\text{ad}}$).

In the multipole expansion of (4.3) only the monopole and dipole terms contribute significantly, the latter being important at high velocities or large impact parameters.

4.3. Molecular Models

In slow collisions of symmetric or nearly symmetric systems the electrons can adjust to the combined field of both atoms. If the collision were to proceed with infinitely small velocity no transitions would be effected. Small deviations from adiabaticity, however, may cause a transfer of the electron to a neighboring molecular orbital thus leading to excitation in the final system. The excitation process depends sensitively on the variation of molecular orbitals as a function of R. For instance, it has been noted very early (Hu 27, Mu 28) that levels can be lifted to higher shells if the nuclei come close to each other. This is the basis of the "promotion model" put forward by Fano and Lichten (Fa 65): Upon adiabatic variation of the internuclear distance the binding energy of an electron can approach that of a higher shell orbital. Thus there is a well-defined region where it can more easily (without the need for large energy transfer) couple to vacant states.

The presence of very high excitation cross sections in slow collisions can be explained in this way. First evidence for quasimolecular effects in heavy-ion collisions was found by Specht *et al.* (Ar 63, Sp 65). Working with fission products as projectiles they observed characteristic variations of hole production as a function of target charge, caused by "level matching." In connection with the test of quantum electrodynamics of strong fields the existence of superheavy quasimolecules had been suggested by the Frankfurt school (see Ra 71), who transplanted the idea of nuclear molecules to the atomic field. In this sense the nuclear two-center shell model is the predecessor of the two-center Dirac equation.

In cases where the promotion mechanism is not operative, direct transitions between distant molecular orbitals or direct ionization to the continuum will prevail. This process will turn out to be important in very heavy collision systems.

4.3.1. Quasimolecular Correlation Diagrams

The calculation of energies and wave functions of an atomic two-center system is the basic problem of quantum chemistry. Due to the high mass ratio between the two types of particles involved, the electronic wave functions can be calculated assuming a fixed position of the nuclei (the Born–Oppenheimer approximation). Nondiagonal coupling terms between electronic and nuclear motion can be treated as perturbations on these states. The quasimolecular theory of electronic excitations in adiabatic atomic collisions is based on the same picture. It is also known under the name PSS method (perturbed stationary states) (Ma 33). As the main difference to chemistry, the internuclear distance now is a given function of time which varies over a wide range of values. It is of paramount importance to construct reliable correlation diagrams depicting the variation of molecular energy eigenvalues as a function of R between the united atom (UA) limit $R \to 0$ and the separated atom (SA) limit $R \to \infty$.

For light collision systems it is sufficient to solve the two-center Schrödinger (TCS) equation

$$[-\tfrac{1}{2}\Delta - Z_1 e^2/r_1 - Z_2 e^2/r_2]\,\phi_n(r) = E_n(r)\,\phi_n(r) \tag{4.10}$$

The simplest example is given by a symmetric two-center one-electron molecule like H_2^+ and its analogs. The problem of solving the TCS partial differential equation is much simplified by the fact that (4.10) can be separated into three ordinary differential equations, if elliptical coordinates $\xi = (r_1 + r_2)/2R$, $\eta = (r_1 - r_2)/2R$, and ϕ are introduced. The resulting problem can be solved with arbitrary accuracy by direct integration, by a series representation, or by expansion in a suitable basis set (Te 30, see also He 65, Ba 68). The problem has two obvious symmetries, which allow for a classifivation of the solutions: The cylindrical symmetry with respect to the internuclear axis leads to a good projection of angular momentum, which can take on the values $\sigma, \pi, \delta, \cdots$ (i.e., $|m| = 1, 2, 3 \cdots$). Furthermore the solutions are parity eigenfunctions (gerade or ungerade states).

Figure 4.2 shows the nonrelativistic one-electron correlation diagram for a symmetric system. The levels are labeled by the state to which they correlate in the limit $R \to 0$. The electron terms of the three most deeply bound shells are seen to vary in a rather complicated way as a function of internuclear distance. Details of the correlation diagram can be understood from the nodal structure of the wave functions. Since the number of nodes is preserved when varying R, correlation rules can be established which determine how the states in the UA and SA limits are connected.

For the SA K-shell the situation is quite clear. Most deeply bound is the even $1s\sigma_g$ state which at small distances is well separated in energy from the other levels (note the logarithmic scale). The $2p\sigma_u$ MO which originates from the odd superposition of SA $1s$ states, however, is antibinding and correlates to the UA L shell. Here it meets with the $2p\pi_u$ and $2s\sigma_g$ MO's which originate from the SA L shell. More drastic examples of the rather counterintuitive process of electron promotion, where the binding energy decreases for approaching nuclei and an electron is lifted to a UA state with larger principal quantum number, can be found in higher shells. In particular, the $4f\sigma_u$ level is even doubly promoted, joining $2p$(SA) with $4f$(UA). In higher shells a complicated pattern of level crossings is observed. It can be explained by (i) the Stark effect at large distances R, (ii) the effect of the quadrupole field near the UA limit, and (iii) the promotion mechanism.

As long as spin–orbit and relativistic effects are neglected, Figure 4.2 is valid for any symmetric one-electron molecule since the TCS Equation (4.10)

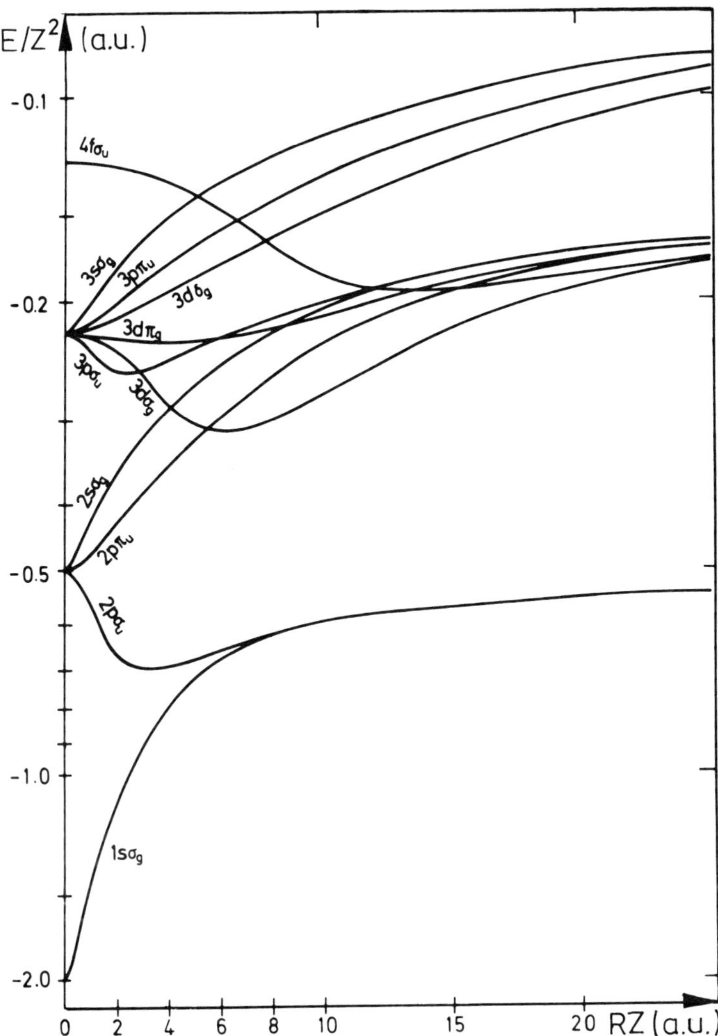

Figure 4.2. Energy terms of the H_2^+ molecule as a function of internuclear separation. Atomic units scaled with Z are used, therefore the figure holds for any nonrelativistic symmetric one-electron molecule.

can be rescaled using $E' = EZ^2$ and $R' = R/Z$. The situation becomes more involved in the case of unequal nuclear charges, $Z_1 \neq Z_2$. Parity no longer is a good quantum number. In the limit $R \rightarrow \infty$ each of the MO's correlates to a definitive atomic state centered at one of the two nuclei instead of a linear combination of atomic wave functions. The general correlation rules are worked out in (Ge 61, Ba 72).

In principle, the one-electron two-center model is a highly idealized approximation. Even though inner-shell excitation can be well understood in an independent particle picture, the effective potential seen by the electron differs from pure Coulomb type due to screening. In quantum chemistry the solution of this problem is relevant mainly at the equilibrium distance of diatomic molecules. The study of complete correlation diagrams for many-electron molecules has received full attention only quite recently. The most elaborate approaches solve the two-center Hartree–Fock equations. The calculations are performed using an expansion in a basis of single- or multicentered orbitals and call for extensive numerical work.

Much of the effort can be saved if the many-body problem is attacked in the quantum statistical approach. This will have particular advantages in heavy systems, since the computing time required to generate the interelectronic potential is independent of the number of electrons. In the Thomas–Fermi model the molecular electrostatic potential $V(r, R)$ (electron–nucleus plus electron–electron interaction) solves the differential equation

$$\nabla^2 V = 8(2/3\pi)^{1/2}(-V)^{3/2} \tag{4.11}$$

Equation (4.11) with the appropriate two-center boundary conditions can be integrated directly by numerical procedures. Instead, the solution may be approximated by an Ansatz introduced very early by Hund (Hu 32)

$$V(\mathbf{r}, R) = \Phi_1(r_1, R)\, Z_1/r_1 + \Phi_2(r_2, R)\, Z_2/r_2 \tag{4.12}$$

(4.12) contains the linear superposition of two screened atomic Coulomb potentials. For the atomic screening functions $\Phi_i(r_i, R)$ analytical approximations are known, the parameters of which now, of course, have to be made R dependent. In their "variable screening model" Eichler and Wille (Ei 75) used a phenomenological interpolation prescription. From the potential (4.12) correlation diagrams can be generated. A large number of systems has been investigated (Ei 75, Ei 76). Dreizler, Gross, and Toepfer (Gr 79, To 79) started from the Thomas–Fermi–Dirac–Weizsäcker model and determined the screening functions $\Phi_i(r_i, R)$ by an optimization procedure. As an example Figure 4.3a shows the correlation diagram for Ne + Ne ($Z = 20$). The energies are in good agreement with the result of the corresponding Hartree–Fock calculation (La 72), Figure 4.3b.

Compared with Figure 4.2 the outer shells are much less strongly bound and the number of level crossings is reduced. Instead, several avoided crossings are observed. This is in accordance with the noncrossing rule of Neumann and Wigner which states that levels having the same symmetry repel each other and cannot become degenerate upon variation of a parameter in the Hamiltonian. In the unscreened case an additional constant

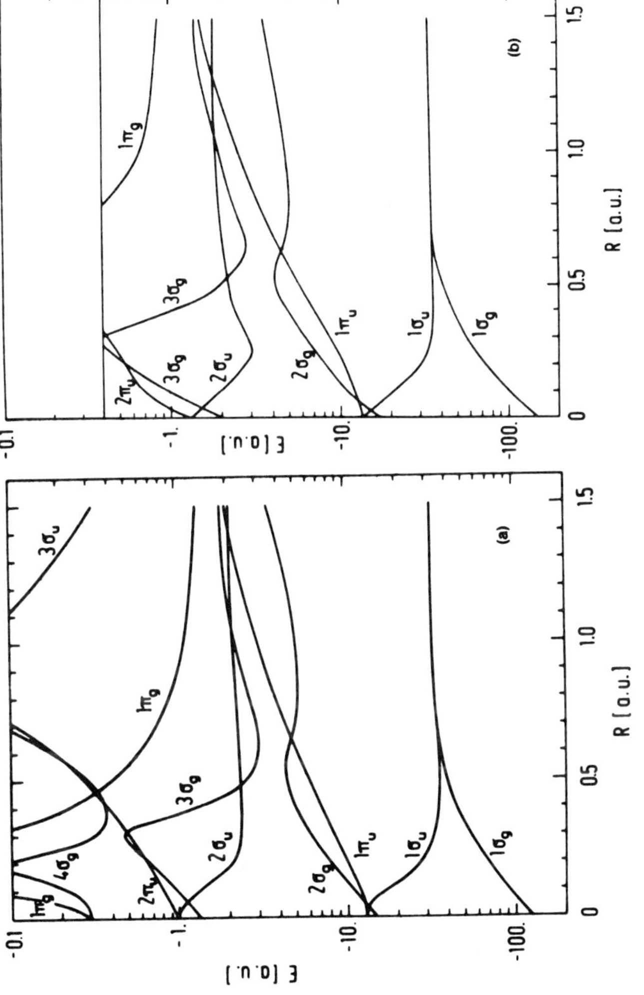

Figure 4.3. Nonrelativistic many-electron correlation diagram for the Ne + Ne quasimolecule. The Thomas–Fermi–Dirac–Weizsäcker model of (To 79), part (a), agrees well with the results of a Hartree–Fock calculation (La 72), part (b). From (To 79).

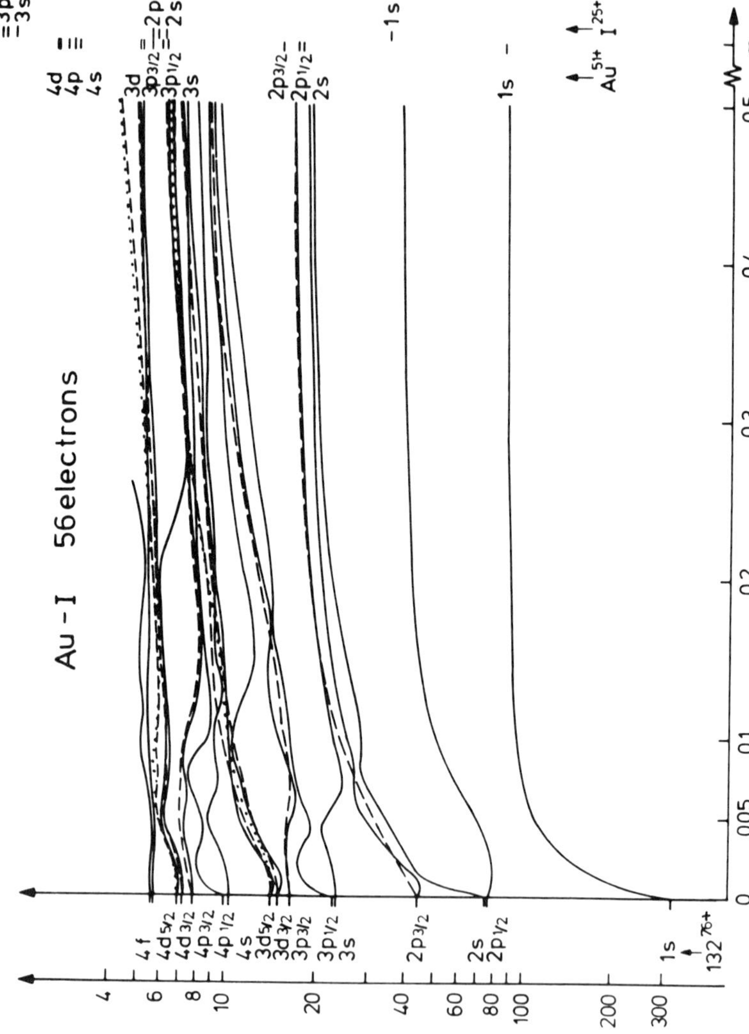

Figure 4.4. Corelation diagram for the $(\mathrm{Au} + \mathrm{I})^{76+}$ system, calculated in the relativistic Hartree–Fock–Slater model. From (Fr 76b).

of motion exists, due to the special nature of the $1/r$ potential, so that more crossings are allowed. The physical relevance of crossings will be discussed in Section 4.3.3. Correlation diagrams for screened symmetric systems exhibit the maximum number of avoided crossings, since the only good quantum number left is m. An example is shown in the Hartree–Fock–Slater correlation diagram of Figure 4.4 calculated by Fricke *et al.* (Fr 76b) for the ionized Au + I quasimolecule ($Z_1/Z_2 = 0.67$). This system is superheavy ($Z = 132$) and has been calculated with the Dirac equation (see below).

4.3.2. *The Two-Center Dirac Equation*

The physics of inner shell excitation in high-Z heavy-ion collisions is governed by relativistic effects. A study of atomic physics in the region of superheavy systems therefore has to be based on the solutions of the two-center Dirac (TCD) equation

$$[\boldsymbol{\alpha} \cdot \mathbf{p} + \beta m - Z_1 e^2/r_1 - Z_2 e^2/r_2]\, \phi_n = E_n \phi_n \tag{4.13}$$

This equation is more difficult to solve than (4.10) since it does not separate in any coordinate system. Furthermore it is dangerous to rely upon a variational principle of Rayleigh–Ritz type since the Dirac Hamiltonian is not bounded from below: Spurous contributions from states of the negative energy continuum can be admixed to a variational wave function.

The TCD equation was first solved by B. Müller *et al.* (Mü 73a, Mü 76b). Working in the elliptical coordinate system he expanded the solution in a set of analytically given basis states and diagonalized the Hamiltonian. Convergence of higher shells was difficult to achieve in this way so that Rafelski and Müller (Ra 76a) subsequently developed a method based on direct numerical integration. Equation (4.13) is equivalent to a set of two partial differential equations in the variables r and θ (the azimuthal angle ϕ is split off easily). (Ra 76a) have chosen to discretize the angular dependence by the use of a multipole expansion

$$\phi_\mu(r) = \sum_\kappa \phi_{\mu\kappa}(\mathbf{r}) = \sum_\kappa \begin{pmatrix} g_{\mu\kappa}(r)\,\chi_\kappa \\ if_{\mu\kappa}(r)\,\chi_{-\kappa} \end{pmatrix} \tag{4.14}$$

where the spinors χ_κ^μ have been defined in (2.5). If also the two-center potential is expanded in multipoles,

$$V(\mathbf{r}, R) = \sum_l V_l(r, R)\, P_l(\cos\theta) \tag{4.15}$$

the radial wave functions are determined by a set of ordinary differential equations

$$\begin{aligned}
\frac{d}{dr} f_{\mu\kappa}(r) &= \frac{\kappa - 1}{r} f_{\mu\kappa}(r) - (E - m)\, g_{\mu\kappa}(r) \\
&\quad + \sum_{s,l} g_{\mu s}(r)\, V_l(r, R)\, A^{\mu}_{\kappa l s} \\
\frac{d}{dr} g_{\mu\kappa}(r) &= -\frac{\kappa + 1}{r} g_{\mu\kappa}(r) + (E + m)\, f_{\mu\kappa}(r) \\
&\quad - \sum_{s,l} f_{\mu s}(r)\, V_l(r, R)\, A^{\mu}_{-\kappa l -s}
\end{aligned} \tag{4.16}$$

The coefficients $A_{\kappa l s} = \langle \chi^{\mu}_{\kappa} |\, P_l(\cos\theta)\, | \chi^{\mu}_{s} \rangle$ are determined from angular momentum algebra. After the sum in (4.14) has been truncated at a sufficiently large angular momentum $j_{\max}$ (typically 5 terms), the set of $2(2j_{\max} + 1)$ coupled differential equations (4.16) is integrated numerically. The energy eigenvalue E_n is determined iteratively to satisfy the boundary conditions at $r \to \infty$ and $r \to 0$.

Typical one-electron correlation diagrams of two very heavy systems, Pb + Pb ($Z = 164$) and U + Cm ($Z = 188$) are presented in Figures 4.5a and 4.5b. The latter quasimolecule becomes supercritical in close collisions. It is the most highly charged system accessible so far in heavy-ion collisions. A comparison with the nonrelativistic case, Figure 4.2, reveals drastic differences. Since relativistic effects act differently on various states, depending on angular momentum, the ordering of levels is changed. All inner shell states gain in binding energy with decreasing internuclear distance. In particular, the states correlating to the UA limit $|\kappa| = 1$, i.e., $ns_{1/2}$ and $np_{1/2}$, are strongly dependent on R (approximately following a power law $E_B \propto R^{-\gamma}$) even at small internuclear distances. The $1s\sigma$ level in Figure 4.5b, for example, triples its binding energy over the last 300 fm of approach, the total variation amounts to a factor 10. Clearly, this is due to the "collapse of the wave functions" discussed in Section 2. The length scale of the problem is no longer governed by the Bohr radius or the Compton wavelength but solely by the value of R. A further difference between nonrelativistic and highly relativistic correlation diagrams is the huge fine structure splitting, which, e.g., removes the degeneracy of $2p_{1/2}\sigma$ and $2s_{1/2}\sigma$ in the UA limit. From this discussion it is clear that characteristic deviations from the nonrelativistic case can be expected for inner shell excitation in close collisions of very heavy systems.

The TCD calculations allow an accurate determination of the critical distance (Mü 73a, Ra 76b) where the binding energy reaches $2mc^2$. These values are decisive for the physics of positron emission in supercritical heavy-ion collisions. Table 1 shows $R_{\rm cr}$ for a number of systems (Be 80b) with total charge between 178 and 190. Since the wave functions are highly localized the finite nuclear size can not be neglected. Equation (4.13) has

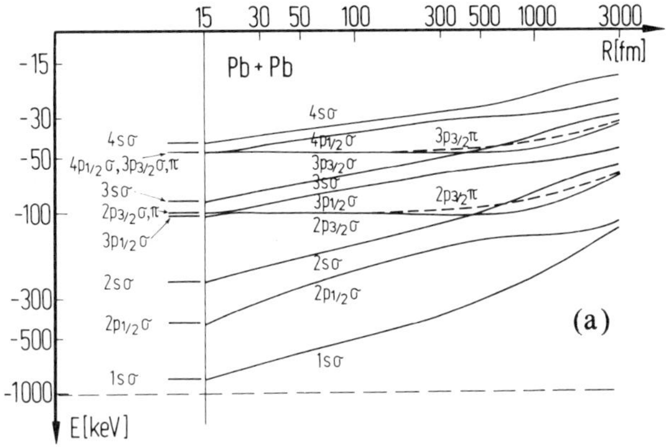

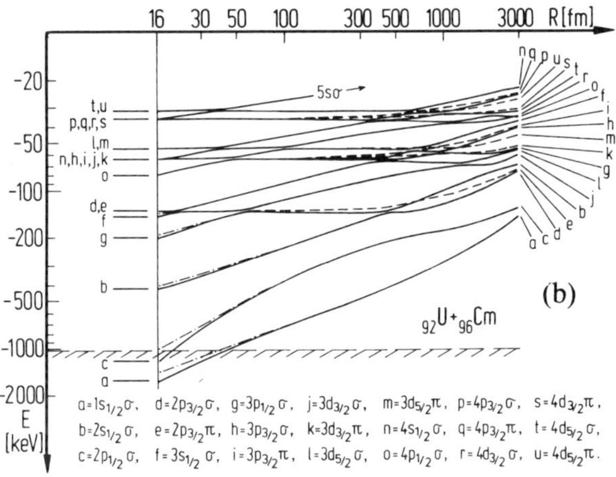

Figure 4.5. (a) Relativistic one-electron correlation diagram of the Pb + Pb quasimolecule showing the lowest σ (solid lines) and π (dashed lines) states. (b) Same as (a) for the very heavy system U + Cm. The dash-dotted lines show the decrease of binding energy caused by the finite extension of the nuclei. From (So 78a, Be 80b).

been solved for the potential generated by two homogeneously charged spheres of radius $R_i = 1.28\ A^{1/3}$; cf. the dashed lines lines in Figure 4.5. For the states $1s\sigma$ and $2p_{1/2}\sigma$ the nuclear volume effect can even exceed 100 keV. The table gives the critical distances both for pointlike and for extended nuclei. In addition the result of Hartree–Fock–Slater calculations in the monopole approximation (cf. Section 4.4.1) is included (Wi 79). The critical distances are reduced due to electron screening. Values for the critical distance also have been calculated by V.S. Popov and collaborators using a variational method. Having eliminated an error in their earlier calculations,

Table 1. Values of the Critical Distance R_{cr} where the 1s State Reaches $E = -mc^2$ for Various Quasimolecular Systems Assuming (1) Pointlike Nuclei, (2) Extended Nuclei, (3) Electron Screening Calculated in the Hartree–Fock–Slater Approximation

System	Z	R_{cr}^{point}	R_{cr}^{ext}	R_{cr}^{HFS}
Pb + Cm	178	26.5	23.2	17.5
Pb + Cf	180	30.0	26.4	20.5
U + U	184	37.2	34.5	26.2
U + Cm	188	4.5	43.0	32.8
U + Cf	190	48.7	47.4	36.3

the values for R_{cr}^{ext} agree well with the results shown in the table (Li 80b). For point nuclei the calculated critical distances of (Li 77) are larger, probably caused by a higher concentration of electron density close to the nuclei.

Relativistic many-electron correlation diagrams have been obtained by Fricke, Rosen *et al.* using the self-consistent field method (Fr 76b, Se 81). A relativistic generalization of the Thomas–Fermi method is developed by Gross and Dreizler (Gr 81). While the innermost states are not very much affected by screening (e.g., about 8% energy shift for $1s\sigma$), higher levels become significantly less bound. Also the positions of avoided crossings and energy minima are altered.

It should be noted that the Hartree–Fock calculations assume total relaxation of the electron shells at each internuclear distance. Even in rather "slow" collisions this assumption is not realistic since, at least, the higher MOs are not adiabatic. Therefore the full self-consistent diagonalization of the electron–electron interaction may be inadequate.

4.3.3. Specific Couplings

The quasimolecular model is particularly successful in situations where the electronic motion is largely adiabatic, except for localized regions of strong coupling. Here the transition probabilities may become large or even close to unity. We will discuss three examples of this kind.

(i) Avoided Crossings. Realistic quasimolecular correlation diagrams—cf. Figures 4.3–4.5—may exhibit a large number of avoided level crossings. In an extremely slow collision the electrons would follow these adiabatic levels, leading to a strikingly different behavior compared to the case of real crossings (Note that the crossings are allowed in one-electron two-center Schrödinger correlation diagrams; cf. Figure 4.2). Clearly, the deviations from a pure $Z_1e^2/r_1 + Z_2e^2/r_2$ type potential (caused by screening, finite nuclear size, spin, and relativity) should not have such a large effect. This anticipation is borne out by an inspection of the time-

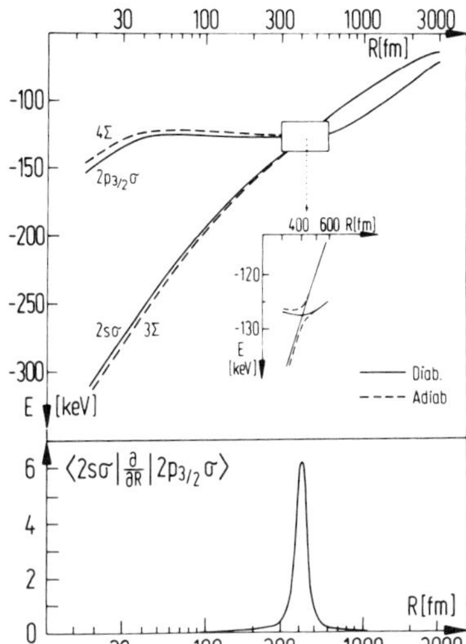

Figure 4.6. Adiabatic quasimolecular states (dashed lines) $2s\sigma$ and $2p_{3/2}\sigma$ in the system Pb + Cm show an avoided crossing near $R = 400$ fm (upper part), which leads to a sharply localized strong radial coupling. This coupling can be eliminated going to diabatic states (full curves) which are allowed to cross; see the inset.

dependent problem. Even at low collision velocities the dynamical wave function will not care much for the small energy gap. Expressed in terms of the adiabatic basis strong transitions are induced by radial coupling.

An example is given in Figure 4.6 for the highly relativistic system Pb + Cm (Be 80b, Ki 80b) which exhibits an avoided crossing between the states 4Σ and 5Σ at $R \simeq 400$ fm. The radial matrix element $\langle 5\Sigma | \partial/\partial R | 4\Sigma \rangle$ is large and strongly peaked at the position of minimum energy difference. Obviously the adiabatic basis states undergo a sudden change. This demonstrates that the wave functions do cross, even if the adiabatic energies are not allowed to do so.

Therefore it is rewarding to introduce a basis set which reflects this behavior and does not give rise to large "unphysical" couplings of the kind shown in Figure 4.6. Such a set of states is called a diabatic basis and has been under discussion for a long time (Li 63, Sm 69). For qualitative considerations diabatization just amounts to the replacement of avoided crossings by true crossings as long as the energy gap is not too large and the collision velocity is not too small. Although the concept is quite clear in practical calculations, a general formal definition of diabatic states presents some problems. A recent detailed discussion is found in (De 81).

A unitary transformation to diabatic states $\Xi_i(R)$, defined by a matrix C

$$\Xi_i = \sum_j C_{ji}(R)\, \phi_j \tag{4.17}$$

serves the purpose to reduce the radial coupling strength. As an extreme choice one may require the dynamic coupling matrix to vanish entirely, i.e. (we restrict our attention to the case of radial coupling only),

$$\langle \Xi_j | \, \partial/\partial R \, | \Xi_i \rangle = 0 \tag{4.18}$$

Insertion of (4.17) leads to the condition (He 79)

$$(d/dR) \, C_{ij} + \sum_k D_{ik} C_{kj} = 0 \tag{4.19}$$

where D is the adiabatic coupling matrix $D_{ik} = \langle \phi_i | \, \partial/\partial R \, | \phi_k \rangle$. The matrix differential equation (4.18) can be integrated with the initial condition $\mathsf{C} \to \mathsf{I}$ for $R \to \infty$. This allows the generation of diabatic states from a given adiabatic set.

Of course a price has to be paid for the simplification contained in (4.18): Since the diabatic states Ξ_i are not exact eigenstates of the Hamiltonian H they will be coupled through off-diagonal potential matrix elements

$$V_{ji} = \sum_k C^{-1}_{jk} E_k C_{ki} \tag{4.20}$$

with the adiabatic energies E_k.

The condition (4.18) should be applied to a limited set of states only since otherwise (the set of states $\langle \Xi_j |$ being complete) it would imply $(\partial/\partial R) \, | \Xi_i \rangle = 0$, i.e., the constancy of Ξ_i. Furthermore the $\partial/\partial R$-operator should be modified by a translational correction (De 81) in the sense of (3.42).

In simple cases a two-state approximation is sufficient. Then Equation (4.19) can be solved immediately. The transformation matrix becomes

$$\mathsf{C} = \begin{pmatrix} \cos\zeta & \sin\zeta \\ -\sin\zeta & \cos\zeta \end{pmatrix} \tag{4.21}$$

with the phase $\zeta = \int_\infty^R D_{12}(R') \, dR'$.

A close avoided crossing is characterized by a sudden increase of ζ by an amount of $\pi/2$. The diabatic states will cross at $\zeta = \pi/4$, which can be seen from the diagonal elements of the potential matrix

$$\mathsf{V} = \begin{pmatrix} E_1 \cos^2\zeta + E_2 \sin^2\zeta & \frac{1}{2}(E_1 - E_2) \sin^2\zeta \\ \frac{1}{2}(E_2 - E_1) \sin^2\zeta & E_1 \sin^2\zeta + E_2 \cos^2\zeta \end{pmatrix} \tag{4.22}$$

This is illustrated in Figure 4.6 for the crossing of the diabatic states $2s\sigma$ and $2p_{3/2}\sigma$ in the Pb + Cm system.

The remaining transition probability between diabatic states can be evaluated with the familiar Landau–Zener model for crossings. If sufficiently extensive coupled-channel calculations are performed, the introduction of diabatic states in principle is not necessary, but it may have computational advantages.

(ii) Rotational Coupling. Apart from level crossings strong transitions can be induced by rotational coupling between states belonging to an atomic multiplet in the limit $R \to 0$. The most prominent example is the process of vacancy creation in the $2p\pi$ level which becomes degenerate with the $2p\sigma$ MO in the UA limit. Since holes are more readily available in the atomic L shell this constitutes the dominant mechanism for K-vacancy production in light collision systems.

In a first approximation only the two states $2p\sigma$ and $2p\pi$ and their rotational coupling have to be taken into account. The two-state problem

$$\begin{aligned} \frac{d}{dt} a_\pi &= \dot{\theta} f(R)\, e^{i\chi_{\pi\sigma}(t)} a_\sigma \\ \frac{d}{dt} a_\sigma &= -\dot{\theta} f(R)\, e^{-i\chi_{\pi\sigma}(t)} a_\pi \end{aligned} \tag{4.23}$$

has to be integrated numerically (Br 72) with realistic rotational coupling matrix elements $f(R) = \langle \phi_\pi | iL_y | \phi_\sigma \rangle$ and phases $\chi_{\pi\sigma} = \int^t dt' \, (E_\pi - E_\sigma)$, using Rutherford trajectories.

Figure 4.7 shows typical $2p\sigma$ vacancy probabilities as a function of impact parameter b. The result can be shown to obey general scaling relations (Ta 76). At low collision velocities $P_{2p\sigma}(b)$ exhibits a broad bump having a maximum at a position which approximately follows $\Delta E b/v \simeq 1$. The inset in Figure 4.7 shows the total transfer cross section which decreases very fast if the impact velocity is to low to reach the strong coupling region. In close collisions (high impact energies) a second, very sharp peak appears which has maximum transition probability one. This peak is correlated to scattering under 90° and can be explained as a kinematic effect. In the UA limit the electronic $2p\sigma$ wave function cannot follow the rapid rotation of the molecular quantization axis but rather will remain fixed in space. Expressed in the rotating basis it will become a superposition of two states, $\phi_{2p\sigma} \cos\theta + \phi_{2p\pi_x} \sin\theta$. Neglecting additional couplings the transfer probability is given just by the kinematical expression (Br 72)

$$P_{2p\sigma}(b) = \sin^2\theta = 4(b^2/a^2)/(1 + b^2/a^2)^2 \tag{4.24}$$

where the parameter a was introduced in (4.6). At $\chi = \pi - \theta = 90°$ c.m. scattering angle ($b = a$) the vacancy transfer will be complete, $P_{2p\sigma} = 1$. The same result is obtained from the coupled equations (4.23) which can be integrated trivially in the UA limit, where $\Delta E = 0$ and $f(R) = 1$. In heavier systems the coupling is reduced due to the fine structure splitting between $2p_{1/2}$ and $2p_{3/2}$ (Ja 80).

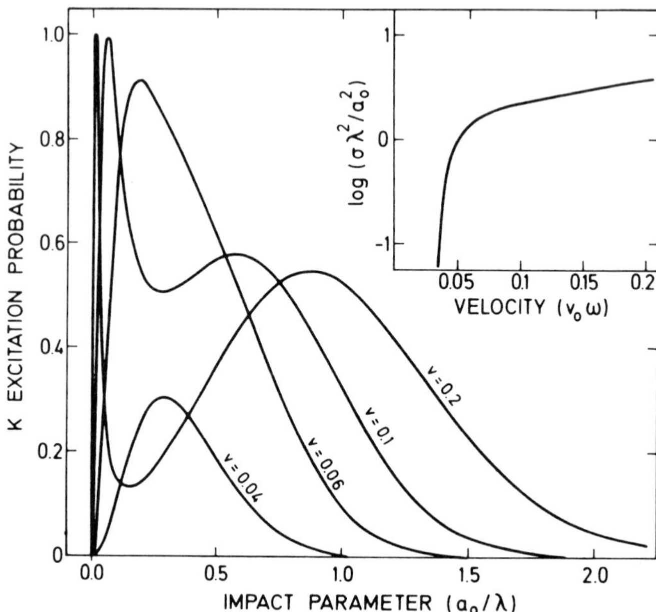

Figure 4.7. Probability for 2pπ–2pσ vacancy transfer due to rotational coupling as a function of impact parameter. The inset shows the integrated cross section as a function of projectile velocity. According to (Ta 76) the lengths and velocities have been expressed in terms of two Z-dependent scaling parameters, λ and ω, to make the curves universal. From (Me 77).

(iii) Vacancy Sharing. High transition probabilities can also be induced at large distances R, if two molecular states become nearly degenerate in the limit $R \to \infty$. This is true for the $1s\sigma$ and $2p\sigma$ states which correlate to the K shells of the heavier (lighter) atom in moderately asymmetric systems. Vacancies observed in the K shell of the heavier partner can originate either from direct $1s\sigma$ excitation or they can be transfered down from the $2p\sigma$ level. The vacancy probability in the latter orbital in general is much larger due to, e.g., the electron promotion mechanism discussed under (ii). Therefore $1s\sigma$ ionization will be masked by the vacancy sharing mechanism if the system is not sufficiently asymmetric.

The sharing is caused by radial coupling between the states $1s\sigma$ and $2p\sigma$. It is expected to occur at an internuclear distance R larger than the atomic radii, where the energy difference is small while coupling matrix elements still are large. Therefore the sharing mechanism is separated from other couplings and a two-state approximation again is meaningful.*

* The vacancy sharing mechanism in some sense is the opposite of a level crossing: While the Landau–Zener model for crossings assumes constant coupling strength and strong variation of the energy difference, in the case of vacancy sharing the roles are reversed.

A very simple analytically solvable model has been given by Demkov (De 64), who studied the process of nonresonant charge transfer. He starts from an expansion of the time-dependent wave function as a linear combination of the two atomic orbitals ϕ_1 and ϕ_2 describing the atomic K shells of the heavy (light) partner,

$$\Phi(t) = a_1'(t)\,\phi_1 + a_2'(t)\,\phi_2 \tag{4.25}$$

After an obvious change in phase the resulting coupled differential equations read

$$\begin{aligned} i\dot{a}_1 &= \tfrac{1}{2}(H_{11} - H_{22})\,a_1 + H_{12}a_2 \\ i\dot{a}_2 &= H_{21}a_1 - \tfrac{1}{2}(H_{11} - H_{22})\,a_2 \end{aligned} \tag{4.26}$$

with $H_{ij} = \langle \phi_i | H | \phi_j \rangle$.

If the relevant values of R are large the energy difference is approximately constant and H_{12} will fall off exponentially, since it involves the overlap of the tails of atomic $1s$ wave functions. With the substitutions (De 64) $\alpha = \frac{1}{2}(H_{11} - H_{22}) = [E_1(\infty) - E_2(\infty)]$ and $H_{12} = H_{21} = \beta e^{-\gamma t}$ (assuming straight line motion $R = vt$) the system (4.26) is solvable exactly. Imposing the initial condition $a_1 = 1/\sqrt{2}$ and $a_2 = -1/\sqrt{2}$ (preparation of a hole in the $2p\sigma$ state at close internuclear distance) Meyerhof (Me 73a) obtained the vacancy sharing ratio (valid for $\beta \gg \gamma$)

$$w = |a_1(\infty)|^2 = 1/(1 + e^{|2x|}) \tag{4.27}$$

with $x = \pi\alpha/\gamma = \pi(E_1 - E_2)/2(2mI)^{1/2}\,v$.

The square root factor can be identified with an averaged reciprocal K-shell radius which is expressed through the atomic binding energies, $I^{1/2} = \frac{1}{2}(E_1^{1/2} + E_2^{1/2})$. For symmetric systems the sharing probability becomes 1/2 since here the K shells correlate to a linear combination of ϕ_{1s} and $\phi_{2p\sigma}$.

The result (4.27) can be qualitatively understood from an argument involving the time–energy uncertainty relation: Transitions will be strongly suppressed unless the product of energy transfer and "interaction time" $\Delta t \simeq R_0/v$ is smaller than 1,

$$\Delta E R_0/v < 1 \tag{4.28}$$

If the interaction region is identified with the K shell radius $R_0 = 1/(2mI)^{1/2}$, the "Massey criterion" Equation (4.28) explains the behavior of (4.27). If a K hole is brought into the collision the sharing region is passed twice, which can lead to interference. The hole transfer probability (De 64) is determined by an expression similar to Equation (4.27) multiplied with an oscillating factor determined by the phase difference $\int_{-\infty}^{\infty} dt' [E_{1s\sigma}(t') - E_{2p\sigma}(t')]$. This effect has been observed experimentally by (Sch 79).

The simple Demkov–Meyerhof formula, Equation (4.27), has been surprisingly successful in explaining the experimental data on vacancy sharing. More refined calculations should be based explicitly on the quasimolecular model. The coupled equations have been integrated numerically using exact binding energies and radial coupling matrix elements (Ma 79a) or improved model couplings H_{ij} (St 80b). The results tend to confirm the analytical result (4.27) even in cases where the assumptions made in deriving this equation are not well justified. Further studies are necessary to investigate the dependence of vacancy sharing on impact parameter and to study the process in very heavy systems. Reliable calculations are required to extract the $1s\sigma$ vacancy probability in nearly symmetric collisions from measured K x-ray data. This quantity is of particular interest in high-Z systems, cf. Section 4.4.

4.4. Electron Excitation in Superheavy Systems

4.4.1. K-Hole Production

From the viewpoint of basic physics the process of inner shell excitation deserves the greatest attention if superheavy collision systems are involved. Here the opportunity is offered to study—to some extent—the physical properties of very strongly bound electronic states, a field which is not accessible by any other method. The ultimate goal is to perform some kind of "spectroscopy of superheavy quasimolecules" by analyzing characteristic features of the excitation process.

If this concept is to hold, the electrons have to respond to the combined charge of both nuclei, i.e., for the inner shells the collision has to be to adiabatic at least to some degree. It was first emphasized in (Ra 71, Ra 72) that superheavy quasimolecules can be formed in close heavy-ion collisions. Due to their strong binding the inner shell electrons will be in relativistic motion. Nuclear velocities of the order $v_\infty \simeq 0.1c$ are sufficient to surmount the Coulomb barrier so that the adiabaticity condition (4.2) holds. Calculations of excitation rates have to be based on the solutions of the two-center Dirac equation (4.13). The theory has been worked out only quite recently (cf. Be 76b, So 79a, So 80, So 81a).

For the process of $1s\sigma$ excitation direct ionization plays a dominant role. As a first step, first-order time-dependent perturbation theory with respect to the molecular basis may be employed

$$P_{1s\sigma}(b) = 2\int_m^\infty dE \left| \int_{-\infty}^\infty dt \, \langle \phi_E(t) | \, \partial/\partial t \, | \phi_{1s\sigma}(t) \rangle \, e^{i(\chi_{1s\sigma} - \chi_E)} \right|^2 \tag{4.29}$$

where the factor $2 = 2(j_i + 1)$ stands for the multiplicity of the initial state.

Since the $1s\sigma$ state is well separated from its neighbors, rotational coupling is weak and can be neglected. The excitation therefore is determined by the matrix elements of the radial coupling operator $\dot{R}\partial/\partial R$ and by the binding energy $E_{1s\sigma}(R)$. In collisions with $(Z_1+Z_2)\alpha > 1$ both quantities are dominated by relativistic effects in the wave functions.

The radial coupling matrix elements can be calculated conveniently using the Hellmann–Feynman identity

$$\langle \phi_f | \, \partial/\partial R \, | \phi_i \rangle = \langle \phi_f | \, \partial V_{TC}/\partial R \, | \phi_i \rangle / (E_i - E_f) \tag{4.30}$$

Figure 4.8a shows the radial coupling matrix elements between $1s\sigma$ and the neighboring $ns\sigma$ states in the U + U quasimolecule (Be 76b). The curves exhibit a remarkable increase at small internuclear distance, following a power law $\propto R^{-1}$. For comparison the figure shows the nonrelativistic $1s\sigma$–$2s\sigma$ matrix element (dash-dotted line), scaled to the U + U system. This curve goes to zero as $R \to 0$. Thus the increase of the TCD matrix elements is clearly caused by the relativistic contraction of the Dirac wave functions and the loss of an atomic length scale other than the internuclear distance.

As it turns, out the structure of quasimolecular orbitals at small R is mainly determined by the mean depth of the potential well created by the two nuclei. Molecular energies and coupling matrix elements can be determined accurately by replacement of the full two-center potential by its

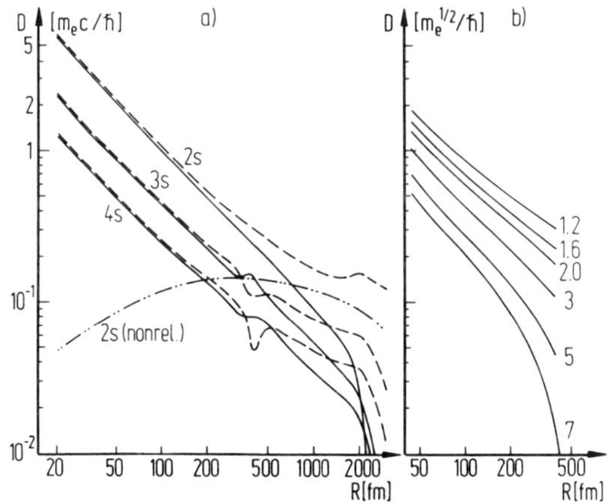

Figure 4.8. (a) Matrix elements for radial coupling joining the $1s\sigma$ with the $2s\sigma$, $3s\sigma$, and $4s\sigma$ levels in U + U as a function of internuclear separation. The solid lines take into account plane wave translation factors in a lowest-order approximation. The dash-dotted line shows the nonrelativistic $1s\sigma$–$2s\sigma$ matrix element scaled to $Z = 184$. (b) Radial coupling matrix elements from the $1s\sigma$ level to the electron continuum, calculated in monopole approximation. From (Be 76b).

monopole part ($l = 0$). This corresponds to the potential of a single blown-up charged spherical shell of radius $R/2$, i.e. (for pointlike charges),

$$V_0(r) = \begin{cases} -(Z_1 + Z_2)\, e^2/r & \text{for} \quad r > R/2 \\ -2(Z_1 + Z_2)\, e^2/R & \text{for} \quad r < R/2 \end{cases} \tag{4.31}$$

Radial coupling matrix elements then are given by simple overlap integrals, restricted to the interior region $r < R/2$:

$$\langle \phi_j | \, \partial/\partial R \, | \phi_i \rangle = 2(E_i - E_j)^{-1} (Z_1 - Z_2)\, e^2/R^2 \int_0^{R/2} r^2 \, dr \, (f_j f_i + g_j g_i) \tag{4.32}$$

For the most relevant states $ns_{1/2}$ and $np_{1/2}$ the resulting energies and coupling matrix elements agree with the true TCD solutions within an accuracy of a few percent for distances up to 500 fm and more (So 78c).

Apart from its general simplicity the monopole approximation has one important technical advantage: It allows an easy evaluation of continuum states. Solutions of the corresponding full TCD problem have not yet been obtained. In contrast, owing to the separability of the two-center Schrödinger equation, nonrelativistic molecular wave functions are available (Po 53, Gr 76a, Po 77). Calculations of ionization in high-Z systems up to now have been made feasible only through reliance on the monopole approximation. The construction of an improved continuum, including appropriate translational corrections, remains a major challenge for further theoretical work. The problem has been discussed in detail by Heinz (He 81). In the monopole approximation the radial coupling matrix elements for ionization, $\langle \phi_{Es\sigma} | \, \partial/\partial R \, | \phi_{1s\sigma} \rangle$, show an increase at small internuclear distance, see Figure 4.8b, just as in the case of bound-bound transitions.

The first valid calculation of $1s\sigma$ excitation in a high-Z heavy-ion collision, U + U, led to remarkable results (Be 76b): Large vacancy creation probabilities in the order of several percent in central collisions close to the Coulomb barrier were predicted. At higher impact parameters or lower bombarding energies the calculated ionization rates decrease very sharply. That this is a truely relativistic effect caused by the strong Coulomb field is underlined by the fact that extrapolations of nonrelativistic calculations had predicted values of $P_{1s\sigma}(0)$ on the order of $10^{-5} \cdots 10^{-6}$; cf. the discussion in (An 77). Obviously a theoretical description has to be based on the Dirac equation from the outset. Although an energy up to 1500 keV has to be transfered to the electron (in the UA limit), the ionization process is effective in producing holes since high Fourier components are present in (4.29) due to the rapid variation of the relativistic coupling matrix elements.

The predictions of (Be 76b), elaborated in (So 78a, So 79a), have been essentially confirmed by careful experimental studies of K-shell excitation in heavy-ion collisions, performed by groups at GSI and Stanford. The absolute values of the calculations, however, fell short of experiment by a factor of about 2 to 5. (Re 79b) noted that this discrepancy can be ascribed to a failure of first-order perturbation theory: Since the radial coupling matrix elements are large, strong virtual excitations are induced during the course of the collision and multistep transitions become important. In (Re 79b) therefore numerical coupled channel calculations have been performed, using the quasimolecular basis in monopole approximation. Up to 8 bound and at least 15 continuum s states were included. According to the many-electron formulation of Section 3.3 single-particle transition amplitudes for various initial conditions had to be calculated. Multistep excitation via intermediate

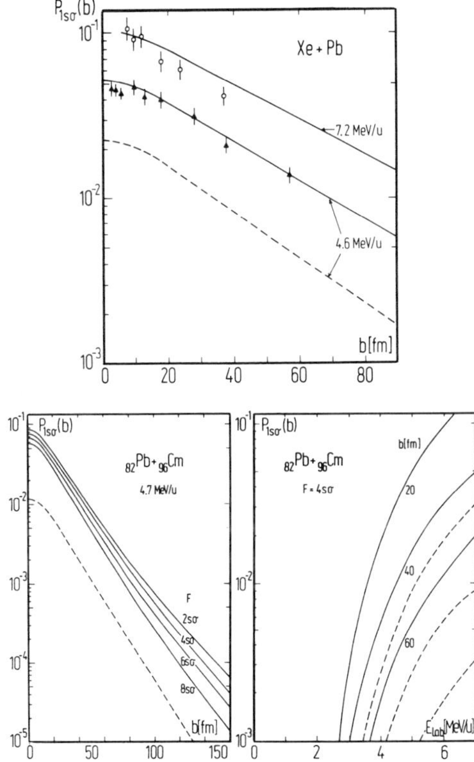

Figure 4.9. (a) Number of $1s\sigma$ vacancies created in Xe + Pb collisions at 4.6 and 7.2 MeV/u laboratory energy. Solid lines: Coupled channel calculations; dashed line: first-order perturbation theory. The experimental data are due to Anholt *et al.* (An 79e). From (So 81a). (b) Same as (a) for the system Pb + Cm at 4.7 MeV/u. The various solid curves are calculated assuming different positions of the Fermi level F as indicated in the figure. From (Re 79b). (c) The number of $1s\sigma$ vacancies in Pb + Cm collisions as a function of bombarding energy for three fixed impact parameters.

bound states was found to enhance the ionization rates while the inclusion of continuum–continuum coupling leads to a rescattering of electrons to higher kinetic energies.

As an example for the outcome of these calculations Figure 4.9 shows $P_{1s\sigma}(b)$ for the "medium heavy" system Xe + Pb ($Z = 136$) at 4.6 and 7.2 MeV/u and the very heavy system Pb + Cm ($Z = 178$) at 4.7 MeV/u (Re 79b, So 80, So 81a). The coupled-channel results (solid lines) are compared with first-order perturbation theory (dashed lines). Figure 4.9b furthermore indicates the dependence on the Fermi level F, i.e., on the availability of initial holes in higher shells. In the superheavy system Pb + Cm the decrease with impact parameter is extremely steep, cf. Figure 4.9b, combined with a strong dependence on bombarding energy, Figure 4.9c. This behavior reflects the large quasimolecular binding energies. In 4.9a experimental points of Anholt *et al.* (An 79e) have been included [earlier experiments were performed by (Gr 77a, Li 78)]. In view of the fact that the theory contains no adjustable parameter (except for the choice of F which is constrained by physical reasoning) the agreement is very satisfying. The same is true for other systems, including Pb + Cm at 5.9 MeV/u (Li 80a), while there is an unexplained discrepancy at lower energies.

The dependence of $P_{1s\sigma}$ for central collisions on the total nuclear charge $Z_1 + Z_2$ is displayed in Figure 4.10 (So 81a). The bombarding energy was chosen to fix the distance of closest approach at values of $R_{\text{min}} = 15$, 20, and 25 fm, respectively. Note that excitation rates in the monopole model depend on the combined nuclear charge only. This is valid if the system is not too asymmetric ($Z_1/Z_2 > 0.6$) and does not account for modifications in the kinematics. The excitation probability exhibits a maximum (So 78b) at

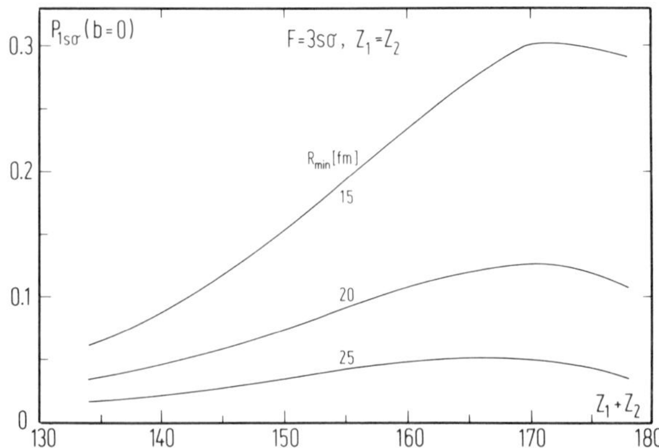

Figure 4.10. The number of $1s\sigma$ vacancies produced in superheavy systems as a function of total nuclear charge $Z_1 + Z_2$, as predicted from coupled-channel calculations. Collisions with distance of closest approach $R_{\text{min}} = 15$, 20, and 25 fm are considered. From (So 81a).

$Z \simeq 170$ reflecting a competition between the relativistic contraction of the wave functions and the need for higher energy transfer due to deeper binding. The latter effect ultimately brings down the curves.

The discussed calculations have been based on one-electron states, neglecting screening. For the inner states in heavy atoms this approximation should not be too bad, since here the electron–nucleus Coulomb force dominates over the electron–electron interaction. To improve the model, coupled-channel calculations have been performed using a basis derived from a monopole Hartree–Fock–Slater potential (deR 84). The rates for $1s\sigma$ excitation are increased only weakly, mainly due to the reduction of binding energies.

The monopole model is less successful in describing $2p_{1/2}\sigma$ excitation in superheavy quasimolecules. Here in addition to the strong radial coupling at small internuclear distances, more subtle molecular effects can play a role. In high-Z systems rotational coupling of the $2p\pi$-$2p\sigma$ type at the UA limit is suppressed due to the fine-structure splitting. However, rotational and radial couplings to various higher levels are possible at intermediate distances, in particular radial coupling to the $2p_{3/2}\sigma$ at about 400 fm (He 78a). To describe such effects the full two-center states have to be used in the coupled-channel calculations.

4.4.2. Delta-Electron Creation

Inner shell hole production due to direct ionization is accompanied by a continuum of "δ electrons" emitted during the collision. Since ionization rates essentially are obtained by an integration over all electron energies, the δ spectrum should even contain more detailed information on the superheavy quasimolecule.

Calculations proceed in the same way as discussed for vacancy production. As an example of an earlier calculation (So 78a, So 79a) using first-order perturbation theory Figure 4.11 shows the differential cross section $d\sigma/dE$ of electron emission in head-on collisions with $R_{min} = 20$ fm for the systems $Z = 140$, 170, and 190. The spectrum extends to high kinetic electron energies; δ electrons of up to 2 MeV energy can be observed with reasonable intensity. It should be noted that the maximum energy which can be transfered to an electron at rest in an elastic collision is of the order 10 keV! High-energy electrons therefore necessarily originate from a spatial region close to the nucleus where they can gather a high initial momentum. The availability of high momentum components in the inner shell wave functions grows strongly with Z in the region $Z\alpha > 1$. This effect is evident in the different slopes of the spectra shown in Figure 4.11 (note that the ordinate scale is highly compressed). The calculation also demonstrates that electrons from the $1s\sigma$ state only make a small contribution to the total

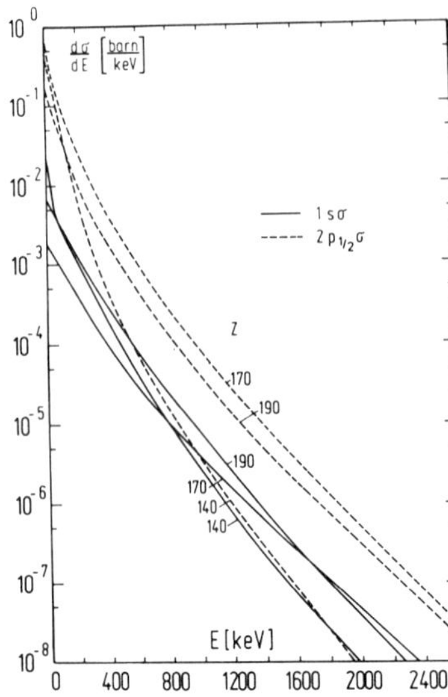

Figure 4.11. Delta-electron spectrum $d\sigma/dE$ calculated in first-order perturbation theory for three superheavy systems with $Z = 140$, 170, and 190, assuming a fixed distance of closest approach in head-on collisions, $2a = 20$ fm. In the heavier systems the contributions of electrons out of the $2p_{1/2}\sigma$ level (dashed lines) are seen to dominate the $1s\sigma$ contribution (full lines).

spectrum even at high energies; coupled-channel calculations do not essentially alter this conclusion. Most of the emission will originate from the $2p_{1/2}\sigma$ and $2s\sigma$ states which combine a high localization of the wave function with a weaker binding energy compared to the $1s\sigma$ level.

δ-electron spectra obtained from coupled-channel calculations again are enhanced in absolute magnitude (So 80, So 81a) compared to perturbation theory (So 78a, Ja 79b), in many cases leading to good agreement with experiments performed at GSI and MPI (Heidelberg) (Bo 78, Gü 81, Ko 81). For example Figure 4.12a shows the spectrum of electrons emitted in 500-MeV collisions of the "medium heavy" ($Z = 135$) system I + Pb (Ko 81), denoted by Σ, together with a spectrum of electrons measured in coincidence with Pb K holes. Apart from many-particle contributions according to (3.56) the latter spectrum should originate from the $1s\sigma$ shell, thus singling out a specific transition (Ko 77). Figure 4.12b shows the result of an elaborate experiment involving the triple coincidence between scattered ion, δ-electron, and Pb K x ray (Ko 81, Gü 81). Both the spectrum, Figure 4.12a, and the impact parameter dependence, Figure 4.12b, of $1s\sigma$ δ-electron emission are explained very closely by the coupled-channel calculations in the monopole model performed by G. Soff. In future experiments δ electrons from the $1s\sigma$ orbital might be singled out by requiring a coincidence with a high-energy MOX ray; cf. Section 5.

In the heaviest quasimolecular system which can be studied without problems due to nuclear background, Pb + Pb, δ electrons have been measured up to 1.8 MeV energy (Ko 81 and references therein). The slope

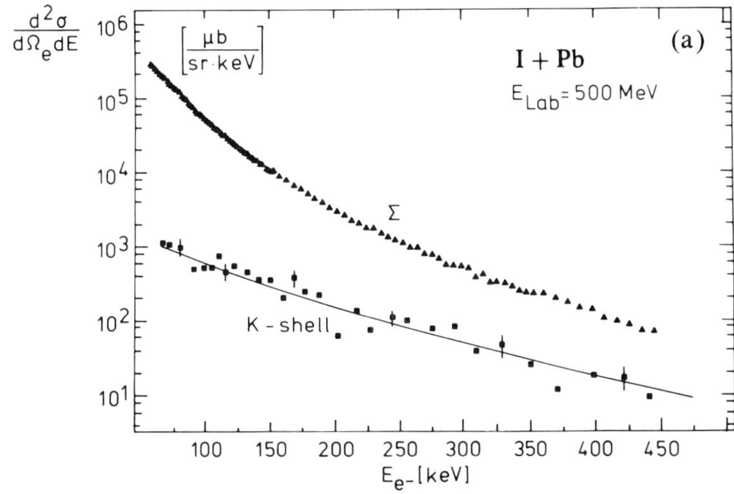

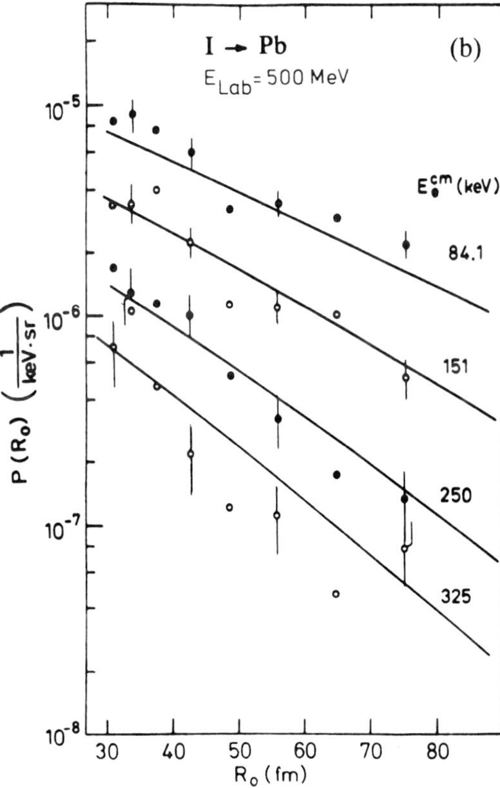

Figure 4.12. (a) Spectrum of δ electrons emitted in I + Pb collisions at 500 MeV bombarding energy obtained from a singles experiment (Σ) and measured in coincidence with a Pb K-hole. The experimental data of Güttner *et al.* (Ko 81) are compared with coupled-channel calculations of Soff. (b) Probability for the emission of δ electrons of a given energy E_e as a functions of the distance of closest approach, compared with theory. From (Ko 81).

and impact parameter dependence again are well described by the coupled-channel theory while there is some discrepancy in absolute magnitude.

The angular distribution of emitted electrons according to the monopole approximation should be isotropic. Its measurement will provide a direct test for the importance of higher multipole components.

Semiquantitative agreement with the measured data can also be obtained using the PBWA theory of (4.4), (4.5). It is necessary, however, to employ the relativistic binding energy and form factor of the united atom configuration. This again demonstrates that δ electrons are direct messengers from the region of strong fields and relativistic binding in a heavy-ion collision.

4.4.3. Spectroscopy of Superheavy Quasimolecules

In "slow" heavy-ion collisions the electrons probe the combined Coulomb potential of both nuclei. As discussed above, the electronic excitation rates become very sensitive to the distortion of the wave functions in this field. This opens the opportunity to gain information on the static properties of the atomic system which is formed transiently during the collision. Since nuclear isotopes with $Z \gtrsim 107$ are not available, the only hope to perform a "spectroscopy" of superheavy atoms rests in the analysis of scattering data.

It is evident from the outset that the high precision of conventional atomic spectroscopy cannot be expected from such studies. The finite, short collision time ($\sim 10^{-19}$–10^{-21} s) translates into an energy spread from ~ 10 keV to 1 MeV via Heisenberg's uncertainty principle. The complex collision dynamics can make a simple interpretation of experimental data rather questionable.

Nevertheless it is instructive to derive analytical expressions which explicitly display how the excitation rates depend on molecular wave functions and transition energies. We are most interested in K shell excitation of superheavy systems. Accoring to the results of the last sections this process is characterized by the following features:

(i) K holes are produced predominantly by direct ionization through radial coupling.

(ii) The excitation originates from small internuclear separations near the distance of closest approach. Here the electron states are well described by the monopole approximation. In superheavy systems with $(Z_1 + Z_2)\,\alpha > 1$ the radial matrix elements for $s_{1/2}$ and $p_{1/2}$ states are large at small R. The relation

$$\langle \phi_E |\, \partial/\partial R \,| \phi_n \rangle = d(E, Z)/R \tag{4.33}$$

where E is the total energy of the emitted electron and $Z = Z_1 + Z_2$ is the total nuclear charge, serves to characterize the coupling in not too asymmetric systems.

(iii) The matrix element is a smooth function of E. The electrons are mainly ionized into states with less than 1 MeV kinetic energy. The transition strength in this range can be approximated by

$$d(E, Z) = d_0(Z)(mc^2/E)^{\gamma/2} \tag{4.34}$$

where γ is a slowly varying function of Z.

Based on these premises Müller *et al.* (Mü 78) derived an approximate scaling law for the $1s\sigma$ ionization probability $P_{1s\sigma}(b)$. The derivation was given in the framework of first-order perturbation theory. As described in Section 5.4.1 this approximation does not give the full size of the excitation probabilities but it is convenient for analytical studies. The amplitude for ionization reads

$$\begin{aligned} a_{n,E} &= -\int_{-\infty}^{\infty} dt\, \dot{R}(t)\langle\phi_E|\, \partial/\partial R\, |\phi_n\rangle \exp\left\{-i\int_0^t dt'\, [E_n(R(t')) - E]/\hbar\right\} \\ &= -d\int_{-\infty}^{\infty} dt\, \frac{\dot{R}(t)}{R(t)} \exp\left[-i\int_0^t dt'\, \Delta E_n(R(t'))/\hbar\right] \end{aligned} \tag{4.35}$$

The solution of this Fourier integral is rendered difficult by the variation of transition energy $\Delta E_n(R(t)) = E_n(R(t)) - E$. According to the assumption (ii) ΔE may be approximately substituted by its value at the distances of closest approach $\Delta E_0 \equiv \Delta E(R_0)$, $R_0 = a\{1 + [1 + (b/a)^2]^{1/2}\}$. The trajectory $R(t)$ describes a Rutherford hyperbola which cannot be expressed in closed form. Usually the parametric representation (4.6) is employed.

An estimate of the integral (4.35) can be obtained from an investigation of the analytic structure of the integrand $R(t)/R(t)$ in the complex time plane. It can be seen that this function has singular points only on the imaginary axis. The zero of $R(t)$ closest to the origin is given by

$$\begin{aligned} \tau_0 = \operatorname{Im}(t_0) &= \frac{a}{v}\left[(\varepsilon^2 - 1)^{1/2} + \frac{\pi}{2} + \sin^{-1}\frac{1}{\varepsilon}\right] \\ &= \frac{1}{v}\left[b + a\left(\pi\text{-}\tan^{-1}\frac{b}{a}\right)\right] \end{aligned} \tag{4.36}$$

where $\varepsilon = (R_0 - a)/a$ is the eccentricity of the orbit.

The integral (4.35) can be closed in the upper half complex t plane, circumventing the singular points on the imaginary axis. The transition amplitude then is given by an integral over the discontinuity of the function

$\dot{R}/R$ across the imaginary time axis from $i\tau_0$ to infinity. Writing $t = i(\tau_0 + x)$ one has

$$a_{n,E} = -d \exp(\tau_0 \Delta E_0/\hbar) \int_0^\infty dx \operatorname{disc} \left[\frac{\dot{R}(x \pm i\varepsilon)}{R(x \pm i\varepsilon)} \right] \exp\left(\frac{-x \Delta E_0}{\hbar} \right) \tag{4.37}$$

The remaining integral can be approximated by the pole contribution at $x = 0$ which yields a factor π. Thus we have

$$|a_{n,E}|^2 \simeq \pi^2 d_0^2 (mc^2/E)^\gamma e^{-2\tau_0 \Delta E_0/\hbar} \tag{4.38}$$

Apart from the energy dependence originating from the matrix element this equation produces an exponential δ-electon spectrum.

In order to obtain the total ionization probability $P(b)$ an integration over energy E must be performed

$$\begin{aligned} P_n(b) &= \int_m^\infty dE \, |a_{n,E}|^2 \\ &= \tfrac{1}{4} D(Z) \, e^{2\tau_0 E_n(R_0)/\hbar} E_\gamma(2\tau_0 mc^2/\hbar) \end{aligned} \tag{4.39}$$

where E_γ is the exponential integral of order γ. Equation (4.39) may be compared with experimental data on ionization; the two required parameters $D(Z) \equiv 4\pi^2 d_0^2 mc^2$ and $\gamma(Z)$ are shown in Figure 4.13.

The ionization rate predicted by Equation (4.39) depends on impact parameter essentially in an exponential manner. Introducing a slowly varying function $f(\tau_0)$ we have

$$\begin{aligned} P_n(b) &\simeq (D/4) f(\tau_0) \, e^{-2E_B \tau_0/\hbar} \\ &\simeq (D/4) f(\tau_0) \, e^{-\pi E_B a/\hbar v} e^{-2(E_B/\hbar v) R_0} \end{aligned} \tag{4.40}$$

where $E_B = mc^2 - E_n(R_0)$ is the electron binding energy. The latter form of

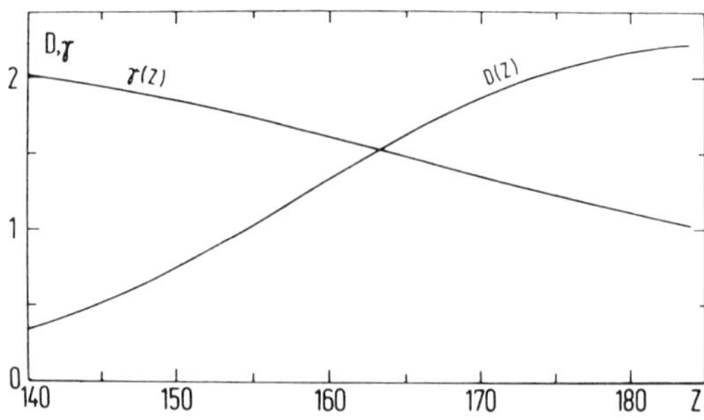

Figure 4.13. Functions $D(Z)$ and $\gamma(Z)$ determining the transiton strength to the continuum. From (Mü 78).

(5.40) is valid only in distant collisions, i.e., for $a/R_0 \ll 1$. The first exponential factor in (4.40) can be identified with the familiar correction factor for Coulomb deflection.

A scaling law of the type (4.39) was first derived in (Mü 78) under more severe approximations, which led to a steeper decrease with impact parameter. Backe (Ba 79a) and Kankeleit (Ka 80) employed a fit to the function $\dot{R}/R$ in the Fourier integral, obtaining a result very similar to (4.36), (4.38). A similar scaling law subsequently has been put forward by Bosch *et al.* (Bo 80). Neglecting the second factor in (4.36) they derived $P(b) \simeq D/2\, \exp[/2(R_0 - a)\, E_B/\hbar v]$ and claimed good agreement with experimental data. Since the true solution of (4.35) underestimates the coupled-channel results typically by a factor of 3–5, this agreement in absolute magnitude is fortuitous, being caused by an incorrect integration over energy E.

Figure 4.14 compares various versions of the scaling law with the exact results of perturbation theory (solid line) and coupled-channel calculations (dotted line). Shown is the $1s\sigma$ ionization probability in Pb + Pb collisions at $2a = 20$ fm (Mü 83).

The experimental data on $1s\sigma$ hole production in superheavy systems can be describe remarkably well by a simple exponential law

$$P_{1s\sigma}(b) = C(Z)\, e^{-2(E_B/\hbar v)R_0} \equiv e^{-2q_0 R_0} \tag{4.41}$$

for a large variety of heavy-ion systems and kinematic parameters (Li 80a). To obtain this result the theoretical binding energies at the distance of closest

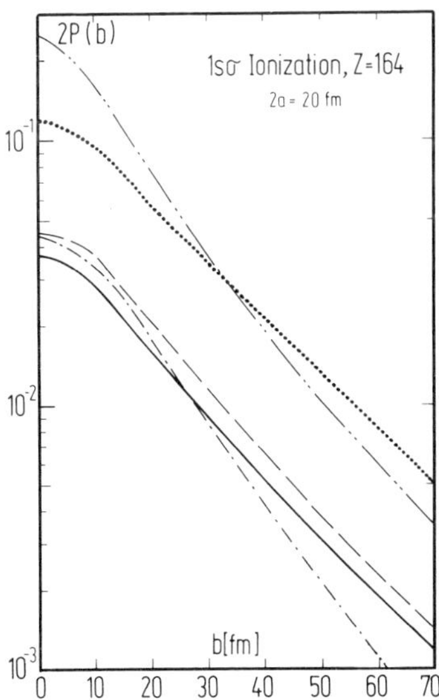

Figure 4.14. Impact parameter dependence of $1s\sigma$ ionization in Pb + Pb collisions at $b = 0$, $2a = 20$ fm, calculated with various analytical approximations. - - -, Equation (5.39); - · - · -, (Mü 78); - ·· -, (Bo 80a) with corrected integration over energy. For comparison the solid and dotted lines give the full numerical results from perturbation theory and coupled-channel integration, respectively.

approach (So 78c) have to be inserted. Vice versa, the binding energy can be deduced from the measured function $P_{1s\sigma}(b)$ (So 78b, Be 78). Relying upon Equation (5.41) Liesen *et al.* (Li 80a, Bo 80) used (1) the slope of $P_{1s\sigma}$ or (2) its variation with impact energy to determine $E_{1s\sigma}(R)$ in various systems in the region $136 < Z < 178$.

Related to this approach is the use of the "Bang–Hansteen scaling law" which was derived in the nonrelativistic SCA theory (Ba 59) and has quite general validity (Ba 79b, Ba 81c): The product of the average impact parameter $\bar{b}$ and the minimum momentum transfer $\bar{q}_0$ is a constant close to one,

$$\bar{b}\bar{q}_0 = C \tag{4.42}$$

A value of $C \simeq 1.1$ was found for the process of electron–positron pair creation (Ar 79). The energy $E_B(\bar{R}_0)$ can be obtained from (5.42) without the knowledge of higher momenta of the impact parameter dependence $P(b)$, (Ar 78).

In general the agreement of quasimolecular $1s\sigma$ binding energies deduced by these methods with the theoretically expected values is remarkably good. In the heaviest system investigated, Pb + Cm, the empirical scaling law (5.41) is violated at small impact parameters (Li 80a). In view of the uncertainties connected with the theoretical justification of the scaling law (5.41), however, it would be premature to conclude that anomalies in the binding energy of very deeply bound electrons have been found.

Perturbation theory undervalues the excitation rates and (to a lesser extent) overestimates the impact parameter dependence. If a simple scaling law including multistep excitation cannot be derived, a "spectroscopy of superheavy quasimolecules" might still be possible, although less elegant: One may introduce arbitrary variations in the quasimolecular energies (and wave functions). The influence of these modifications on the results of coupled-channel calculations for the excitation rates may be studied and compared with experimental data, allowing the determination of the size of the perturbation.

Spectroscopic information may also be obtained from the δ-electron distributions (Gü 81, Ko 81). In the case of electron–hole–particle triple coincidences this has the advantage that both the initial and final state are fixed, giving a definite transition energy $\Delta E(R)$.

4.5. Strong Collisional Magnetic Fields

In addition to the strong Coulomb potential also a magnetic field is created by the fast moving nuclei in a heavy-ion collision. The magnitude of the magnetic field strength B can be estimated easily from Biot-Savart's law.

For example for two colliding nuclei at the distance of closest approach R the value of B at space points along the symmetry axis perpendicular to the scattering plane ($x=z=0$) takes the form $|\mathbf{B}|=\frac{1}{2}(Z_1+Z_2)\,v_\infty R/(R^2+y^2)^{3/2}$. Assuming $Z_1+Z_2=178$, $v_\infty=0.1c$, and $\mathrm{R}=15$ fm the magnetic field strength at the center ($r=0$) reaches 4×10^{15} G. At larger distances the field is of dipole type ($B\propto r^{-3}$); at $r=100$ fm it is still of the order 10^{13} G.

Several years ago Rafelski and Müller (Ra 76a) pointed out that these fields are accessible to experimental observation by studying the excitation of inner shell electrons. Since the quasimolecular $1s\sigma$ orbit in a system with $(Z_1+Z_2)\,\alpha>1$ can shrink to a Bohr radius not larger than 100 fm, inner shell electrons are sensitive to the very strong but highly localized magnetic field B. Since the Bohr magneton is of the order $\mu_B\simeq0.6\cdot10^{-14}$ MeV/G, a very large Zeeman energy splitting can be expected, even approaching the electron rest mass in fast collisions. This is confirmed by quantitative calculations (Ra 76a). The change in binding energies should lead to differences in the ionization probability of spin-up and spin-down electrons. As a consequence the created vacancies as well as the δ-electrons emitted during the collision (Gr 78) should be polarized.

The influence of the magnetic field on inner shell excitation has been evaluated in first-order perturbation theory (Gr 78) and within the coupled-channel approach (So 81b). For consistency with the potential $V(\mathbf{r},t)$ the vector potential $\mathbf{A}(\mathbf{r},t)$ to be inserted in (3.4) should be calculated in the Coulomb gauge. The vector potential created by two charged point sources moving with velocities $v_{\mathrm{i}}(t)$ along the trajectories $R_i(t)$ is known from standard electrodynamics. Neglecting retardation (Ko 78) it takes the form

$$\mathbf{A}_C(r,t)=-\sum_{i=1,2}\frac{Z_ie\mathbf{v}_i}{2\,|\mathbf{r}-\mathbf{R}_i(t)|}-\sum_{i=1,2}\frac{Z_ie\mathbf{v}_i(t)\cdot(\mathbf{r}-\mathbf{R}_i(t)]}{2\,|\mathbf{r}-\mathbf{R}_i(t)|^3}\,[\mathbf{r}-\mathbf{R}_i(t)] \tag{4.43}$$

(4.43) can be split into two parts proportional to the radial and angular component of the nuclear velocity, v_R and v_ϕ. In (So 81b) the vector potential has been expanded in multipoles in analogy with the two-center Coulomb potential (4.15), and the resulting matrix elements of the magnetic interaction Hamiltonian $H_{\mathrm{mag}}=-e\boldsymbol{\alpha}\cdot\mathbf{A}$ between electronic multipole states have been evaluated. In the simplest case (monopole approximation, s states) the spin–flip coupling reduces to

$$\langle s',\pm\tfrac{1}{2}|\,H_{\mathrm{mag}}\,|s,\mp\tfrac{1}{2}\rangle=\pm\frac{i}{3}Ze^2v_\phi\int_0^\infty r^2\,dr\,\frac{r_<}{r_>^2}\,(f'g+fg') \tag{4.44}$$

while the nonflip matrix element vanishes. g and f are the large and small radial wave functions of the Dirac bispinor. Taking $s'=s$ the absolute value of (4.44) gives the Zeeman splitting of an s level. The result is demonstrated in Figure 4.15 for the $1s\sigma$ state in a 5.9 MeV/u collision with $Z_1 + Z_2 = 178$ and impact parameter $b = 20$ fm. In addition to the energy splitting the magnetic field has a similar effect as the potential coupling in (3.16), causing transitions between different basis states. All effects are rigorously included, if the coupled-channel equations are solved with couplings of the type (4.44).

The field (4.43) is expected to cause spin polarization oriented in the direction of the magnetic field strength, i.e., the y axis perpendicular to the scattering plane. To calculate this effect the quantization axis has to be rotated to the y direction. In the second quantized formulation of Section 3.3 the spin polarization of created particles is found to be (So 81b)

$$\begin{aligned}\eta_p &= (N^y_{p\uparrow} - N^y_{p\downarrow})/(N^y_{p\uparrow} + N^y_{p\downarrow}) \\ &= 2\,\mathrm{Im}\left(\sum_{r<F} a^*_{r\downarrow,p\uparrow} a_{r\downarrow,p\downarrow}\right)\bigg/\left(\sum_{r<F} |a_{r\downarrow,p\downarrow}|^2 + \sum_{r<F} |a_{r\downarrow,p\uparrow}|^2\right)\end{aligned} \tag{4.45}$$

Several symmetry properties (cf. Section 3.2.3) of the transition amplitudes $a_{ij}(\infty)$ have been used in deriving (4.45).

Coupled-channel calculations including a large number of bound and continuum states have been performed in (So 81b). Since the nondiagonal magnetic coupling matrix elements (4.44) are smaller than the corresponding radial coupling by about an order of magnitude, the unpolarized excitation rates are not significantly altered, compared with the results of Section 4.3. The polarization effect, however, was found to survive the multistep

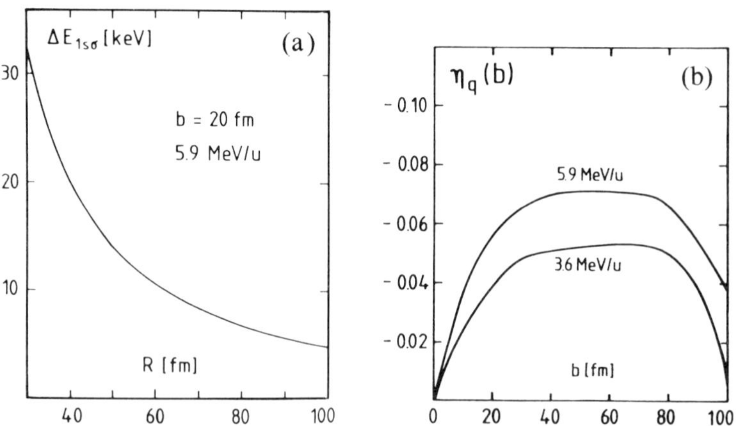

Figure 4.15. (a) Zeeman splitting of the $1s\sigma$ state induced by the strong magnetic field in 5.9 MeV/u Pb + Cm collisions at impact parameter $b = 10$ fm. (b) Spin polarization of electrons caused by the Zeeman energy splitting, drawn as a function of impact parameter for two bombarding energies. From (So 81b).

processes inherent in the coupled-channel approach. As an example, Figure 4.15 shows the spin polarization of created K vacancies in $Z_1 + Z_2 = 178$ collisions at two different impact energies. η_{1s} attains a maximum at impact parameters between $b = 30$ and 80 fm. At $b = 0$ it vanishes, since here $v_\phi = bv_\infty/R^2 = 0$, at large b the magnetic field becomes too weak to influence the ionization process. The spin polarization of high-energy δ electrons emitted out of the K shell is expected to exceed 10%.

These predictions presently are waiting for experimental examination. It should be noted that the strength of the magnetic field tested in a heavy-ion collision is larger by orders of magnitude compared to the macroscopic fields attainable in the laboratory (10^{11} G) or predicted to exist at the surface of neutron stars (10^{12} G).

4.6. High-Energy Collisions

In the previous sections inner shell excitation in "slow" heavy-ion collisions below the nuclear Coulomb barrier has been discussed, mainly in terms of the adiabatic picture. For higher impact energies such a description becomes increasingly problematic. The strong binding in near-symmetric high-Z systems extends the range, where an expansion in the adiabatic basis can be used with some confidence, to energies considerably beyond the Coulomb barrier. Coupled-channel calculations, however, become indispensable since multistep excitations are important and since the transition probabilities become very large.

Soff *et al.* (So 81c) have calculated $1s\sigma$ excitation in Pb + U collisions at bombarding energies from 4.7 up to 300 MeV/u. In these extensive coupled-channel calculations 8 bound states and 60 states in the electron continuum up to 3 MeV kinetic energy were included. Figure 4.16 shows the

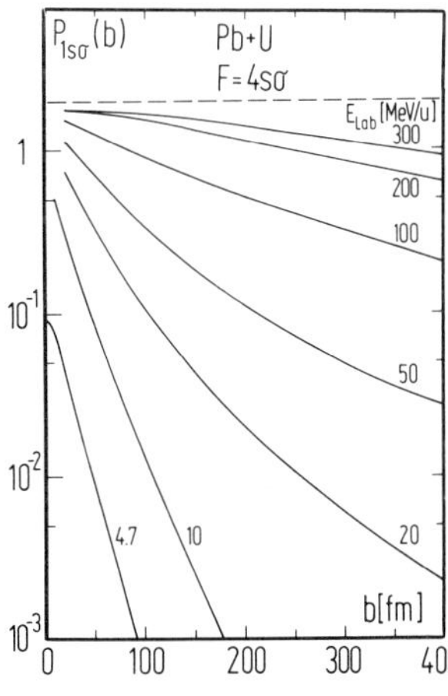

Figure 4.16. Number of $1s\sigma$ vacancies created in Pb + U collisions as a function of impact parameter for various bombarding energies between 4.7 and 300 MeV/u, predicted from coupled-channel calculations in the quasimolecular monopole model. The dashed line indicates the maximum possible value of $P_{1s\sigma}$. From (So 81c).

predicted impact parameter dependence of the $1s\sigma$ vacancy probability $P_{1s\sigma}(b)$ (the region of nuclear reactions at small b has been excluded in the figure). With increasing bombarding energy the slope is greatly reduced in accordance with the scaling rules of 4.4.3 and the values of $P_{1s\sigma}(b)$ come close to unity. At energies $E_{\text{lab}} > 200$ MeV/u a considerable fraction of the K shells is totally depleted even at higher impact parameters. The total $1s\sigma$ vacancy cross section is 10^4 b at 200 MeV/u.

At still higher impact velocities other methods should be employed. Teubner *et al.* (Te 80) calculated K-shell excitation in relativistic U + U collisions in the framework of the classical electron trajectory method; see Section 3.5.4. The relativistic three-body problem with interactions deduced from the retarded Lienard–Wiechert potentials were solved numerically for a large set of initial conditions. Figure 4.17 shows the weighted ionization probability $b\,P(b)$ for a singly occupied K shell. The energies are $E_{\text{lab}} = 1$ GeV/u (dashed curve), 10 GeV/u (full curve), and 100 GeV/u (triangles); some typical statistical error bars are given. The curves indicate that ionization becomes saturated at these highly relativistic impact velocities.

Ionization in asymmetric relativistic collisions can be calculated in Born approximation using the Møller interaction. Using the Coulomb gauge, the interaction between the charged projectile and a target electron can be

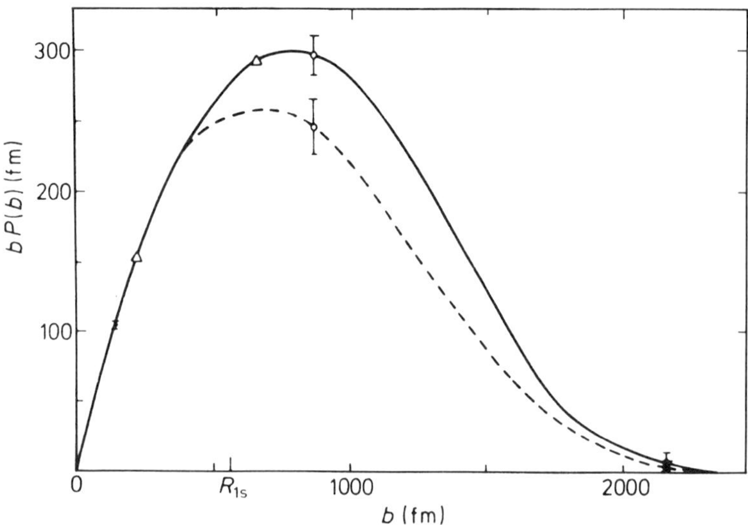

Figure 4.17. Ionization probability weighed with impact parameter for the singly occupied K shell in relativistic U + U collisions, calculated within the classical trajectory model. Dashed line: $E_{\text{lab}} = 1$ GeV/u; solid line: 10 GeV/u; triangles: 100 GeV/u. Some typical statistical error bars of the calculation are shown. From (Te 80).

split into a longitudinal part and a transversal current–current interaction. The latter contribution will eventually become dominant as v approaches the velocity of light. Spin–flip transitions play an important role in the excitation process if the target atom is sufficiently heavy (An 79d). For a detailed discussion see (Mo 80).

5. Molecular Orbital X-Ray Radiation

5.1. Introduction

The classical tool to gain information on the structure of atoms and molecules, well established for over a century, is the method of photon spectroscopy. The analysis of x-ray spectra has led to a remarkably good understanding of the inner shells in heavy atoms. Therefore it would seem natural and very promising to apply this method to a study of the electronic shells in a heavy-ion collision. In particular, in a slow collision x-rays may be created by transitions between adiabatic electron states and thus serve as direct messengers from the transiently formed quasimolecules. Measurement of the energy spectra of x-rays emitted in the collision might even permit a spectroscopy of these elusive objects. For example, high-energy x-rays should reveal the strong binding of the K electrons in a superheavy atomic system.

As it turns out, this aim is difficult to achieve, although the attempts have been successful in part. The problems become evident from a brief glimpse at any molecular correlation diagram. Even for one selected transition the energy difference changes rapidly with internuclear distance, therefore the spectrum will be continuous. Furthermore the information that can be drawn from the shape of the spectrum is also limited. This is due to the short duration of the collision that leads to an energy broadening, in accordance with the Heisenberg uncertainty principle. In the following we will give an account of the theoretical formalism necessary to describe spectra and angular distributions of collisional x-ray radiation. Subsequently the results of some available calculations will be discussed. For a detailed exposition of the history and present status of experiments we refer to the article by Vincent and Greenberg in this volume. Earlier reviews of the field have been given in (Me 77, Mo 78).

5.2. Theory: S-Matrix Formulation

From a theoretical point of view photon emission is easily incorporated in the formalism outlined in Section 3. The whole discussion will be presented in the semiclassical formulation of Section 3.2; see also (Sm 75a,

Br 77). The standard formulation proceeds in the following way: We have the total Hamiltonian

$$H = H_{el} + H_{em} + H_{rad} \tag{5.1}$$

consisting of the electronic Hamiltonian H_{el}, the Hamiltonian H_{em} of the free radiation field, and the radiative Hamiltonian H_{rad}. In the relativistic theory the latter has the minimal coupling form

$$H_{rad} = -e\boldsymbol{\alpha} \cdot \mathbf{A} \tag{5.2}$$

with the vector potential of the quantized radiation field

$$\mathbf{A}(\mathbf{r}, t) = \sum_{\mathbf{k}\lambda} \hat{a}_{\mathbf{k}\lambda} \mathbf{q}_{\mathbf{k}\lambda}(\mathbf{r}, t) + \sum_{\mathbf{k}\lambda} \hat{a}^{+}_{\mathbf{k}} \mathbf{q}^{*}_{\mathbf{k}\lambda}(\mathbf{r}, t) \tag{5.3}$$

$a^{+}_{\mathbf{k}\lambda}$ and $\hat{a}_{\mathbf{k}\lambda}$ are the creation and annihilation operators for a photon having the wave vector k and polarization λ. $\mathbf{q}_{\mathbf{k}\lambda}(\mathbf{r}, t)$ denotes the basis vector of the photon field (normalized to a box of unit volume)

$$\mathbf{q}_{\mathbf{k}\lambda}(\mathbf{r}, t) = \mathbf{q}_{\mathbf{k}\lambda}(\mathbf{r})\, e^{-i\omega t} = \frac{1}{(2\omega)^{1/2}}\, \boldsymbol{\varepsilon}_{\lambda}\, e^{i(\mathbf{k}\cdot\mathbf{r} - \omega t)} \tag{5.4}$$

with the transversal polarization vectors $\boldsymbol{\varepsilon}_{\lambda}$, $\boldsymbol{\varepsilon}_{\lambda} \cdot \mathbf{k} = 0$, $\lambda = 1, 2$.

Now we want to solve the time-dependent Schrödinger equation for the total Hamiltonian (5.1)

$$(H - i\partial/\partial t)\, \Psi_T = 0 \tag{5.5}$$

This differential equation is equivalent to the integral equation

$$\Psi_T(t) = \Phi_i^{(+)}(t)\, |0_k\rangle - i \int_{-\infty}^{t} dt'\, U(t, t')\, H_{rad}\, \Psi_T(t') \tag{5.6}$$

where $\Phi_i^{(+)}$ is a solution to the scattering problem neglecting the radiation field

$$(H_{el} - i\partial/\partial t)\, \Phi_i^{(+)} = 0 \tag{5.7}$$

and $|0_{\mathbf{k}}\rangle$ represents the photon vacuum. The $\Phi_i^{(+)}$ are the dynamical basis states already discussed in Section 3.3. They satisfy boundary conditions of type (3.8) at $t \to -\infty$. The integral kernel contains the time development operator $U(t, t')$ for the uncoupled fields, which solves

$$(H_{el} + H_{em} - i\partial/\partial t)\, U(t, t') = 0 \tag{5.8}$$

with the property $U(t, t) = 1$.

Equation (5.6) has been written in a way which lends itself easily to a perturbation expansion. In first order with respect to the radiation field

$$\Psi_T(t) \simeq \Phi_i^{(+)}(t)\,|0_{\mathbf{k}}\rangle - i\int_{-\infty}^{t} dt'\; U(t,t')\,\hat{H}_{\text{rad}}\,\Phi_i^{(+)}(t)\,|0_{\mathbf{k}}\rangle \tag{5.9}$$

Since the emitted radiation is of low intensity, the first-order expansion (5.9) is totally adequate. At a given time t the probability amplitude to find a photon $|\mathbf{k}\lambda\rangle$ and an electronic configuration which finally ends up in the state ϕ_f is obtained by projection

$$\begin{aligned} C_{k\lambda}^{fi}(t) &= -i\langle \Phi_f^{(-)}(t); \mathbf{k}, \lambda, t \mid \Psi_T(t)\rangle \\ &= -i\int_{-\infty}^{t} dt'\; e^{i\omega t'}\langle \Phi_f^{(-)}(t')|\,(-e\boldsymbol{\alpha})\cdot \mathbf{q}_{k\lambda}^{*}\,|\Phi_i^{(+)}(t')\rangle \end{aligned} \tag{5.10}$$

Thus the intensity of radiation is a Fourier transform of the radiation matrix element taken between the dynamical wave functions $\Phi_i^{(+)}(t)$, integrated forward in time from the initial state ϕ_i, and the dynamical wave function $\Phi_f^{(-)}(t)$, integrated backward in time from the boundary condition ϕ_f. The differential probability for photon emission accompanying the electronic transition $i \to f$ is

$$dP_{\lambda}^{fi}(b) = |C_{\mathbf{k}\lambda}^{fi}|^2\, d^3k = \omega^2\, |C_{\mathbf{k}\lambda}^{fi}|^2 \frac{d\omega}{(2\pi)^3}\, d\Omega_k \tag{5.11}$$

To compare with experiment one has to sum over states that are not resolved. Since in most cases photon polarizations and electron configurations are not measured, a useful quantity is

$$\mathrm{d}^2P/d\Omega_x\, dE_x = \omega^2 \frac{1}{(2\pi)^3} \sum_{f,\lambda} |C_{\mathbf{k}\lambda}^{fi}|^2 \tag{5.12}$$

In analogy with (3.10), (3.11) this result may be used to calculate differential cross sections for x-ray emission just by multiplying with the particle scattering cross section.

The expression (5.10) is sufficiently general to encompass several different physical phenomena. One important process, which we will not pursue in detail here, is radiative electron capture (REC): A hole present in an inner shell of the projectile ion is filled by an outer shell electron of the target atom. Simultaneously a photon is emitted with the mean energy $E_x = \Delta E_{if} + \frac{1}{2}mv^2$, where the latter term describes the energy of the translational motion. REC shows up as a broad x-ray bump at an energy slightly above the (separated atom) characteristic x-ray position (Sch 72). The line shape contains information on the momentum distribution of the target electrons. The process can be calculated (Kl 75) from (5.10) using a Coulomb distorted "traveling orbital" for $\Phi_i^{(+)}$ and a stationary state of the projectile for $\Phi_f^{(-)}$.

In contrast to the REC process, which is characteristic for a nonadiabatic transition (fast nuclear motion), in slow collisions molecular orbitals are formed and radiative transitions between them lead to quasimolecular x-rays (MOX).

The dynamical states $\Phi^{(\pm)}(t)$ required in (5.10) to calculate the radiation amplitudes contain the full solution of the electronic excitation problem and therefore in general have a complicated structure. As discussed in Section 3.2 one usually expands them in a basis ϕ_j

$$\Phi_i^{(\pm)}(\mathbf{r}, t) = \sum_j a_{ij}^{(\pm)}(t) \exp\left(-i \int^t dt' \, E_j\right) \phi_j(\mathbf{r}, t) \tag{5.13}$$

The amplitudes are determined by the coupled-channel equations (3.19) with the boundary conditions $a_{ij}^{(+)}(-\infty) = \delta_{ij}$ and $a_{ij}^{(-)}(\infty) = \delta_{ij}$. Equation (5.13) leads to an expression for the radiation amplitudes in terms of the electron occupation amplitudes $a_{ij}(t)$:

$$\begin{aligned} C_{k\lambda}^{fi}(\infty) = -i \sum_{j,k} \int_{-\infty}^{\infty} dt \exp\left[i \int^t dt'(\omega + \omega_{kj})\right] a_{fk}^{(-)*}(t) \, a_{ij}^{(+)}(t) \\ \times \langle \phi_k | (-e\boldsymbol{\alpha}) \cdot \mathbf{q}_{\mathbf{k}\lambda}^* | \phi_j \rangle \end{aligned} \tag{5.14}$$

with the transition frequency $\omega_{kj} = E_k - E_j$.

Equation (5.14) is easy to read: the electron is excited from state ϕ_i to ϕ_j by any of the couplings (3.16), (3.21), then undergoes a radiative transition to ϕ_k at time t and is transferred finally to the state ϕ_f. The amplitude $C_{\mathbf{k}\lambda}^{fi}$ results from a coherent superposition of transitions between all combinations of intermediate configurations taking place at every instant of time t.

5.3. Theory: Field Theoretical Formalism

The evaluation of (5.14) is rather straightforward if only a single electron (or hole) is present. Except for very light systems in a realistic collision, however, many states are occupied. The distribution of holes turns out of be of great significance for the understanding of experimental results, in particular the anisotropy of the radiation. Although reasonable results have been obtained (without rigorous justification) using a single-particle picture, it is desirable to obtain a formulation which accounts explicitly for many-electron configurations and the Pauli principle.

Equation (5.14) could be employed directly, using many-electron wave functions (Slater determinants) as basis states ϕ_j. For practical applications, however, this procedure is quite infeasible since a forbiddingly large number of configurations can be built even from a modest number of particles. To avoid this problem, Kirsch (Ki 80a) used the methods of quantum field theory to develop an improved theoretical formulation of the MOX radiation. This extends the presentation of Section 3.3, where many-particle aspects of the electron excitation process alone have been treated. Again we will neglect electron–electron two-body interactions.

The set of field theoretical operator equations describing the coupled electron and photon fields has been given in (3.45), (3.46), (3.48), and (3.49). We are now interested in the quanta of the electromagnetic field $\hat{A}_\mu(x)$ in the asymptotic limit $t \to \infty$. In analogy to (3.49), (3.52) let us introduce two expansions of the field operator $A_\mu(x)$ (Ya 50). The inhomogeneous wave equation (3.46) can be solved by

$$\hat{A}_\mu(x) = \hat{A}_\mu^{(+)}(x) + \int d^4y\, D^A(x-y)\hat{\jmath}_\mu(y) \tag{5.15}$$

where $A_\mu^{(+)}$ is called the in-field. Due to the causality property of the advanced photon propagator D^A we have $\hat{A}_\mu(x) \to \hat{A}_\mu^{(+)}(x)$ in the limit† $t \to -\infty$. In the same way an out-field $\hat{A}_\mu^{(-)}(x)$ can be constructed

$$\hat{A}_\mu(x) = \hat{A}_\mu^{(-)}(x) + \int d^4y\, D^R(x-y)\hat{\jmath}_\mu(y) \tag{5.16}$$

with the retarded photon propagator D^R. Now the boundary condition is $\hat{A}_\mu(x) \to \hat{A}_\mu^{(-)}(x)$ for $t \to +\infty$. The operators $\hat{A}_\mu^{(\pm)}(x)$ are solutions of the free wave equation. Therefore they can be quantized by the standard prescription

$$A_\mu^{(\pm)} = \sum_{\mathbf{k}\lambda} [\hat{a}_{\mathbf{k}\lambda}^{(\pm)} q_{\mathbf{k}\lambda\,\mu}(x) + \hat{a}_{\mathbf{k}\lambda}^{(\pm)+} q_{\mathbf{k}\lambda\,\mu}^{*}(x)] \tag{5.17}$$

where $q_{\mathbf{k}\lambda\,\mu}(x)$ is the four-dimensional version of (5.4). Using the Lorentz gauge in the Gupta–Bleuler framework (Kä 72), only transverse polarizations will be observable ($\lambda = 1, 2$); for them $q_{\mathbf{k}\lambda\,0} = 0$. The Heisenberg state $|F\rangle$ is determined by the initial preparation of the system. Assuming that no "in-photons" are present before the collision we have

$$\hat{a}_{\mathbf{k}\lambda}^{(+)} |F\rangle = 0 \tag{5.18}$$

† We do not discuss problems of convergence, which in general have to invoke the adiabatic switching hypothesis.

for all values of **k** and λ. The number of created photons in the mode $(\mathbf{k}, \lambda)$ which is observed after the collision is given by

$$I(\mathbf{k}, \lambda) = \langle F| \hat{a}^{(-)+}_{\mathbf{k}\lambda} \hat{a}^{(-)}_{\mathbf{k}\lambda} |F\rangle \tag{5.19}$$

To obtain an expression for the out-operators in terms of the in-operators, (5.15) and (5.16) may be equated. This leads to

$$\hat{A}^{(-)}_{\mu}(x) = \hat{A}^{(+)}_{\mu}(x) - \int d^4y\, \Delta_0(x-y) \hat{\jmath}_\mu(y) \tag{5.20}$$

where $\Delta_0 = D^A - D^R$ is Schwinger's propagation function for the photon field. Insertion of the Fock expansion (5.17) and projection on the basis $q_{\mathbf{k}\lambda\,\mu}$ leads to (Ki 80a)

$$\hat{a}^{(-)}_{\mathbf{k}\lambda} = \hat{a}^{(+)}_{\mathbf{k}\lambda} - i\hat{h}_{\mathbf{k}\lambda} \tag{5.21}$$

with

$$\hat{h}_{\mathbf{k}\lambda} = \int d^4x\, q^*_{\mathbf{k}\lambda\,m}(x)\, \hat{\jmath}^\mu(x) \tag{5.22}$$

The current operator $\hat{\jmath}^\mu$ is given by (3.48). Therefore the photon source operator $\hat{h}_{\mathbf{k}\lambda}$ can be expressed through a bilinear combination of electron particle and hole operators. Since the action of these operators on the Heisenberg state $|F\rangle$ is known, (3.51), the remainder of the calculation is straightforward. It leads to an expression for the photon number (Ki 80a)

$$\begin{aligned} I(\mathbf{k}, \lambda) &= \langle F| \hat{a}^{(-)+}_{\mathbf{k}\lambda} \hat{a}^{(-)}_{\mathbf{k}\lambda} |F\rangle = \langle F| \hat{h}^+_{\mathbf{k}\lambda} \hat{h}_{\mathbf{k}\lambda} |F\rangle \\ &= \sum_{\substack{t>F \\ i<F}} |H^{fi}_{\mathbf{k}\lambda}|^2 + \left| \tilde{\sum_i} H^{ii}_{\mathbf{k}\lambda} \right|^2 \end{aligned} \tag{5.23}$$

with the abbreviation

$$H^{fi}_{\mathbf{k}\lambda} = -\frac{e}{(2\omega)^{1/2}} \int_{-\infty}^{\infty} dt\, e^{i\omega t} \langle \Phi^{(+)}_f | \varepsilon^*_{\mathbf{k}\lambda\mu} e^{-i\mathbf{k}\cdot\mathbf{r}} \gamma^\mu | \Phi^{(+)}_i \rangle \tag{5.24}$$

The sum $\tilde{\sum}_i$ is defined by

$$\tilde{\sum_i} = \frac{1}{2} \left(\sum_{i<F} - \sum_{i>F} \right)$$

which can be approximated by the sum over all occupied bound states (Hartree–Fock sum) if vacuum polarization effects are neglected. Equation (5.24) can be expressed by the occupation amplitudes $a_{ij}(t)$ in the same way as (5.14). The photon intensity $I(k, \lambda)$ can be identified with the squared radiation amplitude of (5.10) summed over all possible electron configurations

$$I(\mathbf{k}, \lambda) = \sum_f |C^{fi}_{\mathbf{k}\lambda}|^2 \tag{5.25}$$

The expression (5.23) and (5.25) can be shown to be completely equivalent (Ki 80a). Equation (5.23), however, is much easier to evaluate since the states Φ_i now are single-particle wave functions, while they referred to a complete set of N-particle Slater determinants in the last section (for brevity we have used the same notation in both cases). The results (5.23) takes such a simple form since the final electron configuration is not specified. In analogy to (3.56) coincidence rates between a photon and a particle or hole can be evaluated, leading to somewhat lengthy expressions.

5.4. The Angular Distribution

In addition to the energy spectrum a further feature of the MOX radiation can be measured quite easily: the distribution as a function of the photon angle, $d\sigma/dE_x d\Omega_x$. This quantity has turned out to have very interesting properties which we will discuss in the next section.

The ingredients necessary to calculate the angular distribution are contained in (5.23), (5.24). For the radiative matrix element $\boldsymbol{\gamma} \cdot \boldsymbol{\varepsilon}^*_{\mathbf{k}\lambda} e^{-i\mathbf{k}\cdot\mathbf{r}}$ in general the well-known multipole expansion of the radiation field can be used,

$$\boldsymbol{\varepsilon}^*_{\mathbf{k}\lambda} e^{-i\mathbf{k}\cdot\mathbf{r}} = (2\pi)^{1/2}(-)^{\lambda+1} \sum_L \sum_M (-1)^L (2L+1)^{1/2} D^L_{\mu\lambda}(\Omega) \times [\mathbf{A}_{LM}(m) - i\lambda \mathbf{A}_{LM}(e)] \tag{5.26}$$

with the magnetic and electric multipole fields $\mathbf{A}_{LM}(m, e)$. Usually in atomic physics only dipole radiation is important, so that the sum can be restricted to the term $L = 1$. A further significant simplification can be achieved in the long-wavelength approximation, $kr \ll 1$, whereby the plane wave factor is replaced by unity. This approximation is well justified for lighter atoms or outer shells but becomes unreliable for superheavy atomic systems. The use of the correct expression (5.26) is cumbersome since the radiation matrix elements depend explicitly on the wave number k and have to be calculated new for each photon energy. In the long-wavelength limit the dipole radiation matrix element can be rewritten using a commutator trick

$$\langle \phi_f | \boldsymbol{\alpha} | \phi_i \rangle = i\langle \phi_f | [H, \mathbf{r}] | \phi_i \rangle = i(E_f - E_i)\langle \phi_f | \mathbf{r} | \phi_i \rangle \tag{5.27}$$

Let us abbreviate

$$\mathbf{D}^{fi} = \sum_{jk} \int_{-\infty}^{\infty} dt \exp\left[i \int^{t} dt' (\omega - \omega_{jk}) \right] a_{fk}^{(+)*} a_{ij}^{(+)} \mathbf{d}_{kj} \tag{5.28}$$

with the molecular dipole matrix element $\mathbf{d}_{kj} = \omega_{kj} \langle \phi_k | \mathbf{r} | \phi_j \rangle$. This leads to a factorization of (5.14)

$$I(k, \lambda) = \frac{e^2}{2\omega} \sum_{\substack{f>F \\ i<F}} |\boldsymbol{\varepsilon}_{k\lambda} \cdot \mathbf{D}^{fi}|^2 + \left| \tilde{\sum_i} \boldsymbol{\varepsilon}_{\mathbf{k}\lambda} \cdot \mathbf{D}^{ii} \right|^2 \tag{5.29}$$

The result is simplified if (5.29) is summed over polarizations ($\lambda = 1, 2$) and averaged over the azimuthal angle ϕ_x of the photon. This is sufficient to describe singles experiments where the scattering plane is not defined. Likewise nearly all ion–x-ray coincidence measurements performed up to now have used annular particle detectors, thus losing the ϕ_x information. The averaged photon probability is

$$\begin{aligned} d^2P/dE_x \, d\Omega_x(b) &= \frac{\omega^2}{(2\pi)^3} \sum_{\lambda} \frac{1}{2\pi} \int_0^{2\pi} d\phi_x \, I(\mathbf{k}, \lambda) \\ &= \frac{\omega e^2}{(2\pi)^2} \left[A(\omega) + \tfrac{1}{2} B(\omega) \sin^2 \theta_x \right] \end{aligned} \tag{5.30}$$

with the functions

$$A(\omega) = \sum_{\substack{f>F \\ i<F}} |D_{\perp}^{fi}|^2 + \left| \tilde{\sum_i} D_{\perp}^{ii} \right|^2$$

and

$$B(\omega) = \sum_{\substack{f>F \\ i<F}} (2 |D_{\parallel}^{fi}|^2 - |D_{\perp}^{fi}|^2) + \left(2 \left| \tilde{\sum_i} D_{\parallel}^{ii} \right|^2 - \left| \tilde{\sum_i} D_{\perp}^{ii} \right|^2 \right) \tag{5.31}$$

To obtain (5.30) the relation

$$\frac{1}{2\pi} \int_0^{2\pi} d\phi \sum_{\lambda} |\boldsymbol{\varepsilon}_{\lambda} \cdot \mathbf{D}|^2 = |D_{\perp}|^2 + \tfrac{1}{2} \sin^2 \theta_x \, (2 |D_{\parallel}|^2 - |D_{\perp}|^2)$$

has been used. In a c.m. coordinate system with the z axis along the beam direction $\mathbf{v}$, the parallel and orthogonal components are $|D_{\parallel}|^2 = |D_z|^2$ and $|D_{\perp}|^2 = |D_x|^2 + |D_y|^2$. The extension of (5.30) to obtain $d\sigma/d\Omega_x$ is obvious.

In comparison with experiments an anisotropy parameter $\eta(E_x)$ is often used, defined by

$$\eta(E_x) = d\sigma(90°)/d\sigma(0°) - 1 = \tfrac{1}{2} B/A \tag{5.32}$$

Another measure for the angular distribution is the photon alignment A_0, defined by $A_0 = B/(3A + B)$. A full discussion of the polarization phenomena possible for the x-ray radiation that can be emitted in heavy-ion collisions was given in (Br 79).

The matrix elements of the dipole operator $\mathbf{r}$ in (5.28) are best evaluated in the intrisic molecular frame, with the z' axis pointing along the internuclear line. Therefore a transformation has to be performed expressing the intrinsic dipole operator in the inertial c.m. frame. If

$$\theta_N(t) = \int_{-\infty}^{t} dt' \, \dot{\theta}_N(t') = \int_{-\infty}^{t} \frac{bv}{R^2(t')} dt'$$

is the angle of rotation around the $y = y'$ axis we have to insert into (5.28) $\mathbf{d}_{kj}(t) = \mathsf{R}(t) \, \mathbf{d}'_{kj}$ with the matrix

$$\mathsf{R}(t) = \begin{pmatrix} \cos\theta_N(t) & 0 & -\sin\theta_N(t) \\ 0 & 1 & 0 \\ \sin\theta_N(t) & 0 & \cos\theta_N(t) \end{pmatrix} \tag{5.33}$$

The rotation matrix thus introduces an additional time dependence into the Fourier integral (Sm 75b).

5.5. Selected Results

The influence of atomic collisions on the shape of spectral lines was considered by Weisskopf in 1933 (We 33) and (in the classical framework) even earlier by other authors. Today the effect of "collisional line broadening" has been thoroughly investigated owing to its great importance to plasma and astrophysics. These applications mainly concentrate on peripheral collisions. In closer collisions the effect becomes very striking so that it has earned a special name: MOX.

During a close collision the transition energies and strengths of the interpenetrating atoms are greatly modified, leading to the emission of x-ray continua (of comparatively low intensity) distinct from the atomic lines. Such an effect has first been observed by Saris *et al.* (Sa 72) for L radiation in slow Ar + Ar collisions. Subsequently M radiation in I + Au (Mo 72) and K radiation in C + C (Ma 73) and Kr + Kr (Me 73b) were reported. Since that time a large number of MOX spectra has been investigated. The experiments were extended to the K shell of superheavy systems (Fr 75) up to the highest accessible charge $Z_1 + Z_2 = 164$, Pb + Pb (Me 79).

Several theoretical calculations of MOX-radiation have been performed, which, however, still have not led to a complete understanding of the process. Radiative transitions to inner shells are possible only if holes are available that can be filled. Any attempt to perform a calculation of MOX therefore has to consider the mechanism for hole formation first. Two distinct scenarios are possible:

(i) The double collision process: An atomic inner shell vacancy is produced in a first collision. Then in a second close encounter a hole in the MO under consideration is available from the outset. This process requires a high density of atoms; it is important if solid targets are used.

(ii) The one-collision mechanism: The vacancy is created by dynamical excitation and filled by x-ray radiation in one and the same collision. This is the only process available in gas–target experiments. Furthermore, for very heavy projectiles (e.g., U) the lifetime of an atomic K vacancy is too short ($<10^{-17}$ s) to be brought into a second collision.

Of course both alternatives are described by the same theory, in the language of Section 5.3 only the Heisenberg state $|F>$ is defined differently in both cases. The double collision MOX radiation, however, is easier to evaluate. Provided that the MO levels are highly adiabatic the variation of the amplitudes a_{ij} might be neglected entirely. Under the approximation $a_{ij}(t) \simeq \delta_{ij}$ and $a_{fk}(t) \simeq \delta_{fk}$, Equations (5.23) and (5.24) reduce to

$$\frac{dP}{d\omega}(b) = \frac{2}{3}\frac{e^2\omega}{\pi} \sum_{\substack{f>F \\ i<F}} \left| \int_{-\infty}^{\infty} dt \exp\left\{ i \int^t dt' [\omega - \omega_{if}(t')] \right\} \omega_{fi} \langle \phi_f | \mathbf{r} | \phi_i \rangle \right|^2 \tag{5.34}$$

which actually is Weisskopf's formula for line broadening (We 33, Br 74).† The essential features of (5.34) may be illustrated within the stationary phase approximation. It is based on the fact that the Fourier integral contributes mostly at the points $t = \pm t_\omega$, where the phase $\chi_{if}(t) = \int_0^t dt'(\omega - \omega_{if})$ is stationary, i.e., $\omega = \omega_{if}(t_\omega)$. A Taylor expansion of $\chi_{if}(t)$ up to second order leads to

$$\frac{dP}{d\omega} = \frac{4}{3} e^2\omega \sum_{\substack{f>F \\ i<F}} |\mathbf{d}_{fi}(t_\omega)|^2 \, |d\omega_{fi}/dt_\omega|^{-1} \cos^2[\chi_{if}(t_\omega) + \pi/4] \tag{5.35}$$

Ignoring the last factor, the result (5.35) could have been guessed intuitively from the "quasistatic" assumption that an undisturbed emission of photons takes place everywhere along the nuclear trajectory. $dP/d\omega$ is just given by

† Equation (5.34) is not completely correct: the hole has to be created somewhere, therefore the integral should be switched on smoothly at a finite time t^*. A sharp cutoff leads to an unreasonably strong line wing of the separated atom x-ray line (An 76b, He 78b). In addition the radiative decay of the hole could be accounted for by an exponentially decreasing factor.

the rate of spontaneous x-ray emission multiplied by the interval of time available for a transition with $\omega = \omega_{if} \dots \omega_{if} + d\omega$. Calculated for a fixed impact parameter the spectrum (5.35) has an unphysical singularity at $\omega = \omega_{if}(R_{\min})$. Integrated over impact parameter, however, the differential cross section $d\sigma/d\omega$ looks smooth and agrees quite well with the exact solution of the Fourier integral (Ma 74a).

The cosine factor in (5.35) has its roots in quantum mechanics. There is a stationary point of $\chi(t)$ both on the incoming and outgoing branch of the scattering trajectory. The radiation emitted at these two distinct intervals can interfere, leading to an oscillatory MOX spectrum for fixed impact parameter b (Li 74, Ma 74a, Sm 75b). The period of oscillation is determined by the phase difference $\Delta\chi$, thus $\Delta E_x = \pi/t_\omega \simeq \pi v/R_\omega$, where $R_\omega = R(t_\omega)$. Rapid oscillations should occur near the separated atom line, reaching out, with increasing period, up to the UA limit. Integrated over impact parameter the spectrum again looks smooth.

After several unsuccessful attempts, the predicted interference pattern recently was observed in experiments (Ts 83) using a beam of one-electron projectiles impinging on a gas target, e.g. $S^{15} + Ar$. Here a $1s\sigma$-hole is brought into the collision with about 50% probability, thus ensuring the symmetry of the integrand in (5.34) which is essential to get strong interference. In this way the phase integral $\chi_{if}(t_\omega)$ can be measured directly.

The quasistatic model describes the gross features of MOX emission but it fails at high energies E_x, since here the neglected dynamical broadening becomes essential. Experiments usually show structureless exponentially falling spectra, which extend well beyond the UA transition energy and have no definite end point. The high energy tail is explained by solution of the Fourier integral (5.34). Several authors (Ma 74a, Be 75, Mü 75a) have deduced an exponential decrease $d\sigma/dE_x \propto E_x \exp(-E_x/\Gamma)$. The width depends on the molecular transition energies and is proportional to a power of the collision velocity, $\Gamma \propto v^n$, where values of $n = 1/2 \cdots 2/3$ have been predicted. Experimentally the higher value is favored. Very similar shapes have been found for the process of quasimolecular Auger electron emision (Sh 81). An analytical approximation involving complex turning points t_ω which extends the stationary phase approximation beyond the UA transition energy has been considered by (Fr 79).

The shape of the $1s\sigma$ MOX spectrum does not reveal detailed information on the quasimolecular correlation diagram. The situation should be more favorable for $2p\sigma$ MOX caused by transitions filling the $2p\sigma$ molecular orbital (Gi 74, He 76) since the transition energies to this level become stationary at a finite internuclear distance (compare, e.g., Figure 4.4). As a consequence a broad shoulder between the UA K_α and the $1s\sigma$ continuum is observed. For still higher shells $(L, M, \dots)$ maxima in the x-ray spectra are observed which can be identified with transitions where the orbital energies exhibit a shallow minimum as a function of R. The position of these structures scales with Z^2, confirming the quasimolecular origin, cf. (Mo 78).

Semianalytical model solutions of (5.34) were found to explain much of the experimental data (Mü 75b). During the last few years more refined numerical calculations were performed, taking into account the time variation of the amplitudes $a_{ij}(t)$ of (5.14). The spectra of two-collision $1s\sigma$ MOX and one-collision $2p\sigma$ MOX in several light (Br 79) and medium heavy (An 78, An 79b, Jä 81) systems have been studied in detail. For example Jäger *et al.* (Jä 81) performed coupled-channel calculations taking into account the molecular states from $1s\sigma$ up to $3d\sigma$. Figure 5.1 shows the MOX spectrum for 67-MeV Nb + Nb collisions, normalized to the number of produced K holes. The experimental spectrum (Fr 76a) shows a two-component distribution that can be well explained by the two distinct processes of $2p\sigma$ (dotted curve) and $1s\sigma$ MOX (chain curve).

The $1s\sigma$ MOX emission in a superheavy system, Pb + Pb, has been studied by Kirsch (Ki 78, Ki 79). In principle such collisions are of greatest interest in view of a possible spectroscopy of superheavy quasimolecules. Here the one-collision process should prevail and a theoretical model is confronted with the task of describing K hole as well as MOX production. It has been already noted from model calculations for the much less involved p + H system (Th 77) that the one-collision mechanism does not allow for an

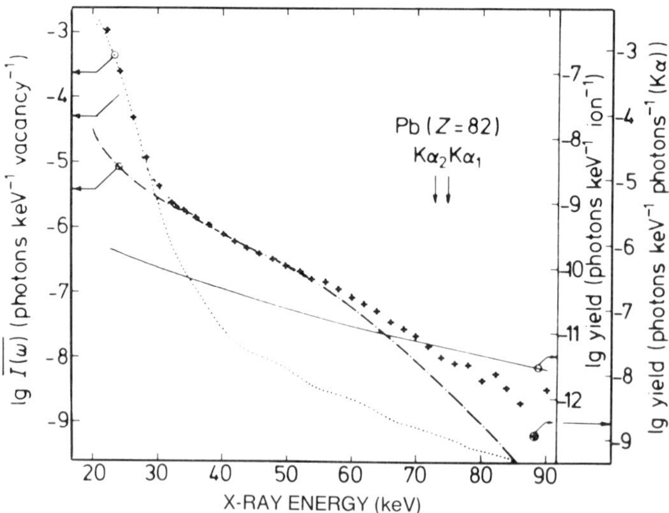

Figure 5.1. Comparison of experimental (Fr 76a) and theoretical (Jä 81) MO x-ray yields for 67-MeV Nb + Nb collisions. Dotted curve: Calculated spectrum of radiation filling $2p\sigma$ vacancies assuming a one-collision process; dash-cotted curve: Calculated two-collision $1s\sigma$ radiation; solid curve: nucleus–nucleus bremsstrahlung of $E2$ type. From (Jä 81).

easy way to relate the MOX emission to a simple quasistatic picture, e.g., by factoring out the K hole probability. The amplitudes $a_{ij}^{(+)}(t)$ of Equation (5.24) are strongly time-dependent oscillating functions, which should not be replaced by an average value.

As discussed in Chapter 4, $1s\sigma$ excitation in high-Z systems is caused by radial coupling to less deeply bound states and by direct ionization to the continuum. The resulting K-hole amplitudes have been employed to calculate the MOX radiation. In (Ki 79) the formulation of Section 5.2 restricted to a one-hole approximation was used, treating the ionization process in first-order perturbation theory. A more recent calculation (Ki 80b) was performed in the many-particle framework of Section 3.3. The coupled-channel problem has been solved with a basis set including the whole K, L, and M shell. The vacancy production rate was scaled to the monopole model result of (Re 79b) to account for excitations to higher states and to the continuum.

As a basic ingredient of this calculation the most important levels of the quasimolecular correlation diagram are shown for the highly relativistic Pb–Pb system in Figure 5.2a, while Figure 5.2b presents the dipole matrix elements of the transitions from various higher states to the $1s\sigma$ orbital. Generally the values decrease at small internuclear distance R, due to the shrinking radius of the shells. There is an interesting exception: transitions

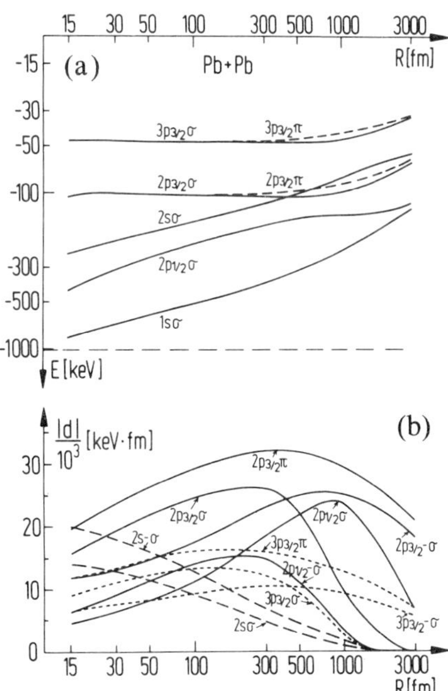

Figure 5.2. (a) States of the quasi-molecule Pb + Pb which have been included in the MOX calculation. (b) Radiative dipole transition matrix elements connecting the $1s\sigma$ orbital of the Pb + Pb quasimolecule with various higher states, drawn as a function of R. Solid lines, $E1$ multipolarity; dashed lines, $M1$ multipolarity.

from the $2s \pm \sigma$ states have a maximum at $R \to 0$, since this $M1$ transition is a purely relativistic effect, strongly enhanced in the high-Z region.

In general, x-ray transition rates in superheavy atomic systems show remarkable differences compared with light atoms. Anholt and Rasmussen (An 74) and Soff and Müller (So 77a) have calculated radiative matrix elements (using the exact multipole expansion) with relativistic Hartree–Fock–Slater wave functions for superheavy atoms. In contrast to the Z^4 scaling in Schrödinger atoms the transition rates exhibit an irregular, sometimes quite erratic behavior with extrema and zeros as a function of Z. This is cased by (i) the shrinking of the electron wave functions and (ii) the the growing magnitude of the "small" component of the Dirac wave function, $f(r)$ in Equation (2.5). Several different multipoles will contribute to a given transition. Normally exotic or forbidden transitions thus can become dominant. For example, the largest x-ray decay width for filling the K shell in a $Z = 173$ atom is associated with $M1$ transitions from the $2s$ initial state. The width is $\Gamma = 226$ eV, which is nearly twice as large as the value for the $E1 + M2$ transition $2p_{3/2}$–$1s$ (So 77a). Thus unusual radiative transition strengths are a further characteristic property of superheavy quasimolecules or atoms; cf. also Section 7.3.

The Pb + Pb $1s\sigma$ MOX yield as a function of x-ray energy calculated by (Ki 80b) is shown in Figure 5.3. The predicated spectrum is compared with the measurement of Meyerhof *et al.* (Me 79, St 80a). At energies beyond 400 keV the agreement in absolute magnitude and slope is remarkably good, while at lower energies the transitions filling higher quasimolecular states (in particular $2p\sigma$) become dominant, which are not calculated.

The model includes radial and rotational coupling between all of the 18 MO of the K, L, and M shell and all allowed $E1$ and $M1$ dipole transitions to the $1s\sigma$ state. The most important single contribution originates from the $2p_{3/2}\pi$ state. Also depicted in Figure 5.3 is the predicted (Re 76) nucleus–nucleus bremsstrahlung (NNB) which is of quadrupole type in a symmetric system and seems to play no role in this collision The situation is different in various lighter or asymmetric systems, where NNB can mask the high-energy tail of the MOX spectrum.

Since they contain more detailed information on the collision mechanism, the study of x-ray–particle–coincidence measurements is highly instructive. For the Pb + Pb system such experiments recently have been performed by K. Stiebing *et al.* (St 81), who find a two-component structure in the impact parameter dependence of $d^2P/dE_x\, d\Omega_x(b)$, attributed to $1s\sigma$ and $2p\sigma$ MOX. Note that such measurements can lead to an experimental determination of the $1s\sigma$ ionization probability even in symmetric collision systems. Again, also for the impact parameter dependence of $1s\sigma$ MOX (at $b < 75$ fm), the agreement between theory (Ki 79) and experiment is quite close. One should keep in mind that the theory has not yet reached its final

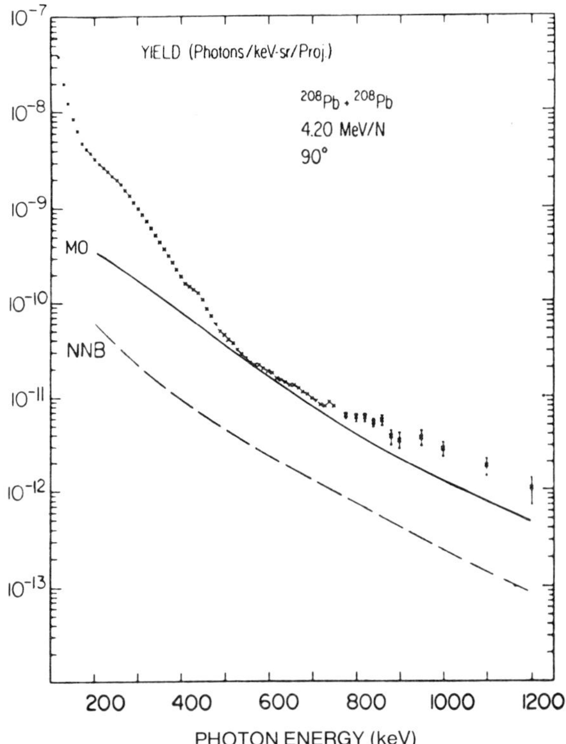

Figure 5.3. Spectrum of quasimolecular $1s\sigma$ radiation in the superheavy system Pb + Pb at 4.2 MeV/u. Solid curve: coupled-channel calculation of Kirsch (Ki 80b); dashed curve: nucleus–nucleus quadrupole bremsstrahlung. The experimental data are from Stoller *et al.* (St 80a).

stage due to several approximations made, in particular the long-wavelength limit of (5.27), which is not well justified at such high x-ray energies. The comparison suggests, however, that the basic mechanism of continuum x-ray emission in heavy-ion collisions is understood and that the effect demonstrates the formation of superheavy quasimolecules.

Despite this success there remains the ambitious goal of performing a true spectroscopy of quasimolecular energy levels. To some extent this indeed has turned out to be feasible through a study of the anisotropy of x-ray radiation. Interest in this quantity was fostered by B. Müller *et al.* (Mü 74), who predicted a positive anisotropy η of the radiation at energies close to the united atom transition energy, based on a model for transitions induced by the rotation of the quasimolecular axis. Although the effect of the rotation later turned out to be more subtle (Gr 76b), experiments by Greenberg *et al.* (Gr 74) and Kraft *et al.* (Kr 74) soon detected a strong positive anisotropy $\eta(E_x)$. In subsequent measurements, most notably the systematic studies of Wölfli and collaborators (St 77), the scaling of the anisotropy maximum with total nuclear charge $Z = Z_1 + Z_2$ was confirmed. Meanwhile x-ray spectra with many distinct peaks in the anisotropy have been measured, where each of them could be assigned to a transition in the quasimolecular correlation diagram.

The theoretical explanation of this effect should be concealed in Equations (5.30), (5.31). Let us ignore, for a moment, the nuclear motion and consider transitions from a p shell to the $1s$ state. In a nonrelativistic system there are three p states, $p\sigma$, $p\pi_x$, $p\pi_y$, oriented along the z', x', and y' axes, respectively. Since $D_\perp = 0$ for the transition $2p\sigma \to 1s\sigma$ and $D_\parallel = 0$ for $2p\pi \to 1s\sigma$, Equations (5.30) and (5.31) lead to the angular distributions $dP \propto \sin^2\theta$ and $dP \propto \frac{1}{2}(1 + \cos^2\theta)$. These are just the familiar radiation patterns for $\Delta m = 0$ and $\Delta m = +1$ dipole radiation, giving a anisotropy η which is large and positive in the former and negative ($\eta = -1/2$) in the latter case. For a completely filled p shell the emission is isotropic. To get a positive anisotropy η an alignment of the p shell is required: The $p\pi$ level should be depopulated compared to the $p\sigma$ state. The existence of such an alignment is not unexpected, since in the limit $R \to \infty$ the $np\pi$ MOs correlate to atomic states less bound than the $np\sigma$ levels, so that more vacancies should be available.

In reality the situation is more complicated than described above, since the molecular axis rotates with time, introducing additional Fourier frequencies in the integral (5.24) via the rotation matrix (5.33). However, it was noted by R. K. Smith *et al.* (Be 76a, Gr 77b) that the influence on the x-ray radiation in part is spurious, canceled by the "electron slip" effect, which is hidden in the amplitudes $a_{ij}(t)$: The strong rotational coupling between the $np\sigma$ and $np\pi$ states in close collisions ($b \simeq a$) leads to transitions with probability close to one; cf. the discussion in Section 4.3.3. The electron cloud remains fixed in space, while the molecular axis changes its orientation. The two extreme cases are illustrated in Figure 5.4 for the $2p\sigma$ state. The dashed line denotes the electron density; the dotted line represents the corresponding radiation pattern.

In a model calculation incorporating "electron slip'" Smith was able to explain the anisotropies measured by Vincent and Greenberg in the Ni + Ni system in terms of a special alignment (Be 76a). Later calculations used the coupled states approach to account for rotational coupling (Br 77, Br 78a, An 78, An 79b). As an example we discuss the result of Anholt (An 78) for the one-collision $1s\sigma$ MOX in the system Ni + Ni. The calculation includes states from $1s\sigma$ up to $3p\sigma$ and their mutual rotational coupling. Radial excitations, however, are neglected. The resulting photon anisotropy parameter A_0 is shown in Figure 5.5a. as a function of x-ray energy. The contributions of individual states taken separately show strong anisotropies

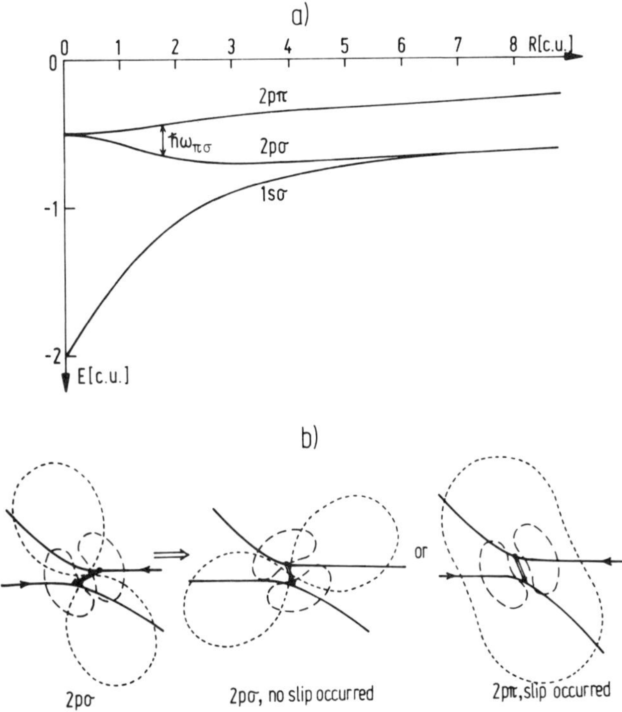

Figure 5.4. (a) Nonrelativistic correlation diagram of symmetric quasimolecules demonstrates the degeneracy of $2p\sigma$ and $2p\pi$ states in the united atom limit. (b) An electron (dashed lines denote the density) initially in, e.g., the molecular $2p\sigma$ state is not likely to follow the nuclear rotation ("no slip"). More often its orientation in space remains fixed, corresponding to a transition to the $2p\pi$ state ("electron slip"). The associated radiation patterns (dotted lines) in the latter case imply a positive photon anisotropy. From (Gr 77b).

which, at high energies, are in agreement with the naive static limit ($A_0 \simeq 1$ for $np\sigma$ and $A_0 \simeq -0.5$ for $np\pi$). For equal population of the substates (no alignment) the predicted anisotropy is small. With the assumption of a depopulation of the $3p\pi$ level, however, the measured anisotropies can be fitted quite well, as shown in Figure 5.5b. The function $\eta(E_x)$ peaks at about the position of the UA K_α and K_β lines, marked by arrows in the figure. At higher energies the anisotropy drops due to an increasing relative contribution of transitions from the $np\pi$ states. In an analogous manner the angular distribution of two-collision $2p\sigma$ MOX has been explained (An 79b), where again a peak at the characteristic transition energy in the quasimolecule was found.

Hartung *et al.* (Ha 79) proposed a simplified model (the kinematic dipole approximation, KDA) where the anisotropy is caused by $\Delta m = 1$ emission from a rotating dipole. The electron dynamics is neglected and the result is

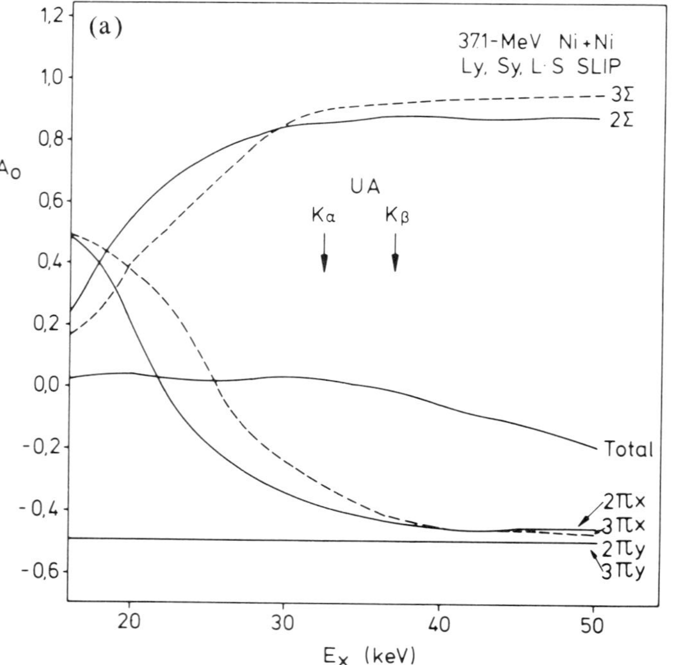

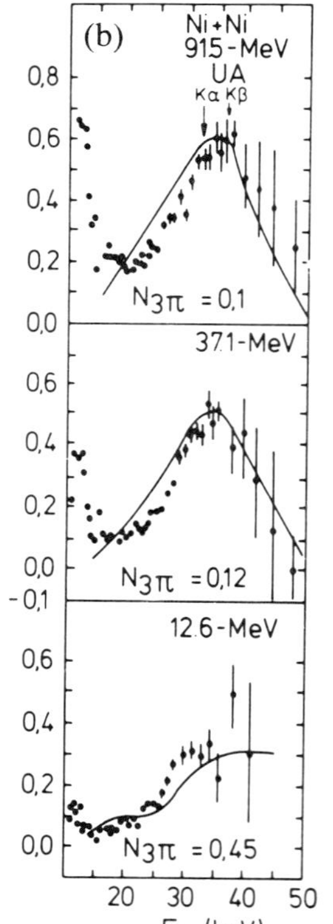

Figure 5.5. (a) Photon alignment parameter A_0 of quasimolecular radiation in 37.1 MeV/u Ni + Ni collisions as a function of photon energy calculated for various transitions ($2p\sigma$, $3p\sigma$, $2p\pi$, $3p\pi$) filling the $1s\sigma$ level. The model includes the "electron slip" effect. (b) Experimentally observed maxima of the x-ray anisotropy can be explained from the model used in (a) under the assumption that the $np\pi$ states are not fully occupied (electron alignment). Fits to the experimental data of Vincent and Greenberg are shown for several bombarding energies. From (An 78).

essentially independent of alignment. In contrast to the alignment model (An 81), the KDA predicts large anisotropies also in the azimuthal angular distribution. This seems to be at variance with recent coincidence measurements (Bü 81).

While the study of maxima in the anisotropy $\eta(E_x)$ was surprisingly successful in light and medium heavy systems, the situation seems to be less encouraging in superheavy quasimolecules. The low-Z results cannot simply be scaled to very heavy systems, since here relativistic and spin–orbit effects become dominant in determining the level structure of the quasimolecule. Experiments in the Pb + Pb system (St 80a) indicate that the MOX anisotropy is nearly zero in the interesting energy region. Therefore it seems unlikely that the MOX spectroscopy of quasimolecules can be extended to the inner shell in very heavy collisions.

6. Positron Creation in Heavy-Ion Collisions

6.1. Introduction

Until recently, theoretical interest in electron–positron pair creation by charged particles was predominantly concentrated on high-energy, relativistic collisions of elementary particles or light nuclei. Here the use of the plane wave Born approximation is justified or the Weizsäcker–Williams method may be applied. In nonrelativistic nuclear collisions it is essential to account correctly for the Coulomb repulsion of the colliding particles, which strongly reduces the cross section (Ba 59). This has been verified by experiments htat failed to detect pair production in slow proton–nucleus collisions (St 58), contrary to the predictions of the Heitler–Nordheim formula (He 34, Li 35).

Renewed interest in the problem has arisen as it was realized that strong electromagnetic fields having interesting physical consequences are generated during a heavy-ion collision. The predicted phenomena have been discussed in Section 2 for the case of static superheavy or supercritical systems. To calculate pair production in high-Z collisions a correct treatment of the distortion of electron and positron states in the Coulomb field of the two nuclei is essential. This provided by the semiclassical quasimolecular model already used for the calculation of inner shell hole and δ-electron production. It was argued in Section 3.3 that positron creation can be described in just the same manner as the lifting of an electron from the "Dirac sea" to an unoccupied electronic state above the Fermi level F.

As a consequence, the rate for direct pair production grows very fast with increasing nuclear charge in the superheavy region. This has to be compared with the $Z_1^2Z_2^2$ dependence deduced from the lowest-order Feynman diagram. Furthermore the role of inner shell bound states (in particular 1s and $2p_{1/2}$) becomes increasingly important if the united nuclear charge approaches Z_{cr}. These may act as intermediate states in multistep excitations or (in the still hypothetical collisions of fully stripped ions) be the dominant final states for the created electron.

In the case of subcritical collisions ($Z_1 + Z_2 < 173$) the same formalism as for inner shell excitation can be used. If supercritical collisions ($Z_1 + Z_2 > 173$) are to be treated, however, additional problems arise, since the adiabatic 1s state vanishes from the discete spectrum of the Dirac operator and turns into a resonance in the antiparticle continuum. A method to describe the time-dependent resonance will be described below.

6.2. Dynamical Theory of Supercritical Collisions

6.2.1. *Discussion of the Problem*

Normally the energy eigenvalues of the adiabatic electronic bound states of the Dirac equation are confined to the gap region $-mc^2 < E < mc^2$. In supercritical collisions, however, the deepest bound state joins the lower continuum and becomes a resonance. In the static limit, a hole brought into this state will decay spontaneously by positron emission leaving a stable filled atomic K shell. The lifetime of the resonance is of the order of 10^{-19} s and therefore considerably larger than the collision time ($\simeq 2 \times 10^{-21}$ s for U + U collisions with $R < R_{cr}$). Excitations induced by the nuclear motion will be of eminent importance. Therefore a formalism is required which describes dynamical excitations and at the same time accounts for the resonance character of the supercritical state.

The coupled differential Equation (3.19) are not directly applicable to this situation: In the region $|t| < t_{cr}$ the 1$s\sigma$ state together with its amplitude disappears from the set of discrete states. Instead, the radial coupling matrix elements involving the lower continuum develop very strong and narrow (few keV) maxima near the (time-dependent) position of the resonance. These couplings are not suitable for numerical treatment. Several ways to circumvent this problem can be imagined.

In the quasistatic approximation it is assumed that an undisturbed decay of the resonance takes place at each point of the trajectory (Pe 73, Ge 73, Po 73, Po 74, Ma 75b). The transition rate is proportional to the decay width $\Gamma(t)$, which is determined parametrically by the nuclear motion. This approximation produces unacceptable energy spectra and is generally insufficient, since coherence effects of hole excitation and positron emission and, most important, the consequences of dynamical broadening may not be neglected. The first calculation that went beyond the static approximation was performed by K. Smith *et al.* (Sm 74).

Alternatively one might try to use a subcritical basis: Problems associated with the dynamical treatment of the resonance might be avoided by expansion in a basis which remains subcritical throughout the whole collision. This is most easily achieved if one uses not the adiabatic eigenstates of $H(R(t))$ but those of a modified Hamiltonian $H(\rho(t))$, where $\rho(t)$ describes a trajectory satisfying $\rho > R_{cr}$. The function $\rho(t)$ may be chosen arbitrarily, the simplest choice being $\rho(t) = R(t)$ for $|t| > t_0$ and $\rho(t) = R_0$ for $|t| < t_0$. If $R_0 = R_{cr}$ this corresponds to a switching from the adiabatic basis in the subcritical region to a frozen basis ϕ_k^0 in the supercritical domain.

If the colliding nuclei approach to distances $R(t)$ closer than R_0, there will arise potential couplings due to the operator $\Delta H = H(R(t)) - H(R_0)$. This description in principle is correct and the couplings are easily numerically managable. In accordance with the discussion of Section 3.5.2 it turns out, however, that it is very difficult to achieve completeness in the modified basis ϕ_k^0. The nondiagonal matrix elements of ΔH decrease only very slowly with increasing energy separation between initial and final state so that a forbiddingly large number of channels would have to be included in the calculation.

6.2.2. *A Projection Method for the Supercritical Resonance State*

The methods used to describe excitations in supercritical collisions discussed in the last section had serious disadvantages, making them unsuitable for numerical calculations. Now we will develop an approach (Re 80b, Re 81a) which avoids these difficulties and moreover has heuristic value for the interpretation of the positron creation process. We start from the observation that the continuum wave function of the supercritical system at resonance energy $E_p = E_{res}$ is quite similar to the discrete $1s\sigma$ state in the subcritical case. In addition to a strongly localized density distribution

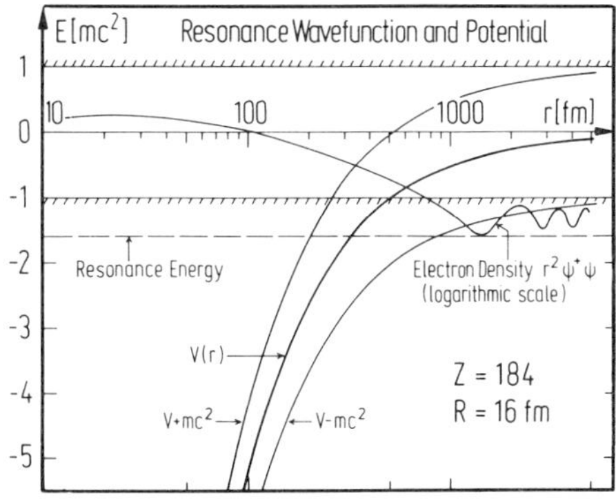

Figure 6.1. The potential well $V(r)$ in a U + U quasimolecule at nuclear contact. Also shown are the borders of the gap $V(r) \pm mc^2$, the energy of the $1s\sigma$ resonance, and the radial density of the positron continuum wave function at resonance energy. The density is drawn on a logarithmic scale covering five orders of magnitude. From (Re 81a).

having the extension of the atomic K shell, the former exhibits an oscillating tail (small in amplitude) reaching out to infinity (see Figure 6.1). This structure reflects the occurrence of a tunneling process through the gap separating the particle and antiparticle solutions of the Dirac equation (cf. the problem of Klein's paradox quoted in Section 2). Apart from the asymptotic behavior the $1s\sigma$ wave function retains much of its identity. Many properties, e.g., the radial matrix elements for ionization, may be continued smoothly to the supercritical region just by neglecting the tail of the wave function.

This procedure can be put on a firmer basis. In a first step a "quasibound" resonance state $|\Phi_R\rangle$ is defined as a reasonable approximation to the $1s\sigma$ state in the absence of a penetrable tunneling barrier. In a second step a new positron continuum $|\tilde{\phi}_{E_p}\rangle$ has to be constructed which excludes the resonance, is orthogonal to $|\Phi_R\rangle$, and preserves completeness. This is achieved with a projection operator technique which had been developed for applications in nuclear physics. The resulting set of modified stationary states will be used as a basis for expanding the time-dependent wave function in analogy to Equation (3.17).

We start from the assumption that a "physically reasonable" wave function $|\Phi_R\rangle$ describing the bound-state properties of the resonance has been found. This state should be normalized

$$\langle \Phi_R \mid \Phi_R \rangle = 1 \tag{6.1}$$

and orthogonal to the states $|\phi_\alpha\rangle$ outside the positron continuum

$$\langle \phi_\alpha \mid \Phi_R \rangle \simeq 0 \tag{6.2}$$

where $|\phi_\alpha\rangle$ denotes all bound states and the electron continuum. The resonance now is to be extracted from the positron continuum $|\phi_{E_p}\rangle$ resulting in a modified continuum $|\tilde{\phi}_{E_p}\rangle$ orthogonal to the bound state

$$\langle \tilde{\phi}_{E_p} | \Phi_R \rangle = 0 \tag{6.3}$$

which still spans the same subspace.

This problem was solved in nuclear physics in connection with the continuum shell model (Wa 70, Ba 71, Mi 78). The formalism applied there makes use of the projection operators P and Q introduced by Feshbach (Fe 62a) projecting on "open" and "closed" channels. In the subcritical case they will be defined to project on the space of continuum positron states

$$P_0 = \int_{-\infty}^{-m} dE_p \, |\phi_{E_p}\rangle\langle\phi_{E_p}| \tag{6.4}$$

and on its complement

$$Q_0 = \sum_n |\phi_n\rangle\langle\phi_n| + \int_m^\infty dE_e\, |\phi_{E_e}\rangle\langle\phi_{E_e}|$$

$$\equiv \sum_\alpha |\phi_\alpha\rangle\langle\phi_\alpha| \tag{6.5}$$

$|\phi_{E_p}\rangle$ and $|\phi_\alpha\rangle$ are eigenstates of the Hamiltonian H,

$$(E_p - H)\,|\phi_{E_p}\rangle = 0 \quad \text{and} \quad (E_\alpha - H)\,|\phi_\alpha\rangle = 0 \tag{6.6}$$

so that P_0 and Q_0 satisfy the usual relations of orthogonal projection operators, $P_0^2 = P_0$, $Q_0^2 = Q_0$, $P_0 Q_0 = 0$.

If the strength of the potential exceeds the critical value, the $1s\sigma$ state becomes a resonance in the positron continuum thus entering P_0 space. The aim now is to transfer the bound state contribution represented by the wave function Φ_R to Q space. We define the new projection operators

$$Q = \sum_\alpha |\phi_\alpha\rangle\langle\phi_\alpha| + |\Phi_R\rangle\langle\Phi_R| \quad .$$

$$P = 1 - Q := \int_{-\infty}^{-m} dE_p\, |\tilde\phi_{E_p}\rangle\langle\tilde\phi_{E_p}| \tag{6.7}$$

Under the assumption (6.2) the operators Q, P again are orthogonal projectors. If $|\Phi_R\rangle$ was chosen judiciously the newly defined modified continuum $|\tilde\phi_{E_p}\rangle$ will no longer show resonance behavior. The $|\tilde\phi_{E_p}\rangle$ are eigenstates of the Hamiltonian restricted to the subspace P, i.e.,

$$(E_p - \mathrm{PHP})\,|\tilde\phi_{E_p}\rangle = 0 \tag{6.8}$$

Using (6.6), (6.7), and the orthogonality relations (6.2), (6.3) this equation may bej transformed to a more explicit form

$$(E_p - H)\,|\tilde\phi_{E_p}\rangle = -\,\langle\Phi_R|\,H\,|\tilde\phi_{E_p}\rangle\,|\Phi_R\rangle \tag{6.9}$$

The modified continuum states satisfy the original Dirac equation supplemented by an inhomogeneous term containing an integral over the solutions $|\tilde\phi_{E_p}\rangle$. Fortunately the kernel is separable so that (6.9) can be solved easily. The formal solution of Equation (6.9) can be given using the Green's function, $(E - H)\,G = 1$. Alternatively, the defining Equation (6.9) can be solved directly by a straightforward integration procedure as reported by Wang and Shakin (Wa 70).

The projection formalism is easily extended to the case of several resonances. In the present context this is not required since in heavy-ion collisions apart from the $1s\sigma$ only the $2p_{1/2}\sigma$ state becomes supercritical (at $Z \sim 185$). In symmetric systems and generally if the monopole approximation is used the $p_{1/2}\sigma$ and $s_{1/2}\sigma$ continua do not couple and can be treated independently. The $2s\sigma$ state dives at $Z \sim 245$, which is far too high to be reached in a collision of two heavy ions.

The resonance state $|\Phi_R\rangle$ and its associated modified positron continuum $|\tilde{\phi}_{E_p}\rangle$ have to be determined at all values of the internuclear distance $R < R_{\rm cr}$. They will be used as a basis to expand the time-dependent wave function in complete analogy to the subcritical case, i.e.,

$$\Phi_i(t) = \int_m^\infty dE_e\, a_{i,E_e} \phi_{E_e} e^{-iE_e t} + \sum_{n>1s} a_{i,n} \phi_n e^{-i\chi_n}$$
$$+ a_{i,1s} \Phi_R e^{-i\chi_{1s}} + \int_{-\infty}^{-m} dE_p\, a_{i,E_p} \tilde{\phi}_{E_p} e^{-iE_p t} \qquad (6.10)$$

If the resonance energy E_{1s} is identified with $\langle \Phi_R | H | \Phi_R \rangle$ and the orthogonality relation (6.2) is fulfilled, we obtain the old set of coupled differential equations. There arises, however, one important modification: Since the resonance state $|\Phi_R\rangle$ is not an eigenstate of H it has an additional interaction with the continuum. The time-derivative coupling matrix elements $1s\sigma \to E_p$ must be replaced by

$$\langle \phi_{E_p} | \partial/\partial t | \phi_{1s} \rangle \to \langle \tilde{\phi}_{E_p} | \partial/\partial t | \Phi_R \rangle - i \langle \tilde{\phi}_{E_p} | H | \Phi_R \rangle \qquad (6.11)$$

A hole prepared in $|\Phi_R\rangle$ therefore will decay by spontaneous positron emission in addition to the dynamically induced transitions described by the $\partial/\partial t$ coupling. In the static limit, $R(t) = \text{const}$, this would lead to an exponential decay with the width

$$\Gamma = 2\pi \, |\langle \tilde{\phi}_{E_p} | H | \Phi_R \rangle|^2 \qquad (6.12)$$

where E_p is taken at the position of the resonance.

The described formalism thus has led to the emergence of "induced" and "spontaneous" positron couplings, the latter resulting from the presence of an unstable state, $|\Phi_R\rangle$, in the expansion basis. It is difficult, however, to draw simple conclusions from this fact. Both coupling matrix elements enter via their Fourier transforms, depending on the time development of the heavy-ion collision. Their contributions have to be added coherently so that in a given collision there is no physical way to distinguish between them.

The projection method for constructing a modified basis starts from the quasibound state $|\Phi_R\rangle$ which is supposed to have the properties of a $1s\sigma$ bound state. A straightforward prescription to generate Φ_R starts from the positron continuum wave function ϕ_{E_p} at the energy of the resonance $E_p = E_{\text{res}}$, cutting it off at large distances $r > r_c$

$$\Phi_R(r) = C\phi_{E_{\text{res}}}(r)\,\theta(r_c - r) \tag{6.13}$$

The normalization constant C is determined such that (6.1) is fulfilled. The cutoff wave function (6.13) has been employed to describe resonances in nuclear physics.

The cutoff distance, r_c, is a somewhat arbitrary parameter which should be chosen to lie within the region of the tunneling barrier defined by $r_- < r < r_+$ with $E_{\text{res}} - V(r_\pm) \pm mc^2 = 0$. At larger distances r the wave function ϕ_{E_p} begins to oscillate. This is illustrated in Figure 6.1, where the density of the positron wave function at $E = E_{\text{res}}$ is shown together with the gap of the Dirac equation and the nuclear Coulomb potential $V(r)$. The system considered is U + U at internuclear distance $R = 16$ fm, i.e., at nuclear contact. This and all the following calculations are performed in the monopole approximation.

To avoid problems associated with the discontinuity of the wave function a modified cutoff procedure has been adopted in (Re 80b). $|\Phi_R\rangle$ has been defined as an eigenfunction of the Dirac–Hamiltonian with the modified potential

$$\tilde{V}(r) = \theta(r_c - r)\,V(r) + \theta(r - r_c)\,V(r_c) \tag{6.14}$$

For distances $r < r_c$ $\tilde{V}$ agrees with the old potential V; at large distances the potential is kept fixed so that the 1s energy remains inside the gap region. This artifice produces a smoothly decreasing tail of Φ_R, while the wave function agrees with $\phi_{E_{\text{res}}}$ in the interior region.

Using the smoothly cut off resonance wave function Φ_R the modified continuum $\tilde{\phi}_{E_p}$ has been calculated by numerical integration of the inhomogeneous Dirac Equation (6.9). As an example, Figure 6.2 shows the large and small components of ϕ_{E_p} at resonance energy and the cutoff wavefunction Φ_R (dotted lines). The lower part of the figure displays the modified continuum $\tilde{\phi}_{E_p}$ which shows no resonance behavior.

The modified continuum $\tilde{\phi}_{E_p}$ can be shown to have the usual orthonormality properties. Its phase shift, however, is changed by a counter term which cancels the steep variation near the position of the resonance and leaves a smooth nonresonating phase. For the U + U system at $R = 16$ fm, Figure 6.3 demonstrates the absence of any structure in the phase shift δ, while $\tilde{\delta}$ sharply increases by π at $E = E_{\text{res}}$. Thus the resonance has been completely eliminated from the continuum. Also shown in Figure 6.3 is the "resonance excess" of the wave function ϕ_{E_p} defined as the ratio between the maximum and asymptotic value of $u_1(r)$, the large component of the radial wave function. The narrow Breit–Wigner-like maximum of this ratio charac-

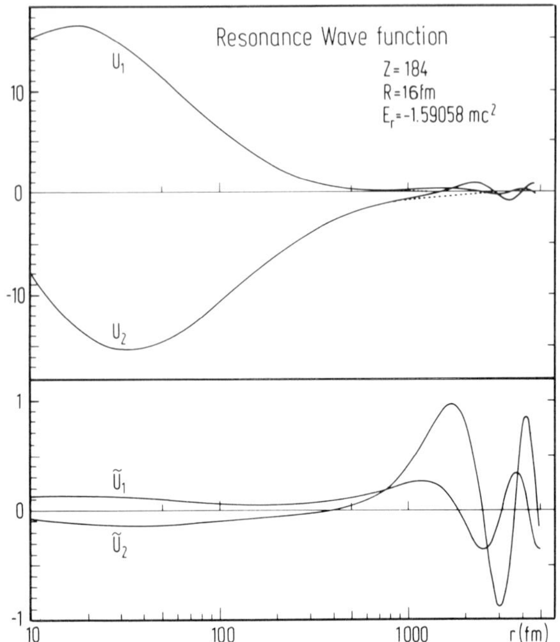

Figure 6.2. Upper part: The large and small components $u_1(r)$ and $u_2(r)$ of the $s_{1/2}$ continuum ϕ_{E_p} as in Figure 6.1. The dashed lines indicate the resonance wave function Φ_R (not normalized). Lower part: The modified continuum wave function $\tilde{\phi}_{E_p}$ at the resonance energy. From (Re 81a).

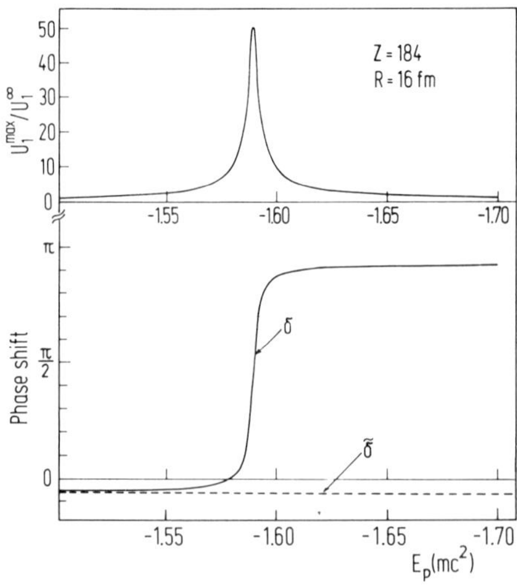

Figure 6.3. Analysis of the $1s\sigma$ resonance in the quasimolecule U + U at $R = 16$ fm. Upper part: The ratio of the maximum and asymptotic value of u_1 Lower part: The phase shift of ϕ_{E_p} and $\tilde{\phi}_{E_p}$, i.e., of the original and modified continuum. From (Re 81a).

terizes a sharp resonance; its high value is material for the concept of defining a quasibound state by cutting off ϕ_{E_p}.

Particularly important are the results for the decay width Γ of the state $|\Phi_R\rangle$ as expressed by the squared nondiagonal matrix element of the Hamiltonian at energy $E = E_{\text{res}}$; cf. Equation (6.12). The values of Γ were found to be largely independent of the details of the cutoff procedure for Φ_R and to be in full agreement with the results of a phase-shift analysis of the old continuum ϕ_{E_p}; cf. Section 2.3.

Figure 6.4 shows the curves $\Gamma(R)$ for the systems U + Th ($Z = 182$), U + U ($Z = 184$), and U − Cm ($Z = 188$). The calculation of decay widths has been extended beyond the distance of nuclear contact. This region is important if positron creation accompanying nuclear reactions is considered; cf. Section 7.5. A geometry of two overlapping charged spheres with normal nuclear density has been assumed in the calculation of the Coulomb potential $V_{l=0}(r)$. Note the strong decrease of Γ with growing two-center distance and the high Z dependence. This implies that it will be necessary to study close collisions of very heavy systems if any effects of the spontaneous decay of the resonance are to be observed. In a realistic collision the values of Γ are further reduced due to electron screening (deR 84).

Since the decay width $\Gamma = 2\pi |\langle \tilde{\phi}_{E_{\text{res}}} | H | \Phi_R \rangle|^2$ provides an adequate description of the positron emission process only in the limit of infinite collision time, the coupling matrix elements $\langle \tilde{\phi}_{E_p} | H | \Phi_R \rangle$ and $\langle \tilde{\phi}_{E_p} | \partial/\partial R | \Phi_R \rangle$ have to be studied in more detail. The values of $\langle \tilde{\phi}_{E_p} | \partial/\partial R | \Phi_R \rangle$ join

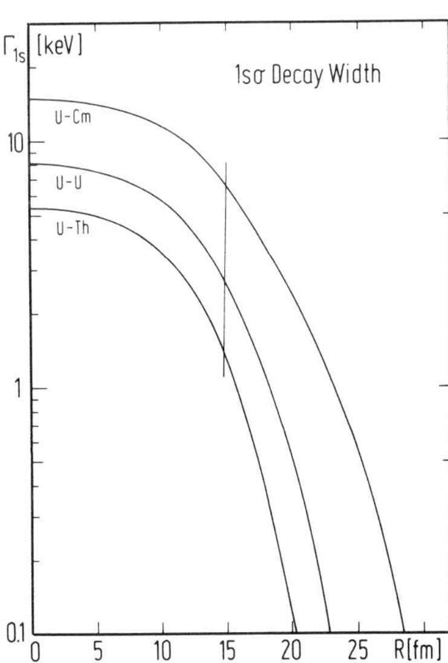

Figure 6.4. Decay width Γ of the $1s\sigma$ resonance as a function of internuclear distance for the systems $Z = 182$, 184, 188. The calculation is extended to the region of nuclear contact (left of the vertical line) where a geometry of two spheres overlapping without compression has been assumed. From (Re 81b).

smoothly with $\langle \phi_{E_p} | \partial/\partial R \, | \Phi_R \rangle$ at the critical distance $R = R_{cr}$. The structure of the dynamic coupling matrix elements, however, is altered. While the matrix elements increase monotonically in the subcritical region, they reach a maximum and fall off again at small internuclear distance R. Except for these matrix elements joining Φ_R, all further couplings to the modified continuum $\tilde{\phi}_{E_p}$ show no structure, again demonstrating that the resonance has been successfully extracted from the continuum.

6.3. Results

The formalism developed in Section 3 and in the last section has been applied to calculate positron emission in various heavy-ion collisions. The required amplitudes entering (6.10) were calculated both in time-dependent perturbation theory up to second order and by numerical solution of the coupled differential equations. In the following we will first briefly discuss the perturbative results. The main emphasis will be placed on the subsequent presentation of the coupled-channel calculations for positron creation in both sub- and supercritical collisions.

In order to arrive at numerical results, the same approximations as in Section 4 have been applied to the semiclassical quasimolecular model. The calculations are restricted to $\kappa = -1$ and $\kappa = +1$ states ($ns_{1/2}$ and $np_{1/2}$). Both sets are decoupled since they have different parity. They are expected to be the dominant channels on theoretical grounds, since in the superheavy systems under consideration the wave functions wih $|\kappa| = 1$ are most severely distorted by the strong potential, leading to large coupling matrix elements.

Pair production in high Z heavy-ion collisions was first calculated in (So 77b). In first-order perturbation theory (direct pair creation caused by radial coupling between the electron and positron continua) rather high production rates and, above all, a very strong dependence on nuclear charge, following roughly $P^{e^+} \propto (Z_1 + Z_2)^{18}$, was predicted. This effect is due to the deformation of the continuum states at small internuclear distance. It has been termed "shake-off of the vacuum polarization charge," since the same combination of continuum wave functions that is responsible for the pair creation enters in the expression for the static vacuum polarization.

Inner shell states do not contribute in first order since they are initially occupied. As the investigation of the role which they play in the positron production process is of major interest, the analysis has to be pushed to higher (at least second) order. In (Re 78) pair excitation by direct and two-step transitions via the six innermost $|\kappa| = 1$ bound states was considered. The main results are as follows:

(i) $s(\kappa = -1)$ and $p_{1/2}(\kappa = +1)$ waves contribute roughly equally to positron production.

(ii) Although being of higher order, the contributions of the $1s$ and $2p_{1/2}$ states are comparable in magnitude to the direct pair creation process. They grow particularly fast with nuclear charge and constitute the largest single components for systems heavier than $Z_1 + Z_2 \sim 175$.

(iii) The kinematical characteristics of direct and two-step excitations do not differ much. In both cases the same amount of energy has to be transferred in a similar region of space and time. This has been essentially confirmed by analytic calculations (Ba 81c) based on united atom wave functions.

The perturbative approach describes much of the physics involved in the excitation process. To obtain quantitative results, however, multi step transitions should be taken into account. As discussed in Section 4 for the case of K-hole excitation coupled-channel calculations have been performed including the positron continuum (Re 81a).

We will analyze the results obtained for the four different heavy-ion collision systems Pb + Pb, Pb + U, U + U, and U + Cf with the total charges $Z_1 + Z_2 = 164$, 174, 184, and 190, respectively. The latter two systems are supercritical. The impact energy per nucleon was kept constant at $E_{lab} = 5.9$ MeV/u, corresponding to a nuclear velocity of about $v/c \simeq 0.11$.

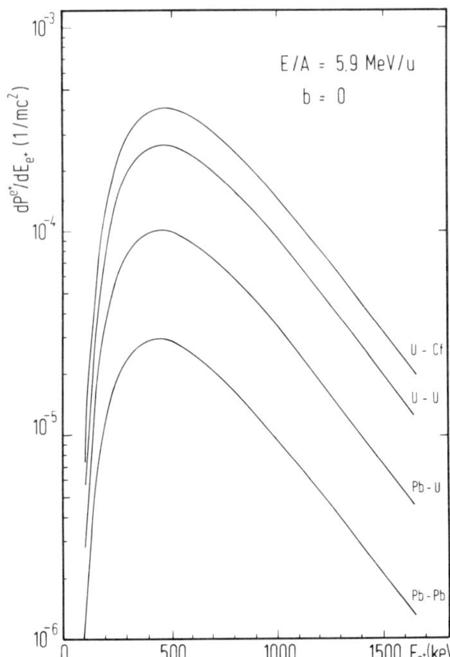

Figure 6.5. Spectra of positrons emitted in central collisions at 5.9 MeV/u lab. energy for systems with united nuclear charge $Z = 164$, 174, 184, and 190 as predicted by coupled-channel calculations. From (Re 81a).

Figure 6.5 shows the calculated energy spectra of positrons dP/dE_p produced in 5.9 MeV/u head-on collisions. As is well known, the emission of low-energy positrons is suppressed by Coulomb repulsion while at high energies the spectra fall off exponentially in analogy to the spectra of δ electrons. The shapes do not differ qualitatively from the perturbative results. Obviously they are nearly independent of the charge of the collision system, having a maximum at about 450 keV kinetic energy. Electron screening will shift this value to lower energies by about 50 keV (deR 84).

Figure 6.6a gives the impact parameter dependence of positron production. The abscissa is the distance of closest approach R_{min}, which is related to impact parameter b by $R_{\text{min}} = a\{1 - [1 + (b/a)^2]^{1/2}\}$. A nearly perfect exponential decrease of the curves is noted. The high value of the slope is understandable in terms of the energy transfer required to produce a pair. Therefore the impact parameter dependence will be much weaker (Re 81a) in collisions of bare nuclei, where the gap between the lowest empty state ($1s\sigma$) and the positron continuum becomes small or even vanishes. The predicted values of $Pe^+(R_{\text{min}})$ in this situation are shown in Figure 6.6b.

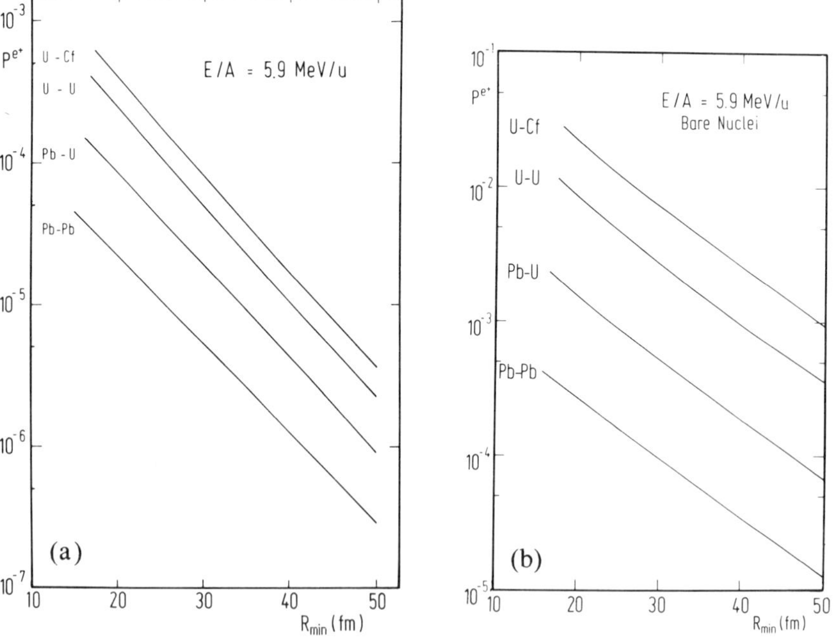

Figure 6.6. (a) Emission probability of positrons in 5.9-MeV/u collisions as function of the distance of closest approach R_{min}, assuming occupied inner shells prior to the collisions. (b) Same as (a) for collisions of totally stripped heavy ions. The large enhancement is mainly due to the contribution of the $1s\sigma$ state. From (Re 81a).

The rates for positron production increase very fast with total nuclear charge, flattening somewhat for the highest Z values. If parametrized by a power law $(Z_1 + Z_2)^n$ the power takes values between 20 and 13. If the distance of closest approach R_0 is kept fixed instead of the impact velocity, n becomes still larger.

In collisions of fully stripped ions positron production is increased by up to two orders of magnitude. Here also the Z dependence is extremely steep ($n \simeq 29$). Mainly responsible for this effect is the contribution of the 1s state, which in normal collisions is suppressed due to the small K-vacancy probability. If the K shell is empty it becomes the dominant final state for pair production. This clearly reflects the strong coupling between the 1s state and the antiparticle continuum which it approaches and even enters in supercritical systems. Experiments with fully stripped uranium ions would require an acceleration to relativistic energies and subsequent deceleration (or a colliding beam setup) and will not be feasible in the near future.

6.4. Discussion

The results displayed in the last section demonstrate remarkable features of the pair production mechanism in "slow" collisions of highly charged nuclei. In particular, the large excitation rates and their high sensitivity on total nuclear charge in the region $Z_1 + Z_2 \simeq Z_{\mathrm{cr}}$ are characteristic for the action of the time-dependent strong Coulomb field.

One question must be studied in more detail: What is the influence of the "diving" of the 1s level in supercritical collisions on positron production? The energy specta and impact parameter dependence depicted in Figures 6.5 and 6.6 have shown that theory does not predict any drastic change of observables at the border of the supercritical region but rather a smooth increase of production rates with Z. If the "spontaneous" coupling $\langle\tilde{\phi}_{E_p}|\, H \,|\Phi_R\rangle$ is left out of the calculation, however, the resulting positron production is reduced and the spectrum is distorted. An explanation for this result can be gained by looking at the structure of the coupling matrix elements as a function of time. In Figure 6.7 the values of $\langle\tilde{\phi}_{E_p}|\, \dot{R}\partial/\partial R \,|\Phi_R\rangle$ and $\langle\tilde{\phi}_{E_p}|\, H \,|\Phi_R\rangle$, $E_p = -2m$, are drawn for a head-on U + Cf collision with $2a = 18$ fm. For comparison the corresponding radial coupling in a Pb–Pb collision is included in the graph. Obviously, the "spontaneous" coupling is compensated by a corresponding reduction of the "1s-induced" contribution. Both couplings have to be added coherently [with a relative phase factor of i, cf. (6.11)], leading to excitation rates which do not differ qualitatively from the subcritical results.

This observation is in general agreement with the notion of a "dynamical width" which may be associated with a collision time τ by the uncertainty principle, $\Gamma_{\mathrm{dyn}} \simeq h/\tau$, and which leads to a smooth transition

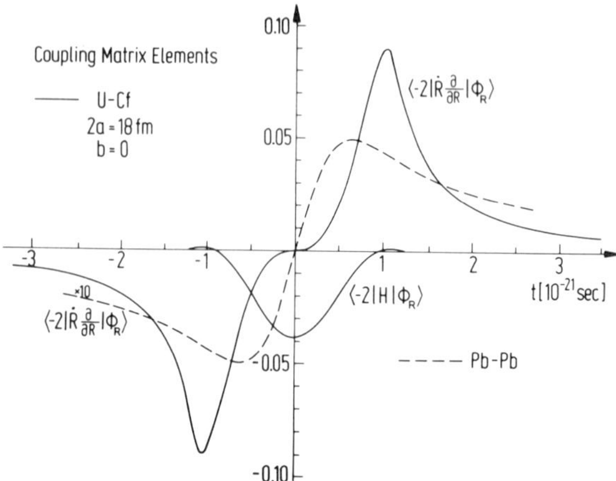

Figure 6.7. Matrix elements of the "spontaneous" and "induced" emission of positrons out of the $1s\sigma$ state, drawn as a function of time in central collisions of U + Cf, $2a = 18$ fm. For comparison the dashed line shows the radial coupling in a Pb + Pb collision (multiplied by a factor of 10). From (Re 81a).

between sub- and supercritical collisions. The shortness of τ prevents any sudden threshold effects and in particular precludes the existence of sharp structures in the positron spectra near the "diving energy" $E_{1s}(R_{\min})$. A proof of supercriticality of a given collision according to these calculations, therefore, will not be possible by the observation of qualitative features in the excitation rates. Rather, a quantitative analysis is needed, including the rate of $1s\sigma$ vacancy formation, which is sensitive to the binding energy. However, a unique signature for spontaneous positron production can be gained in collisions with prolonged interaction time; cf. Section 7.

Since beams of very heavy ions at energies close to the Coulomb barrier have become available at GSI (Darmstadt), a number of experiments have been performed to study positron production in highly charged collision systems (Ba 78, Ko 79). The search for the such processes has been largely successful.

A major problem in analyzing the experiments lies in the background originating from nuclear processes (Ob 76). Already well below the barrier the nuclei can be excited by Coulomb excitation. Photons with energy greater than 1022 keV can undergo pair conversion. This background process can be estimated by measuring the γ spectrum and folding with the theoretically known (Sch 78, Sch 81) pair conversion coefficients. Figure 6.8a shows that the measured positron production (Ba 81a) in light systems can be explained solely by nuclear conversion. For $Z_u > 160$, however, an excess is observed in the experiment. This is in accordance with the predicted steep rise of "atomic" positron production; cf. Figure 6.8b.

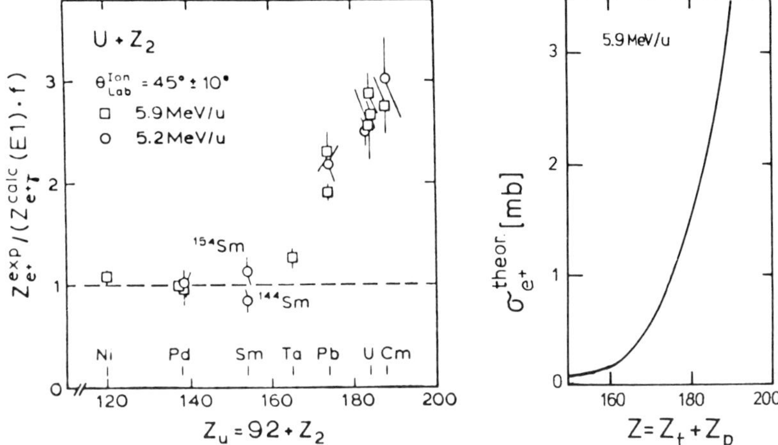

Figure 6.8. In the left part the ratio between measured positron production rates and the pair conversion rates expected from nuclear γ-ray spectra is displayed as a function of the united charge $Z = 92 + Z_2$ (Ba 81). The rise at $Z > 160$ cannot be explained by nuclear processes. The right part shows the predicted cross section of atomic positron emission in 5.9-MeV/u collisions.

Further experiments have studied in detail the dependence on impact parameter and bombarding energy (Ba 78, Ko 79). As an example Figure 6.9 shows $P^{e^+}(R_{\min})$ for the systems Pb + Pb, Pb + U, U + U, and U + Cm (Ba 78, Ba 81b). The scattered particle was detected in a fixed angular

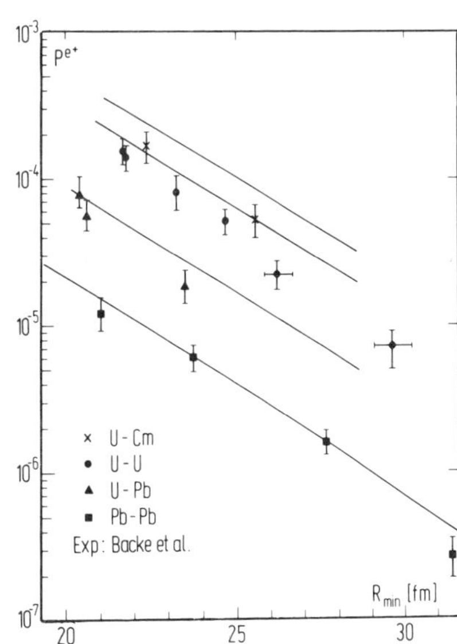

Figure 6.9. Positron emission probability for various heavy-ion collision systems, drawn as a function of the distance of closest approach. The scattered ion is detected in an angular window $\theta_{\text{lab}} = 45° \pm 10°$. Experimental data taken from Backe *et al.* (Ba 77, Ba 81).

window $\theta_{lab} = 45° \pm 10°$ so that the various values of R_{min} were obtained by variation of the collision energy. The theoretical values are symmetrized and averaged over the region of impact parameters defined by the experimental angular window. In general the agreement is quite good, especially for the Pb + Pb system. In the heavier systems theory has a tendency to overestimate the measured data; also the slopes are somewhat steeper than predicted. The general trends can be described by scaling properties of the kind introduced in Section 4.4.3.

The energy spectra of positrons have been studied by three independent groups, H. Backe *et al.* (Ba 81a,b), H. Bokemeyer, J. S. Greenberg *et al.* (Bo 81), and C. Kozhuharov, P. Kienle *et al.* (Be 81, Ki 81). While the shape of dP/dE_{e+} in some cases was found to be in good agreement with predictions (Ba 81a), remarkable structures have been observed recently, (Be 81, Bo 81) which may be associated with nuclear processes; cf. Section 7.5.

The experiments performed so far have convincingly established the predicted strong increase of positron production in collisions with very high total nuclear charge and its concentration on close collisions. A more detailed discussion of experimental results is presented in the article of Vincent and Greenberg.

7. Atomic Physics and Nuclear Reactions

7.1. Introduction

The physics of electronic excitations in ion–atom collisions can be extended to a new "degree of freedom" if energies above the nuclear Coulomb barrier are investigated. While below the barrier the nuclear motion was determined by pure Coulomb scattering, now the nuclear interaction modifies the collision. In the semiclassical picture, at nuclear contact attractive (and dissipative) forces begin to act and the trajectory $\mathbf{R}(t)$ deviates from the usual Rutherford orbit. The ejectiles of the reaction may differ from the incoming configuration with respect to charge, kinetic energy, and angular momentum. Even in an elastic collison a time delay can be introduced, separating the incoming and outgoing branches of the trajectory. It is quite obvious that all of the electronic excitation processes discussed in this article will be modified in the presence of a nuclear reaction.

The resulting phenomena can be divided into two classes:

(1) United atom effects are characteristic of the formation of an atom with nuclear charge $Z_1 + Z_2$. Such processes become prominent when a sufficiently long-lived compound nucleus (CN) is formed; their intensity

increases with the CN lifetime. Two kinds of transitions can be expected in the united atom system: x-ray radiation (or the competing emission of Auger electrons) and, in supercritical systems, spontaneous positron emission.

(2) Interference effects are caused by the change of phase relations between excitations that take place in the incoming and outgoing branches of the trajectory. Inner shell hole production as well as δ-electron, x-ray, and induced positron emission can be influenced in this way.

Most of the phenomena have come under closer investigation only recently. The aim in these studies is twofold: From the viewpoint of atomic physics, information on details of the electronic excitation mechanism may be gained, which is not accessible through subbarrier collisions, e.g., the amplitude of vacancy formation at distance of closest approach in central collisions, $a_{ij}(b=0, t=0)$. Most interesting in the present context is the prospect of forming a superheavy quasiatom for some prolonged interaction time and of studying its properties. This might give a chance to reduce the influence of collision broadening, which has severely hindered attempts to deduce atomic properties from measurements of quasimolecular transition probabilities. In particular, the process of spontaneous positron production could be identified unambiguously in this way (Ra 78b).

As a second objective also information on the nuclear reaction mechanism can be obtained. Attempts to deduce CN lifetimes in proton-induced reactions already have proved successful in several cases. Using heavy ions as projectiles, a variety of reaction types is known, which are associated with prolonged interaction times. Lighter systems like $^{12}C + {}^{12}C$ show indications for the formation of nuclear molecules. For heavier systems fusion–fission processes and deep-inelastic reactions are of interest. The latter process is dominant in collisions of heavy or intermediate projectiles above the Coulomb barrier. Deep inelastic scattering has become one of the best-studied processes in nuclear heavy-ion physics. Various microscopic and macroscopic models have been proposed to explain a wide range of phenomena. As the basic assumptions of these models are not accessible to direct experimental verification, it would be of great importance to observe the course of the reaction "from the outside." The effects discussed in this chapter may provide an independent "atomic clock" which monitors the time scale of the nuclear collision.

7.2. Interference Effects

From a large-scale point of view, which is adequate or a description of atomic processes, and in the semiclassical picture a nuclear reaction is characterized by a "standstill" of the nuclear motion, separating the incoming and outgoing branches of the nuclear trajectory. Not surprisingly, atomic excitations that can take place all along the trajectory will be influenced by this change since the phase relations are affected.

Such an effect was first postulated by Eisberg *et al.* (Ei 60) for the process of nuclear bremsstrahlung. Radiation is associated with the cessation and initiation of the nuclear current at the beginning or end of the reaction. Both contributions can interfere with a relative phase $\exp(iE_x T)$. This interference leads to a structure in the bremsstrahlung spectrum, from which the CN lifetime τ_{CN} in a compound elastic reaction can be deduced. Such an experiment has been performed successfully (Ma 76).

The same idea can be applied to electronic excitations. In the following the nuclear reaction will be described by joining together three branches of a classical trajectory: two Rutherford orbits (straight lines will suffice for many cases) valid in the regions $t<0$ or $t>0$, and a section with constant $R=R_0$ in the intermediate region $0 \leqslant t \leqslant T$. To simplify the discussion let us assume that first-order perturbation theory is valid. Then the electronic transition amplitude will read

$$a_{ij}(\infty) = -i \int_{-\infty}^{\infty} dt\, M_{ji} \exp[-i(\chi_i - \chi_j)] \tag{7.1}$$

The time integral can be split into three parts: $\int_{-\infty}^{\infty} = \int_{-\infty}^{0} + \int_0^T + \int_T^{\infty}$. If the phases $\chi_i(t)$ of (3.17) are defined symmetric with respect to $t=0$, (7.1) leads to

$$\begin{aligned} a_{ij}(\infty) &= [a_{ij}(0) - H_{ji}(0)/\omega_{ij}] - \exp(-i\omega_{ij} T)\,[a_{ij}'^{*}(0) - H_{ji}(0)/\omega_{ij}] \\ &\equiv b_{ij}(0) - \exp(-i\omega_{ij} T)\, b_{ij}'^{*}(0) \end{aligned} \tag{7.2}$$

Here the second integral has been performed analytically and the third one was transformed using the symmetry relation (3.27) of the coupling matrix under time reflection. ω_{ij} denotes the transition frequency in the united atom limit, $\omega_{ij} = E_i - E_j$. The amplitude $a_{ij}'(0)$ has to be calculated integrating forward in time, but using the kinematical parameters (kinetic energy, angular momentum) of the outgoing branch of the trajectory. In general a_{ij}' will also depend on the scattering angle θ.

According to the assumption $R=0$ only the potential coupling acts during nuclear contact. The term $H_{ji}(0)/\omega_{ij}$ causes a rearrangement of the electronic states towards the adiabatic limit. If a molecular or UA basis is used, $H_{ji}(0)$ does vanish and the amplitudes $a_{ij}(t)$ remain constant during the reaction. Equation (7.2) reduces to

$$a_{ij}(\infty) = a_{ij}(0) - \exp(-i\omega_{ij} T)\, a_{ij}'^{*}(0) \tag{7.3}$$

A generalization of (7.3) which is exact to all orders is given by (Mü 80a)

$$a_{ik}(\infty)=\sum_{jl} a_{ij}(0)\, A_{jl}\, a'_{kl}(0) \tag{7.4}$$

where the matrix A of the time evolution operator is diagonal, if the adiabatic basis is used

$$A_{jl}=\delta_{jl}\exp[-iE_j(0)\ T] \tag{7.5}$$

Equations of type (7.4) have been used in (Mü 80a) to study the influence of variations in the trajectory $\mathbf{R}(t)$ on excitation rates.

Equation (7.2) was first studied by Ciocchetti and Molinari (Ci 63, Ci 65), who calculated ionization in light-particle-induced elastic reactions and suggested the measurement of time delays. Observable quantities involve the absolute square of (7.3). Interference effects are expected if the factor ωT is of the order one. In this case the excitation process is sensitive to the phase angle of $a_{ij}(0)$, a quantity which otherwise is not observable. Large values of ωT effectively lead to a loss of coherence between the two partial amplitudes.

Equation (7.2) is based on a semiclassical, sequential picture of the scattering process. As shown by Feshbach and Yennie (Fe 62b) for the case of nuclear bremsstrahlung, the discussion can be extended to a fully quantal description. In the distorted wave Born approximation (DWBA) [see (Bl 81, Fe 81)] the inelastic scattering amplitude for atomic excitation is

$$f_{ij}=(\mu/2\pi)(k_f/k_i)^{1/2}\,\langle \Psi^{(-)}_{\mathbf{k}_f}\phi_f\,|\,H_{\text{int}}\,|\,\Psi^{(+)}_{\mathbf{k}_i}\phi_i\rangle \tag{7.6}$$

Excitation cross sections are obtained by squaring (7.6) and summation over the relevant final states. The nuclear wave functions $\Psi^{(\pm)}_k$, satisfying incoming (outgoing) boundary conditions, can be approximated by their asymptotic form $(R\to\infty)$ which is determined by the nuclear scattering amplitude $f(\theta,E)$. This leads to

$$f_{ij}=f(\theta,\ E-\omega_{ij})\,b_{ij}(0)-f(\theta,E)\,b'^{*}_{ij}(0) \tag{7.7}$$

where θ and E are the nuclear scattering angle and kinetic energy. This result has a particularly lucid interpretation: The electronic excitation may either take place on the incoming branch, in which case only the energy $E-\omega_{ij}$ is available for the nuclear reaction. Otherwise the nuclear scattering occurs at energy E and the electron is excited on the way out. The structure of (7.7) is very similar to (7.2). For slowly varying $f(\theta,E)$ the relative phase in (7.7), $f(\theta,E)/f(\theta,E-\omega)$ is consistent with the classical expression $e^{-i\omega T}$ in (7.2), if T is replaced by the proper quantum mechanical "delay time" $\tau_{\text{QM}}=-i(d/dE)\ln f(E)$, (Bl 78).

Blair *et al.* (Bl 78) have measured K-shell excitation across a narrow proton-induced nuclear resonance [rapid variation of $f(E)$, i.e., long delay time]. The interference effects of (7.7) have been clearly identified, but the measurement seems to be not consistent with the phase angle of $a_{ij}(0)$ obtained from SCA theory. In another experiment (Du 80) an interference effect was observed, although $\omega T \ll 1$, contrary to theoretical expectations (Fe 81, Bl 81).

The theory of excitations in delayed nuclear collisions, in its semiclassical version, has been applied to heavy-ion collisions. Preferably, coincidence measurements with reaction products should be performed, specifying those parameters in the outgoing channel (kinetic energy loss, deflection angle) which are favorable for a reaction with long interaction time.

Assuming $a_{ij} = a'_{ij}$, which is valid if energy loss and excitation of higher multipoles are neglected, the spectrum of electrons emitted out of the K shell in perturbation theory can be written (Re 79c, So 79b)

$$dP/dE(T) = 4\,|a_{1s,E}(0)|^2 \sin^2\{[E - E_{1s}(0)]\,T/2 + \arg[a_{1s,E}(0)]\} \quad (7.8)$$

The observation of oscillations in the δ-electron spectrum with a width ε therefore would allow us to deduce in an unambiguous way the nuclear reaction time $T = 2\pi\hbar/\varepsilon$. Figure 7.1 shows the electron spectrum for a 7-MeV/u Xe + Pb collision assuming several values of the nuclear delay time (So 79b). The calculations were performed in the coupled-channel approach, which demonstrates that multistep transitions do not destroy the effect.

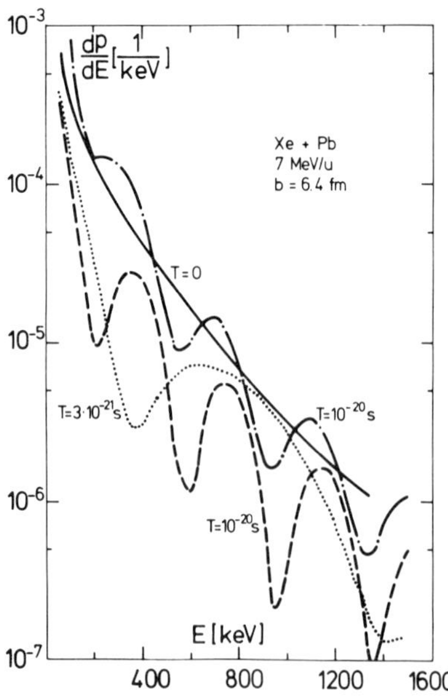

Figure 7.1. Coupled-channel calculation for the energy spectrum of δ electrons emitted out of the $1s\sigma$ state in Xe + Pb collisions above the Coulomb barrier (7 MeV/u) assuming a Coulomb trajectory (solid line, $T = 0$) and two fixed nuclear delay times $T = 3 \times 10^{-21}$ and 10×10^{-21} s. The dash-dotted line represents the total δ-electron spectrum ($T = 10^{-20}$ s) originating from the states $1s\sigma$, $2s\sigma$, and $3s\sigma$. From (So 79b).

The spectrum (7.8) still has to be folded with a lifetime distribution function $\rho(T)$ associated with the nuclear reaction

$$\overline{dP/dE} = \int_0^\infty dT\, \rho(T)\, dP/dE(T) \tag{7.9}$$

The predicted oscillations can survive this averaging, if $\rho(T)$ is peaked at a finite value of T. This should be a realistic assumption for the case of deep inelastic collisions, if favorable kinematic conditions are selected.

A quantity less sensitive to the folding (7.9) is the K-vacancy creation rate P_{1s}, which is obtained by integrating (7.8) over electron energy. P_{1s} also contains information on the nuclear reaction and could be used to measure the time scale (An 79a). It is obvious from Figure 7.1 that destructive interference prevails for $T > 0$. The resulting vacancy probabilities in a deep inelastic collision therefore should be reduced compared to the case of pure Coulomb scattering. Figure 7.2 demonstrates this reduction according to coupled-channel calculations of (Mü 80b) for the systems Pb + Pb and U + U. Due to the strong increase of binding energies in superheavy systems, in U + U collisions a small time delay $T \simeq 10^{-21}$ s should be sufficient to produce a drastic reduction. Since both $P_{1s\sigma}$ and $P_{2p\sigma}$ are affected, vacancy sharing in the symmetric system has no adverse effect.

Up to now time delay effects in deep inelastic heavy-ion reactions have

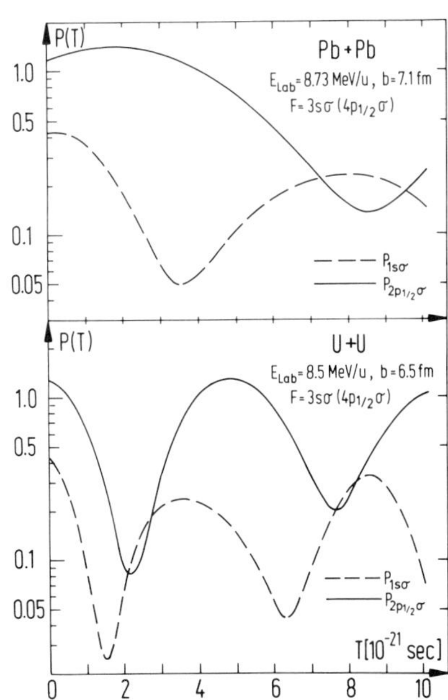

Figure 7.2. The vacancy productions rates in the quasimolecular $1s\sigma$ (dashed lines) and $2p_{1/2}\sigma$ (solid lines) levels as a function of nuclear delay time T in Pb + Pb and U + U collisions above the Coulomb barrier. From (Mü 80b).

not been ientified experimentally, mainly due to the conversion of high-multiplicity nuclear γ rays. If successful, the method promises valuable independent information on the mechanism of heavy-ion reactions.

7.3. Compound Nucleus X-Rays

In a collision of nuclei with sufficiently high energy there is a certain probability of exciting an inner shell vacancy simultaneously with the formation of a compound nucleus. The resulting resonance of the combined atom–nucleus system may decay in two distinct ways. Either the CN decays first, leaving a vacancy in the separated system, which then is filled by an x-ray transition corresponding to an atom with charge, say, Z_2'; or, alternatively, the vacancy can decay while the compound nucleus still is formed, leading to the emission of a photon with the increased energy characteristic for the atom $Z_1 + Z_2$. The ratio of intensities of both x-ray lines should be a measure for the CN lifetime τ_{CN} in comparison with the radiative lifetime τ_x of the vacancy. Simple reasoning leads to the probability P_x of CN x-ray emission (neglecting effects of the fluorescence yield)

$$P_x = P_K(0)\,\tau_{\text{CN}}/(\tau_{\text{CN}} + \tau_x) = P_K(0)\,\Gamma_x/(\Gamma_{\text{CN}} + \Gamma_x) \tag{7.10}$$

where $P_K(0)$ is the probability for vacancies created in the incoming part of the trajectory. Equation (7.10) can be shown to be valid under quite general assumptions (McV 81) as long as the electronic and nuclear decay are decoupled. In the case of strongly overlapping nuclear resonances the width Γ has to be replaced by a fluctuation length derived from the autocorrelation function of the nuclear scattering amplitude (Br 80).

The application of (7.10) to measure compound nuclear lifetimes τ_{CN} was first suggested some time ago by Gugelot (Gu 62). Recently Chemin *et al.* (Ch 79) and Röhl *et al.* (Rö 79) succeeded in determining τ_{CN} for elastic proton scattering. They found times of the order of 10^{-17} s. In principle in reactions of this type lifetimes in the interval 10^{-15}–10^{-18} s could be experimentally accessible via CN x-rays (Me 80).

From a theoretical point of view, CN emission is just a special case of the molecular orbital x-ray mechanism discussed in Section 5. In the semiclassical approximation the nuclear reaction is characterized by a special type of trajectory $R(t)$. The formation of a compound nucleus in this picture corresponds to an interval $0 < t < T$ during which the nuclei are totally amalgamated, $R = 0$ or at least $R(t) = \text{const.}$. MO x-rays are continuously emitted along the whole trajectory. With increasing reaction time T emission at the UA transition energy is gradually enhanced until, in the limit $T \gg \tau_x$, the characteristic x-ray line of the atom with charge $Z_1 + Z_2$ becomes predominant in the spectrum.

The effect of time delay on MOX emission can be calculated in a similar way as discussed in the last section for the case of inner shell excitation, using a schematic nuclear trajectory. Anholt (An 79c) has calculated x-ray spectra for the symmetric heavy-ion collisions Ni + Ni and the superheavy system Yb + Yb, taking the CN lifetime as a free parameter. Figure 7.3 shows the differential photon probability dP/dE_x predicted for central 200-MeV Ni + Ni collisions assuming the double-collision mechanism for $1s\sigma$ vacancy creation. Assigned to the different curves is the parameter $\omega_{K\alpha}T$, the product of UA transition frequency and nuclear delay time. In the absence of a reaction, the spectrum shows the customary smooth decrease characteristic for MOX emission. For CN lifetimes satisfying $\omega_{K\alpha}T > 20$ the UA K_α line of the compound nucleus Ba at 32 keV can be

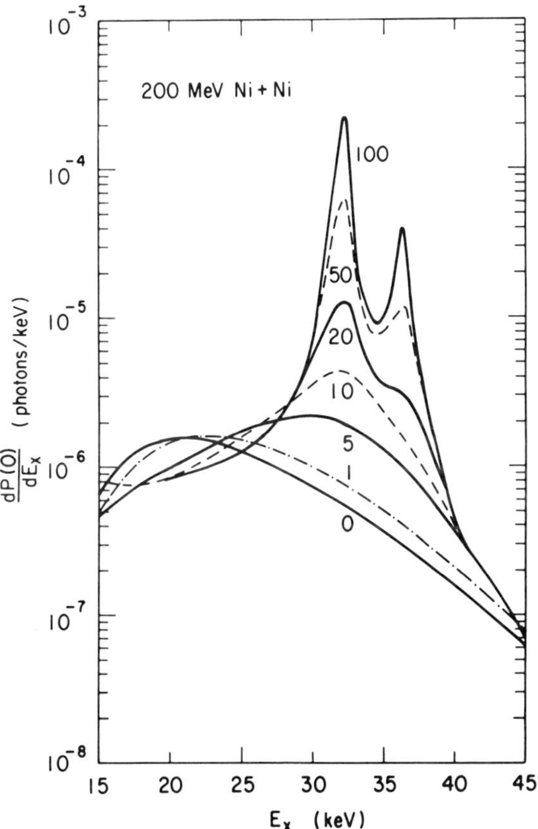

Figure 7.3. Quasimolecular x-ray spectra emitted in a 200-MeV Ni + Ni fusion–fission reaction calculated for various values of the compound nucleus lifetime τ. The parameters denote the product $\omega_{K_\alpha}\tau$ of CN lifetime and UA K_α transition frequency. From (An 79c).

clearly distinguished. At still higher values of T the reduction of Heisenberg broadening allows us to separate the K_α and K_β lines.

Reaction times of the order of 10^{-19} s are required to observe distinct CN x-ray lines in superheavy systems. This effect might be used to detect and identify the transient formation of superheavy atoms (Ki 83). Fusion–fission reactions, observable in systems with $Z_u < 140$, should provide sufficiently long reaction times (An 79c). Severe background problems up to now have obstructed experimental attempts to detect the effect.

7.4. Positron Creation

With respect to fundamental importance, the spontaneous emission of positrons is the most interesting process, which can benefit from time delay in nuclear reactions (Ra 78b, Re 81b). As discussed in Section 6, in subbarrier collisions no clear threshold effects indicating supercriticality are expected. If the time scale can be extended at the distance of closest approach, however, positron creation will be enhanced in a characteristic way. In the language of Section 6.2 this is caused by the decay matrix element $\langle \tilde{\phi}_{E_p} | H | \Phi_R \rangle$. Using the schematic model of Section 7.2 for the nuclear trajectory, the influence of the delay is easily estimated. The integral $\int_0^T$ now makes an important contribution. Disregarding for a moment excitations in the region outside $0 < t < T$ the number of positrons with energy E reads

$$\frac{dP^{e^+}}{dE} \simeq P_{1s}(0) \frac{\Gamma}{2\pi} T^2 \left[\frac{\sin(E - E_{1s})\, T/2}{(E - E_{1s})\, T/2} \right]^2 \tag{7.11}$$

The spectrum of positrons thus ultimately develops a line at the position of the resonance, E_{1s}, which becomes increasingly narrow with growing delay time. The width of the line is $\Delta E_{1/2} \simeq 5.6\hbar/T$, as long as $\Gamma T \ll \hbar$.

Detailed dynamical calculations of positron production in various sub- and supercritical systems have been performed using the schematic nuclear trajectory for the nuclear reaction (Re 81b). As an example, Figure 7.4 shows the positron spectra predicted for Pb + Pb and U + U collisions above the Coulomb barrier assuming delay times of $T = 0$, 3, 6, and 10×10^{-21} s. In subcritical collisions (Figure 7.4a) the delay causes modulations in the positron spectrum in much the same way as predicted for δ-electron emission in deep inelastic heavy-ion collisions.

In addition to the oscillatory interference patterns an enhancement of positron production in time-delayed supercritical collisions is observed if the binding energy of the lowest bound state(s) exceeds the value $2mc^2$. Figure 7.4b demonstrates that for long delay times a distinct peak in the positron spectrum is predicted at the location of the supercritical bound state

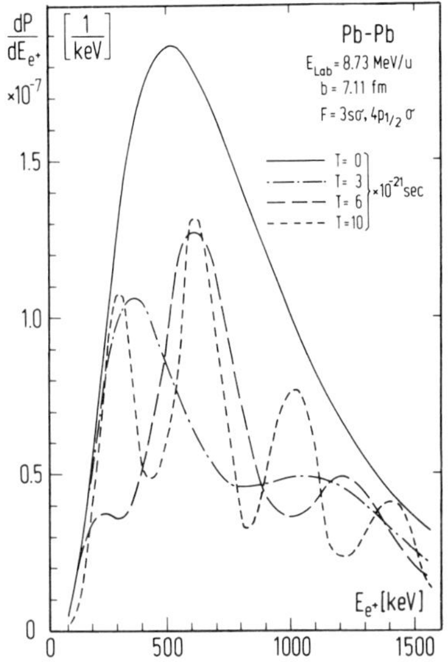

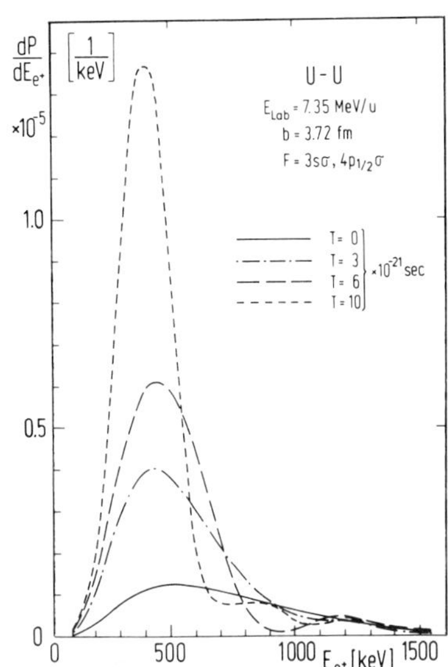

Figure 7.4. Calculated positron spectra assuming grazing Coulomb collisions (solid lines) or nuclear reactions leading to delay times of $T = 3\times$, $6\times$, and 10×10^{-21} s using a schematic model for the trajectory. Whereas for lighter systems (part a, Pb + Pb) only modulations of the spectrum are expected, in supercritical systems (part b, U + U) for sufficiently large times a distinct peak builds up at the position of the $1s\sigma$ resonance. From (Re 81b).

resonance (binding energy minus $2mc^2$) due to the spontaneous pair-creation mechanisms. A detailed analysis reveals that the line due to spontaneous emission of positrons in delayed collisions emerges gradually as $Z = Z_1 + Z_2$ exceeds Z_{cr}.

Figure 7.5 shows the total emission probability for positrons (integrated over positron energy E_{e^+}), for the four systems Pb + Pb, Pb + U, U + U, and U + Cm [solid lines: s state ($\kappa = -1$), dashed lines: $p_{1/2}$ state ($\kappa = +1$) contributions]. Simple schematic trajectories have been employed which allow for an arbitrary nuclear reaction time. The variation of P^{e^+} with delay time T is of particular interest. While in the subcritical systems $P^{e^+}(T)$ remains roughly unchanged (or is even reduced by up to a factor of 2 due to destructive interference), the probability rises approximately in linear relation with T for the supercritical systems. This reflects the increasing contribution of the spontaneous decay of the $1s$ resonance. At small values of T interference effects are seen to prevail.

The discussion up to now was quite schematic. To analyze a given experiment, however, the nuclear trajectories employed should be consistent with the elastic and inelastic heavy-ion scattering data. An alternative approach would be to treat the nuclear motion in a semiphenomenological way by solving the classical equations of motion including a real

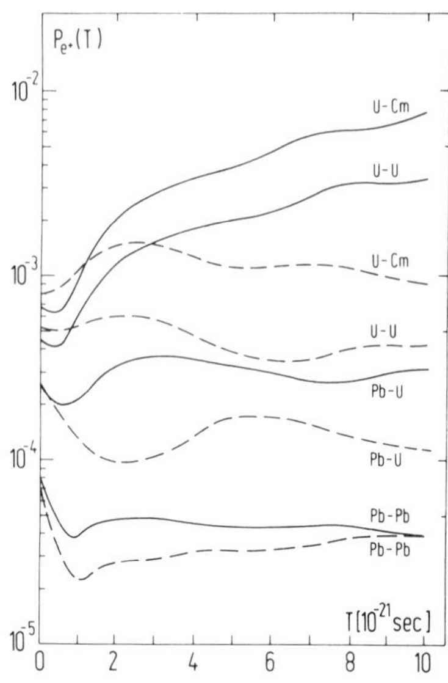

Figure 7.5. Total probability for positron emission calculated as a function of nuclear reaction time for the angular momentum channels $p_{1/2}$ (dashed lines) and $s_{1/2}$ (solid lines). In the supercritical systems U + U and U + Cm the latter contribution is found to increase strongly with T. Simple schematic trajectories are used. From (Re 81b).

nucleus–nucleus potential and frictional forces. Many models with different degrees of refinement have been discussed in the literature. Trajectories calculated with several different classical friction models have been employed to calculate positron emission (Re 81b). Figure 7.6a shows the employed trajectories, compared to the ordinary Rutherford orbit, for a head-on U + U collision with $E_{\text{Lab}} = 7.5$ MeV/u. The modified positron spectra calculated with two of the models are shown in Figure 7.6b. After incoherent summation of $s_{1/2}$ and $p_{1/2}$ partial waves, an enhancement of the maximum and a drift towards lower energies is observed. However, the relatively small time delay ($\Delta T < 10^{-21}$ s) introduced by the friction models, constructed to describe deep-inelastic scattering, is not large enough to cause the unambiguous emergence of a line due to the spontaneous emission process.

Experiments performed by Backe *et al.* (Ba 81b) on U + U and U + Cm collisions well above the Coulomb barrier seem to indicate an enhancement of dP/dE_{e^+} at lower kinetic energies in qualitative agreement with the prediction of Figure 7.6b. Further investigations along these lines seem to be quite promising, both to establish the mechanism of positron production and to deduce the nuclear reaction time scale.

In addition one might speculate on a more pronounced manifestation of time delay effects in supercritical collisions. As demonstrated by Rehm *et al.* (Re 79a) in lighter collision systems, very long reaction times may be obtained at energies slightly above the Coulomb barrier. If this trend persists at even lower bombarding energies, i.e., if a nuclear reaction mechanism involving very long delay times exists in the vicinity of the Coulomb barrier,

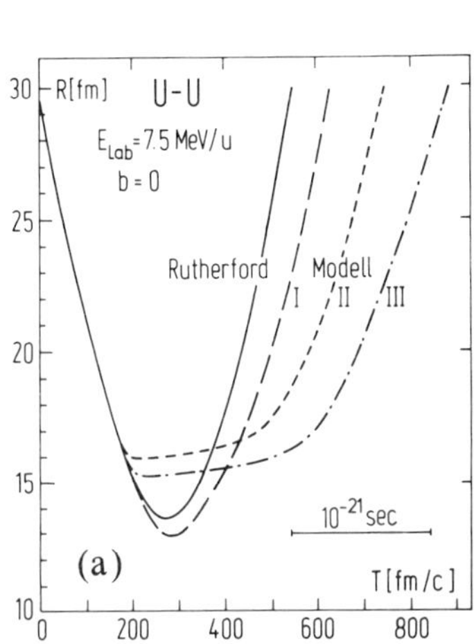

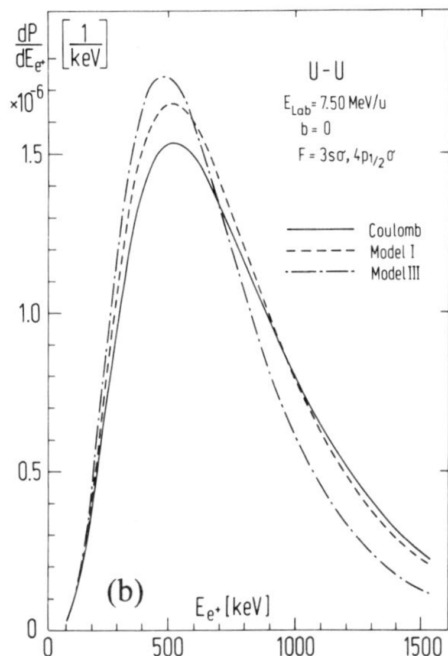

Figure 7.6. (a) Nuclear trajectories for head-on U + U collisions at 7.5 MeV/u lab energy calculated from Coulomb interaction alone (solid line) or obtained in the framework of various classical frictions models. (b) Spectra of positrons obtained from coupled-channel calculations using several of the trajectories depicted in part (a). From (Re 81b).

then the following scenario is conceivable for those collisions: At a given scattering angle a small fraction of the detected ions may result from a nuclear reaction associated with small impact parameter collisions, while the dominant contribution to the scattering cross section is determined by Rutherford scattering. As positron production is very strongly enhanced for reaction times larger than 10^{-20} s, however, a peak superimposed on the smooth spectrum of positrons, emitted in the much more frequent "distant" Coulomb collisions, could be observable. A rough estimate shows that for long delay times a peak may be prominent even if the differential reaction cross section is much less than 1 % of the Rutherford cross section (Re 81b).

Indications for such structures have been found in the experiments of the groups of P. Kienle *et al.* (Be 81) and of J. S. Greenberg *et al.* (Bo 81) for U + Th, U + U, and U + Cm collisions at energies close to the Coulomb barrier. The spectra might be explained if the presence of a small component of nuclear reactions with very long interaction time of $\simeq 4 \times 10^{-20}$ s is assumed. This value is needed to produce narrow structures in the spectrum.

Background effects like nuclear conversion, in particular of E0 type (So 81d), have to be ruled out by further experiments. If the results are confirmed, they would constitute the experimental evidence for the predicted process of spontaneous positron production. At the same time the observation of very long reaction times would have far-reaching impact on the theory of nuclear structure and nuclear reactions in the domain of superheavy elements.

Note added in Proof

The text and references for the "Note Added in Proof" are located at the end of this volume on pages 490 to 492.

References

(Aa 76) O. Aashamar and L. Kocbach, *Z. Phys.* **A279**:237 (1976).

(Ab 66) R. Abrines and I. C. Percival, *Proc. R. Soc. London* **88**:861, 873 (1966).

(Ak 65) A. I. Akhiezer and N. B. Berestetzkii, *Quantum Electrodynamics*, Interscience, New York (1965).

(Am 78) P. A. Amundsen, *J. Phys. B: At. Mol. Phys.* **11**:3197 (1978).

(An 74) R. Anholt and J. O. Rasmussen, *Phys. Rev. A* **9**:585 (1974).

(An 76a) J. U. Andersen, E. Laegsgaard, M. Lund, and C. D. Moak, *Nucl. Instrum. Methods* **132**:507 (1976).

(An 76b) R. Anholt, *J. Phys. B: At. Mol. Phys.* **9**:L249 (1976).

(An 77) R. Anholt and W. E. Meyerhof, *Phys. Rev. A* **16**:190 (1977).

(An 78) R. Anholt, *Z. Phys.* **A288**:257 (1978).

(An 79a) R. Anholt, *Phys. Lett.* **88B**:262 (1979).

(An 79b) R. Anholt, *Z. Phys.* **A291**:289 (1979).

(An 79c) R. Anholt, *Z. Phys.* **A292**:123 (1979).

(An 79d) R. Anholt, *Phys. Rev. A* **19**:1004 (1979).

(An 79e) R. Anholt, W. E. Meyerhof, and Ch. Stoller, *Z. Phys.* **A291**:287 (1979).

(An 81) R. Anholt, *Phys. Lett.* **84A**:62 (1981).

(Ar 63) P. Armbruster, E. Roeckl, H. J. Specht, and A. Vollmer, *Z. Naturforsch.* **9a**:1301 (1964).

(Ar 78) P. Armbruster, H.-H. Behncke, S. Hagmann, D. Liesen, F. Folkmann, and P. H. Mokler, *Z. Phys.* **A288**:277 (1978).

(Ar 79) P. Armbruster and P. Kienle, *Z. Phys.* **A291**:399 (1979).

(Ba 58) D. R. Bates and R. McCarroll, *Proc. R. Soc.* London **A245**:311 (1958).

(Ba 59) J. Bang and J. M. Hansteen, *Kgl. Dan. Vid. Selsk., Mat. Fys. Medd.* **31**, Vol. 13 (1959).

(Ba 68) D. R. Bates and R. H. G. Reid, *Adv. At. Mol. Phys.* **4**:13 (1968).

(Ba 71) H. W. Barz, I. Rotter, and J. Höhn, *Phys. Lett.* **37B**:4 (1971).

(Ba 72) M. Barat and W. Lichten, *Phys. Rev. A* **6**:211 (1972).

(Ba 73) G. Basbas, W. Brandt, and R. H. Ritchie, *Phys. Rev. A* **7**:1971 (1973).

(Ba 78) H. Backe, L. Handschug, F. Hessberger, E. Kankeleit, L. Richter, F. Weik, R. Willwater, H. Bokemeyer, P. Vincent, Y. Nakayama, and J. S. Greenberg, *Phys. Rev. Lett.* **40**:1443 (1978).

(Ba 79a) H. Backe, in *Trends in Physics* (M. M. Woolfson, ed.). Adam Hilger Ltd., Bristol (1979), p. 445.

(Ba 79b) J. Bang and J. M. Hansteen, *Phys. Lett.* **72A**:218 (1979).

(Ba 81a) H. Backe, in *Present Status and Aims of Quantum Electrodynamics*, Lecture Notes in Physics, Vol. 143, Springer, Berlin (1981), p. 277.

(Ba 81b) H. Backe, W. Bonin, E. Kankeleit, M. Krämer, R. Krieg, V. Metag, P. Senger, N. Trautmann, F. Weik, and J. B. Wilhelmy, in: *Proc. of the NATO Advanced Studies Institute "Quantum Electrodynamics of Strong Fields"* (Lahnstein, 1981), Plenum Press, New York (1983), p. 107.

(Ba 81c) J. Bang and J. M. Hansteen, *Phys. Scr.* **22**:609 (1981).

(Ba 81d) M. Bawin and J. Cugnon, *Phys. Lett.* **107B**:257 (1981).

(Be 63) F. Beck, H. Steinwedel, and G. Süssmann, *Z. Phys.* **171**:189 (1963).

(Be 75) H. D. Betz, F. Bell, H. Panke, W. Stehling, E. Spindler, and M. Kleber, *Phys. Rev. Lett.* **34**:1256 (1975).

(Be 76a) W. Betz, G. Heiligenthal, J. Reinhardt, R. K. Smith, and W. Greiner, *Proc. 9th Int. Conf. on the Physics of Electronic and Atomic Collisions, Seattle* (J. S. Risley and R. Geballe, eds.), University of Washington Press, Seattle (1976), p. 531.

(Be 76b) W. Betz, G. Soff, B. Müller, and W. Greiner, *Phys. Rev. Lett.* **37**:1046 (1976).

(Be 78) H.-H. Behncke, D. Liesen, S. Hagmann, P. H. Mokler, and P. Armbruster, *Z. Phys.* **A288**:35 (1978).

(Be 80a) R. L. Becker, A. L. Ford, and J. F. Reading, *J. Phys. B: At. Mol. Phys.* **13**:4059 (1980).

(Be 80b) W. Betz, Ph.D. thesis, Frankfurt (1980).

(Be 81) E. Berdermann, F. Bosch, M. Clemente, F. Güttner, P. Kienle, W. König, Ch. Kozhuharov, B. Martin, B. Povh, H. Tsertos, W. Wagner, and Th. Walcher, *GSI Scientific Report 1980*, Darmstadt, March (1981), p. 128.

(Bj 65) J. D. Bjorken and S. D. Drell, *Relativistic Quantum Fields*, McGraw-Hill, New York (1965).

(Bl 78) J. S. Blair, P. Dyer, K. A. Snover, and T. A. Trainor, *Phys. Rev. Lett.* **41**:1712 (1978).

(Bl 81) J. S. Blair and R. Anholt, *Phys. Rev. A* **25**:907 (1982).

(Bo 78) F. Bosch, H. Krimm, B. Martin, B. Povh, Th. Walcher, and K. Traxel, *Phys. Lett.* **78B**:568 (1978).

(Bo 80) F. Bosch, D. Liesen, P. Armbruster, D. Maor, P. H. Mokler, H. Schmidt-Böcking, and R. Schuch, *Z. Phys.* **A296**:11 (1980).

(Bo 81) H. Bokemeyer, K. Bethge, H. Folger, J. S. Greenberg, H. Grein, A. Gruppe, S. Ito, R. Schulé, D. Schwalm, J. Schweppe, N. Trautmann, P. Vincent, and M. Waldschmidt, in: *Proc. of the NATO Advanced Studies Institute "Quantum Electrodynamics of Strong Fields"* (Lahnstein, 1981), Plenum Press, New York (1983), p. 273.

(Br 72) J. S. Briggs and J. H. Macek, *J. Phys. B: At. Mol. Phys.* **5**:579 (1972).

(Br 74) J. S. Briggs, *J. Phys. B: At. Mol. Phys.* **7**:47 (1974).

(Br 75) J. S. Briggs, *J. Phys. B: At. Mol. Phys.* **8**:L485 (1975).

(Br 76) J. S. Briggs, *Rep. Prog. Phys.* **39**:217 (1976).

(Br 77) J. S. Briggs and K. Dettmann, *J. Phys. B: At. Mol. Phys.* **10**:1113 (1977).

(Br 78a) J. S. Briggs and K. Taulbjerg, in: *Structure and Collisions of Ions and Atoms* (I. A. Sellin, ed.), Springer, Berlin (1978), p. 105.

(Br 78b) S. J. Brodsky and P. J. Mohr, in: *Structure and Collisions of Ions and Atoms* (I. A. Sellin, ed.), Springer, Berlin (1978), p. 3.

(Br 79) J. S. Briggs, J. H. Macek, and K. Taulbjerg, *J. Phys. B: At. Mol. Phys.* **12**:1457 (1979).

(Br 80) P. von Brentano and M. Kleber, *Phys. Lett.* **92B**:5 (1980).

(Bu 79) E. H. S. Burhop, *Adv. At. Mol. Phys.* **15**:329 (1979).

(Bu 81) C. Burnap, H. Brysk, and P. F. Zweifel, *Nuovo Cimento* **64B**:407 (1981).

(Bü 81) P. Bürgy, G. Bonani, E. Morenzoni, M. Stöckli, Ch. Stoller, M. Suter, and W. Wölfli, *Z. Phys.* **A300**:105 (1981).

(Ca 50) K. M. Case, *Phys. Rev.* **80**:797 (1950).

(Ch 68) I. M. Cheshire, *J. Phys. B: At. Mol. Phys.* **1**:482 (1968).

(Ch 76) K.-T. Cheng and W. R. Johnson, *Phys. Rev. A* **14**:1943 (1976).

(Ch 79) J. F. Chemin, S. Andriamonje, J. Roturier, B. Saboya, J. P. Thibaud, S. Joly, S. Plattard, J. Uzureau, H. Laurent, J. M. Maison, and J. P. Shapiro, *Nucl. Phys.* **A331**:407 (1979).

(Ci 63) G. Ciocchetti, A. Molinari, and R. Malvano, *Nuovo Cimento* **40B**:69 (1965).

(Ci 65) G. Ciocchetti and A. Molinari, *Nuovo Cimento* **29**:1262 (1965).

(De 64) Yu. N. Demkov, *Sov. Phys. JETP* **18**:138 (1964).

(De 71) A. M. Desiderio and W. R. Johnson, *Phys. Rev. A* **3**:1267 (1971).

(De 81) J. B. Delos, *Rev. Mod. Phys.* **53**:287 (1981).

(deR 84) T. H. J. de Reus, J. Reinhardt, B. Müller, W. Greiner, G. Soff, and U. Müller, *J. Phys. B: At Mol. Phys.* **17**:615 (1984).

(Di 34) P. A. M. Dirac, *Proc. Camb. Phil. Soc.* **30** (1934).

(Do 71) H. G. Dosch, J. H. D. Jensen, and V. F. Müller, *Phys. Norvegica* **5**:2 (1971).

(Du 80) W. Duinker, J. van Eck, and A. Niehaus, *Phys. Rev. Lett.* **45**:2102 (1980).

(Ei 60) R. M. Eisberg, D. R. Yennie, and D. H. Wilkinson, *Nucl. Phys.* **18**:338 (1960).

(Ei 75) J. Eichler and U. Wille, *Phys. Rev. A* **11**:1973 (1975).

(Ei 76) J. Eichler, U. Wille, and B. Fastrup, *Phys. Rev. A* **14**:707 (1976).

(Ei 81) D. Eichenauer, N. Grün, and W. Scheid, *J. Phys. B: At. Mol. Phys.* **14**:3929 (1981).

(Er 71) G. W. Erickson, *Phys. Rev. Lett.* **27**:780 (1971).

(Fa 61) U. Fano, *Phys. Rev.* **124**:1866 (1961).

(Fa 65) U. Fano and W. Lichten, *Phys. Rev. Lett.* **14**:627 (1965).

(Fe 62a) H. Feshbach, *Ann. Phys. (*N.Y.*)* **19**:287 (1962).

(Fe 62b) H. Feshbach and D. R. Yennie, *Nucl. Phys.* **37**:150 (1962).

(Fe 81) J. M. Feagin and L. Kocbach, *J. Phys. B: At. Mol. Phys.* **14**:4349 (1981).

(Fo 77) A. L. Ford, E. Fitchard, and J. F. Reading, *Phys. Rev. A* **16**:133 (1977).

(Fr 75) W. Frank, P. Gippner, K.-H. Kaun, H. Sodan, and Yu. P. Tretyakov, *Phys. Lett.* **58B**:41 (1975).

(Fr 76a) W. Frank, P. Gippner, K.-H. Kaun, P. Manfrass, and Yu. P. Tretyakov, *Z. Phys.* **A277**:333 (1976).

(Fr 76b) B. Fricke, T. Morovic, W.-D. Sepp, A. Rosen, and D. E. Ellis, *Phys. Lett.* **59A**:375 (1976).

(Fr 77) B. Fricke and G. Soff, *At. Data and Nucl. Data Tables* **19**:83 (1977).

(Fr 79) W. Fritsch and U. Wille, *J. Phys. B: At. Mol. Phys.* **12**:L335 (1979).

(Fu 51) W. M. Furry, *Phys. Rev.* **81**:115 (1951).

(Fu 73) L. P. Fulcher and A. Klein, *Phys. Rev. D* **8**:2455 (1973).

(Gä 80) P. Gärtner, J. Reinhardt, B. Müller, and W. Greiner, *Phys. Lett.* **95B**:181 (1980).

(Gä 81) P. Gärtner, U. Heinz, B. Müller, and W. Greiner, *Z. Phys.* **A300**:143 (1981).

(Ge 61) S. S. Gershtein and V. D. Krivchenkov, *Sov. Phys. JETP* **13**:1044 (1961).

(Ge 69) S. S. Gershtein and Y. B. Zeldovich, *Lett. Nuovo Cimento* **1**:835 (1969).

(Ge 73) S. S. Gershtein and V. S. Popov, *Lett. Nuovo Cimento* **6**:593 (1973).

(Gi 74) P. Gippner, K. H. Kaun, F. Stary, W. Schulze, and Yu. P. Tretyakov, *Nucl. Phys.* **A230**:509 (1974).

(Gr 65) T. A. Green, *Proc. Phys. Soc.* **86**:1017 (1965).

(Gr 74) J. S. Greenberg, C. K. Davis, and P. Vincent, *Phys. Rev. Lett.* **33**:473 (1974).

(Gr 76a) P. T. Greenland and W. Greiner, *Theor. Chim. Acta* **42**:273 (1976).

(Gr 76b) M. Gros, B. Müller, and W. Greiner, *J. Phys. B: At. Mol. Phys.* **9**:1849 (1976).

(Gr 77a) J. S. Greenberg, H. Bokemeyer, E. Emling, E. Grosse, D. Schwalm, and F. Bosch, *Phys. Rev. Lett.* **39**:1404 (1977).

(Gr 77b) M. Gros, P. T. Greenland, and W. Greiner, *Z. Phys.* **A280**:31 (1977).

(Gr 78) W. Greiner, B. Müller, and G. Soff, *Phys. Lett.* **69A**:27 (1978).

(Gr 79) E. K. U. Gross and R. M. Dreizler, *Phys. Rev. A* **20**:1798 (1979).

(Gr 81) E. K. U. Gross and R. M. Dreizler, in: *Proc. of the NATO Advanced Studies Institute "Quantum Electrodynamics of Strong Fields"* (Lahnstein, 1981), Plenum Press, New York (1983), p. 383.

(Gu 62) P. C. Gugelot, in *Direct Interactions and Nuclear Reaction Mechanisms* (E. Clementel and C. Villi, eds.), Gordon and Breach, New York (1962), p. 382.

(Gü 81) F. Güttner, W. König, B. Martin, B. Povh, H. Skapa, J. Soltani, Th. Walcher, F. Bosch, and C. Kozhuharov, *Z. Phys.* **A304**:207 (1982).

(Gy 74) M. Gyulassy, *Phys. Rev. Lett.* **33**:921 (1974).

(Gy 75) M. Gyulassay, *Nucl. Phys.* **A244**:497 (1975).

(Ha 81) A. Hansen and F. Ravndal, *Phys. Scr.* **23**:1036 (1981).

(Ha 75) J. M. Hansteen, *Adv. At. Mol. Phys.* **11**:299 (1975).

(Ha 79) H. Hartung and B. Fricke, *Phys. Lett.* **71A**:335 (1979).

(He 34) W. Heitler and L. Nordheim, *J. Phys. (*Paris*)* **5**:449 (1934).

(He 65) K. Helfrich and H. Hartmann, *Theor. Chim. Acta* **3**:21 (1965).

(He 76) K.-H. Heinig, H.-U. Jäger, H. Richter, and H. Woittenek, *Phys. Lett.* **60B**:249 (1976).

(He 78a) G. Heiligenthal, W. Betz, G. Soff, B. Müller, and W. Greiner, *Z. Phys.* **A285**:105 (1978).

(He 78b) K.-H. Heinig, H.-U. Jäger, L. Münchow, H. Richter, H. Woittenek, *J. Phys. B: At. Mol. Phys.* **11**:115 (1978).

(He 79) T. G. Heil and A. Dalgarno, *J. Phys. B: At. Mol. Phys.* **12**:L557 (1979).

(He 81) U. Heinz, W. Greiner, and B. Müller, *Phys. Rev. A* **23**:562 (1981).

(Ho 81) M. Horbatsch and R. M. Dreizler, *Z. Phys.* **A300**:119 (1981).

(Hu 27) F. Hund, *Z. Phys.* **40**:742 (1927).

(Hu 32) F. Hund, *Z. Phys.* **77**:12 (1932).

(Hu 40) F. Hund, *Z. Phys.* **117**:1 (1940).

(Hü 81) A. Hühnergarth, N. Grün, and W. Scheid, GSI scientific report 1980, p. 149 (1981).

(Ja 79a) D. H. Jakubassa, *Z. Phys.* **A290**, 13 (1979).

(Ja 79b) D. H. Jakubassa, *Z. Phys.* **A293**, 281 (1979).

(Ja 80) D. H. Jakubassa and K. Taulbjerg, *J. Phys. B: At. Mol. Phys.* **13**:757 (1980).

(Jä 81) H. U. Jäger, P. G. Gippner, K. H. Heinig, H. Richter, N. F. Truskova, and H. Woitennek, *J. Phys. B: At. Mol. Phys.* **14**:701 (1981).

(Kä 72) G. Källen, *Quantum Electrodynamics*, Springer Verlag, Berlin (1972).

(Ka 80) E. Kankeleit, *Nukleonika* **25**:253 (1980).

(Ki 78) J. Kirsch, W. Betz, J. Reinhardt, G. Soff, B. Müller, and W. Greiner, *Phys. Lett.* **72B**:298 (1978).

(Ki 79) J. Kirsch, W. Betz, J. Reinhardt, B. Müller, W. Greiner, and G. Soff, *Z. Phys.* **A292**:227 (1979).

(Ki 80a) J. Kirsch, B. Müller, and W. Greiner, *Z. Naturforsch.* **35a**:579 (1980).

(Ki 80b) J. Kirsch, Ph. D. thesis, Frankfurt (1980).

(Ki 80c) C. L. Kirschbaum and L. Wilets, *Phys. Rev. A* **21**:834 (1980).

(Ki 81) P. Kienle, in *Proc. of the NATO Advanced Studies Institute "Quantum Electrodynamics of Strong Fields"* (Lahnstein, 1981), Plenum Press, New York (1983), p. 293.

(Ki 83) J. Kirsch, B. Müller, and W. Greiner, *Phys. Lett.* **94A**:151 (1983).

(Kl 29) O. Klein, *Z. Phys.* **53**:157 (1929).

(Kl 75) M. Kleber and D. H. Jakubassa, *Nucl. Phys.* **A252**:152 (1975).

(Kl 78) M. Kleber, *J. Phys. B: At. Mol. Phys.* **11**:1069 (1978).

(Kl 79) M. Kleber and K. Unterseer, *Z. Phys.* **A292**:311 (1979).

(Ko 77) C. Kozhuharov, P. Kienle, D. H. Jakubassa, and M. Kleber, *Phys. Rev. Lett.* **39**:540 (1977).

(Ko 78) D. Kolb, W.-D. Sepp, B. Fricke, and T. Morovic, *Z. Phys.* **A286**:169 (1978).

(Ko 79) C. Kozhuharov, P. Kienle, E. Berdermann, H. Bokemeyer, J. S. Greenberg, Y. Nakayama, P. Vincent, H. Backe, L. Handschug, and E. Kankeleit, *Phys. Rev. Lett.* **42**:376 (1979).

(Ko 80) L. Kocbach, J. M. Hansteen, and R. Gundersen, *Nucl. Instrum. Methods* **169**:281 (1980).

(Ko 81) C. Kozhuharov, in *Electronic and Atomic Collisions*, Proceedings of the XII ICPEAC, Gatlinburg, Tennessee (S. Datz, ed.), North-Holland, Amsterdam (1982), p. 179.

(Kr 74) G. Kraft, P. H. Mokler, and H. J. Stein, *Phys. Rev. Lett.* **33**:476 (1974).

(Kr 81) J. Krause and M. Kleber, in: *Proc. of the NATO Advanced Studies Institute "Quantum Electrodynamics of Strong Fields"* (Lahnstein, 1981), Plenum Press, New York (1983), p. 489.

(La 72) F. P. Larkins, *J. Phys. B: At. Mol. Phys.* **5**:571 (1972).

(Le 63) J. M. Levy-Leblond, *J. Math. Phys.* **4**:776 (1963).

(Le 69) H. Levy II and W. R. Thorson, *Phys. Rev.* **181**:244, 252 (1969).

(Le 71) C. F. Lebeda, W. R. Thorson, and H. Levy, *Phys. Rev. A* **4**:900 (1971).

(Li 35) E. Lifshitz, *Phys. Z. Sowjetunion* **7**:385 (1935).

(Li 63) W. Lichten, *Phys. Rev.* **131**:229 (1963).

(Li 74) W. Lichten, *Phys. Rev. A* **9**:1458 (1974).

(Li 77) V. I. Lisin, M. S. Marinov, and V. S. Popov, *Phys. Lett.* **69B**:141 (1977).

(Li 78) D. Liesen, P. Armbruster, H.-H. Behncke, and S. Hagmann: *Z. Phys.* **A288**:417 (1978).

(Li 80a) D. Liesen, P. Armbruster, F. Bosch, S. Hagmann, P. H. Mokler, H. J. Wollersheim, H. Schmidt-Böcking, R. Schuch, and J. B. Wilhelmy, *Phys. Rev. Lett.* **44**:983 (1980).

(Li 80b) V. I. Lisin, M. S. Marinov, and V. S. Popov, *Phys. Lett.* **91B**:20 (1980).

(Lü 81) H. J. Lüdde and R. Dreizler, *J. Phys. B: At. Mol. Phys.* **B14**:2191 (1981).

(Ma 33) H. S. W. Massey and R. A. Smith, *Proc. R. Soc. London* **A142**:142 (1933).

(Ma 73) J. R. MacDonald, M. D. Brown, and T. Chiao, *Phys. Rev. Lett.* **30**:471 (1973).

(Ma 74a) J. H. Macek and J. S. Briggs, *J. Phys. B: At. Mol. Phys.* **7**:1312 (1974).

(Ma 74b) H. S. W. Massey and H. B. Gilbody, in *Electronic and Ionic Impact Phenomena*, Vol. IV, Clarendon, Oxford (1974).

(Ma 75a) D. H. Madison and E. Merzbacher, in *Atomic Inner-Shell Processes* (B. Craseman, ed.), Vol. I, Academic Press, New York (1975), p. 1.

(Ma 75b) M. S. Marinov and V. S. Popov, *Sov. J. Nucl. Phys.* **20**:641 (1975).

(Ma 76) C. Maroni, I. Massa, and C. Vannini, *Nucl. Phys.* **A273**:429 (1979).

(Ma 79a) A. Macias, A. Riera, and A. Salin, *J. Phys. B: At. Mol. Phys.* **12**:447 (1979).

(Ma 79b) V. Maruhn-Rezwani, N. Grün, and W. Scheid, *Phys. Rev. Lett.* **43**:512 (1979).

(McV 81) K. W. McVoy, X. T. Tang, and H. A. Weidenmüller, *Z. Phys.* **A299**:195 (1981).

(Me 58) E. Merzbacher and H. W. Lewis, in *Encyclopedia of Physics* (S. Flügge, ed.), vol. XXXIV, p. 166, Springer, Berlin (1958).

(Me 73a) W. E. Meyerhof, *Phys. Rev. Lett.* **31**:1341 (1973).

(Me 73b) W. E. Meyerhof, T. K. Saylor, S. M. Lazarus, W. A. Little, and B. B. Triplett, *Phys. Rev. Lett.* **30**:1279 (1973).

(Me 77) W. E. Meyerhof and K. Taulbjerg, *Ann. Rev. Nucl. Sci.* **27**:279 (1977).

(Me 79) W. E. Meyerhof, D. L. Clark, C. Stoller, E. Morenzoni, W. Wölfli, F.

Folkmann, P. Vincent, P. H. Mokler, and P. Armbruster, *Phys. Lett.* **70A**:303 (1979).

(Me 80) W. E. Meyerhof, J. F. Chemin, R. Anholt, and Ch. Stoller, *Trans. NY Acad. Sci. Ser. II* **40**:134 (1980).

(Mi 78) M. Micklinghoff, *Nucl. Phys.* **A295**:228 (1978).

(Mo 65) N. S. Mott and H. S. W. Massey, *The Theory of Atomic Collisions*, 3rd ed., Oxford, London (1965).

(Mo 72) P. H. Mokler, H.-H. Stein, and P. Armbruster, *Phys. Rev. Lett.* **29**:827 (1972).

(Mo 74) P. J. Mohr, *Ann. Phys. (N.Y.)* **88**:26, 52 (1974).

(Mo 77) B. L. Moiseiwitsch, *Rep. Prog. Phys.* **40**:843 (1977).

(Mo 78) P. H. Mokler and F. Folkmann, in: *Structure and Collisions of Ions and Atoms* (I. A. Sellin, ed.), Springer, Berlin (1978), p. 201.

(Mo 80) B. L. Moiseiwitsch, *Adv. At. Mol. Phys.* **16**:281 (1980).

(Mu 28) R. S. Mulliken, *Phys. Rev.* **32**:186 (1928).

(Mü 72a) B. Müller, H. Peitz, J. Rafelski, and W. Greiner, *Phys. Rev. Lett.* **28**:1235 (1972).

(Mü 72b) B. Müller, J. Rafelski, and W. Greiner, *Z. Phys.* **257**:62, 183 (1972).

(Mü 73a) B. Müller, J. Rafelski, and W. Greiner, *Phys. Lett.* **47B**:5 (1973).

(Mü 73b) B. Müller, J. Rafelski, and W. Greiner, *Nuovo Cimento* **A18**:551 (1973).

(Mü 74) B. Müller, R. K. Smith, and W. Greiner, *Phys. Lett.* **49B**:219 (1974).

(Mü 75a) B. Müller, *Proc. 9th Int. Conf. on the Physics of Electronic and Atomic Collisions, Seattle* (J. S. Risley and R. Geballe, eds.), University of Washington Press, Seattle (1976).

(Mü 75b) B. Müller and J. Rafelski, *Phys. Rev. Lett.* **34**:349 (1975).

(Mü 76a) B. Müller, *Ann. Rev. Nucl. Sci.* **26**:351 (1976).

(Mü 76b) B. Müller and W. Greiner, *Z. Naturforsch.* **31a**:1 (1976).

(Mü 78) B. Müller, G. Soff, W. Greiner, and V. Ceausescu: *Z. Phys.* **A285**:27 (1978).

(Mü 80a) B. Müller and V. E. Oberacker, *Phys. Rev. C* **22**:1909 (1980).

(Mü 80b) U. Müller, J. Reinhardt, G. Soff, B. Müller, and W. Greiner, *Z. Phys.* **A297**:357 (1980).

(Mü 81) B. Müller, in: *Proc. of the NATO Advanced Studies Institute "Quantum Electrodynamics of Strong Fields"* (Lahnstein, 1981), Plenum Press, New York, (1983), p. 41.

(Mü 83) B. Müller, J. Reinhardt, W. Greiner, and G. Soff, *Z. Phys.* **A311**:151 (1983).

(Nu 76) R. W. Numrich and J. S. Dahler, *J. Phys. B: At. Mol. Phys.* **9**:L403 (1976).

(Ob 76) V. Oberacker, G. Soff, and W. Greiner, *Nucl. Phys.* **A259**:324 (1976).

(Pa 78a) M. Pauli and D. Trautmann, *J. Phys. B: At. Mol. Phys.* **11**:667 (1978).

(Pa 78b) M. Pauli, F. Rösel, and D. Trautmann, *J. Phys. B: At. Mol. Phys.* **11**:2511 (1978).

(Pe 73) H. Peitz, B. Müller, J. Rafelski, and W. Greiner, *Lett. Nuovo Cimento* **8**:37 (1973).

(Pi 69) W. Pieper and W. Greiner, *Z. Phys.* **218**:327 (1969).

(Pi 77) R. D. Piacentini and A. Salin, *J. Phys. B: At. Mol. Phys.* **10**:1515 (1977).

(Po 45) I. Pomeranchuk and J. Smorodinsky, *J. Phys. USRR* **9**:97 (1945).

(Po 53) G. Poots, U. Öpik, and D. R. Bates, *Proc. Phys. Soc. (London)* **A66**:1113 (1953).

(Po 71) V. S. Popov, *Sov. J. Nucl. Phys.* **12**:235 (1971).

(Po 73) V. S. Popov, *Sov. J. Nucl. Phys.* **17**:322 (1973).

(Po 74) V. S. Popov, *Sov. Phys. JETP* **38**:18 (1974).

(Po 77) L. I. Ponomarev, T. P. Puzynina, and L. N. Somov, *J. Phys. B: At. Mol. Phys.* **10**:1335 (1977).

(Ra 71) J. Rafelski, L. Fulcher, and W. Greiner, *Phys. Rev. Lett.* **27**:958 (1971).

(Ra 72) J. Rafelski, B. Müller, and W. Greiner, *Lett. Nuovo Cimento* **4**:469 (1972).

(Ra 74) J. Rafelski, B. Müller, and W. Greiner, *Nucl. Phys.* **B68**:585 (1974).

(Ra 76a) J. Rafelski and B. Müller, *Phys. Rev. Lett.* **36**:517 (1976).

(Ra 76b) J. Rafelski and B. Müller, *Phys. Lett.* **65B**:205 (1976).

(Ra 78a) J. Rafelski, L. P. Fulcher, and A. Klein, *Phys. Rep.* **38C**:227 (1978).

(Ra 78b) J. Rafelski, B. Müller, and W. Greiner, *Z. Phys.* **A285**:49 (1978).

(Ra 79) J. Rankin and W. R. Thorson, *Phys. Rev. A* **18**:1990 (1979).

(Re 69) D. Rein, *Z. Phys.* **221**:423 (1969).

(Re 73) J. F. Reading, *Phys. Rev. A* **8**:3262 (1973).

(Re 76) J. Reinhardt, G. Soff, and W. Greiner, *Z. Phys.* **A176**:285 (1976).

(Re 77) J. Reinhardt and W. Greiner, *Rep. Prog. Phys.* **40**:219 (1977).

(Re 78) J. Reinhardt, V. Oberacker, B. Müller, W. Greiner, and G. Soff, *Phys. Lett.* **78B**:183 (1978).

(Re 79a) K. E. Rehm, H. Essel, K. Härtel, P. Kienle, H. J. Körner, P. Sperr, and W. Wagner, *Physs. Lett.* **86B**:251 (1979).

(Re 79b) J. Reinhardt, B. Müller, W. Greiner, and G. Soff, *Phys. Rev. Lett.* **43**:1307 (1979).

(Re 79c) J. Reinhardt, B. Müller, W. Greiner, and G. Soff, *Z. Phys. A292*:211 (1979).

(Re 80a) J. F. Reading and A. L. Ford, *Phys. Rev. A* **21**:124 (1980).

(Re 80b) J. Reinhardt, B. Müller, and W. Greiner, in *Coherence and Correlation in Atomic Collisions* (H. Kleinpoppen and J. F. Williams, eds.), Plenum, New York (1980), p. 331.

(Re 81a) J. Reinhardt, B. Müller, and W. Greiner, *Phys. Rev. A* **24**:103 (1981).

(Re 81b) J. Reinhardt, U. Müller, B. Müller, and W. Greiner, *Z. Phys.* **A303**:173 (1981).

(Ri 75) G. A. Rinker and L. Wilets, *Phys. Rev. A* **12**:748 (1975).

(Ri 78) T. H. Rihan, N. S. Aly, E. Merzbacher, B. Müller, and W. Greiner, *Z. Phys.* **A285**:397 (1978).

(Rö 79) S. Röhl, S. Hoppenau, and M. Dost, *Phys. Rev. Lett.* **43**:1300 (1979).

(Sa 31) F. Sauter, *Z. Phys.* **69**:742 (1931).

(Sa 72) F. W. Saris, W. F. van der Weg, T. Tawara, and R. Laubert, *Phys. Rev. Lett.* **28**:717 (1972).

(Sa 81) K. R. Sandhya Devi and S. E. Koonin, *Phys. Rev. Lett.* **47**:27 (1981).

(Sch 40) L. I. Schiff, H. Snyder, and J. Weinberg, *Phys. Rev.* **57**:315 (1940).

(Sch 51) J. Schwinger, *Phys. Rev.* **82**:664 (1951).

(Sch 69) S. B. Schneidermann and A. Russek, *Phys. Rev.* **181**:311 (1969).

(Sch 72) H. W. Schnopper, H. D. Betz, J. P. Delvaille, K. Kalata, A. R. Sohval, K. W. Jones, and H. E. Wegner, *Phys. Rev. Lett.* **29**:898 (1977).

(Sch 77) G. B. Schmid, *Phys. Rev. A* **15**:1459 (1977).

(Sch 78) P. Schlüter, G. Soff, and W. Greiner, *Z. Phys.* **A286**:149 (1978).

(Sch 79) R. Schuch, G. Nolte, H. Schmidt-Böcking, and W. Lichtenberg, *Phys. Rev. Lett.* **43**:1104 (1979).

(Sch 81) P. Schlüter, G. Soff, and W. Greiner, *Phys. Rep.* **75**:327 (1981).

(Se 73) V. Sethu-Raman, W. R. Thorson, and C. F. Lebeda, *Phys. Rev. A* **8**:1316 (1973).

(Se 81) W.-D. Sepp, B. Fricke, and T. Morovic, *Phys. Lett.* **81A**:258 (1981).

(Sh 73) R. Shakeshaft, J. Macek, and E. Gerjuoy, *J. Phys. B: At. Mol. Phys.* **6**:794 (1973).

(Sh 81) A. P. Shergin, R. Mann, and H. F. Beyer, *Z. Phys.* **A302**:191 (1981).

(Sm 69) F. T. Smith, *Phys. Rev.* **179**:111 (1969).

(Sm 74) K. Smith, H. Peitz, B. Müller, and W. Greiner, *Phys. Rev. Lett.* **32**:554 (1974).

(Sm 75a) K. Smith, B. Müller, and W. Greiner, *J. Phys. B: At. Mol. Phys.* **8**:75 (1975).
(Sm 75b) R. K. Smith and W. Greiner, *Spontaneous and Induced Radiation from Intermediate Molecules*, unpublished report (1975).
(So 73) G. Soff, J. Rafelski, and W. Greiner, *Phys. Rev. A* **7**:903 (1973).
(So 74) G. Soff, B. Müller, and J. Rafelski, *Z. Naturforsch.* **29a**:1267 (1974).
(So 77a) G. Soff and B. Müller, *Z. Phys.* **A280**:243 (1977).
(So 77b) G. Soff, J. Reinhardt, B. Müller, and W. Greiner, *Phys. Rev. Lett.* **38**:592 (1977).
(So 78a) G. Soff, W. Betz, B. Müller, W. Greiner, and E. Merzbacher, *Phys. Lett.* **65A**:19 (1978).
(So 78b) G. Soff, B. Müller, and W. Greiner, *Phys. Rev. Lett.* **40**:540 (1978).
(So 78c) G. Soff, W. Betz, J. Reinhardt, and J. Rafelski, *Phys. Scr.* **17**:417 (1978).
(So 79a) G. Soff, W. Greiner, W. Betz, and B. Müller, *Phys. Rev. A* **20**:169 (1979).
(So 79b) G. Soff, J. Reinhardt, B. Müller, and W. Greiner, *Phys. Rev. Lett.* **43**:1981 (1979).
(So 80) G. Soff, J. Reinhardt, B. Müller, and W. Greiner, *Z. Phys.* **A294**:137 (1980).
(So 81a) G. Soff, B. Müller, and W. Greiner, *Z. Phys.* **A299**:189 (1981).
(So 81b) G. Soff, J. Reinhardt, and W. Greiner, *Phys. Rev. A* **23**:701 (1981).
(So 81c) G. Soff, J. Reinhardt, and W. Greiner, *Phys. Lett.* **83A**:158 (1981).
(So 81d) G. Soff, P. Schlüter, and W. Greiner, *Z. Phys.* **A303**:189 (1981).
(So 82) G. Soff, P. Schlüter, B. Müller, and W. Greiner, *Phys. Rev. Lett.* **48**:1465 (1982).
(Sp 65) H. J. Specht, *Z. Phys.* **185**:301 (1965).
(St 58) W. E. Stephens and H. Staub, *Phys. Rev.* **109**:1196 (1958).
(St 77) Ch. Stoller, W. Wölfli, G. Bonani, M. Stöckli, and M. Suter, *J. Phys. B: At. Mol. Phys.* **10**:L347 (1977).
(St 80a) Ch. Stoller, E. Morenzoni, W. Wölfli, W. E. Meyerhof, F. Folkmann, P. Vincent, P. H. Mokler, and P. Armbruster, *Z. Phys.* **A297**:93 (1980).
(St 80b) N. Stolterfoht, *J. Phys. B: At. Mol. Phys.* **13**:L559 (1980).
(St 81) K. E. Stiebing, K. Bethge, W. Schadt, H. Schmidt-Böcking, F. Bosch, D. Liesen, D. Maor, P. H. Mokler, P. Vincent, S. Hagmann, and R. Schuch, GSI scientific report 1980, Darmstadt (1981), p. 126.
(Ta 75) K. Taulbjerg, J. Vaaben, and B. Fastrup, *Phys. Rev. A* **12**:2325 (1975).
(Ta 76) K. Taulbjerg, J. S. Briggs, and J. Vaaben, *J. Phys. B: At. Mol. Phys.* **9**:1351 (1976).
(Ta 77) K. Taulbjerg, *J. Phys. B: At. Mol. Phys.* **10**:L341 (1977).
(Te 30) E. Teller, *Z. Phys.* **61**:458 (1930).
(Te 80) E. Teubner, G. Terlecki, N. Grün, and W. Scheid, *J. Phys. B: At. Mol. Phys.* **13**:523 (1980).
(Th 69) W. R. Thorson and H. Levy, *Phys. Rev.* **181**:230 (1978).
(Th 77) W. R. Thorson and J. H. Choi, *Phys. Rev. A* **15**:550 (1977).
(Th 78) W. R. Thorson and J. B. Delos, *Phys. Rev. A* **18**:117, 135 (1978).
(Th 79) J. Theis, J. Reinhardt, and B. Müller, *J. Phys. B: At. Mol. Phys.* **12**:L479 (1979).
(Th 81) J. Theis and B. Müller, *Phys. Rev. A* **24**:89 (1981).
(Ts 83) I. Tserruya, R. Schuch, H. Schmidt-Böcking, J. Barrete, Wang Da-Hai, B. M. Johnson, K. W. Jones, and M. Meron, *Phys. Rev. Lett.* **50**:30 (1983).
(To 79) A. Toepfer, E. K. U. Gross, and R. M. Dreizler, *Phys. Rev. A* **20**:1808 (1979).
(Ue 35) E. A. Uehling, *Phys. Rev.* **48**:55 (1935).
(Va 81) J. Vaaben and K. Taulbjerg, *J. Phys. B: At. Mol. Phys.* **14**:1815 (1981).
(Vo 61) V. V. Voronkov and N. N. Koleznikov, *Sov. Phys. JETP* **12**:136 (1961).

(Wa 70) W. L. Wang and C. M. Shakin, *Phys. Lett.* **32B**:421 (1970).
(We 33) V. Weisskopf, *Phys. Z.* **34**:1 (1933).
(We 58) F. G. Werner and J. A. Wheeler, *Phys. Rev.* **109**:126 (1958).
(Wi 54) E. H. Wichmann and N. M. Kroll, *Phys. Rev.* **96**:232 (1954).
(Wi 79) K.-H. Wietschorke, B. Müller, W. Greiner, and G. Soff, *J. Phys. B: At. Mol. Phys.* **12**:L31 (1979).
(Wi 80) T. G. Winter and G. J. Hatton, *Phys. Rev. A* **21**:793 (1980).
(Ya 50) C. N. Tang and D. Feldman, *Phys. Rev.* **79**:972 (1950).
(Ze 72) Ya. B. Zeldovich and V. S. Popov, *Sov. Phys. Usp.* **14**:673 (1972).

2

JACK S. GREENBERG (top) is Professor of Physics and a member of the Wright Nuclear Structure Laboratory at Yale University. He received a B.Eng. and M.Sc. from McGill University and a Ph.D. from Massachusetts Institute of Technology. Before joining the Faculty at the Yale University Physics Department, he was a Rutherford Memorial Fellow at the Swiss Federal Institute of Technology (ETH), Institute of Physics in 1955–1956. He served as Director of Graduate Studies in the Physics Department at Yale University from 1967–1969. In 1969, he was a Senior Faculty Fellow at the Weizmann Institute of Science, and in 1971 he held a Guest Professorship at Erstes Physikalishes Institute, University of Köln. He was a recipient of the Senior U.S. Scientist Award of the Alexander Von Humboldt Foundation in 1976–1977 and in 1984, served as Guest Professor and Consultant at the Gesellschaft für Schwerionenforschung in Darmstadt, Federal Republic of Germany. He is a Fellow of the American Physical Society and his research has been in nuclear, particle and atomic physics.

PAUL VINCENT (bottom) is currently a Physics Division staff member in the relativistic heavy-ion research group at Brookhaven National Laboratory. He graduated Phi Beta Kappa from Hamilton College in 1972, and received his Ph.D. in experimental heavy-ion atomic physics from the Yale Wright Nuclear Structure Laboratory in 1977. Following an appointment as guest scientist at GSI Darmstadt, West Germany, where he pursued studies of x-ray and positron production in high energy, heavy-ion collisions, Vincent was Assistant Professor of Physics at Rutgers University from 1980–1984.

High-Energy Atomic Physics—Experimental

J. S. GREENBERG AND P. VINCENT

1. Introduction

1.1. Scope of Review

These volumes are compelling testimony to the rapid development of heavy-ion collisions as penetrating probes of nuclear structure and as sources of information on nuclear dynamics. But with equal emphasis, the availability of heavy-ion projectiles has had a major impact on other fields of physics. One such particularly fruitful overlap is with atomic physics. Atomic collisions with light projectiles such as photons, electrons, and protons have always been important probes of atomic structure. Most recently, the introduction of a greatly expanded pool of heavy-ion projectiles and the ability to accelerate these to high energies has further extended this field and, more significantly, introduced unique opportunities to study new phenomena. Indeed, during the last decade, paralleling the development of heavy-ion accelerators, a clearly distinguishable field of high-energy atomic and molecular physics has emerged. Recent conferences reflect its scope (Od 80, Gr 81, Cr 82, Da 82).

The pool of available projectiles now spans the range of the Periodic Table from protons to uranium with energies from keV to GeV. The diversity in scattering systems that can be studied is clearly enormous and the field presently reflects this diversity. Broadly speaking the problems being pursued involve attempts, on the one hand, to understand the physics of the ion–atom

J. S. GREENBERG • A. W. Wright Nuclear Structure Laboratory, Yale University, New Haven, Connecticut 06511. P. VINCENT • Department of Physics, Brookhaven National Laboratory, Upton, New York 11973.

collision *per se* and, on the other hand, to exploit the collisions as an environment and a diagnostic tool to study questions of fundamental interest and novel aspects of atomic physics. The latter can only be addressed experimentally with the unique conditions generated in ion–atom collisions.

Among the former type of investigations are attempts to pursue our understanding of the complex scattering problem, begun with light projectiles, over a wide range of conditions. In this context a large effort is going into the study of electron excitation, ionization, charge transfer mechanisms, and radiative processes. As will be discussed in this article, each of these delineates many interesting phenomena as a diversity of scattering conditions are achieved from the many possibilities to choose systems and collision kinematics.

The utilization of high-energy heavy-ion beams to provide an environment to study problems of fundamental interest also is proving to be very rewarding. For many years experiments investigating the relativistic properties of atomic structure have continued to furnish important probes into the range of validity of the theory of quantum electrodynamics (QED). Generally the growth of the sensitivity of these tests has relied on a slow but progressive improvement in the experimental accuracy, taking advantage of an evolving technology. In the main the focus of experiments has been the hydrogen atom because of its simplicity. Heavy-ion beams are providing an important extension to this work by making it feasible to explore increasingly higher Z one- and two-electron type atoms. The simplicity of the hydrogen atom is retained, but such systems open up to experimental scrutiny the interesting subject of radiative corrections in strong fields (Mo 81), and with this the possibility of testing the standard formulation of QED in strong fields (Br 78).

But in addition to these more traditional avenues of investigation, a new dimension is introduced by the availability of superheavy collision systems. The fact that it is possible to bring together two uranium atoms within internuclear separations of nuclear dimensions has provided the opportunity to explore atomic physics in a domain where relativistic effects are no longer a perturbative consideration but, instead, constitute the prominent features dominating the binding energy and spatial behavior of atomic states (Re 85).

The feature of special interest, from which this relativistic behavior evolves, is the presence of extreme electric and magnetic fields that penetrate the space in the immediate vicinity of the two nuclear charges when they are close together well within the innermost orbits of the electron cloud. Here electric fields achieve intensities of $>10^{19}$ V/m for short periods of $\sim 2 \times 10^{-21}$ s if the scattering is Rutherford, or even longer if giant nuclear composites are made. Our special interest in these strong fields is connected with the opportunity to observe for the first time an elemental property of the vacuum associated with quantum field theory; the theory allows for the spontaneous creation of real particles in strong *static* external fields[†] (Kl 29, Sa 32, We 33, He 36, Hu 40, Sc 51, Vo 61, Be 63, Ge 69, Pi 69, Do 71, Mu 76, Re 77, Ra 78). The normal vacuum is unstable in such a *supercritical* field and decays into a new state that contains real particles. Of course, this process may occur for any elementary particle field. However, it is difficult to envision achieving the necessary physical situation except for the electron–position field. As will be demonstrated, in this case the formation of short-lived superheavy quasiatoms in a collision between heavy atoms (Ge 70, Ra 71) or the formation of a giant superheavy nucleus (Re 81) can offer the suitable conditions. In the presence of a supercritical electric field the instability of the vacuum appears as the *spontaneous emission of positrons* (Vo 61, Ge 69, Pi 69).

The search for this fundamental process and the experimental effort to study the novel aspects of the atomic physics associated with superheavy collision systems will be at the focus of this chapter. In this context more specifically we will be considering the investigations that have been addressing a number of the central questions concerning (1) the utilization of superheavy collision systems to extract information on superheavy atomic species, (2) the possible existence of electron binding energies in atoms which can exceed the supercritical value of twice the electron mass, (3) the possibility of employing collision systems such as U + U and U + Cm to observe the spontaneous decay of the supercritically bound state, and (4) the possibility of exploiting atomic processes, such as positron emission, as probes of exotic nuclear properties such as the formation of superheavy nuclear species.

Although our main concentration is the recent work with superheavy systems, we will be considering also many aspects of ion–atom collision experiments carried out with light and medium weight collision systems. They play a pivotal role in understanding the ionization and radiation processes in heavy-ion collisions as well as in developing the basis for the molecular model of ion–atom collisions, which is one of the essential ideas associated with the physics of the superheavy collision systems.

Of course, the expansion of the atomic collision field with the introduction of many projectile species has been accompanied by the development of appropriate experimental techniques to gather and handle this new information. Particularly in experiments involving heavy collision systems, x rays, electrons, and positrons are emitted in a high-energy range not usually encountered in low-energy atomic collision studies. Moreover, the experiments often involve the detection and identification of energetic heavy ions in a background sometimes containing copious fluxes of lighter

[†] For review of QED of strong fields see (Gr 81, Br 78, Re 85) and (Mu 76, Re 77, Ra 78). These contain more references to early literature and historical developments.

particles. Typically multicoincidence and multiparameter data accumulation is involved. Therefore, techniques have evolved to accommodate the new requirements. Since the radiations and experimental conditions in many ways duplicate those encountered in nuclear physics, the detection systems in the two fields are often similar. When beneficial to the discussion, this article will discuss some of the more specialized experimental arrangements.

For orientation to the content of this article, let us review first some of the ideas and intuitive physical pictures which have been at the focus of developments in ion–atom collision studies during the past decade.

1.2. Excitation Mechanisms

In large part the current ion–atom collision studies at high bombarding energy are centered around an interest in the physics of inner shells. Of course, the accumulation of information on inner-shell processes, both excitation and radiative decay, has a long and distinguished history. It is only necessary to recall the essential role played by such information in developing the fundamental ideas underlying quantum theory (Ja 66, Me 76). The pioneering work on characteristic x rays by G. G. Barkla (Ba 11) and H. G. J. Moseley (Mo 14) provided demanding tests for the newly developing theory, and further experiments, such as the observations by Siegbahn (Si 20) of x-ray satellites, posed new challenges to understand the residual interactions between inner-shell electrons outside the confines of a theory of hydrogenic atoms.

Heavy-particle-induced excitation and ionization of inner-shell electrons also contributed to the evolution of the new quantum physics in its early growth period (Me 58). Some of the first inquiries into atomic excitations using collisions with nuclear projectiles can be traced to Rutherford and colleagues (Ch 13, Ch 14, Ru 13), and many aspects of this subject were pursued before World War II. Particularly, pioneering experiments such as those carried out by W. M. Coates (Co 34) in 1934 using mercury beams were especially prophetic in anticipating questions of the most current interest regarding inner shells. With keen insight he suggested an essential role for a "temporarily formed quasimolecule" to account for his observations of characteristic and continuous x rays. As noted above and discussed in detail below, this idea, resurrected and developed in detail 30 years later (Fa 65), forms the basis of and motivation for much of the current research with heavy-ion–atomic interactions.

Some selected data can be used to illustrate some of these main avenues of inquiry and also indicate some of the principal developments in the field which opened up the possibility of addressing the questions associated with strong binding in inner shells and the QED of strong fields. As a historical note we mention that during the 1940s there was a lagging interest in atomic collisions which were then viewed as basically understood and, in fact, were overshadowed by exciting developments in nuclear physics. Renewed interest in this subject was activated in the 1950s when new accelerators with high-energy beams and later in the 1960s a new variety of projectiles began to produce data with unexplained phenomena. The bulk of the data from heavy-ion bombardment dates to the 1970s (Ga 73, Ke 73, Cr 75, Cr 82, Me 76c, Me 77, Mo 78, Ri 78, Od 80, Gr 81, Da 82).

One such interesting set of data is shown in Figure 1.1 (Br 70). It compiles *K*-shell ionization cross sections for various targets excited by the light projectiles H and He and by some heavier ions. Guided by a universal representation obtained from scaling laws for Coulomb ionization (Ma 75), a reduced cross section and a reduced velocity is chosen for the ordinate and abscissa, respectively. Irrespective of considering any details, the main features of such a plot clearly illustrate that quite different mechanisms must be responsible for ionization by the light H and He projectile moving with high velocities on the one hand, and by the heavier ions associated with low velocities on the other hand. In fact these data reflect the principal mechanisms represented by the two extreme theoretical models for ion–atom encounters.

One of these essentially involves the classic problem of direct transfer of momentum to the bound electrons of the target by the incident projectile due to the Coulomb interaction between the projectile and target electrons. The projectile's role here is merely that of an isolated charged particle. Such direct Coulomb excitation is expected to dominate when the encounter is sufficiently swift and for light-ion projectiles (Ma 75). More specifically, the velocity of the projectile, v_{ion}, must approach or exceed the electron's velocity, v_e, while the projectile charge, Z_1, is small compared with the target nuclear charge, Z_2, so that the electron cloud is only slightly disturbed and perturbation theory is valid.

Due to many experiments yielding data such as shown in Figure 1.1 and complimentary theoretical studies, collisions by fast structureless particles such as protons and alpa particles are now well understood. Review articles by Madison and Merzbacher (Ma 75) and Garcia *et al.* (Ga 73) provide a comprehensive account of these studies. Indeed, a variety of theoretical approaches are available presently to describe data such as shown on the right-hand side of Figure 1.1. Thus the application of the Born approximation (Me 58, Me 76, Ga 73, Ma 75) and the semiclassical impact parameter method (Ba 59, Ha 75, Ha 75b) have led to a basic understanding of the Coulomb ionization processes. Although the simplest models have been challenged by experiments and their limitations noted, the incorporation of refinements in the theory which include the effects of the increased binding produced by the projectile, polarization, relativistic corrections and the Coulomb deflection leads to an impressive description of the experimental

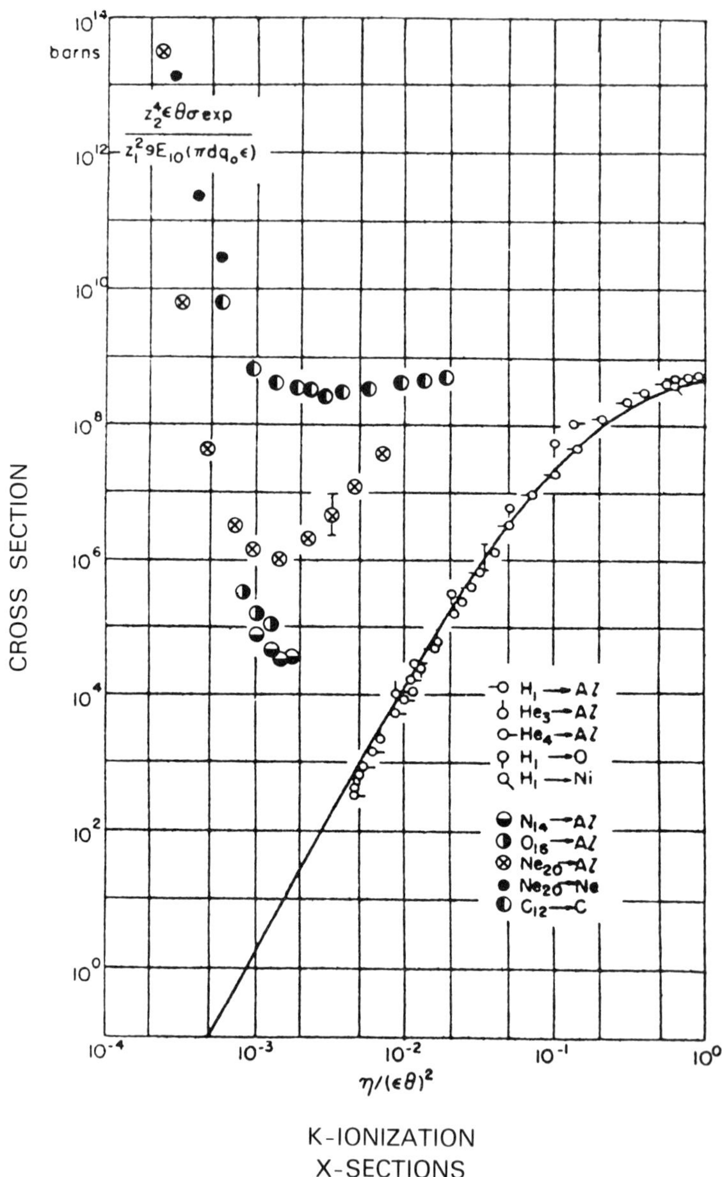

Figure 1.1. Reduced K-shell ionization cross sections as a function of the reduced velocity for various targets excited by the ions indicated in the figure (Br 70). The solid line represents the theoretical K-shell Coulomb ionization cross sections in the Born approximation. The sources of the data are listed in (Br 70).

data (Br 66b, Ba 71, Ba 71b, Ba 73, Ba 73b, Am 75, Cl 75, An 77, An 78c, An 78d). With the theory in hand, the light projectiles can now be used as a sensitive probe of the wave functions in the inner confines of the atoms. For example, a now classic illustration of the effectiveness of these probes is found in the detection of the prominent dip in the 2*s* ionization cross section produced by the radial node in the 2*s* wave function (Da 82).

The data shown on the left-hand side of Figure 1.1, representing collisions where $Z_1 \simeq Z_2$ and $v_{\text{ion}} \ll v_e$, however, leads us into a different regime of ion–atom collisions which offers the opportunity to explore new aspects of the collision system. With the measured cross sections in these cases differing from the simple Coulomb ionization predictions by as much as a factor of 10^{14}, new physical consequences are clearly implied. Indeed, an explanation for these data is found in the formation of the transient quasimolecules visualized by Coats (Co 34).

1.3. Quasimolecular Phenomena

The simple Coulomb excitation description of ionization by heavy projectiles colliding at low velocities falls short for several reasons. In such slow encounters the quasiadiabatic effect of the projectile on the target electrons can no longer be treated as a small correction to the direct projectile-electron Coulomb interaction. Instead it is the dominant effect. The distortion energy associated with the presence of the projectile nuclear charge well within the core of the target atom can be comparable to and even exceed the ionization energy. Thus the overlapping electron clouds experience the combined influence of the two nuclear centers. If the system develops slowly enough, the exchange interactions among the electrons contribute to excitations in addition to Coulomb excitation.

Qualitatively the role of the quasimolecule in the excitation and ionization processes can be readily understood. The basis of quasimolecule formation is in the disparity between the nuclear collision velocity and the velocities of the orbiting electrons. Thus for the fast-moving inner-shell electrons the Coulomb potential produced by the colliding nuclei varies sufficiently slowly so that the electrons can relax adiabatically to the nuclear motion. At small internuclear separations, well within their orbiting radii where the electrons cannot distinguish between the two nuclear centers, they act as if they are bound by the combined Z of the nuclear charges. Asymptotically the quantum numbers are, therefore, those of the united quasiatom at one extreme and those of the separated atoms at the other. In the intermediate region, between the very distant and very close internuclear separations, the electrons evolve through a series of quasimolecular states in the two-center field as the internuclear distance decreases and subsequently increases again. The detailed evolution of these molecular orbitals with time provides the mechanisms for excitation and ionization.

As will be discussed in more detail in Sections 2 and 3, an important characteristic property of the transient molecular orbitals is a promotion or demotion of the principal quantum number and binding energy in passing from the separated atom to the united atom limits. In their evolution with changing internuclear separation, molecular orbitals of very different principal quantum number in the separated atom limit may cross or approach very closely in binding energy at particular internuclear distances. The decreased binding energy or the close proximity of molecular orbitals at finite internuclear separations can lead to excitation induced by a departure from complete adiabaticity due to radial variation or rotation of the internuclear axis with time. Upon separation, the electrons, thus, may not necessarily return to their original atomic states having crossed into available empty orbits, and the atom may then be left in an excited configuration leading to x-ray emission after the ions separate. This mechanism of electron excitation by promotion was first proposed by Fano and Lichten (Fa 65) and marked the current revival of the molecular formulation of atomic collisions. The molecular model explains the large energy transfer even for slow ions, the large cross sections corresponding to geometric atomic dimensions, the threshold behavior of these cross sections, and the now classic discovery by Armbruster (Ar 62, Ar 64) and Specht (Sp 65) of the resonance behavior in the production of K, L and M x rays when levels of the target and projectile match. In the latter work, especially, may be found the first indication of the new phenomena which has motivated much of the current interest.

While the studies of characteristic x-ray production provide this indirect information on quasimolecular phenomena, more direct signatures for quasimolecule formation are supplied by other radiations from heavy-ion collisions. For example as illustrated in Figure 1.2 the x-ray spectra not only consist of characteristic lines but also of x-ray continua. The characteristic lines are associated with the isolated atomic or ionic species after the collision. On the other hand the continuum radiations are not characteristic of the separated atoms but may originate from the collision complex. Figure 1.2 illustrates mechanisms leading to two continuum radiations which fall into this category and which will be of particular interest to our discussion. As observed for characteristic x-ray emission, both atomic and molecular mechanisms can be responsible for the emission of the continuum x rays depending largely on the ratio of the ion and orbiting electron velocities, v_{ion}/v_e.

For example radiative electron capture (REC) (Sc 72, Ki 73, Me 77, Mo 78) of target electrons into projectile vacancies is associated with atomic processes and high collision velocities, while x rays emitted during radiative

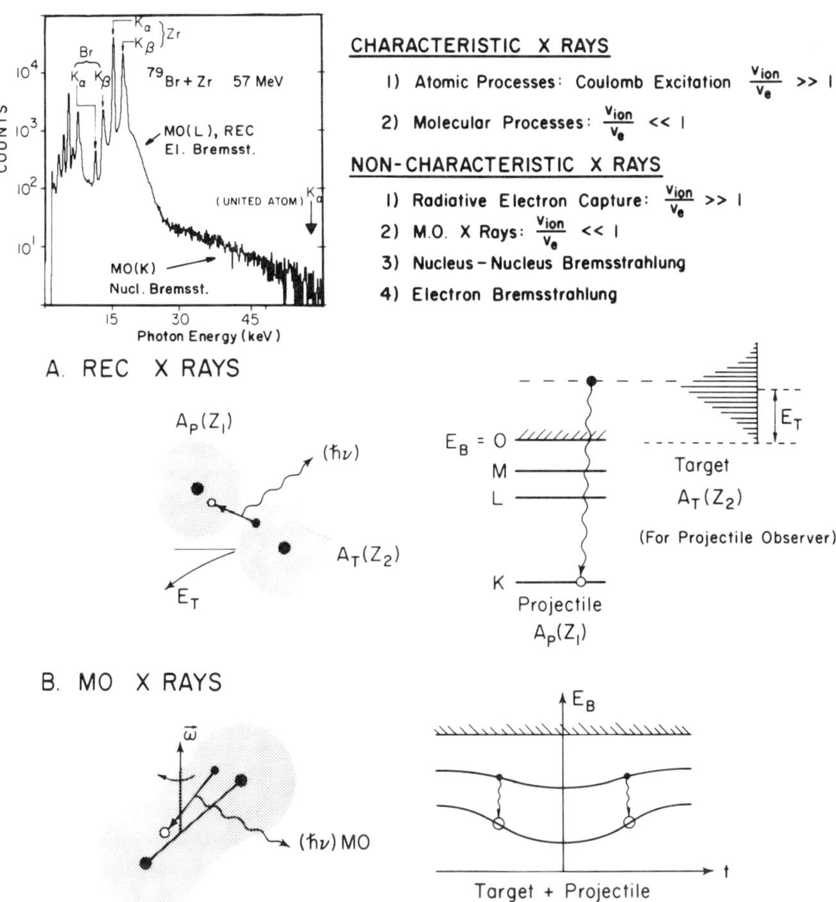

Figure 1.2. Characteristic and continuum x rays emitted from Br + Zr collisions. The figure illustrates two mechanisms, radiative electron capture (REC) and molecular-orbital x-ray emission (MO) that contribute prominently to the continuum x rays.

transitions between molecular orbitals (MO x rays) (Sa 72, Me 77, Mo 78) represent the other extreme. REC transitions are between atomic states, and the emission spectrum is determined predominantly by the momentum change occurring during the transfer of the electron. The spectrum, therefore, reflects the momentum distribution of the electrons in the initial target atomic state. On the other hand, since MO x-ray emission originates from transitions between the molecular orbitals, in principle, it can yield information on the structure of the collision complex and on the time evolution of these states. As such it provides the most direct demonstration of electronic quasimolecule formation.

It bears emphasis that the x-ray spectrum from heavy-ion collisions is rich in information on many radiative processes. In addition to the ones already mentioned, there is radiative ionization (Ja 75, An 76b), nucleus–nucleus bremsstrahlung (Re 76, Tr 76), secondary electron bremsstrahlung (Fo 74, Is 76), inverse bremsstrahlung (Sc 74), and radiation accompanying other electron processes. Some of these have been discussed in a recent review by Meyerhof and Taulberg (Me 77). Although of intrinsic interest, unfortunately they constitute troublesome backgrounds in some of the studies of primary interest to this article such as MO x-ray emission. In this context, they are discussed in Section 3.

The x-ray spectrum is not the only source of information on the quasimolecule and on ion–atom collisions in general. Delta electrons emitted during the ionization process and positron creation in superheavy collision systems are important experimental approaches to study the united atom characteristics of superheavy collision systems. In detecting the delta electrons we acquire one more differential piece of information not available in vacancy production studies via x-ray deexcitation alone, since the energy transfer is now more clearly defined. Section 4 will discuss how the presence of high energies in the delta-electron spectrum demonstrates the existence of high momentum components of the inner-shell wave functions which, in turn, reflect the relativistic shrinkage of the atomic space by the deep binding potentials in superheavy collision systems. The role played by the action of the combined charges $(Z_1 + Z_2)$ to produce a central potential which governs the electronic motion becomes very apparent from the study of dynamic aspects of delta-electron and positron emission. These points will be dealt with in Sections 4 and 5.

Therefore, as will become evident from such studies, the molecular picture not only provides a convenient set of basis states with which we describe the atomic excitations in slow ion–atom collisions, but the electronic quasimolecule becomes the focus of attention itself with access to new physics. In a broad sense, mapping the spectroscopy of quasimolecular levels is a central task of high-energy atomic physics since it forms the foundation for confronting the solutions to the two-center Dirac equation as a function of internuclear separation on which many of the theoretical predictions for the scattering problem are based. This mapping is clearly not the usual challenge encountered in conventional spectroscopy since in this case a time dependence is an inherent feature of the energy level structure. Nevertheless, a spectroscopy of sorts is attainable in superheavy systems, and we will discuss how some of their interesting spectroscopic features are reflected in inner-shell vacancy production and in the emission of MO x rays, delta electrons, and positrons.

1.4. Superheavy Quasiatoms

Probably most current interest in the quasimolecule, on a fundamental level, reflects its potential role as a vehicle which can provide the conditions for observing the instability of the electron–positron vacuum by spontaneous positron emission (Ge 70, Ra 71). The reasons for investigating this process have been mentioned briefly above. We also outlined some of the associated interesting questions to be explored. For further orientation to the experiments that will be discussed, let us briefly consider some qualitative aspects that clarify the conditions for an unstable vacuum and indicate how these conditions can be met with the formation of superheavy quasimolecules. For a comprehensive discussion of the theory and of the historical development of the interest in strong field phenomena, we refer the reader to chapter 1 in this volume by Reinhardt and Greiner (Re 85).

The source of the vacuum instability in the electron–positron field is traced to the behavior of the Dirac energy-level spectrum when the binding energy exceeds $2mc^2$ (Pi 69, Ge 69). Such strong binding can only occur for atoms with $Z\alpha > 1$. We can venture into this domain, well beyond the reach of stable atomic species, with a superheavy quasiatom formed momentarily in the united atom limit of a U + U collision where the nuclei almost touch. An examination of the development of the quasimolecular energy-level spectrum for U + U collisions with internuclear separation, or equivalently with time, shows why. Figure, 1.3 for example, displays such a spectrum based on the one-electron solutions to the two-center Dirac equation (Mu 73, Mu 76b). It bears emphasis that these calculations, incorporating complete adiabaticity for the energy levels, are calculated for the diatomic molecule frozen at each internuclear separation. Also shown in Figure 1.3 are calculations of binding energies and radial expectation values for the most tightly bound $1s\sigma$ orbital as a function of internuclear separation for a number of superheavy collision systems with combined charge Z (De 81). These calculations display some of the very unusual properties predicted for the exotic atoms represented by the quasiatoms formed in the collision.

For example, the binding energy of the $1s$ state increases very rapidly as the $Z\alpha = 1$ boundary is passed. In the early development of relativistic quantum mechanics $Z\alpha = 1$ appeared to be an apparent boundary due to the well-known singularity in the Sommerfeld fine-structure formula for point nuclei $j = 1/2$ solutions to the Dirac equation for this charge. Although this ambiguity has been removed by considering charge centers of finite dimensions, the onset of the rapid increase of the binding energy for the $1s$ state beyond $Z\alpha = 1$ still marks the remnants of the effect of a collapsing wave function mentioned below.

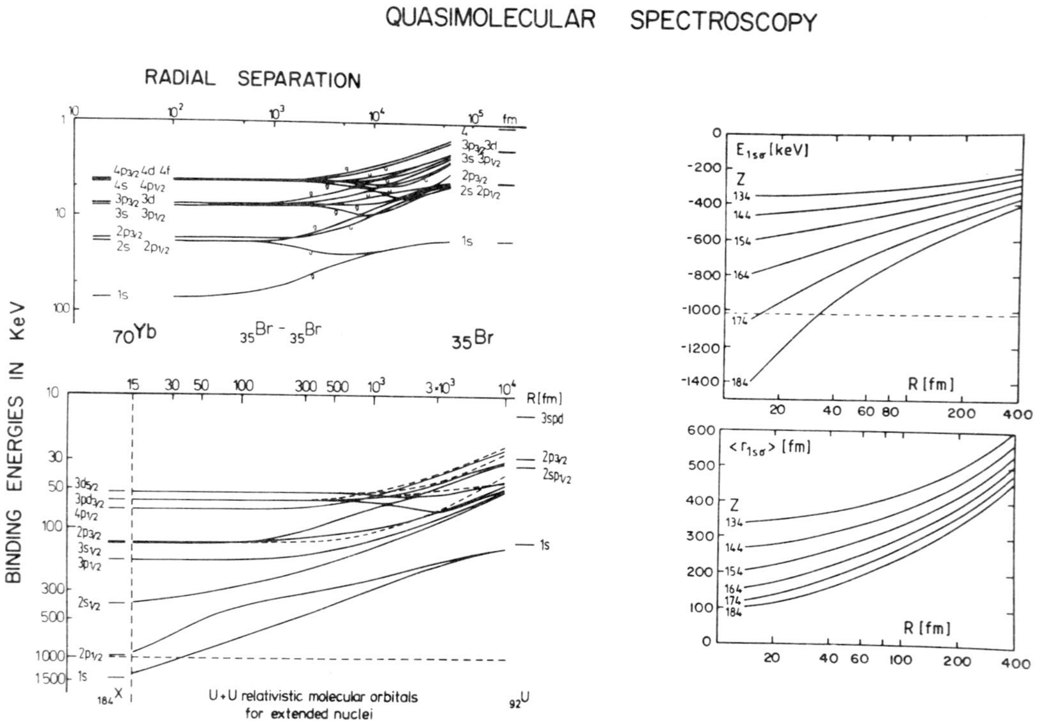

Figure 1.3. The development of the quasimolecular energy level spectra for Br + Br and U + U collisions with internuclear separation R. The spectra are based on the one-electron solutions of the two-center Dirac equation (Mu 73, Mu 76b). The comparison of the light and superheavy systems illustrates some of the relativistic features. Also shown separately are binding energies and radial expectation values for the $1s\sigma$ orbital as a function of R for a number of superheavy collision systems (De 81).

The binding energy of the electron soon exceeds its own rest mass at $Z \simeq 150$ and continues to increase even more precipitously toward $-mc^2$ as the charge is increased even further. The relativistic effects which are responsible for this rapid growth of binding energy with Z also lead to severe further modifications of the energy-level structure so that it no longer bears much resemblance to a nonrelativistic hydrogenlike structure. For instance, the fine-structure splitting of the $2p_{3/2}$–$2p_{1/2}$ states becomes very pronounced, achieving values in excess of mc^2. Another prominent new feature is the interchange of the $2p_{1/2}$ and $2s_{1/2}$ energy levels and the appearance of a large energy splitting between them where in light atoms a near degeneracy exists. The latter property particularly reflects the relativistic spinor character of the wave functions.

In fact, the dominant influence of relativistic effects in superheavy atoms is even more profoundly exhibited by the behavior of the wave functions. Figure 1.3 illustrates that the $1s_{1/2}$ wave function collapses around the nuclear center to dimensions well within the Compton wavelength for $Z\alpha > 1$. A measure of the size of this effect is obtained by noting that for Pb $\langle r_{1s} \rangle \simeq 750$ fm while for $Z = 170$ $\langle r_{1s} \rangle \simeq 120$ fm. But more dramatic is the prediction (De 81) that for $Z = 170$ the relativistically calculated electron density at the nucleus exceeds the corresponding nonrelativistic value by more than three orders of magnitude. This enhancement particularly illustrates the unusual nature of the electronic inner-shell structure of superheavy atoms. Indeed, it is the contraction of the wave function together with the accompanying rapid variation of the binding energy of the inner-shell molecular orbitals with decreasing internuclear separation, as displayed in Figure 1.3, that focus the mechanisms of inner-shell excitation and positron emission at the turning points of the ion–atom collision trajectories. This focussing of phenomena in the immediate vicinity of the two nuclear charges when they are close together, in turn, allows the probing of the properties of the quasiatom in the united atom limit and effectively excludes complications that could be introduced by the outer electronic structure. As will be demonstrated, it is the source of the spectroscopy of quasiatoms referred to above.

1.5. The Unstable Vacuum

But probably the most novel feature of superheavy atoms occurs when the binding energy reaches and exceeds $2mc^2$. The required critical charge, Z_{cr}, is predicted to be 173 for $1s$ states (Po 71, So 74, Wi 79). [Earlier studies of this question are considered in (Sc 40, Po 45, We 58, Vo 61, Pi 69, Re 69.)] The theory describing the physics as we cross this boundary is considered in detail in Chapter 1 by Reinhardt and Greiner in this volume and also in a number of review papers (Re 77, Ra 78, Mu 76, Br 78). From these considerations we can qualitatively envision the measurable effects associated with the physical significance of reaching and passing the critical charge.

The theory demonstrates that for finite nuclei there are no undue difficulties in tracing the energy levels, as Z increases, to the value Z_{cr} where they merge with the negative energy continuum. At this threshold theory predicts (Mu 72, Mu 72b, Ze 72) that the bound state ceases to exist as it joins the negative energy continuum, but instead develops into a resonance shared over the continuum states (Figure 1.4). The width of this resonance grows with increasing penetration into the continuum as Z increases beyond Z_{cr}. The appearance of this resonance reflects the crucial aspect of the physics since it marks a decaying state and thus an instability.

The physical process that follows is suggested by considering the smallest energy required to create an electron–positron pair in the vicinity of such an overcritically charged bare nucleus. Below the critical charge energy must be fed into the atomic system to create pairs. Figure 1.4 illustrates that the energy required is the difference between $2mc^2$ and the binding energy of the $1s$ state. It decreases as Z increases, and eventually at the critical charge it changes sign. Therefore, this boundary marks the threshold where it becomes energetically favorable to create an electron–positron pair provided the $1s$ state is unoccupied by an electron. The spontaneous emission of such a pair is not forbidden by any conservation law, and thus a free positron escapes and an electron remains bound under supercritical binding conditions ($>2mc^2$).

There is, therefore, an essential difference between a hydrogenlike atom with $Z < Z_{cr}$ and a superheavy atom with $Z > Z_{cr}$. For, as depicted in Figure 1.4, ionization of an electron from the former leads to a stable ionized atom, while permanent ionization of the inner-shell electrons of the superheavy atom with $Z > Z_{cr}$ is not a stable state. After K-shell ionization, an atom with $Z > 173$ will always spontaneously surround itself with two electrons. In the process positrons are emitted with a rather well-defined kinetic energy centered at a value equal to the excess of the binding energy beyond $2mc^2$. This two-electron state becomes the lowest energetically stable state for this atom. Calculations show that for $Z \simeq 180$ it forms on a time scale of 10^{-19} s, corresponding to the width of the resonance state of $\sim$10 keV (Mu 72, Mu 72b, Ze 72).

Of course as the central charge is increased arbitrarily—so that successive bound states join the negative energy continuum—successive phase transitions occur in which the vacuum increases its negative charge, as illustrated schematically in Figure 1.4. The vacuum thus breaks down in supercritical fields and the new ground state is necessarily charged ["charged vacuum" (Fu 73, Ra 74, Mu 73b)]. It is stabilized by the Pauli principle.

Calculations of the interaction of the electron with the radiation field do not seem to alter these predictions, as discussed in detail by Reinhardt and

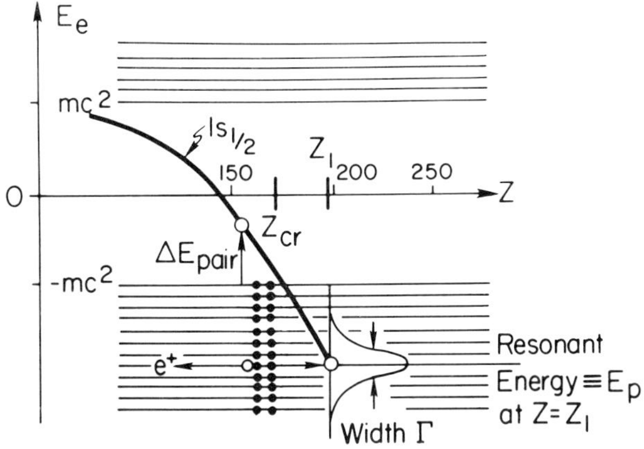

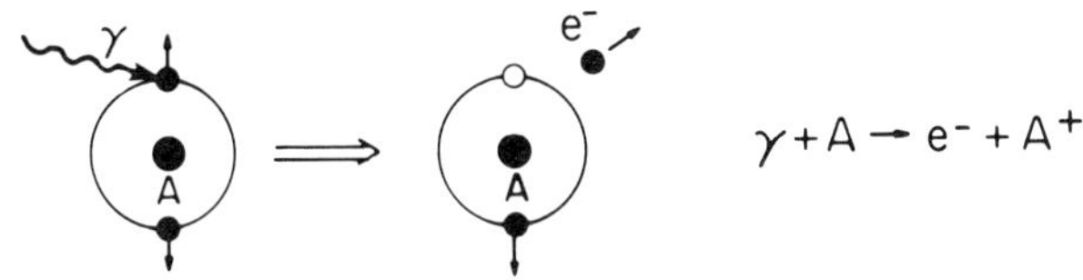

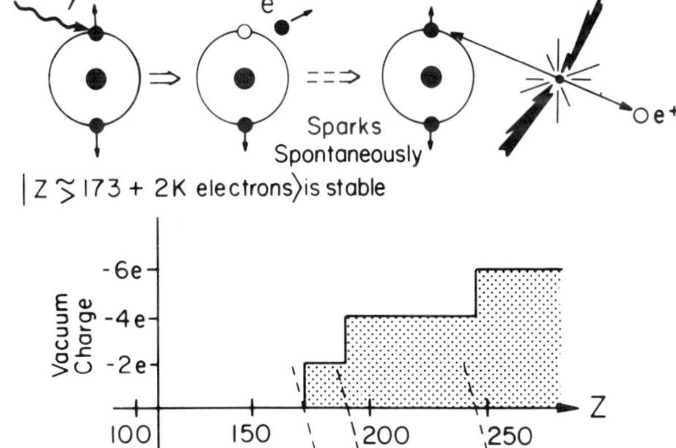

Figure 1.4. A schematic illustration of the diving of the 1*s* atomic state with increasing Z and its transformation into a decaying resonance state when the binding energy exceeds $2mc^2$. For $Z > Z_{cr}$ the ionized $1s_{1/2}$ state is unstable to spontaneous positron emission. As successive vacant bound states join the negative energy continuum, the vacuum increases its negative charge.

Greiner (Re 85). They show that the self-energy correction decreases the binding by only ~11 keV (So 82) for $Z = 170$, while vacuum polarization enhances the binding of the 1*s* electron by a similar amount (So 74, Gy 74). Spontaneous positron emission is predicted to be, therefore, a unique signature for establishing the change in the ground state of QED as the supercritical field threshold is passed. As previously mentioned, it is the search for this process that has motivated much of the experimental interest in ion–atom collisions of superheavy systems since stable nuclei with $Z > 173$ are not provided by nature.

It can be readily visualized how the formation of superheavy quasimolecules can simulate the conditions for observing spontaneous positron emission. If a vacancy is produced in the $1s\sigma$ molecular orbital by ionization at a suitable prior time, then spontaneous positron emission can occur during the fraction of the collision time when the $1s\sigma$ binding exceeds $2mc^2$ at the quasiatomic limit. Therefore it would seem that our inability, presently, to produce a true superheavy nucleus with a sufficiently large atomic number can be circumvented by probing into the electron structure of heavy-ion collision systems. But, unfortunately, the quasiatom endures only for a fleeting moment. The duration of this moment turns out to be a critical parameter in the hunt for spontaneous positron emission since the time evolution of the collision influences both the spontaneous and other positron production processes. This important consideration is discussed in more detail in Section 5.

It bears noting that naively the collision of two massive uranium atoms with GeV energies would not be expected to provide the most accommodating conditions to yield the detailed information needed for exploring fundamental aspects of atomic structure and QED. On the scale usually associated with the energy and momentum transfer encountered in classical atomic phenomena, such encounters have to be considered very violent events. There is a deep interpenetration of the electronic structures of the target and projectile atoms, accompanied by extensive electronic excitations. Copious emission of x rays and electrons as well as positron creation testify to the strong perturbation of the electron–positron field. Nuclear surfaces may also overlap leading to nuclear reactions and the additional emission of gamma-rays, internal conversion electrons and positrons, x rays and nuclear constituents. The emitted spectrum of radiations is thus quite complex. The salient property that makes these collisions tractable experimentally for obtaining information on the quasiatom is the focusing by relativistic effects of the interesting processes around the nuclear core of the quasiatom at the distance of closest approach. As noted above the reason behind this focusing is the rapid spatial collapse of the inner-shell wave functions accompanied by the rapid increase in binding energy, both of which are characteristic of the superheavy systems.

1.6. Recent Experiments

The experimental effort to study quasimolecular phenomena has evolved over a decade of interesting measurements. Particularly since April 1976, when the first uranium beams were accelerated at the UNILAC in Darmstadt to energies sufficient to reach the Coulomb barrier for the heaviest nuclear species, a large quantity of data have been accumulated which provide evidence for the formation of quasimolecules in superheavy collision systems and information on the behavior of their electronic structure. The questions associated with the search for the decay of the vacuum and, more generally, with the physics related to positron creation processes in heavy-ion collisions are being addressed by several experiments. Although the experimental picture pertinent to answering these questions is yet to be completed, a level of achievement has been reached which indicates that interesting results may not only be emerging on the electrodynamic aspects of the questions but also on nuclear physics. A particularly exciting recent development is the observation of narrow peaks in the positron spectra from U + Cm, U + U, Th + Cm, and U + Th collisions near the Coulomb barrier. The origin of these peaks cannot be identified with the short collision times associated with dynamic mechanisms of positron production involving Coulomb trajectories only. The possibility that these peaks may be signatures for the spontaneous decay of the vacuum following the formation of metastable superheavy nuclear complexes is one of the interesting options under investigation.

1.7. Organization of Discussion

To discuss these developments together with other related investigations over the past decade, this chapter has been organized along the following sequence of topics:

Since the quasimolecular concept is an essential ingredient throughout our discussion, it is treated first in Section 2. We have also mentioned that the experimental approaches to investigating superheavy quasiatoms involve the study of the dynamic mechanisms leading to the excitation and deexcitation of the electronic structure. Particularly, access to information on strong-binding effects in quasimolecules all start with the ionization process involving the inner shells. While the ionization process produces delta-electron emission, the deexcitation of the vacancy can occur via the emission of characteristic x rays, molecular orbital x rays, and positrons. All of these provide experimental probes into the inner-shell structure. Therefore, we will consider in successive sections the experiments involving the detection of characteristic x rays, molecular orbital x rays, delta electrons, and positrons. Section 5 deals more specifically with the recent measurements to identify spontaneous positron emission. Here we will discuss in some detail the experimental approaches, the early measurements, the most recent detection of peak structures in the positron spectra, and consider the origin of these peaks. The chapter concludes with an outlook for further investigations and challenges for nuclear physics.

2. The Quasimolecule

In many respects the electronic quasimolecule is the unifying concept of our discussion. It forms the basis for exploring much of the data on x-ray emission in the lighter collision systems, and is the unique source of information on the spectroscopy of superheavy atoms; it is the only possible vehicle presently available for exploring the QED of strong fields experimentally. We have already mentioned the inception of the idea by Coates (Co 34), its revival in 1965 by Fano and Lichten (Fa 65), and early work by Armbruster (Ar 62, Ar 64) and Specht (Sp 65). Other experimental contributions which have developed the molecular description of atomic collisions are considered throughout this article. With regard to those most closely related to the evolution towards investigations of superheavy collision systems and the QED of strong fields, we may mention the first convincing observation and identification of MO *K* x rays by Saris *et al.* in 1972 (Sa 72), the subsequent evidence for quasimolecule formation through the observation of MO x rays from the *M* (Mo 72, Ju 74) and *K* (Ma 73, Me 73b, Da 74) shells, the discovery of the emission anisotropy for *K* MO x-rays (Gr 74) and the use of this anisotropy to identify the emitter as the center-of-mass system (Me 75, Fo 76, Vi 78), the first observation of *K* MO x rays from superheavy systems (Me 79, St 80), the isolation of selected MO transitions (Ob 80), the establishment of large *K*-vacancy production probabilities and their concentration at small internuclear separations (Gr 77, Be 79b), and the use of *K*-vacancy production to demonstrate the large binding energies in superheavy quasiatoms (Gr 77, Li 80c, Bo 80). Of course, the suggestion, independently by Rafelski, Fulcher, and Greiner (Ra 71) and Gershtein and Zel'dovich (Ge 70), that quasimolecule formation may provide the suitable environment for observing the instability of the vacuum in supercritical fields marked a milestone in this field.

Much of the progress in the quantitative development of the quasimolecular model is due to the availability of x-ray data from heavy-ion collisions. For extensive discussions of the results of experiments performed prior to 1975 we refer the reader to the review articles by Meyerhof and Taulberg (Me 77) or Mokler and Folkmann (Mo 78). Although some of the salient features of these earlier data will be mentioned in Section 3, the discussion in this chapter focuses principally on results of experiments performed after 1975 with emphasis on most recent results. In this section, particularly, the principal features of the quasimolecular description of atomic collisions are summarized. Aspects of relevant quasimolecular correlation diagrams for light and heavy encounters are discussed and contrasted. Excitation processes in the quasimolecule are then treated with discussion of coupling mechanisms among bound and continuum states in relativistic and nonrelativistic situations. This leads us to an interesting scaling law for excitation in the relativistic quasimolecule which becomes a useful tool for understanding the principal underlying physics associated with atomic excitations in superheavy collision systems.

2.1. General Features of the Quasimolecular Model

The viability of the quasimolecular description of high-energy, heavy-ion collisions rests centrally on the condition that the ratio of the inner-shell electron velocities to that of the moving projectile is much greater than unity. A qualitative measure of the velocity and distance scales involved in such collisions is obtained by comparing the distance of closest internuclear separation, R_0, the mean radius, r_K, of the atomic K shell, the projectile velocity, v_p, and the velocity of the bound K-shell electron v_k. For 70-MeV, Ni + Ni collisions $R_0 = 30$ fm, $r_K = 3000$ fm, $v_p/c = 0.05$, and $v_K/c = 0.2$. For 1-GeV Pb + Pb collisions, $R_0 = 20$ fm, $r_K = 800$ fm, $v_p/c = 0.1$, and $v_K/c = 0.6$. These numbers indicate that collisions deep within the atomic K shell are possible using projectiles whose motion is slow relative to that of the K-shell electrons. Such conditions are favorable for the atomic electrons to adjust their orbits to the combined projectile–target charge, leading to the formation of quasimolecular orbitals.

A key simplification to calculating the experimental cross sections and excitation probabilities in such adiabatic collisions is the use of the Born–Oppenheimer approximation (Ke 73) where the nuclear and atomic electron motions are decoupled and the total wave function is written as a product of the nuclear and electron components:

$$\psi_i(\mathbf{R}, \mathbf{r}_i, t) = \psi_N(\mathbf{R}, t)\, \psi_e(\mathbf{r}_i, t) \tag{2.1}$$

Here $\mathbf{R}$ is the internuclear separation, $\mathbf{r}_i$ is the coordinate of the ith electron, and t is the time. For nearly all the collisions of interest to this discussion, the nuclear motion is well described by the classical motion of Rutherford scattering since the Bohr–Sommerfeld parameter (Bo 48)

$$\eta = \frac{Z_p Z_T e^2}{\hbar v} \tag{2.2}$$

is much greater than unity. Under such conditions, a classical impact parameter can be used to describe the nuclear scattering trajectories and it becomes convenient to employ the impact parameter formalism in calculations. For small impact parameters an unscreened point nuclear potential is often adequate, and the relationship between the classical impact parameter, b, and the center-of-mass scattering angle, $\theta_{\text{c.m.}}$, is given by the familiar Rutherford relationship:

$$b = \frac{Z_p Z_T e^2}{2E_{\text{c.m.}}} \cot(\theta_{\text{c.m.}}/2) \tag{2.3}$$

where Z_p and Z_T are the projectile and target nuclear charges, and $E_{\text{c.m.}}$ the center-of-mass collision energy. For large impact parameters, particularly for higher Z collisions, the influence of electron screening on the nuclear trajectories must be taken into account (Co 80). In many cases this problem is satisfactorily treated with the use of a screened internuclear potential of the kind suggested by Bohr (Bo 48):

$$V(R) = \frac{Z_p Z_T e^2}{R} e^{-R/a} \tag{2.4}$$

with

$$a = 0.88\, a_0/(Z_p^{2/3} + Z_T^{2/3})^{1/2} \tag{2.5}$$

where $a_0 = 5.29 \times 10^{-9}$ cm is the first Bohr radius for hydrogen.

The electronic motion has been treated in various approximation schemes. In some cases of inner-shell excitation phenomena, the effects of electron–electron interactions can be safely neglected and the electronic wave function is well described by the solutions to the one-electron two-center Schrödinger or Dirac equations. The influence of many electron effects are calculated using Hartree–Fock or variational techniques (Br 73b, Ta 75). With the time development of the electron wave function given by the solution to the equation for the electronic Hamiltonian H_e,

$$i\hbar \frac{\partial \psi_e(\mathbf{r}, t)}{\partial t} = H_e \psi_e(\mathbf{r}, t) \tag{2.6}$$

the adiabatic tracking of the electrons with the nuclear motion in the molecular orbital picture can be used to expand the electron wave function $\psi_e(\mathbf{r}, t)$ in a complete set of molecular orbital basis functions:

$$\psi_e(\mathbf{r}, t) = \sum_k a_k(t)\, \phi_k[\mathbf{r}, R(t)] \exp\left[(-i/\hbar)\int_{-\infty}^{t} dt' \langle \phi_k | H_e | \phi_k \rangle\right] \tag{2.7}$$

For nonrelativistic cases, ϕ_k solve the two-centered Schrödinger equation:

$$\left(\frac{-\nabla_r^2}{2m} - \frac{Z_p e^2}{|\mathbf{r} - \mathbf{R}/2|} - \frac{Z_T e^2}{|\mathbf{r} + \mathbf{R}/2|}\right) \phi_k(\mathbf{r}, \mathbf{R}) = E_k \phi_k(\mathbf{r}, \mathbf{R}) \tag{2.8}$$

where the origin has been chosen halfway between the projectile and target nuclei. For high-Z collisions, where relativistic effects become important, ϕ_k solve the two-center Dirac equation:

$$\left[\boldsymbol{\alpha} \cdot \mathbf{p} + \beta m - \frac{Z_p e^2}{|\mathbf{r} - \mathbf{R}/2|} - \frac{Z_T e^2}{|\mathbf{r} + \mathbf{R}/2|}\right] \phi_k(\mathbf{r}, \mathbf{R}) = E_k \phi_k(\mathbf{r}, \mathbf{R}) \tag{2.9}$$

2.1.1. Molecular Orbital Diagrams

The solutions to Equation (2.8) or (2.9) generate sets of two-center electronic eigenfunctions for every internuclear separation R. The evolution of each wave function ϕ_k as a function of R maps the development of the atomic states in the collision constructed quasimolecule. Particularly the variation of the molecular binding energies, E_k, with changes in R is conveniently displayed in molecular orbital diagrams. (An example has already appeared in Figure 1.3.) In evolving from the separated atom limit ($R = \infty$) to the united atom limit ($R = 0$), most orbitals will increase in binding as R decreases, but some orbitals will decrease in binding for decreasing R. The topography of the R-dependent molecular orbital levels, and their proximity to unoccupied neighboring states and to the continuum, provide important information on the probability of electronic transitions during the course of the collision.

Without solving Equations (2.8) and (2.9) explicitly, one can attempt to determine the correlation between the limiting separated and united atom cases, i.e., to ascertain which separated atom levels correlate to which united atom states. Such correlation diagrams are used in molecular physics to study the structure of diatomic molecules (He 65). In those systems, adiabatic molecular orbitals calculated for fixed nuclei of separation R are good basis states for the molecular Hamiltonian. However, for high-energy atomic collisions, where the interesting region of small internuclear distances is attained using relatively high-velocity projectiles, dynamic effects become an important consideration.

A principal result of incorporating such dynamic effects into the correlation diagram is the changed behavior of the molecular orbitals at avoided level crossings (Ke 73). The Wigner–Neumann noncrossing rule states that, in general, two identical energy eigenvalues cannot be achieved by varying one real parameter (Vo 29). In the nonrelativistic two-center

Coulomb problem, an additional constant of motion exists and the Wigner–Neumann rule is inapplicable (Co 67, He 72, Ge 61). Therefore, for pointlike nuclei the one-electron two-centered Schrödinger equation exhibits solutions where molecular orbitals of like symmetry may become degenerate in energy and cross each other at some internuclear separations. Where departures from the pure, nonrelativistic, two-center point charge potential occur, such as in many-electron atoms, the Wigner–Neumann rule prevents orbitals of like symmetry from crossing (Vo 29). For example, in symmetric quasimolecules (same target and projectile), parity is a good quantum number and states of the same parity and (nonrelativistically) orbital angular momentum projection along the internuclear axis cannot cross. This avoidance of crossings can be qualitatively viewed as an effective repulsion between the two states of like symmetry at the crossing radius. As a result of this repulsion electronic charge is redistributed, the wave functions change, and the binding energies of the repelled orbitals veer away from each other.

If the energy separation between repelled states is small and the time spent by the quasimolecule at the crossing radius short, charge redistribution may not have sufficient time to occur and the states may cross. Viewed in another way, transitions can occur at the avoided crossing between the two states (Ze 32, La 32). In cases where this transition probability is large, adiabatic molecular orbitals fail to appropriately describe the evolution of the many-electron collision system, and dynamically modified orbitals (i.e., diabatic orbitals), are necessary.

Work on the use of such dynamically modified orbitals was pioneered by Fano, Lichten, and Barat (Li 63, Li 67, Li 76, Fa 65, Ba 72b). As a result of their studies, these authors proposed that in a collision the electron wave functions at avoided crossings will evolve along the level whose molecular wave function preserves the number of radial nodes in the initial wave function. Therefore, a set of diabatic levels may be defined with each level conserving the number of radial nodes in its wave function as R changes from infinity to zero. A diabatic correlation diagram may, therefore, be generated showing the molecular connections between the separated and united atom limits.

Figure 2.1 illustrates the diabatic correlation diagram for a nonrelativistic, slightly asymmetric collision system (Ba 72b). The subscript "a" denotes the heavier and "b" the lighter of the two collision partners. The diabatic molecular states are labeled (nonrelativistically) with the principal quantum number and orbital angular momentum of the united atom state. The σ (solid lines), π (dashed lines), and δ (dotted lines) denote the projection of the orbital angular momentum along the internuclear axis ($\sigma = 0, \pi = 1, \delta = 2$, etc.). The figure illustrates the connection of the $1s\sigma$ molecular orbital to the K shell of the heavier collision partner, and of the $2p\sigma$ orbital to that of the lighter collision partner. There are numerous

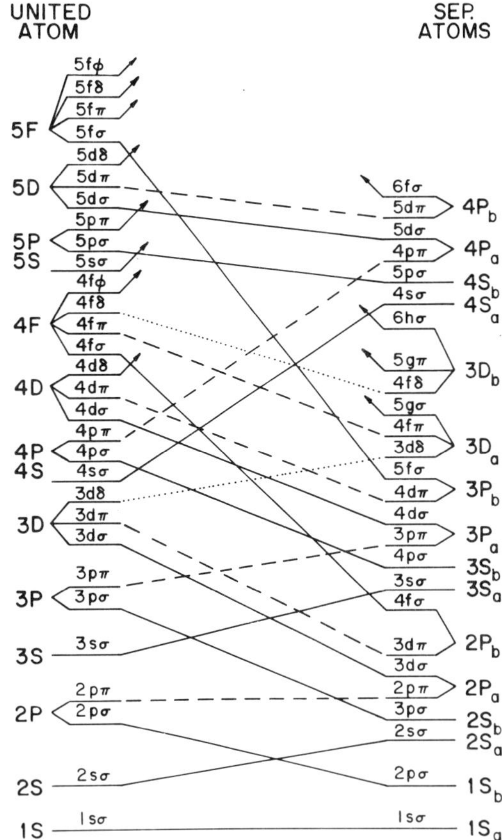

Figure 2.1. Quasimolecular correlation diagram for a slightly asymmetric collision system. The K shell of the heavy collision partner (subscript "a" in the figure) connects to the $1s\sigma$ orbital and that of the lighter collision partner (subscript "b") to the $2p\sigma$ molecular level (Ba 72b).

crossings between diabatic states of like symmetry (i.e., $2p\sigma$ and $2s\sigma$) which would adiabatically repel each other at avoided crossings.

Clearly, calculations of inner-shell excitation in heavy-ion encounters within the molecular orbital model require detailed knowledge of the two-center electron wave functions versus R, including many-electron effects and screening. The solution to the time-independent Schrödinger equation for such modified potentials will yield adiabatic basis states exhibiting avoided level crossings (Br 73b, Ta 75, Ei 76). An example of such adiabatic molecular orbitals is given in Figure 2.2 for the Ne + O quasimolecule (Ta 75). The adiabatic states are labeled with their orbital angular momentum projection along the internuclear axis. In this case there is an avoided level crossing between the second and third σ states (2σ and 3σ) at small internuclear separations.

Figure 2.3 plots the adiabatic electronic 2σ and 3σ wave functions along the internuclear axis in the vicinity of the crossing radius $R_C = 0.0575$ a.u.

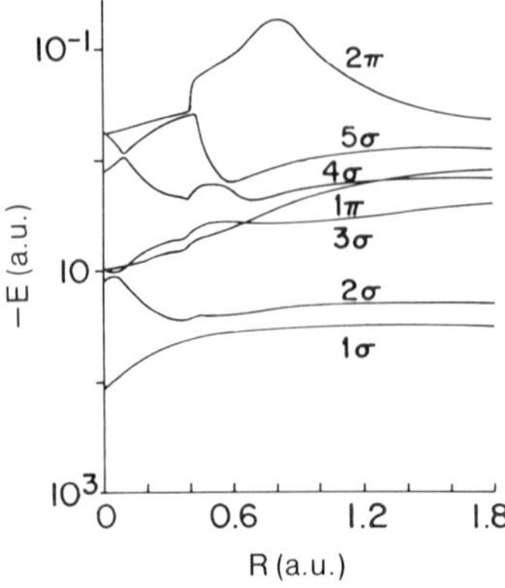

Figure 2.2. Quantitative calculation of the adiabatic molecular-orbital binding energies versus internuclear separation R for the Ne + O quasimolecule (Ta 75).

(Ta 75). At $R = 0.055$, the 2σ wave function exhibits "s"-like character and the 3σ "p"-like properties. At $R = 0.060$, the 2σ character has changed to "p"-like and the 3σ to "s"-like. This rapid change in the wave functions with R causes large couplings between them at the avoided crossing. The diabatic states would simply map 2σ for $R < R_C$ into 3σ for $R > R_C$ ($2s\sigma$ diabatic molecular orbital) and 3σ for $R < R_C$ into 2σ for $R > R_C$ ($2p\sigma$ diabatic molecular orbital). The diabatic orbitals are labeled with the principal quantum number and orbital angular momentum of the united atom level to which they correlate. The resulting diabatic molecular level diagram is shown in Figure 2.4 (Br 75). The diabatic $2p\sigma$ and $2s\sigma$ states cross at $R = R_C$.

It must be emphasized that the choice between adiabatic and diabatic orbitals is a choice of basis states in which the true electron wave function is to be expanded [equation (2.7)]. In principle, any complete set will do. However, in practical calculations only truncated sets of basis states can be employed. More importantly, the physics of the collision problem and the extent to which the inner-shell electrons are "really" bound in molecular

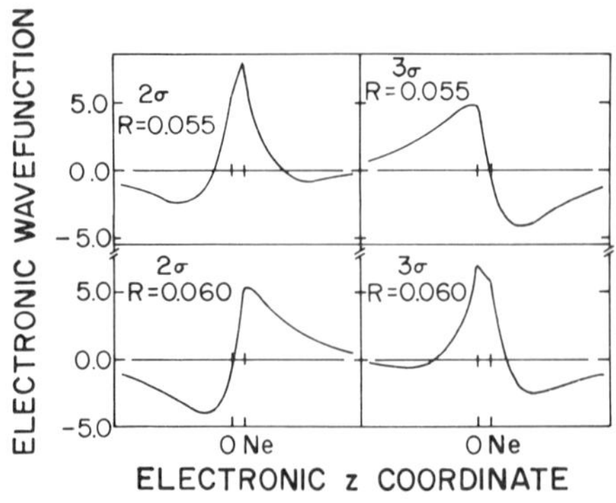

Figure 2.3. Radial adiabatic molecular 2σ and 3σ wave functions along the internuclear axis in the vicinity of a crossing radius for the Ne + O quasimolecule (Ta 75).

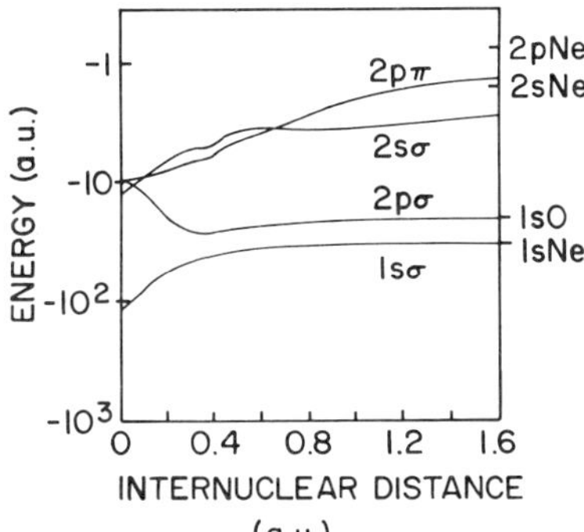

Figure 2.4. Diabatized molecular Ne + O orbitals (Ta 75).

orbitals depend on how successfully measurements of various inner-shell phenomena can be calculated with truncated molecular orbital bases. The overwhelming success of the molecular orbital model lies in its ability to describe wide ranges of measured inner-shell excitation phenomena using only a small number of molecular orbital states.

Discussion up to this point has been limited to essentially nonrelativistic cases. For high-Z collisions the two-centered Dirac equation must be solved. A more detailed example than that given in Figure 1.3 of the adiabatic energy-level diagram calculated with the self-consistent relativistic Dirac–Fock method is shown in Figure 2.5 (Se 81). The molecular levels are labeled with the principal quantum number, orbital angular momentum, and total angular momentum ($\mathbf{j} = \boldsymbol{l} + \mathbf{s}$) of the united atomic orbital to which they correlate, as well as with the projection m_j of j along the internuclear

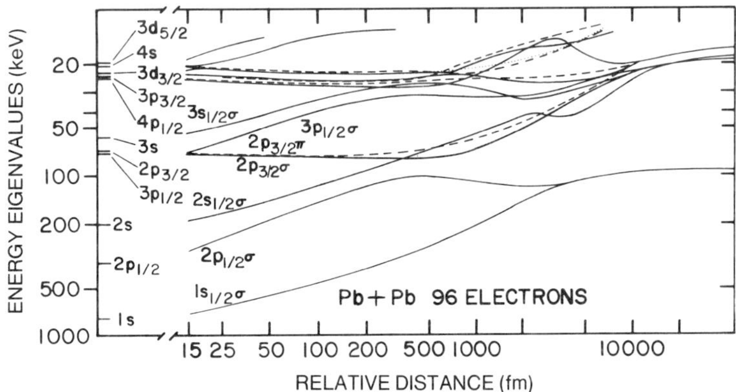

Figure 2.5. Relativistic Dirac–Fock calculation of molecular orbitals in the Pb + Pb quasimolecule calculated for 96 electrons. The left-hand labels indicate the united atom state to which the orbitals correlate (Se 81).

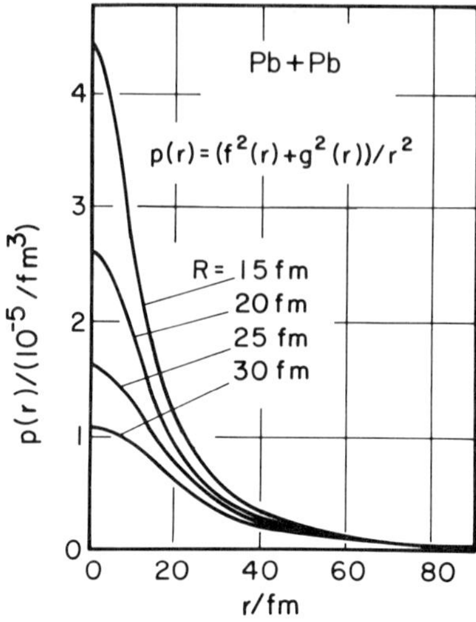

Figure 2.6. Electron density distribution for the $1s\sigma$ state in the Pb + Pb quasimolecule at various R (So 79).

axis. The notation $2p_{1/2}\sigma$, for example, indicates $j = 1/2$, $m_j = 1/2$. σ, π, and δ now correspond to $m_j = 1/2$, $3/2$, and $5/2$, respectively.

In the introduction it was pointed out that a most important feature of such superheavy collision systems is simply the magnitude of the binding energy that can be achieved. The other interesting feature noted was the strong localization of the highly relativistic electron wave functions near the nucleus and the important role played by the finite nuclear size due to the close overlap of atomic and nuclear dimensions. But, in addition, it was also emphasized that key features of the high-Z quasimolecules are the rapid changes of both the binding energy and electron density at the nucleus with decreasing internuclear separation. For example, Figure 2.6 indicates how the electron density grows with decreasing R, in fact, increasing as $1/R^2$, and Figure 2.5 displays how the innermost orbitals continue to change energy at increasingly accelerated rates as R decreases. These properties lead to important consequences for the physics of these collisions, producing large ionization of the inner shells at small internuclear separations and through these properties facilitate experimental study of the highly bound state wave functions. However, as will be discussed, the fact that ionization occurs so readily implies that dynamic effects on the molecular wave functions are large, eventually limiting the spectroscopic information that can be derived from measurements on these systems.

2.2. Coupling Mechanisms

The time development of the electron wave function couples quasimolecular states to each other and to the continuum. Although diabatizing the adiabatic orbitals eliminates or minimizes spurious couplings due to level crossings, large couplings among diabatic states will nevertheless exist, producing transitions among these orbitals and between bound and continuum levels. Treatment of these couplings originates with the substitution of the molecular basis state expansion for the electron wave function [Equation (2.7)] into the time-dependent Schrödinger or Dirac equation [Equation (2.6)] leading to the following infinite set of coupled-channel equations for the quasimolecular orbital occupation amplitudes (Ke 73):

$$\begin{aligned} i\hbar \dot{a}_m(t) = & \sum_{k\neq m} a_k(t)\langle\phi_m| H_e |\phi_k\rangle \\ & \times \exp\left[-(i/\hbar)\int_{-\infty}^{t} dt'(\langle\phi_k| H_e |\phi_k\rangle - \langle\phi_m| H_e |\phi_m\rangle)\right] \\ & - i\hbar \sum_{k\neq m} a_k(t)\, \dot{R}\langle\phi_m| \partial/\partial R |\phi_k\rangle \\ & \times \exp\left[-(i/\hbar)\int_{-\infty}^{t} dt'(\langle\phi_k| H_e |\phi_k\rangle - \langle\phi_m| H_e |\phi_m\rangle)\right] \\ & + \sum_{k\neq m} a_k(t)\langle\phi_m| \boldsymbol{\omega}\cdot\mathbf{j} |\phi_k\rangle \\ & \times \exp\left[-(i/\hbar)\int_{-\infty}^{t} dt'(\langle\phi_k| H_e |\phi_k\rangle - \langle\phi_m| H_e |\phi_m\rangle)\right] \end{aligned} \quad (2.10)$$

The three terms represent different mechanisms leading to transitions among the bound and continuum states.

The first is a potential coupling term. It is nonzero when the expansion basis does not exactly diagonalize the electronic Hamiltonian. If, for example, the basis states are chosen as one-electron two-center eigenfunctions, the actual Hamiltonian, H_e, will include screening corrections with, in general, nonzero matrix elements between the unscreened molecular basis states. For most cases of interest to inner-shell excitation in energetic atomic collisions, potential coupling between molecular basis states is small compared to couplings represented by the other two terms. A notable exception to this statement is the case of spontaneous positron emission, which, as we will see in Section 5, can be viewed as originating from a potential coupling matrix element.

The second and third terms in Equation (2.10) are dynamic couplings originating from the operator,

$$\frac{\partial}{\partial t} = \dot{R}\frac{\partial}{\partial R} + \dot{\theta}\frac{\partial}{\partial \theta} \quad (2.11)$$

The radial coupling matrix element involving $\dot{R}\,\partial/\partial R$ couples molecular states of the same spin, orbital angular momentum projection along the internuclear axis, and (for symmetric systems) parity. In relativistic systems where m_j, the projection of the total (spin + orbital) angular momentum, is a good quantum number, states of the same m_j value are coupled. In lighter collision systems the radial coupling term is effective in coupling states at avoided crossings and in coupling the $2p\sigma$ and $1s\sigma$ orbitals in asymmetric quasimolecules at large internuclear separations. For such asymmetric systems, the $2p\sigma$ orbital connects to the K shell of the lighter and the $1s\sigma$ orbital to that of the heavier collision partner. Vacancies generated in the $2p\sigma$ orbital can radially couple down to the $1s\sigma$ state during the outgoing portion of the collision. This "vacancy sharing" process is a dominant mechanism for heavy collision partner $(1s\sigma)$ K-shell excitation in lighter systems (De 64, Ol 72, Me 73). However, for sufficiently asymmetric encounters the energy gap between the $1s\sigma$ and $2p\sigma$ levels increases, thereby reducing the radial coupling strength. For very asymmetric systems, direct radial coupling between the $1s\sigma$ level and the continuum eventually dominates heavy collision partner K-shell excitation.

For high-Z collisions, the rapid shrinking of the $1s_{1/2}\sigma$ radial wave function with decreasing internuclear separation R (see Figures 2.5 and 2.6) leads to dramatic increases in the strengths of the radial coupling matrix elements. Even in slightly asymmetric encounters, direct radial coupling of the $1s_{1/2}\sigma$ state to unoccupied higher-lying molecular orbitals and to the continuum is responsible for substantial heavy collision partner K-vacancy production (Re 85). $2p_{1/2}\sigma$ excitation also proceeds partially by radial coupling. As in the lighter encounters, vacancy sharing of these $2p_{1/2}\sigma$ holes dominates $1s_{1/2}\sigma$ excitation in symmetric and nearly asymmetric collision systems.

The third term in Equation (2.10) represents rotational coupling. This operator couples states of the same spin whose orbital angular momentum projections differ by one unit. Where parity is a good quantum number, only states of the same parity are coupled. For relativistic systems, states whose m_j values differ by one unit are coupled.

In lighter-Z symmetric or nearly symmetric systems, the $2p\sigma$ orbital becomes degenerate with the $2p\pi$ level in the united atom limit (see Figures 2.1 and 2.4). If the $2p\pi$ state is vacant prior to the collision, or if a vacancy is generated in this level during the early stages of the collision (or in a prior collision), this vacancy will strongly rotationally couple to the $2p\sigma$ orbital. This $2p\pi$–$2p\sigma$ coupling provides an "exit channel" for $2p\sigma$ electrons, thereby leading to vacancies in the K shell of the lighter collision partner (Li 67, Li 76). Studies have shown that rotational coupling followed by $2p\sigma$–$1s\sigma$ vacancy sharing is the dominant process leading to K-shell excitation in lighter adiabatic symmetric and asymmetric collisions (see

Section 3). For higher-Z encounters, the large relativistic splitting between the $2p_{1/2}\sigma$ and $2p_{3/2}\pi$ orbitals inhibits their rotational coupling (see Figure 2.5). $2p_{1/2}\sigma$ excitation in such systems is, therefore, a combination of radial and rotational coupling.

In Section 3, examples of the statements made in the preceding paragraphs will be given, and extensive discussion of the various coupling mechanisms leading to inner-shell excitation in heavy-ion encounters will be presented. At this time we illustrate some of the above remarks with Figures 2.7 and 2.8. Figure 2.7 graphs examples of the radial matrix elements $\langle Ns\sigma|\,\partial/\partial R\,|\,1s\sigma\rangle$ for $1s_{1/2}\sigma$ excitation to higher-lying bound states in the U + U quasimolecule (Be 76). For comparison purposes, a scaled nonrelativistic $\langle 2s\sigma|\,\partial/\partial R\,|\,1s\sigma\rangle$ matrix element is also shown. The figure illustrates how the relativistic properties of the superheavy quasimolecule lead to a dramatic increase in the magnitude of radial coupling, particularly at small internuclear separations.

Figure 2.8 compares and contrasts calculated energy distributions of electrons ionized into the continuum from the $1s\sigma$ and $2p\sigma$ states in $H^+ + H$ (Se 73) and in the $Z_{\text{united}} = 190$ (e.g., Am + Am) quasimolecules (So 79). The adiabaticity of these encounters is given qualitatively by the ratios of the projectile velocities to those of the separated atom K-shell electrons. In both collision systems, the differential cross sections for delta-electron emission decrease nearly exponentially with electron kinetic energy. Whereas the $2p\sigma$ cross sections exceed the $1s\sigma$ yields by roughly a factor of 500 in the $H^+ + H$ encounters, $2p_{1/2}\sigma$ excitation is only a factor of roughly 5 larger than $1s_{1/2}\sigma$ ionization in the superheavy quasimolecule. This contrasting

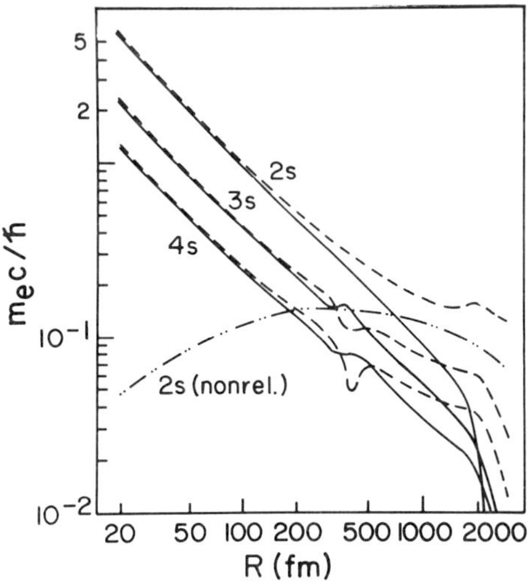

Figure 2.7. Radial coupling matrix elements between the $1s\sigma$ state and the $2s\sigma$, $3s\sigma$, and $4s\sigma$ orbitals in U + U versus internuclear separation. The dashed lines and solid lines are without and with translation factor corrections, respectively. The dash-dotted line shows a nonrelativistic $1s\sigma$–$2s\sigma$ matrix element scaled to the U + U case (Be 76).

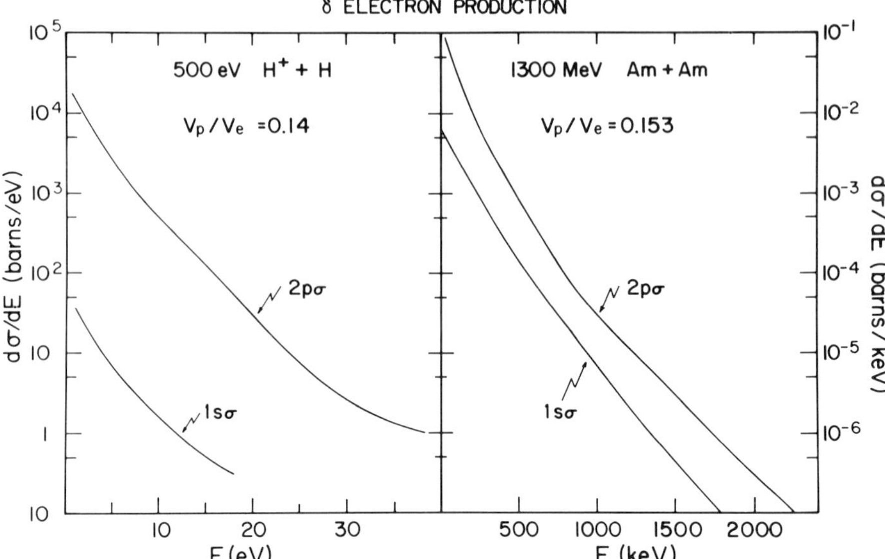

Figure 2.8. Comparison of δ electron spectra from $H^+ + H$ and Am + Am collisions. The $H^+ + H$ calculation includes both radial and rotational coupling, and the Am + Am involves radial coupling only (Se 73, So 79).

behavior illustrates the large increase in radial coupling of the $1s_{1/2}\sigma$ orbital to higher-lying states in the relativistic systems.

A general point which is intrinsic to the quasimolecular description of electron excitations concerns the role of translation factors. Whenever electrons are bound in molecular orbitals, they travel with a net translational velocity given by the velocity of the projectile–target center-of-mass systems in the laboratory. However, at large internuclear separations, each electron will connect to a state in either the projectile or target atom, and travel, in general, at velocities in the laboratory which are not equal to the center-of-mass velocity. Failure to make the proper theoretical connection between these two traveling frames of reference will introduce spurious couplings and excitations in molecular orbital calculations due to the intrinsic inability of the molecular wave functions to exhibit the correct asymptotic properties at large R. Since the transition from the "quasimolecule" to the "atom" is an adiabatic one, care must be taken in making the connection between these two regimes. Modifications to the molecular wave functions such that they exhibit the proper asymptotic dependence are achieved by translation factors. We will not consider this rather complicated topic here, but merely point out that its proper treatment is quite important in both low- and high-Z encounters (Va 81, He 81, Ki 81). The reader is referred to the theoretical treatment of high-energy atomic collisions in this series for further discussion (Re 85).

2.3. $1s\sigma$ Excitation in Very-Heavy Collision Systems

It has been previously noted that a central motivation for studying high-energy heavy-ion encounters is to determine the extent to which such collisions can be used to transiently form a united quasiatom where the K-shell binding energy exceeds $2mc^2$. In this connection, especially, the behavior of the $1s\sigma$ orbital in high-Z systems and its properties are of principal concern. We shall see that the excitation processes leading to x ray, delta-electron, and positron emission associated with this state provide the evidence for the formation of quasiatoms. Moreover, it turns out that these processes all exhibit common characteristic features. These features are compactly expressed in the form of a scaling law derived in the context of first-order time-dependent perturbation theory (Mu 78). Particularly the scaling law clarifies the relationship between the characteristic momentum transfer necessary for $1s\sigma$ excitation and the characteristic distance scale associated with such momentum transfer. Furthermore, comparison between coupled-channel calculations and the results of perturbation theory provides insight into the physical processes governing the behavior of this most deeply bound molecular orbital.

Studies have shown that for $1s\sigma$ excitation, potential and rotational coupling matrix elements can be neglected compared to that of radial coupling (Re 85). The first-order transition probability from the initial $1s\sigma$ orbital to some final state of energy E is given by

$$a(E) = -\int_{-\infty}^{\infty} dt\, \dot{R}(t)\langle\phi_E|\,\partial/\partial R\,|\phi_{1s\sigma}\rangle \exp\left(-(i/\hbar)\int_{-\infty}^{t} dt'\,[E_{1s\sigma}(R(t')) - E]\right) \tag{2.12}$$

where we have characterized the final state by its total relativistic energy E, and have taken $\langle\phi_{1s\sigma}|\,H_e\,|\phi_{1s\sigma}\rangle \equiv E_{1s\sigma}$, and $\langle\phi_E|\,H_e\,|\phi_E\rangle \equiv E$ [see Equation (2.10)]. The behavior of the excitation matrix element $\langle\phi_E|\,\partial/\partial R\,|\phi_{1s\sigma}\rangle$ is responsible for the scaling features of all $1s\sigma$ excitation phenomena. We turn now to discussion of its properties.

Perhaps the most significant aspects of $1s\sigma$ radial coupling matrix elements are their strong increases with decreasing internuclear separation in general (Re 85), and, in particular, their strong peaking near the point of closest approach of the two nuclei. This behavior provides the opportunity of introducing two key simplifications in evaluating the excitation probabilities with Equation (2.12).

One of these exploits the strong enhancement of the excitation probability at small internuclear separations to only use the monopole term of the two-center Coulomb potential, $V_{\mathrm{TC}}(\mathbf{r}, \mathbf{R})$, to approximate the binding energies of, and transition matrix elements between, quasimolecular states. Expressing V_{TC} in a multipole expansion (So 78b, Re 85),

$$V_{\mathrm{TC}}(\mathbf{r}, \mathbf{R}) = -\frac{Z_T e^2}{|\mathbf{r} - \mathbf{R}/2|} - \frac{Z_p e^2}{|\mathbf{r} + \mathbf{R}/2|} = \sum_{l=0}^{\infty} P_l(\cos\theta)\, V_l \qquad (2.13)$$

the monopole term is

$$\begin{aligned} V_0 &= -\frac{(Z_T + Z_p)\, e^2}{r}, \qquad r > R/2 \\ &= -\frac{2(Z_T + Z_p)\, e^2}{R}, \qquad r \leqslant R/2 \end{aligned} \qquad (2.14)$$

Retaining only the monopole term simplifies the wave functions and particularly provides access to continuum wave functions (Re 85) which are presently not available for the two-center problem.

The successful use of this approximation has been extremely important to the development of the theory for superheavy collision systems. In particular, calculations of coupled-channel effects have been facilitated by the simplified matrix elements given by the monopole potential.

We illustrate these latter remarks with the results of Figure 2.9, where exact solutions to the one-electron two-centered Dirac equation energy eigenvalues (solid lines), are compared with the approximate results of the monopole approximation for the Pb + Pb quasimolecule (So 78b). In the region of greatest importance to inner-shell excitation and positron emission ($R < 200$ fm), the difference in binding energy is less than 1%. The properties of the quasimolecular orbitals at small R are therefore mainly determined by the mean depth of the nuclear Coulomb potential.

Within the context of this approximation and assuming pointlike nuclei, the radial matrix element in Equation (2.12) can be written (Mu 78)

$$\langle \phi_E |\, \partial/\partial R \,| \phi_{1s\phi} \rangle = \frac{d(E, Z)}{R} \qquad (2.15)$$

where $d(E, Z)$ is a smoothly varying function of the combined nuclear charges $Z_T + Z_p = Z_u$ and the final state energy E. This $1/R$ radial dependence of the excitation matrix element, which reflects the $1/R^2$ dependence of the electron density of superheavy systems, is directly responsible for the scaling behavior of all $1s\sigma$ excitation phenomena. The amplitude for exciting an electron from the $1s\sigma$ state to the continuum then becomes (Mu 83)

$$a(E) = -d \int_{-\infty}^{\infty} dt\, \frac{\dot{R}(t)}{R(t)} \exp\left(\frac{i}{\hbar} \int_{-\infty}^{t} dt' \{E - E_{1s\sigma}[R(t')]\} \right) \qquad (2.16)$$

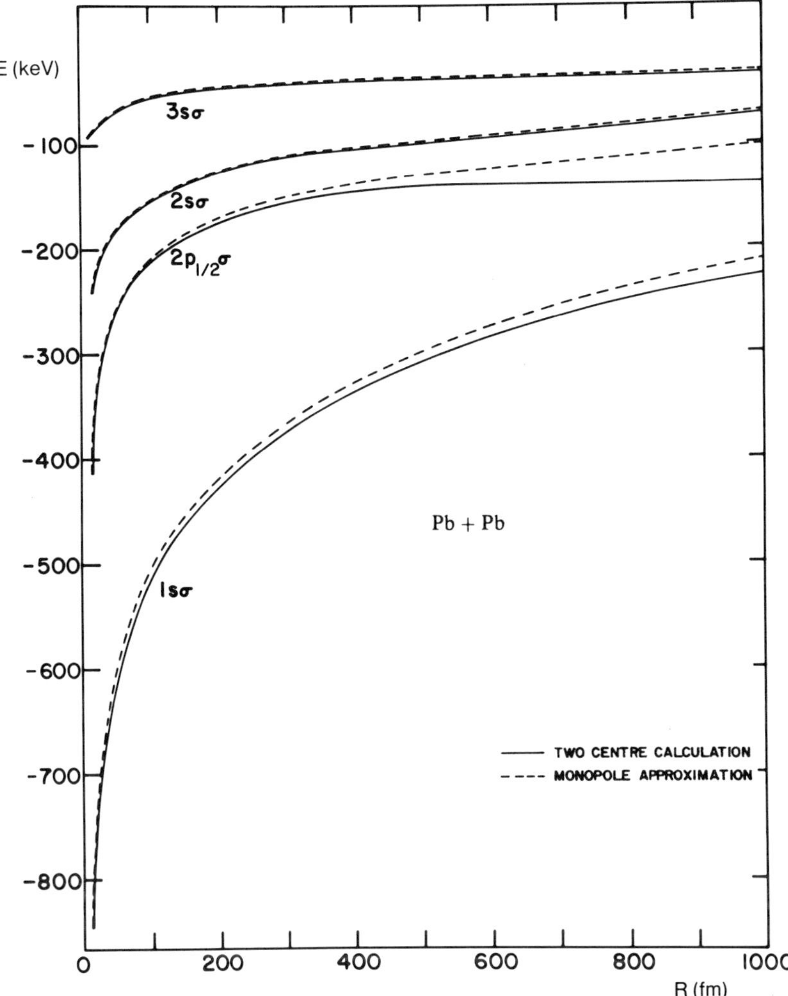

Figure 2.9. Comparison of four lowest σ state binding energies in the Pb + Pb quasimolecule versus R as calculated in the monopole approximation (dashed lines) and exact two-center models (solid lines) (So 78b).

which leads to an expression for the $1s\sigma$ excitation probability in heavy-ion Rutherford scattering encounters,

$$|a(E)|^2 = \pi^2 d_0^2(Z) \left(\frac{mc^2}{E}\right)^{\gamma(Z)} \exp - \left\{\frac{2\tau_0}{\hbar}\,[E - E_{1s\sigma}(R_0)]\right\} \tag{2.17}$$

where,

$$d_0^2(Z) = d^2(E, Z) \left(\frac{E}{mc^2}\right)^{\gamma} = \frac{D(Z)}{4\pi^2 mc^2} \tag{2.18}$$

represents the strength of the radial coupling matrix element. This decoupling of the E and Z dependences is valid for electrons ionized into states with less than 1 MeV kinetic energy. $D(Z)$ and $\gamma(Z)$ are both slowly varying functions of Z which may be evaluated by employing the parametrization (Mu 78, Mu 83)

$$\begin{aligned} D(Z) &= 1.2 + 0.054(Z-158) - 0.000015(Z-158)^3 \\ \gamma(Z) &= 2.08 - 0.024(Z-140) \end{aligned} \tag{2.19}$$

Taking advantage of the second key feature of the matrix elements mentioned above—their peaking at the distance of closest approach—and using the property that $\dot{R}/R$ assumes its largest value near this turning point, $E_{1s\sigma}(R_0)$ is evaluated at the distance of closest approach,

$$R_0 = a + (a^2 + b^2)^{1/2} \tag{2.20}$$

where

$$a = \frac{Z_p Z_T e^2}{2E_{\text{c.m.}}} \tag{2.21}$$

for impact parameter b and center-of-mass energy $E_{\text{c.m.}}$. The time,

$$\tau_0 = \frac{1}{v}\left[b + a\left(\pi - \tan^{-1}\frac{b}{a}\right)\right] \tag{2.22}$$

is characteristic for Rutherford scattering at incident projectile velocity v and represents a measure of the interval which contributes principally to the integral of Equation (2.16) (Ka 80, Mu 83).

Some of the physical consequences that follow from Equation (2.17) are illustrated by the probability for emitting delta electrons with kinetic energy E_δ (Mu 78, Mu 83),

$$\begin{aligned} \frac{dP(R_0, E_\delta)}{dE_\delta} = \frac{1}{2} D(Z) \frac{1}{mc^2}\left(\frac{mc^2}{mc^2 + E_\delta}\right)^{\gamma(Z)} \\ \times \exp - \left\{\frac{2\tau_0}{\hbar}\left[E_\delta + |E^B_{1s\sigma}(R_0)|\right]\right\} \end{aligned} \tag{2.23}$$

In deriving Equation (2.23), one uses $dP/dE = 2\,|a(E)|^2$, where the factor of 2 stems from the presence of two electrons in the $1s\sigma$ orbital. $E_\delta = E - mc^2$ and $E^B_{1s\sigma}(R_0) = E_{1s\sigma}(R_0) - mc^2$ are the delta-electron kinetic and $1s\sigma$ binding energies, respectively.

The magnitude of the delta-electron excitation probability of Equation (2.23) is governed by the matrix element and kinematic factors preceding the exponential term, and the shape of the electron kinetic energy distribution is exponentially decreasing with increasing energy transfer $\Delta E = E_\delta + |E^B_{1s\sigma}(R_0)|$. Such exponential dependence on energy transfer is typical of all Coulomb ionization phenomena in superheavy collision systems including positron emission, and reflects the kinematic relationship between the distance scale over which excitation takes place and the momentum transfer necessary to induce the excitation. This relationship is illustrated by evaluating Equation (2.23) under several simplifying assumptions. We first approximate (Bo 80, Mu 83)

$$\tau_0 \cong \frac{R_0}{v} \tag{2.24}$$

and note that

$$\hbar q_{\min}(R_0) = \frac{E_\delta + |E^B_{1s\sigma}(R_0)|}{v} \tag{2.25}$$

is simply the minimum momentum transfer necessary to ionize the electron to a final state of energy E_δ. Making these substitutions in Equation (2.23) yields

$$\frac{dP(R_0, q_{\min})}{dE_\delta} = C(Z, E_\delta) \exp\left[-2R_0 q_{\min}(R_0)\right] \tag{2.26}$$

Next we approximate $R_0 \cong b$ and define an average impact parameter,

$$\langle b(q_{\min})\rangle \equiv \frac{\displaystyle\int_0^\infty b \frac{dP(b, q_{\min})}{dE_\delta} b\, db}{\displaystyle\int_0^\infty \frac{dP(b, q_{\min})}{dE_\delta} b\, db} \tag{2.27}$$

Evaluating Equation (2.27) with Equation (2.26), leads to the result

$$\langle b\rangle q_{\min} \cong 1 \tag{2.28}$$

This result can be qualitatively compared to the general statement of the uncertainty principle, $\Delta b\, \Delta q_{\min} \geqslant 1$. Therefore, the existence of a conjugate relationship between a characteristic length and a characteristic momentum scale in Equation (2.28) can simply be viewed as a restatement of the uncertainty principle. However, the fact that the product of $\langle b\rangle$ and $q_{\min}$ approximately equals one is a property of the radial coupling matrix element itself. Bang and Hansteen have, in fact, shown that Equation (2.28) is a common feature of all $1s\sigma$ excitation phenomena across the entire Periodic Table (Ba 59, Ba 79).

This discussion will be elaborated in Sections 3 and 4 in connection with specific data on x-ray and delta-electron emission. In particular, with regard to the latter, we will see how the high-energy delta electrons reflect the high-momentum components of the initial state wave function. In this respect it becomes evident that the quasiatomic wave functions for the Br + Br system shown in Figure 1.3 are quite different from those for U + U. In considering the various data we will also see that the characteristic length scale associated with inner-shell atomic phenomena in very heavy-ion collisions is roughly equal to nuclear dimensions. Under such circumstances, strong couplings between nuclear and atomic degrees of freedom can be expected. These large overlaps between the united atom and its nucleus constitute a unique physical situation qualitatively different from that present in even the heaviest of the stable elements. It is precisely this unexplored domain of strongly overlapping nuclear and atomic degrees of freedom where new physical phenomena have been sought. In the following pages, we will develop this theme with discussions of available experimental data.

3. X-Ray Spectroscopy

3.1. Introduction

X-ray measurements are responsible for the bulk of experimental data on quasimolecular collisions. The cross sections and impact parameter dependences for x-ray emission reflect the properties of quasimolecule formation in heavy-ion collisions where vacancies generated in inner shells decay principally by x-ray emission. Furthermore, the availability of good-resolution, high-efficiency, solid-state x-ray detectors has made x-ray measurements rather straightforward. For the heaviest collision systems some data have been gathered on delta-electron emission, and extensive studies of positron emission continue to be pursued. However, in lighter collisions, nearly all information about the quasimolecule has been extracted from x-ray experiments.

From a historical perspective, the first data on quasimolecule formation were obtained using lighter collision systems. These studies served to solidify the two-center Schrödinger equation formulation of quasimolecular processes, and they have provided the essential framework within which heavy-system relativistic effects are understood. The manner in which inner-shell vacancy production processes scale with united atom Z_u has played a central role in our understanding of the quasimolecule. Only since the late 1970s have data from the heaviest collision systems been obtained. Therefore, most of our understanding of adiabatic heavy-ion encounters stems from extensive measurements on lighter systems. Studies of x-ray cross sections and impact parameter dependences, as a function of the projectile charge state and target thickness, have illustrated the effect of outer-shell vacancies on inner-shell processes and highlighted the importance of nonmolecular excitations in adiabatic encounters. In general, efforts to define as precisely as possible the initial and final states associated with excitation processes have led to improved quantitative understanding of quasimolecular encounters.

X-ray studies have also mapped the behavior of $1s\sigma$ excitation and, to a lesser extent, $2p_{1/2}\sigma$ vacancy production in the heaviest collision systems. K x-ray decays of these vacancies provide a clear and experimentally accessible signal of the time evolution of the most deeply bound quasimolecular states during the course of the collisions. In fact, x-ray studies were first to establish that inner-shell excitation was large in heavy collision systems and that it was concentrated at small impact parameters (Gr 77). Subsequent investigations of delta-electron (see Section 4) and positron (Section 5) emission deepened understanding of strong binding effects. Particularly for the positron measurements, where the existence of a vacancy in the $1s\sigma$ orbital at small internuclear separations is a prerequisite for spontaneous positron emission, aspects of $1s\sigma$ vacancy formation, clarified through x-ray studies, have yielded essential information.

In addition to characteristic x-ray detection, MO x-ray studies have provided complementary information on the physics of vacancy production and decay in heavy-ion encounters. Since these transitions take place during the course of the collision itself, they provide a signal which directly examines the structure of the combined projectile–target molecular system. Studies of collision broadening of the MO x-ray radiation have led to improved understanding of dynamic aspects of quasimolecule formation. Furthermore, the directional anisotropy of MO x-ray emission from nonrelativistic encounters has suggested a semiempirical scaling law relating the x-ray energies associated with a distinctive peak in the anisotropy to the binding energy of the united-atom K shell. Unfortunately, the directional anisotropy is not a prominent feature of relativistic quasimolecular systems, and a straightforward extension of the scaling behavior characteristic of lighter quasimolecules to the most interesting very-heavy-ion encounters does not appear feasible at this time.

Heavy-ion collisions lead to the production of other kinds of x-ray intensities of various, sometime exotic, origins. Although some of these sources of x-ray production provide little useful information on strong binding phenomena in quasimolecules, they are, nevertheless, interesting in their own right. In the Introduction we mentioned such additional sources of x-ray generation which include radiative electron capture, electron and nuclear bremsstrahlung, and two-electron–one-photon x-ray emission. All these x-ray processes are associated with high-energy heavy-ion collision phenomena and have provided interesting results for atomic physics.

In this x-ray section, we will discuss the various means by which x-ray processes provide information on reaction mechanisms active in heavy-ion encounters. In Section 3.2, some experimental aspects of x-ray detection are presented and discussed. In Section 3.3, x-ray processes in nonrelativistic collision systems are reviewed with initial discussion of characteristic K x-ray results, and subsequent presentation of MO x-ray data. Section 3.4 treats relativistic encounters, comparing and contrasting these results with findings at lower Z. Section 3.5 delineates other x-ray processes characteristic of high-energy atomic collisions.

3.2. Experimental Considerations

Typical x-ray spectra from adiabatic heavy-ion encounters are illustrated in Figure 3.1 where data from Ni + Ni collisions at 64.8 and 12.6 MeV are shown (Vi 77, Vi 78). These data were taken with an intrinsic germanium x-ray detector employing a 0.86-mm Al absorber. Visible at photon energies above the characteristic K x-ray lines of Ni are the continuum MO x rays which extend beyond the united-atom K x-ray transition energy of barium. The small peak visible at 15 keV is not due to pulse pile-up effects but, instead, to two-electron–one-photon transitions to a

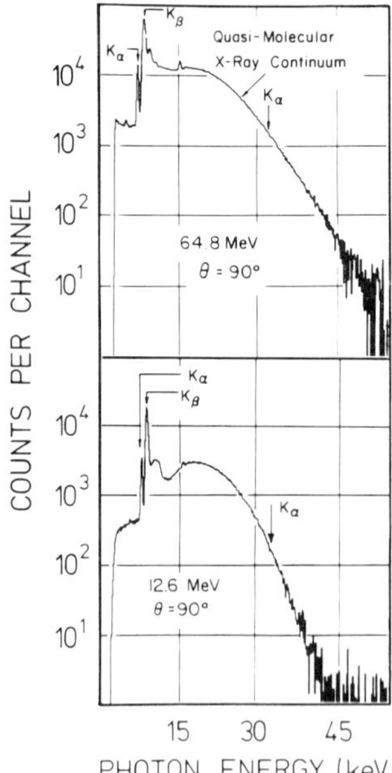

Figure 3.1. X-ray spectra from Ni + Ni collisions taken with an intrinsic germanium x-ray spectrometer. A thick Al absorber has been used (Vi 77, Vi 78).

doubly vacant Ni atomic K shell. These transitions will be discussed in Section 3.5 below.

The Al absorber serves to attenuate characteristic K x-ray radiation by a factor of roughly 5×10^5. Therefore, the spectral distributions shown span more than 10 orders of magnitude in intensity. The discrete K_α and K_β x-ray lines provide global information on K-vacancy generation during the course of these encounters. Comparisons between K_α and K_β strengths yield relative information on L- and M-shell vacancy configurations at the time of x-ray emission. The continuum transitions reflect properties of the molecular orbitals during the course of the collisions, and their dependence on incident projectile energy provides additional information on the role of dynamic effects on the quasimolecular levels.

From an experimental point of view, study of such spectra necessitates the use of a x-ray spectrometer with high resolution and detection efficiency throughout the x-ray energy range from several to roughly 100 keV. These requirements are well satisfied by planar intrinsic germanium or lithium drifted silicon detectors. In addition to good resolution and detection efficiency, the planar crystal shape of these x-ray detection systems provides fast, uniform pulse rise times resulting in good timing resolution in coincidence experiments. When combined with surface barrier or gas avalanche particle detectors, timing resolutions between 10 and 20 ns are achievable. The Si(Li) spectrometers generally provide better energy resolution than the Ge detectors ($\sim$150 eV at 5.9 keV versus $\sim$250 eV at 5.9 keV), but have full-energy peak efficiencies which decrease rapidly with increasing x-ray energy due to the strong Z dependence of the photoelectric absorption cross section. Therefore, Si(Li) spectrometers are used for measurements below roughly 30 keV, and Ge detectors for a broader range of x-ray energies up to 100 keV. Below one or two keV, absorption of x rays in beryllium entrance windows and/or detector dead layers in both systems limits their detection efficiencies.

In cases where only modest resolution but large detection efficiences are required, the relatively inexpensive NaI crystal can be used. However, at x-ray energies below 10 keV, the limited light output of these crystals results in poor energy and time resolution. For very low-energy x rays (<3 keV), gas proportional counters are used (Ri 80).

The data of Figure 3.1 were taken with an intrinsic germanium spectrometer 5 mm in thickness. Absorption due to the 0.86-mm Al filter was 50% at 20 keV, decreasing rapidly above this energy and increasing rapidly below. Without such absorption, pulse pile-up would contribute significantly to the continuum x-ray band. An example of x-ray data obtained from an asymmetric collision system using this same detector is shown in Figure 3.2 (Vi 77, Vi 78). Here a sandwich of 0.0843-mm Cu and 0.892-mm Al was employed, with the Al absorber preventing fluorescent Cu x rays from

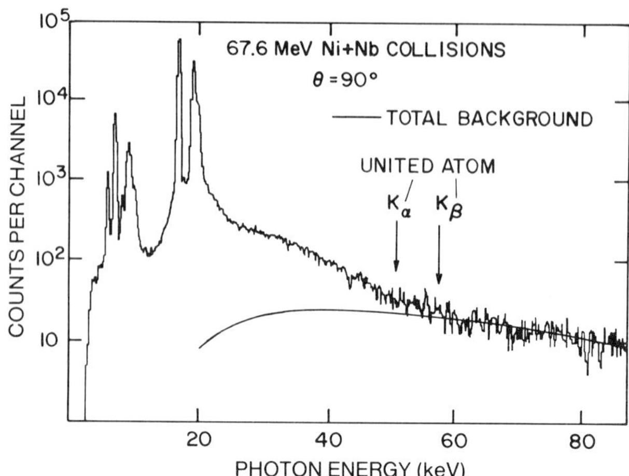

Figure 3.2. X-ray spectra from Ni + Nb collisions. Cu and Al absorbers have been employed (Vi 77, Vi 78).

reaching the detector. The upper peaks near 20 keV are Nb K x rays, and the lower peaks are Ge K x-ray escape peaks from the Nb K x rays. The Ni lines are strongly absorbed and are not seen. The solid black line in the figure indicates the presence of nucleus–nucleus bremsstrahlung background. In contrast to the Ni + Ni case, bremsstrahlung x rays dominate in the vicinity of the united atom MO x-ray limit. This is due to decreased $1s\sigma$ vacancy production in asymmetric collisions and increased background yields from dipole bremsstrahlung radiation. These points will be discussed in greater detail in Section 3.5.

Figure 3.3 shows the intrinsic full energy peak detection efficiency of the 5-mm-thick Ge x-ray detector (Vi 77), which was determined using a variety of readily available calibration standards (Ri 80b). Calculation of the efficiency can be performed with the use of one or two adjustable parameters (i.e., the Ge dead layer thickness and crystal charge collection efficiency) (Ha 73). The detection efficiency is quite high over a broad range of x-ray energies, making these detectors ideally suited for x-ray measurements above several keV.

In measurements where an absorber is needed to attenuate unwanted low-energy x rays, a pure metal should be chosen. Such chemically pure filters eliminate problems due to fluorescence of contaminating elements and allow calculation of absorption based on measurements of the foil thickness. In general, best results are obtained for determination of absorber corrections by combining calculation (St 70) and measurement of the absorption.

In experiments where coincidences with scattered particles are not performed, normalization of the measured yields can be carried out using an

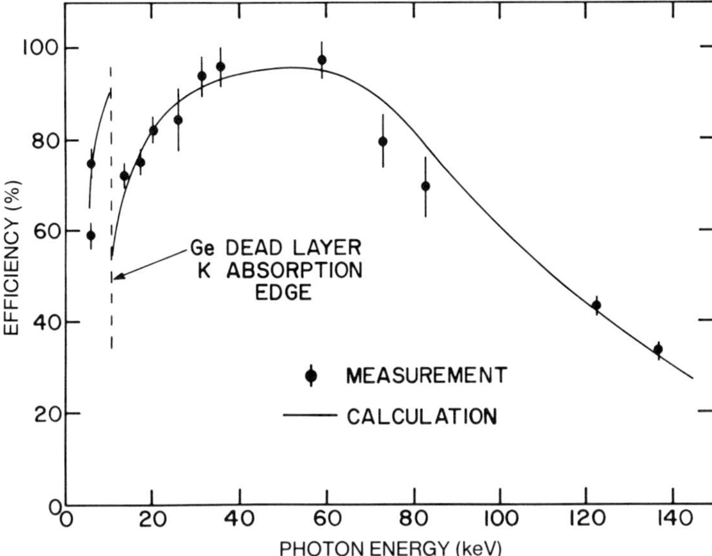

Figure 3.3. Full energy peak efficiency for an intrinsic germanium x-ray detector versus energy. Calculation (solid lines) agrees well with measurements (Vi 77).

electron suppressed Faraday cup and beam current integration or by normalization to Rutherford scattering in a monitor surface barrier detector. Electron suppression is effected with small permanent magnets or suppression voltages of at least several hundred volts. Figure 3.4 shows an example of a beam line used for such measurements. Here magnets are used to prevent slit generated electrons from reaching the chamber Faraday cup and to block target electrons from leaving.

Targets in high-energy atomic physics experiments are an important consideration. The complications arise for principally two reasons: the large size of many atomic collision cross sections and the dependence of these cross sections on the atomic configuration of the projectile. Since the atomic cross sections depend on the charge state of the incident projectile, care must be exercised in choosing a proper, pure charge state for the incident beam. Since metastable atomic configurations can be excited by stripping or entrance window foils, the projectile may not always be in its ground state. If the atomic configuration of the projectile is to be maintained throughout the collision, extremely thin targets are needed. This requirement usually results in the use of gas targets at pressures of 10^{-3} Torr or less for precise atomic physics measurements. However, even with the use of such targets, it is often not possible to maintain single-collision conditions for the cross sections of interest. This follows since, even for studies of inner-shell excitation

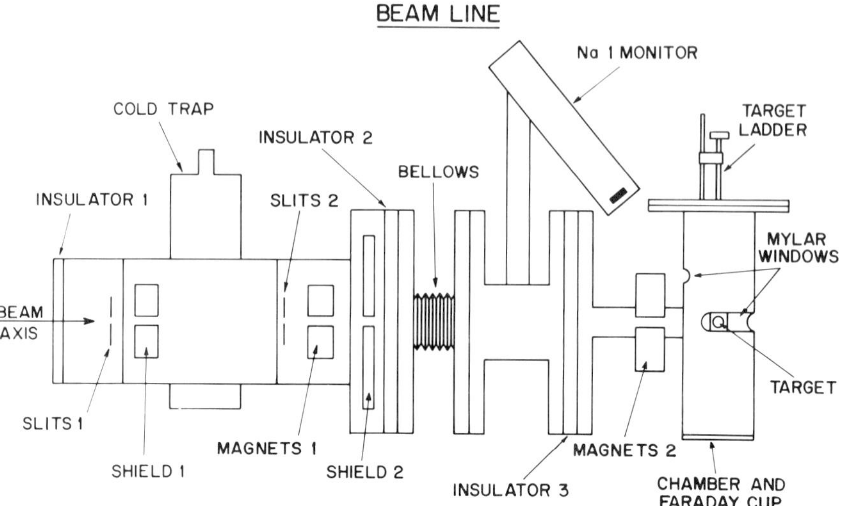

Figure 3.4. Exemplary beam line used for K x-ray and MO x-ray cross section measurements (Vi 77).

phenomena, the vacancy configuration of the outer shells during the collision has significant effects and collision cross sections for loosely bound atomic electrons are very large. Furthermore, gas targets are limited in their range of Z, can be difficult to prepare experimentally, or simply are not thick enough to produce high enough counting rates to do the experiment at hand. For these reasons foil targets are often used in atomic physics experiments. An example of an experimental setup of a gas target system is given in Figure 3.5 (Wa 78, Ri 80).

In experiments where thin foils are used, foil thicknesses from a few $\mu g/cm^2$ to foils thick enough to stop the incident beam can be employed. However, strictly speaking, such measurements only produce determinations of yields, not cross sections, since multiple collisions are always occurring. Even under conditions where changes in the beam energy of the incident projectile while traversing the target are negligible, target thickness effects can result in cross sections for inner-shell phenomena which differ easily by factors of 2 or more. The question of target thickness effects can therefore be quite subtle, and each measurement must be considered on an individual basis (Gr 80).

Vacancy lifetimes in atoms decrease rapidly with increasing electron binding. Cross sections also shrink with decreasing atomic radii. For these reasons, the remarks made above concerning target thickness and multiple collision effects are particularly important in low-Z collisions and outer-shell phenomena. For K-shell excitation processes at high Z, such effects are

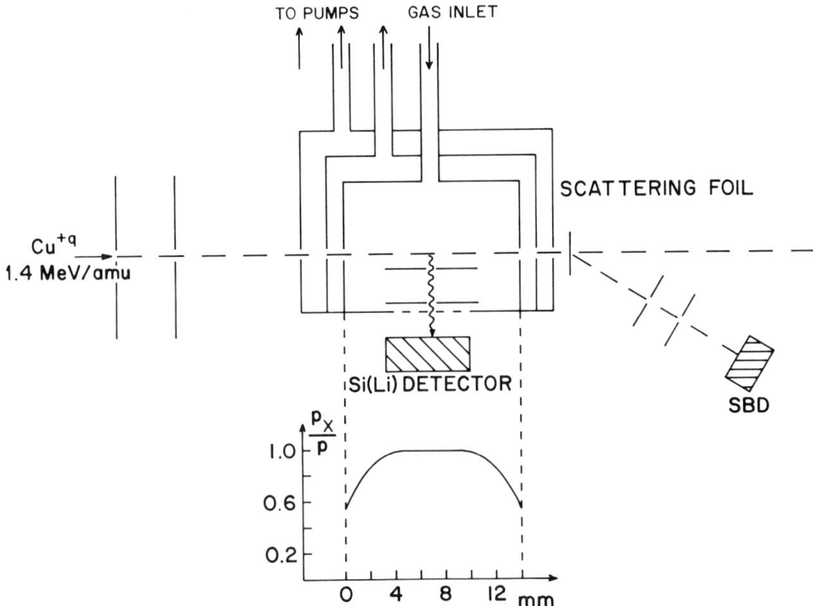

Figure 3.5. Exemplary experimental arrangement for x-ray measurements using gas targets. The calculated pressure distribution P_x in the gas is drawn versus gas cell length. P is the measured pressure in the middle of the cell. A surface barrier detector (SBD) monitors the beam current (Wa 78, Ri 80).

expected to become less important. However, there is, as yet, little data available from superheavy collision systems concerning charge state and multiple collision phenomena. Calculations indicate that significant amounts of K-shell excitation occur to empty bound states (So 83). Therefore, the role of outer-shell atomic configurations even in the heaviest possible collision systems may be significant.

Isotopic purity of the target material may not be required in these experiments. An exception to this rule is the case of MO x-ray measurements, where backgrounds from nucleus–nucleus bremsstrahlung can be bothersome. (See Figure 3.2.) For such studies, use of isotopically pure, identical projectile–target combinations eliminates the prominent dipole term in the nucleus–nucleus bremsstrahlung multipole expansion, leaving only small quadrupole contributions. These points will be discussed further in Section 3.5.

Extremely important information has been gained from coincidence measurements between x rays and scattered projectiles. These studies of the impact parameter dependences of atomic excitations have provided valuable detailed differential information. An example of an experimental arrangement is shown in Figure 3.6 (Sc 82). Even for studies of K-shell excitation, atomic radii are very large compared to the size of the nucleus. For this reason, measurements are often made at scattering angles of less than one degree in the laboratory. Clearly, very well collimated beams and large target-to-detector distances are needed for such studies. Multiple scattering of the projectiles can also occur at sufficiently small scattering angles leading to the use of thin targets for studies involving large impact parameters (Co 80).

Conversion from scattering angle to classical impact parameter can be

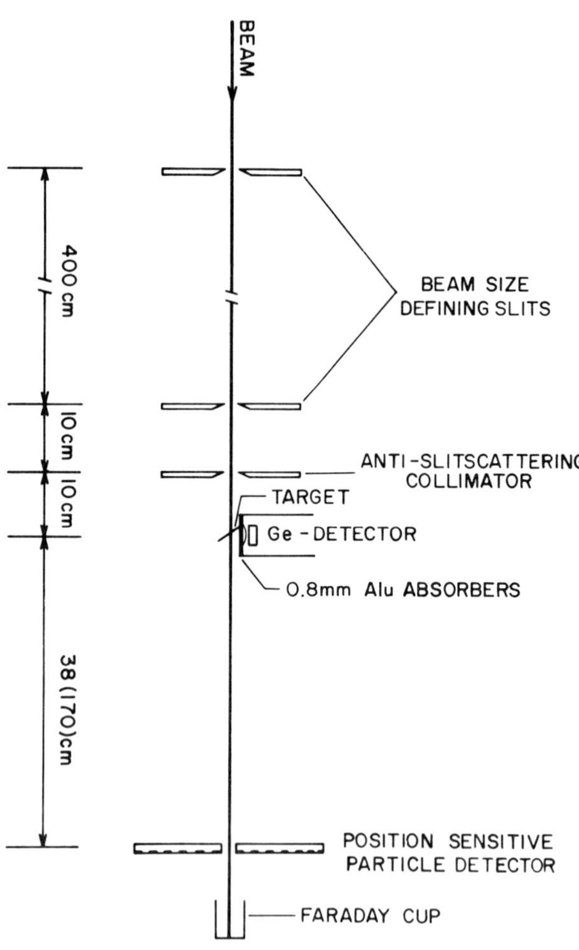

Figure 3.6. Experimental arrangement for x-ray, particle coincidence studies (Sc 82).

carried out using the Rutherford scattering relationship [Equation (2.3)] for large-angle scattering, providing the incident projectile energy is well below the Coulomb barrier. For small-angle scattering, screened Coulomb potentials [Equations (2.4) and (2.5)] must be used and multiple scattering effects contended with. Although sometimes complicated, these problems can usually be satisfactorily solved (Co 80).

In symmetric collisions, only the Rutherford trajectory corresponding to 45° scattering in the laboratory is well defined. At all other scattering angles, kinematically identical contributions from close and distant encounters will be measured. Because of the rapid decrease in Rutherford scattering cross section with increasing scattering angle, distant collisions (forward scattering) dominate the symmetrized scattered particle yields. However, in measurements of impact parameter dependences for various inner-shell excitation phenomena, enhanced probabilities for atomic excitation in close collisions can lead to important contributions from backward scattering events at forward laboratory angles. In lighter collision systems, these recoil contributions from close collisions can usually be safely neglected. However, the very steep impact parameter dependences associated with relativistic encounters lead to large contributions from close collisions. In such cases, only symmetrized production probabilities can be determined. These problems can be circumvented to some extent through the use of asymmetric collision partners. In such cases, particle detector resolution must be sufficiently good to kinematically distinguish close from distant encounters.

In experiments using light projectiles, surface barrier detectors can be used for particle detection. For heavier ions, rapid radiation damage of these detectors necessitates the implementation of gas avalanche or plastic scintillator detectors. There has been considerable development of gas detection systems for use in heavy-ion experiments, and x-ray/particle timing resolution using either the surface barrier or gas avalanche detector is almost always limited by the x-ray detector used. These avalanche particle detectors are operated at gas pressures of typically 10 Torr and therefore require thin entrance windows and appropriate vacuum pumping (Ga 77b).

High-resolution x-ray measurements can be performed using crystal diffraction spectrometers (Ho 80). As has been pointed out earlier, even measurements of inner-shell phenomena are often strongly influenced by outer-shell vacancies. High-resolution studies of K x-ray satellite structure in high-energy atomic collisions have provided valuable information on the atomic configuration at the time of K-shell deexcitation. Measurable differences in the K x-ray satellite structure have been found not only for different projectile–target combinations, but also for different chemical compounds containing the same excited atomic element (Ho 80). Further discussion of this sensitivity of inner-shell excitation to outer-shell phenomena follows below.

3.3. X Rays from Lighter Collision Systems

As noted previously, quasimolecular systems can be broadly characterized as relativistic or nonrelativistic in character. In lighter collision systems, molecular binding energies much less than the electron rest mass energy are achieved, and the two-center Schrödinger equation provides an adequate description of the data. This description becomes increasingly inadequate for high-Z encounters where relativistic effects eventually dominate the behavior of the inner quasimolecular orbitals. Although the boundary between light and heavy collisions is not well defined, we have found organization of the data by united atom Z (Z_u) convenient, and have generally taken $Z_u < 100$ to be light and $Z_u > 100$ to be heavy.

Due to the limited amount of x-ray data at very high Z_u, the bulk of information on the behavior of the quasimolecule comes from lighter systems. Although the degree to which results at lower Z_u can be scaled to high-Z_u encounters is not always clear, we have noted that data from lighter collision systems, nevertheless, constitute the framework within which quasimolecular phenomena across the Periodic Table are discussed. We, therefore, initiate the discussion of x-ray producing processes in heavy-ion collisions with considerations of lighter-Z_u results.

Our discussion of these systems begins with a look at discrete K x-ray transitions followed by an examination of MO x-ray measurements. In keeping with our principal interest in high-binding phenomena, processes associated with the excitation and decay of the innermost quasimolecular $1s\sigma$ and $2p\sigma$ orbitals occupy the main focus of the discussion.

3.3.1. K X Rays

Vacancies generated in the $1s\sigma$ and $2p\sigma$ molecular orbitals generally survive the encounters and lead to separated atom K x-ray radiation. In this section, we organize our discussion of these K x rays into two broad categories: cross-section measurements, and impact-parameter-dependent excitation probabilities. Cross-section data are discussed from an experimental viewpoint. The experimental parameters which can be varied include projectile and target Z, projectile energy, target thickness, and projectile charge state. Our discussion centers on how the variation of these parameters leads to a better understanding of $1s\sigma$ and $2p\sigma$ excitations. A treatment of impact parameter dependence studies shows that agreement between cross-section data and theory does not necessarily mean that all aspects of the relevant excitation processes are well understood. Impact parameter regions of excitation which do not make dominant contributions to the total excitation cross section may contain serious discrepancies with theory which do not effect the integrated x-ray yield. Therefore, the differential features of impact parameter dependence studies provide detailed information on inner-shell excitation processes, which, in turn, leads to a deeper theoretical understanding of these collisions.

3.3.1.1. Cross-Section Measurements. Large amounts of information on mechanisms responsible for K-shell excitation cross sections in symmetric and asymmetric collision systems have been obtained in recent years. The cases which have been particularly important to the atomic physics of the very-heavy collision systems include all aspects of quasimolecular collision phenomena and studies of heavy projectiles in solid targets. We first review the Z dependence of the measured cross sections before discussing selected examples of their dependence on projectile charge state and target thickness.

3.3.1.1a. Z dependences. In Figure 3.7 (Me 76b) total production cross sections (ordinate) for K vacancy generation in 202-MeV Kr collisions with various targets (abscissa) are shown. Triangles represent the projectile and

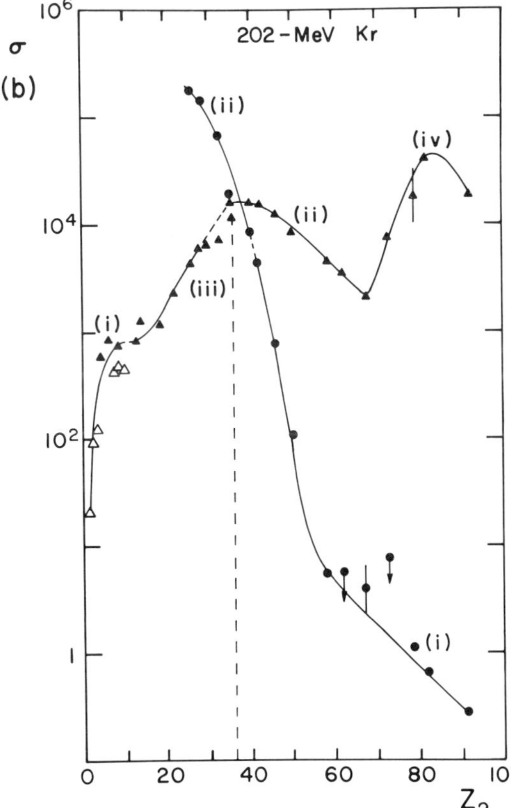

Figure 3.7. Projectile and target K vacancy cross sections in 202-MeV ^{84}Kr collisions. Symbols: △, ▲, thick target extracted projectile cross sections; ●, thick target extracted target cross sections. Curves guide the eye (Me 76b).

circles the target excitations. The dashed vertical line is at $Z_{\text{target}} = Z_{\text{projectile}}$. Adopting the description used by Meyerhof (Me 76b), four regions of excitation are distinguished in the figure.

Region i: Region i refers to excitation of the high-Z collision partner in very asymmetric collisions. For light targets, this corresponds to projectile excitation and for heavy targets, target ionization. For very asymmetric collisions between low Z targets and projectiles, a perturbation theory approach has been successful in describing excitations of the high-Z collision partner (Me 58, Ba 59). In these calculations, the lower-Z collision partner induces ionization by perturbing the atomic wave function of the K-shell electron of the high-Z partner. As the Z of the lighter collision partner increases, this perturbation becomes larger and larger, eventually leading to a breakdown of the approximation. Incorporation of relativistic effects becomes important in Region i at high Z (Ja 57, Am 76).

Alternatively, "Region i" collisions can be described within the context of the molecular orbital picture. We recall that for symmetric collisions the K shells of the projectile and target atoms each connect to both $2p\sigma$ and $1s\sigma$ orbitals. However, for asymmetric collisions (see Figure 2.1) the $1s\sigma$ orbital connects to the K shell of the heavy collision partner and the $2p\sigma$ level to that of the lighter collision partner. The binding energy separation between these two molecular orbitals inhibits their mixing, and heavy collision partner K x rays can therefore be said to originate from the $1s\sigma$ orbital. This molecular orbital approach is essential for nearly symmetric encounters where neither collision partner can be considered as a perturbation acting on the atomic wave functions of the other.

Regions ii and iii: These regions of nearly symmetric encounters are dominated by excitations of the quasimolecular $2p\sigma$ orbital followed by vacancy sharing on the outgoing part of the collision between the light ($2p\sigma$) and heavy ($1s\sigma$) collision partner (Me 77c). The $2p\sigma$ vacancies themselves can be generated by direct excitation of the $2p\sigma$ electron to an empty bound state or to the continuum. Alternatively, rotational coupling of the $2p\sigma$ to a vacant $2p\pi$ orbital will also create $2p\sigma$ holes. In cases where a $2p\pi$ vacancy is present in the ingoing part of the collision, the rotational coupling process dominates (in lighter systems). For nearly symmetric collisions, $2p\sigma$–$1s\sigma$ vacancy sharing is the dominant means of generating high-Z K vacancies.

Region iv: This is the range of collisions where the K-shell binding energy of the lighter collision partner matches that of the L shell ($2p\sigma$) of the higher Z partner. Large increases in the K-vacancy cross sections of the lower Z partner then occur. A number of explanations for this phenomenon have been proposed, all within the molecular orbital picture (Ba 72b, Me 78). In these cases, the K shell of the lower Z partner is either connected with or strongly coupled to orbitals whose binding energies decrease sharply with decreasing internuclear separation ($3d\sigma$ orbital). Electrons in such states have large ionization probabilities, thereby accounting for the observed increase in K vacancy production.

If we confine ourselves to processes involving the highly bound $2p\sigma$ and $1s\sigma$ molecular orbitals, direct and rotational $2p\sigma$ excitation, $2p\sigma$ vacancy sharing, and direct $1s\sigma$ excitation are the topics of central concern (Region i, ii, and iii). Furthermore, the quantity of theoretical interest is the cross section for vacancy production in the various collision systems. Experiment determines the cross section for K x-ray production or, for very light systems, K Auger electron excitation. Therefore, conversion of measured K x-ray data into K-vacancy cross sections requires knowledge of the K-shell fluorescent yields. Calculations and measurements of these yields when the atom has all electrons except a single K-shell vacancy (neutral atom fluorescent yields) have been carried out (Ba 72). However, for symmetric or nearly symmetric collisions, excitation of the K shell is usually accompanied by multiple excitations of the outer shells. The atoms are rarely, if ever, "neutral" at the time of x-ray emission. Determination of these charge-state-dependent fluorescent yields can be a problem. Since fluorescent yields increase with increasing atomic Z, the problem is particularly serious in the lighter Z collision systems. Statistical arguments can be used to estimate the yields (Mc 69, La 71, La 71b) or, in some cases, theoretical calculations have been carried out (Bh 73, Bh 75, Ch 75).

Region i: We turn our attention first to a discussion of direct excitation phenomena (Region i in Figure 3.7) (Ba 73, Ba 78d, An 77). As a first example, Figure 3.8 summarizes results of target K-shell excitation using protons as projectiles (Ba 73). The data are plotted in the form of scaled cross sections (ordinate) versus scaled projectile energies (abscissa) and are compared to plane-wave Born approximation (PWBA) calculations for direct excitation of the electrons from their initial atomic $1s$ orbital. In the figure, $\theta_K = \hbar\omega_{2K}/Z_{2K}^2 R$, where $\hbar\omega_{2K}$ is the binding energy of the target K shell, $Z_{2K} = (Z_2 - 0.3)$ the effective screened target K shell Z, and $R = 1/2$ a.u. $= 13.6$ eV. θ_K is therefore a dimensionless quantity which measures the nonhydrogenic aspects of the K-shell ionization energy. θ_K varies from 0.6 to 0.9 throughout the range of collision systems plotted in Figure 3.8. ε is a binding energy correction factor which modifies θ_K in order to correct for the increased target binding caused by the presence of the projectile inside the atomic target K shell during the collision. $\eta_K = v_1^2/v_{2K}^2$, where v_1 is the projectile velocity, and v_{2K} the Bohr velocity of the screened projectile K-shell electron, is proportional to the incident projectile energy. a_0 is the Bohr radius. The function $E_{10}(x)$ is the exponential integral of order 10, and derives from corrections to straight line trajectories due to nuclear Coulomb deflection of the projectile. The argument of the exponential integral function, $x = \pi d q_{\min}$, contains the variables d, which is one half the distance of closest approach in a head-on collision, and $q_{\min} = \omega_{2K}/v_1$.

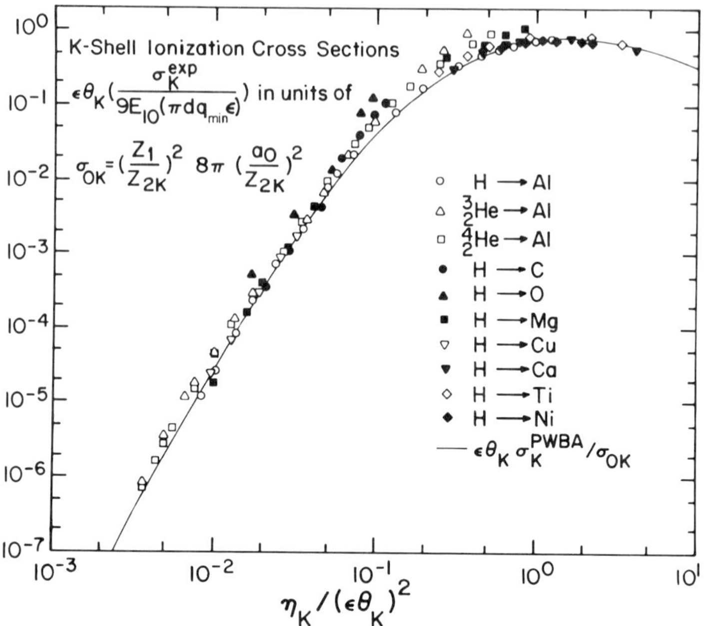

Figure 3.8. Universal graph of K-shell ionization cross sections. Solid curve PWBA prediction (Ba 73).

Corrections to the PWBA calculations of Figure 3.8 due to nuclear Coulomb scattering and target K electron binding are important because excitation usually takes place during collisions which deeply penetrate the target K shell. This is true since the minimum momentum transfer necessary to ionize the K shell is given by [see also Equation (2.25) and related discussion]

$$\hbar q_{\min} = \frac{\hbar \omega_{2K}}{v_1} \tag{3.1}$$

For slow projectiles $q_{\min}^{-1} \ll a_{2K}$, where a_{2K} denotes the target K-shell atomic radius. This presence of the ionizing projectiles deep inside the target K shell leads to an effective increase in the target atom's binding energy. In Figure 3.8, the poorer agreement between measurement and theory at higher values of the abscissa coordinate is due to polarization of the target K shell in the field of the projectile, and to contributions from projectile electron capture.

Extensions of the studies shown in Figure 3.8 to collisions involving heavier projectiles have been made. One example is given in Figure 3.9 (Pr 80) where results for the projectile energy dependence of sulfur K-shell

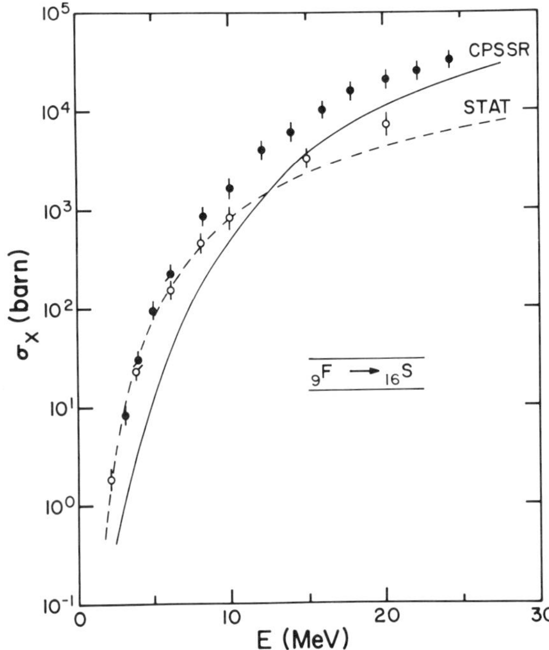

Figure 3.9. Sulfur K x-ray production cross sections. Open circles are gas target measurements and solid circles thin solid target measurements. The solid curve is calculated direct ionization cross sections, the dashed curve the predictions of the statistical model (Pr 80).

excitation in F + S collisions are presented. The data involving these heavier projectile are plotted in comparison (solid line) to Coulomb-perturbed stationary-state calculations (CPSS) (Ba 78d) with relativistic corrections (CPSSR) (Am 76b). These calculations include nuclear Coulomb deflection and they incorporate binding energy and target polarization effects as an intrinsic part of the PSS formalism. They are supplemented by a correction for relativistic effects.

Also shown in the figure as a dashed curve is the result of statistical model calculations (Mi 67, Br 76). These calculations are essentially made within the framework of the molecular orbital picture. However, in contrast to a simple rotational coupling mechanism for $2p\sigma$ excitation, which is essentially a two-state calculation, the statistical model views $2p\sigma$ excitation as a random walk process through which the $2p\sigma$ electron travels upwards in binding energy through a complicated series of molecular orbital crossings and couplings. It is the very complexity of these many couplings which allows the ionization process to be described as a diffusion of the electron "upward" in energy space to unoccupied molecular orbitals or to the continuum. Using this model, $2p\sigma$ ionization cross sections have been calculated. A vacancy sharing calculation (see discussion below) gives the respective low- and high-Z K-vacancy fractions.

The CPSSR calculations are moderately successful in reproducing the beam energy dependence of the measured data in Figure 3.9, but are too low in absolute magnitude; the agreement is, in general, better for increasingly asymmetric collisions and higher bombarding energies. At low beam energies, the statistical model does quite well, but falls below the measured cross sections for bombarding energies above 10 MeV. Taken together, these results demonstrate the importance of molecular excitation processes under conditions where the collisions can be considered adiabatic (low energies).

Alternative descriptions of direct excitation phenomena in asymmetric collision systems have been successfully given using the binary encounter (Ga 73, Ha 73b) and semiclassical approximations (Ba 59, Ch 69, Ha 75) (BEA and SCA, respectively). Within the BEA context, the K-shell excitation cross section is written as (Ga 73, Ha 73b),

$$\sigma = \frac{Z_p^2}{U_K^2} f\left(\frac{E}{\lambda U_K}\right) \tag{3.2}$$

where Z_p is the projectile atomic number, U_K is the binding energy of the target K shell, E is the projectile energy, λ is the ratio of the projectile mass to that of the electron, and f is a universal function. A plot of $U_K^2\sigma/Z_p^2$ versus $E/\lambda U_K$, therefore, yields a universal curve. This description is appropriate for highly asymmetric collision systems involving light projectiles.

In more symmetric adiabatic collisions, molecular orbitals will be formed. In these encounters, inner-shell electrons are not bound by U_K at the time of their excitation, but rather by the binding energy of either the $1s\sigma$ or $2p\sigma$ molecular orbital at the internuclear distance where excitation occurs. Since excitation usually takes place deep within the atomic K shells, the binding energy at the time of excitation can be approximated by the binding energy of the united atom $1s\sigma$ and $2p\sigma$ states for heavy and light collision partner excitation, respectively.

Figure 3.10 (Fo 76b) shows results of scaled cross-section measurements versus a scaled BEA bombarding energy for a variety of collision systems. In the figure, the U_K binding energy used for very asymmetric cases has been replaced with the value U_{ua} indicating the united atom binding energy of the appropriate ($1s\sigma$ or $2p\sigma$) molecular orbital. The solid line is the modified BEA prediction. Agreement between experiment and theory is quite good when the united atom molecular binding energies are used to scale the cross sections.

The results presented in Figure 3.10 provide an intuitively pleasing unification of the atomic and molecular views of these collisions.

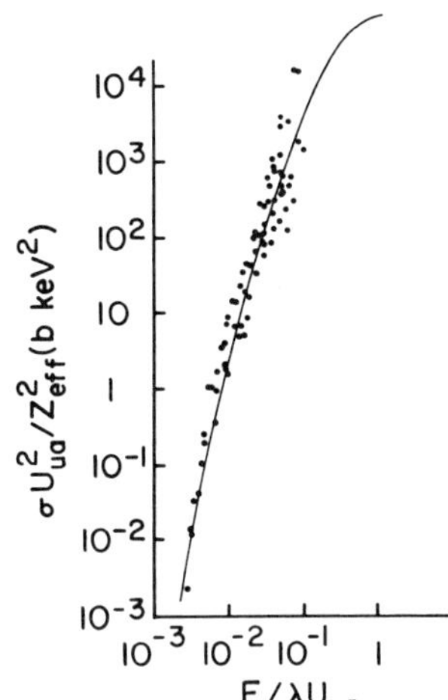

Figure 3.10. Comparison of reduced *K*-vacancy production cross sections with BEA calculations using the united atom binding energy (solid curve) (Fo 76b).

Nevertheless, from a theoretical point of view, the scaling procedure is without justification, since it implies the equivalence of both the lighter and heavier collision partners, while the BEA formulation uses the target nucleus to define the electron velocity distribution and binding energy and the projectile nucleus to transfer momentum and energy to that target electron (Ga 73, Ha 73b). However, a theoretical justification for such a description has been suggested by the results of Briggs (Br 75b), who showed that excitation in slow collisions can be described as Coulomb ionization from united atom molecular orbitals by the coherent field of both nuclei. We also make use of this description to consider data on delta-electron emission in Section 4.

Regions ii and iii: We now turn to a treatment of regions ii and iii of Figure 3.7. In these regions, a full molecular orbital description is likely to be most successful. In fact, some of the most compelling evidence in support of the molecular orbital model itself has come precisely from studies of *K*-vacancy production in adiabatic low-*Z* symmetric collision systems.

Figure 3.11 shows results of K Auger electron studies in Ne^+ on Ne collisions (St 75) where data, taken over a wide range of collision energies, are compared to calculations of $2p\sigma$–$2p\pi$ molecular orbital rotational coupling. An initial vacancy in the Ne projectile (Ne^+) is considered to have a certain statistical probability (Br 72) (see Section 3.3.1.1b below) of being found in that $2p\pi$ molecular orbital which rotationally couples to the $2p\sigma$ level. Simple multiplication of ab initio two-state rotational coupling calculations (Br 73c) by this statistical factor gives the cross section for *K*-vacancy production. The solid line in Figure 3.11 represents the results of such a calculation. There is striking agreement in magnitude and shape

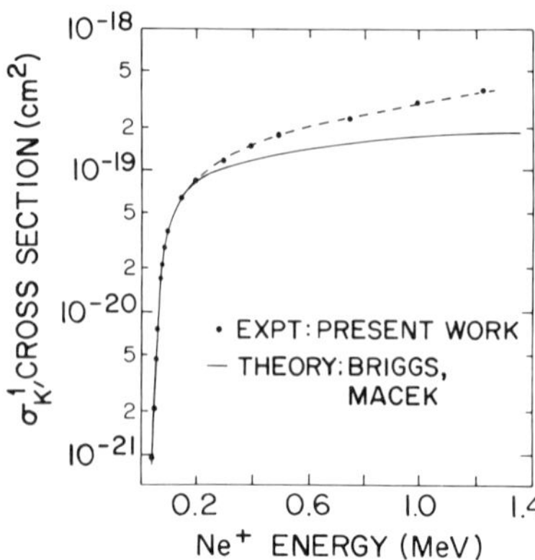

Figure 3.11. Total cross sections for K-shell ionization in $Ne^+ + Ne$ collisions. Dashed line guides the eye. Solid line, predictions of rotational coupling calculations (St 75).

between the energy-dependent K-shell cross section calculations and measurement up to 250 keV projectile energy. The increase in experiment over theory at higher energies has been interpreted as additional velocity-dependent excitation of the $2p\pi$ orbital during the ingoing portion of the collision leading to a greater probability that the $2p\pi$ orbital is vacant and, therefore, larger subsequent rotationally excited $2p\sigma$ electron yields (Fa 74).

Many additional measurements of K-shell excitation in nearly symmetric collision systems have been carried out. In solid targets, multiple collision effects can be of some importance (Me 77c). In these multiple collisions, atomic $2p$ vacancies generated in an initial encounter survive long enough to be carried into a second collision where they first correlate to the $2p\pi$ orbital before rotationally coupling to the $2p\sigma$ level. This process competes with direct excitation of the $2p\sigma$ orbital itself, which, in turn, can be estimated within the context of a modified BEA calculation (Fo 76b). At higher projectile energies in solid targets, both processes occur, and evaluation of the precise magnitude of each of them is problematic. Nevertheless, it has become clear that the concept of the molecular orbital itself must be incorporated into any description of K-shell excitation in regions ii and iii.

Heavy collision partner K-shell excitation ($1s\sigma$ orbital) in nearly symmetric encounters is strongly influenced by the transfer of copiously produced lighter collision partner ($2p\sigma$ orbital) K vacancies. A formulation of this process in energetic atomic collisions was given by Meyerhof (Me 73), who showed, following the results of the charge transfer theory of Demkov (De 64), that the vacancy transfer probability, w is given by

$$w = \frac{1}{1 + \exp(2x)} \tag{3.3}$$

where

$$2x = \frac{\pi\,|(I_1)^{1/2} - (I_2)^{1/2}|}{(\frac{1}{2}mv_i^2)^{1/2}} \cong \frac{\pi\,|Z_1 - Z_2|}{v_i/v_0} \tag{3.4}$$

I_1 and I_2 are the K-shell binding energies of the heavier and lighter partners, m the electron mass, v_i the incident projectile velocity, and $v_0 = e^2/\hbar$ the Bohr velocity. In deriving Equation (3.3), Meyerhof assumed that, since $2p\sigma$ vacancies are, in general, generated during the course of the collisions, transfer is restricted to the outgoing portion of the encounters (one passage transfer). Furthermore, sharing is assumed to occur at internuclear distances large compared to the K-shell radii, so that the binding energies for the low-Z ($2p\sigma$) and high-Z atomic ($1s\sigma$) levels are well approximated by their separated atom K-shell values.

Comparison between the predictions of Equation (3.3) and experiment is made by expressing the K-vacancy production cross sections of the lighter (σ_L) and heavier (σ_H) collision partners as (Be 79b)

$$\begin{aligned} \sigma_L &= (1 - w)\,\sigma_{2p\sigma} + w\sigma_{1s\sigma} \\ \sigma_H &= w\sigma_{2p\sigma} + (1 - w)\,\sigma_{1s\sigma} \end{aligned} \tag{3.5}$$

where $\sigma_{2p\sigma}$ and $\sigma_{1s\sigma}$ are the cross sections for the direct production of vacancies in the $2p\sigma$ and $1s\sigma$ orbitals, respectively. Equation (3.5) allows for two-way transfer of vacancies between the $1s\sigma$ and $2p\sigma$ levels. If $\sigma_{1s\sigma} \ll \sigma_{2p\sigma}$, i.e., if direct $1s\sigma$ excitation is much less than $2p\sigma$ rotational coupling excitation,

$$\frac{\sigma_H}{\sigma_L} \cong \frac{w}{1 - w} \tag{3.6}$$

and comparison between the predictions of Equation (3.3) and experiment is straightforward. Figure 3.12 illustrates the results of such a comparison. The solid line in the figure is the prediction of Equations (3.3) and (3.4). Also shown (dotted line) is the two passage transfer prediction (ingoing plus outgoing portions of the collision contribute to transfer) derived by Demkov (De 64). The agreement between the data and one passage theory is excellent over a wide range of w and x values. As we shall see in subsequent discussions, the success of this vacancy-sharing formulation has greatly facilitated extraction of direct $1s\sigma$ vacancy production probabilities from studies of K x-ray generation in asymmetric collision systems across the Periodic Table.

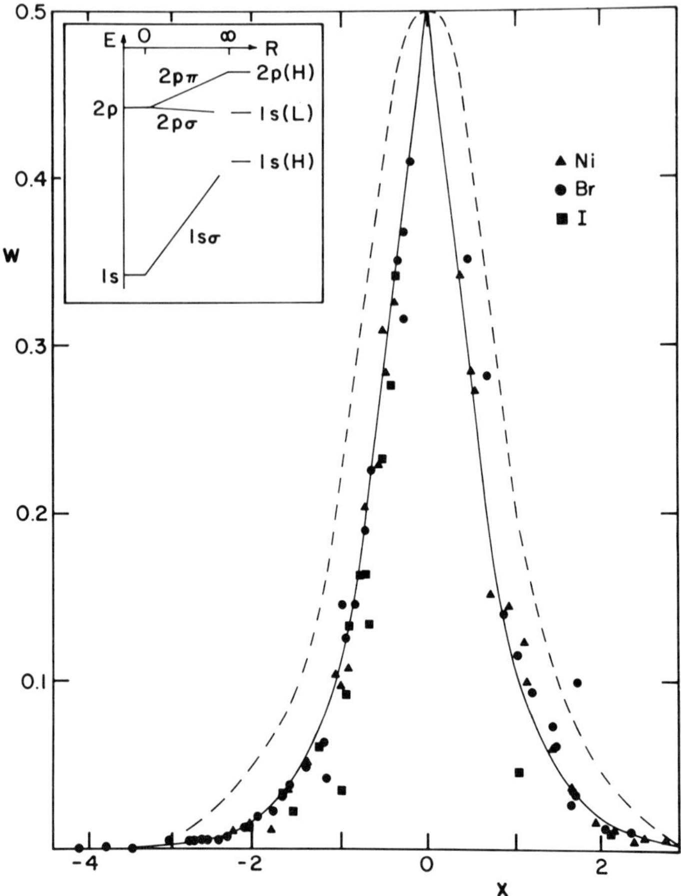

Figure 3.12. *K*-vacancy transfer probability *w* versus the universal parameter *x*. Solid line, theory of Equations (3.3) and (3.4). In the inset, the gap between the $1s\sigma$ and $2p\sigma$ molecular orbitals is schematically shown. Dashed line, two passage transfer theory (Me 73).

3.3.1.1.b. Charge state and target thickness dependences: We now leave our discussion of the *Z* dependence of total cross sections and turn to considerations of charge state and target thickness dependence data. The information provided by studies of the *Z* dependence of heavy-ion *K* x-ray cross sections was instrumental in delineating molecular aspects of vacancy production and in illustrating the relationship between $1s\sigma$ and $2p\sigma$ molecular orbitals in heavy versus light collision partner *K* x-ray emissions. However, more demanding tests of the molecular picture are made by experiments which precisely specify the initial and final states of the atoms involved. Better definition of initial and final states is given by measurements of the dependence of K x-ray yields on incident projectile charge state.

As the projectile charge state is increased, the atomic orbitals of the incident ion are stripped from the outside of the atom inwards. Except for the possible production of metastable states, the stripped ions quickly settle into their ground-state configuration, with the lowest-lying orbitals filled. For sufficiently high charge states, vacancies are brought into the innermost quasimolecular orbitals at the beginning of the collision. The evolution of these vacancies throughout the course of the encounters can then be studied with K x-ray measurements.

For nearly symmetric collision systems, lighter collision partner K x rays originate from $2p\sigma$ orbital vacancies and those of the heavy collision partner from $2p\sigma$–$1s\sigma$ vacancy sharing. The $2p\sigma$ vacancies, in turn, are mostly produced by $2p\pi$, $2p\sigma$ rotational coupling. In slightly asymmetric systems (see Figure 2.1) the $2p\pi$ orbital connects to the $2p$ atomic orbital of the heavier collision partner. Of the six $2p$ orbitals, two connect to the $3d\sigma$ state, and four to the $2p\pi$ level. Two of these $2p\pi$ orbitals (the $2p\pi_x$ orbitals) rotationally couple to the $2p\sigma$ level during the collision. Therefore, based on statistical considerations alone, the number of $2p\pi_x$ vacancies brought into a collision is just $N_\pi = N/3$, where N is the number of initial $2p$ vacancies in the high-Z collision partner. For symmetric collisions, there are 12 $2p$ electrons in the initial state and $N_\pi = N/6$, where N now signifies the sum of initial projectile plus target $2p$ vacancies. These considerations give

$$\sigma_{2p\sigma} = N_\pi \sigma_{\text{rot}} \tag{3.7}$$

where σ_{rot} is simply the cross section for $2p\pi$–$2p\sigma$ rotational coupling per $2p\pi_x$ vacancy.

These concepts can be tested with studies of the dependence of $\sigma_{2p\sigma}$ on incident projectile charge state. Clearly, gas targets must be used for such studies in order to ensure single collision conditions. Figure 3.13 shows results of experiments using Ne and N projectiles on various gas targets (Fo 77). In the figure the ratios of summed projectile plus target K-vacancy cross sections to the predictions of Equation (3.7) are plotted versus incident bombarding energy. In these studies, scaled values of σ_{rot} were used (Ta 76). If experiment and theory are in agreement, these ratios should be unity. In Figure 3.13 the different charge state results tend to form a single curve, but only for the Ne $\rightarrow O_2$ and Ne $\rightarrow$ Ne cases is the observed ratio approximately equal to unity. The increase in σ_{exp} with bombarding energy observed in the Ne $\rightarrow N_2$ and Ne $\rightarrow CH_4$ results could, in part, be explained by additional velocity-dependent excitations of the $2p\pi_x$ orbital during the ingoing part of the collisions (Fa 74). However, such an explanation does not account for the N $\rightarrow CH_4$ and N $\rightarrow N_2$ results where the observed ratios are actually less

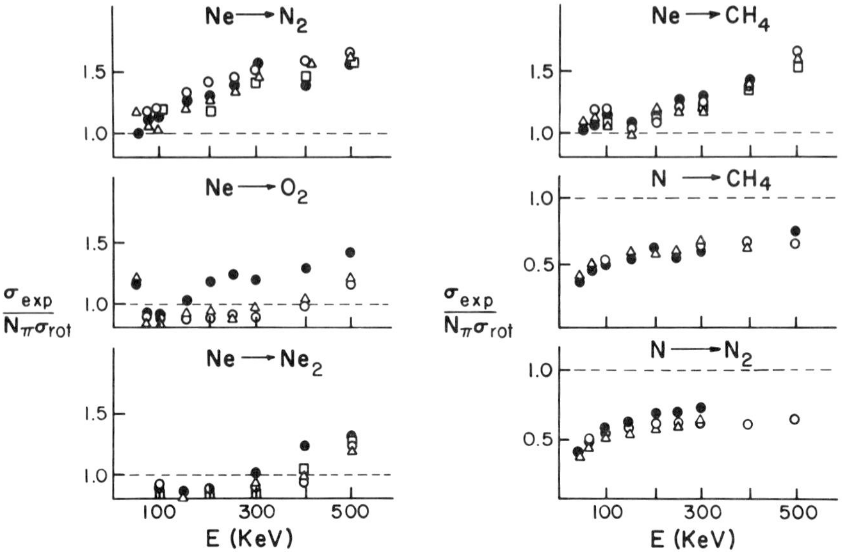

Figure 3.13. Comparison between K-shell vacancy cross sections and theory. The ratios $\sigma_{exp}/N_\pi \sigma_{rot}$ should equal 1 for rotational coupling excitations, where σ_{exp} are the experimental cross sections, σ_{rot} the theoretical rotational coupling cross section per $2p\pi_x$ vacancy, and N_π the initial statistical number of $2p\pi_x$ vacancies. The experimental values are obtained from 1^+(●), 2^+(○), 3^+(△), and 4^+(□) projectiles (Fo 77).

than unity. Based on these data, the authors of Figure 3.13 have suggested the possible inapplicability of the scaled σ_{rot} cross sections (Ta 76) in these cases due to insufficiently accurate evaluation of the $2p\pi$–$2p\sigma$ energy difference and its dependence on internuclear separation (Fo 77).

Figure 3.14 provides a second example of the use of charge state dependence data to test the quasimolecular picture of atomic collisions (Le 81). In the figure total K-shell (projectile plus target) excitation cross sections for

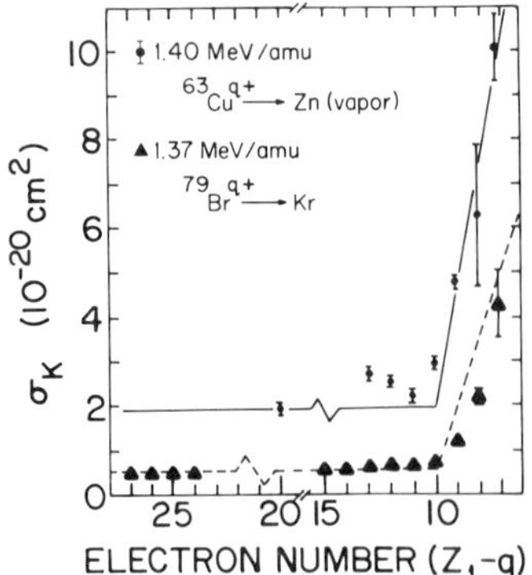

Figure 3.14. Projectile plus target K-excitation cross sections for near symmetric collisions. ● 1.4 MeV/amu $Cu^{q+} \rightarrow Zn$ (vapor); ▲ 1.37 MeV/amu $Br^{q+} \rightarrow Kr$. Abscissa is electron number $Z_1 - q$. For $Z_1 - q < 10$ the solid and dashed lines are rotational coupling predictions for Cu → Zn and Br → Kr, respectively (Le 81).

the nearly symmetric $^{63}Cu \rightarrow Zn$ (vapor) and $^{79}Br \rightarrow Kr$ collisions at 1.40 and 1.37 MeV/amu, respectively, are shown versus the number of electrons present in the incident projectile. As the L shell of the projectile becomes partially vacant, strong increases in the K-shell cross sections are observed. The solid and dashed lines are the predictions of $2p\pi$–$2p\sigma$ rotational coupling calculations for $Cu \rightarrow Zn$ and $Br \rightarrow Kr$, respectively.

The nonzero cross section for K-shell excitation in cases where the projectile L shell is initially filled is interpreted (Le 81) as $2p\pi$–$2p\sigma$ rotational coupling (Ta 76, Sc 80) following the generation of dynamically induced $2p\pi$ vacancies during the ingoing part of the collision (Le 77). The results of Figure 3.14 therefore imply that this dynamic $2p\pi$ excitation process is nearly independent of the initial projectile charge state. As the L shell opens, statistical arguments are combined with scaled rotational coupling calculations to compute increases in K-vacancy production, as was discussed above in connection with the results of Figure 3.13. An additional small complication associated with the data of Figure 3.14 is introduced by the presence of the vacancies in the L shell of the lighter collision partner. $2p$ orbitals of the lighter collision partner connect to the $4f\sigma$ and $3d\pi$ molecular orbitals (see Figure 2.1). In the theoretical calculations of Figure 3.14, the probability of coupling these vacancies to the $2p\pi$ orbitals was calculated assuming $3d\pi$–$2p\pi$ radial coupling as the dominant transfer mechanism (Le 77). Overall agreement between measurement and calculation in Figure 3.14 is good.

Figures 3.13 and 3.14 have illustrated the role that charge dependent studies can play in testing the predictions of the rotational $2p\pi$–$2p\sigma$ coupling process for K-shell excitation. Further discussion of this coupling mechanism is given in Section 3.3.1.2 below in connection with impact parameter dependence studies. Although discrepancies sometimes as large as a factor of 2 between absolute cross sections and the predictions of the rotational coupling description have been obtained, the vast majority of measurements involving this excitation process provide overwhelming proof of its essential validity and, therefore, for the underlying quasimolecular model of atomic collisions.

As the preceding discussion has shown, inner-shell excitation processes are effected by the vacancy configuration of the outer shells, both prior to and during the course of heavy-ion collisions, since the Pauli principle prohibits excitation of inner-shell electrons to occupied outer orbitals. Therefore, a complete description of K-shell excitation in heavy-ion encounters involves a full treatment of both outer- and inner-shell vacancy production processes. However, in energetic collisions, the weakly bound outer orbitals may not have sufficient time to form molecular levels, making an atomic rather than a quasimolecular description of their behavior more appropriate. Charge state dependence measurements can be utilized to study such nonmolecular aspects of heavy-ion encounters.

Figure 3.14 illustrated enhanced probabilities for K-shell excitation in nearly symmetric collisions involving projectiles with L-shell vacancies. In Figure 3.15 (Gr 76) large increases in target K x-ray production are shown in asymmetric collisions involving projectiles with K-shell vacancies. In this figure normalized Cu K x-ray yields from 60-MeV Cl $\rightarrow$ Cu collisions are exhibited as functions of incident Cl charge state, q, and target thickness. We postpone discussion of the target thickness dependence until later in this section, focusing attention at this time on the q dependence of the K x-ray yields.

The increase in target K x-ray production can be interpreted as reflecting K-shell, K-shell coupling between the target and projectile atoms. Since only s states are involved, radial coupling must be examined. By estimating the radial coupling probability between the separated atomic K shells using the Demkov charge transfer theory (De 64, Me 73), the authors of Figure 3.15 were able to account for the observed sharp increase in target K x-ray yield (Gr 76, Gr 80, Ma 80).

Extensions of these charge transfer studies to comprehensive measurements of the Z dependence of not only single, but also double K-shell to K-shell transfer have been carried out by Hall, Richard *et al.* (Ha 81). In these experiments, bare 5 MeV/amu projectiles impinged on 1 μg/cm^2 Ti foils. The transfer of one (two) target K-shell electrons was detected through the observation of the various Ti K x-ray satellite (hypersatellite) transitions using a Si(Li) detector. The resulting single and double K-shell to K-shell transfer cross sections versus incident projectile Z are given in Figure 3.16. The solid lines in the figure are calculations of the transfer cross sections based on the two-state atomic expansion model of Lin *et al.* (Li 78, Li 80). On the basis of the excellent agreement between theory and experiment, the authors of Figure 3.16 conclude that the double vacancy transfer process could make significant contributions to total K-vacancy

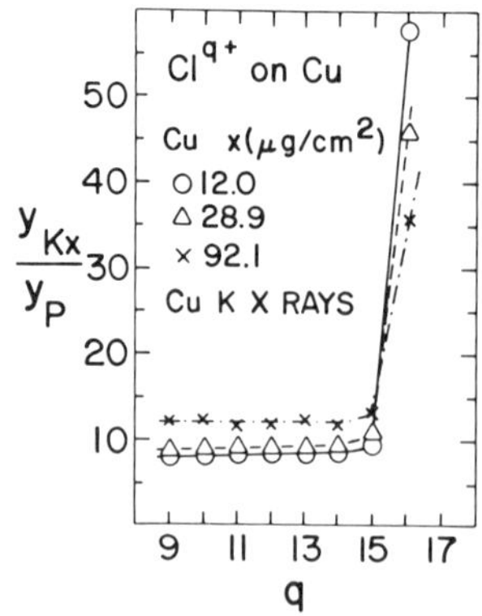

Figure 3.15. Cu K x-ray yield per scattered ion (y_{Kx}/y_p) versus projectile charge and target thickness for 60-MeV Cl on solid Cu targets. The scattered ions, measured at a laboratory scattering angle of 30°, normalize the x-ray yields (Gr 76).

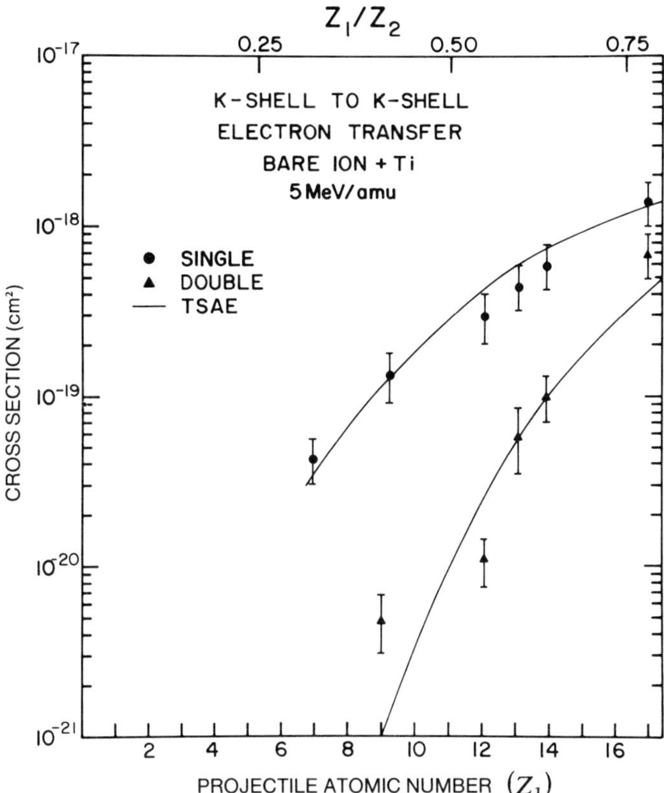

Figure 3.16. Projectile Z dependence of single and double K-shell to K-shell electron transfer for bare ions on Ti. Solid curves are two-state atomic expansion calculations (Ha 81).

production in nearly symmetric collision systems, particularly at beam velocities approaching those of the K-shell atomic electrons (Ri 80, Ha 81).

The data of the Figure 3.15 displayed different K x-ray yields for different target thicknesses. For low projectile charge states, the yields increased with target thickness, and for projectiles with a K vacancy, the thin target yields exceeded those from the thick target. These results reflect the importance of multiple collision effects in thin solid foils. Study of these target thickness dependent processes have provided valuable insight into the dynamic evolution of projectile and target atomic configurations during energetic heavy-ion collisions.

We illustrate these effects with the example of Figure 3.17 where Cu K-shell yields as a function of target thickness are given for various charge state Al, S, and Si ions in thin solid Cu targets (Ga 77). Depending on the incident ionic charge state, target K-shell excitation either increases or

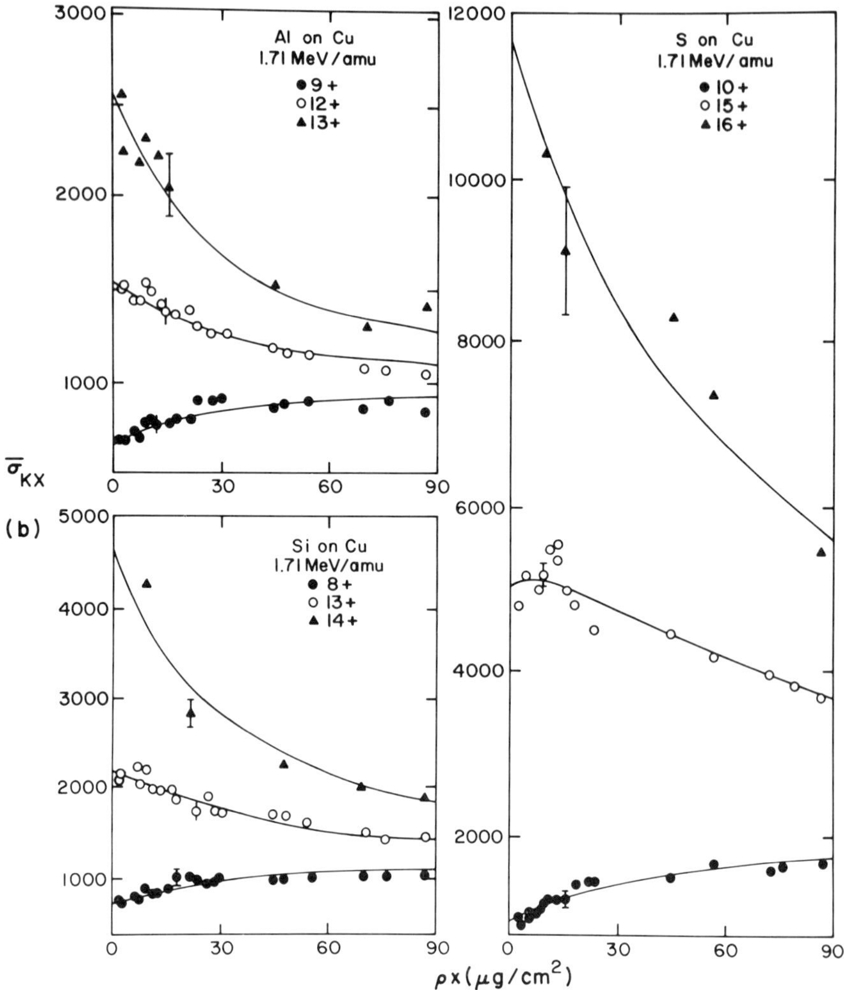

Figure 3.17. Effective Cu K x-ray cross section versus target thickness for Al, Si, and S ions on thin solid Cu targets. Solid lines are calculations of three-component model (Ga 77).

decreases with target thickness. These results have been analyzed using a three-component model of K-vacancy production (Al 58, Be 73, Gr 76, Gr 76b, Gr 80, Ho 75). The model allows for an enhancement in target vacancy production by projectiles with a K vacancy over that from projectiles without a K vacancy by a factor α. The enhancement for projectiles carrying two K vacancies is parametrized by a factor β. Changes in the projectile K-vacancy fractions as the target is traversed explain the observed target thickness dependence. Therefore, in, for example, the

Al + Cu data of Figure 3.17, the 13^+ Al cross sections decrease with target thickness due to quenching of the projectile K vacancies as the target is traversed. The increase in the Al 9^+ cross sections with target thickness reflects the excitation of Al K-shell vacancies in the target.

The solid lines in Figure 3.17 are the results of least-squares fits to the data of the several parameters comprising the three-component model. These parameters include the various projectile K-shell charge changing cross sections as a function of target thickness and the α and β enhancement factors. Once these parameters are determined, they can be compared to theoretical calculations. Results of such a comparison suggest that projectile K-vacancy quenching is due primarily to charge exchange (Ni 67) from all shells of the target atom to the projectile. Incident ion vacancy production in the target is in rough agreement with PWBA direct Coulomb ionization by Cu of the projectile K shell. The enhancement factors α and β can be successfully described by a vacancy-sharing calculation of transfer from the 1s state of the projectile directly to the target K shell. Results of these and other studies indicate that, when only one projectile vacancy is involved, a simpler two-component model may be used instead of the three-component description given here (Be 73, Ho 75).

We, therefore, see that cross-section studies of K-ray production have provided guidelines for the applicability of the molecular orbital picture of atomic collisions. In general, when the projectile velocity is slow compared to the Bohr velocity associated with any particular atomic orbital, that orbital can be expected to evolve along molecular levels. Therefore, in general, adiabatic heavy-ion collisions will exhibit inner-shell excitations governed by molecular processes while having outer-shell ionization cross sections whose behavior is essentially atomic in character. Since outer-shell vacancy configurations affect inner-shell excitation, both atomic and molecular reactions must, in principle, be accounted for in complete treatments of K-shell excitation phenomena. Studies of the Z, charge state, and target thickness dependences of K x-ray production in lighter systems have provided experimental tests of various theoretical atomic and molecular excitation mechanisms and, as such, have established the framework within which all heavy-ion encounters are discussed and understood.

Before closing this section on K x-ray cross sections, we give two additional examples of the importance of simultaneous inner- and outer-shell excitation phenomena in heavy-ion collisions. Figure 3.18 (Wa 74) displays results of high-resolution (Wa 77, Be 81) crystal spectrometer measurements of K_α x rays in collisions of various projectiles with Al targets. As the Z of the projectile increases, the satellite structure becomes more and more complicated. Each peak corresponds to an additional L-shell vacancy at the time of x-ray emission. Clearly, K-shell excitation with heavy ions is accompanied by multiple L-shell vacancy production.

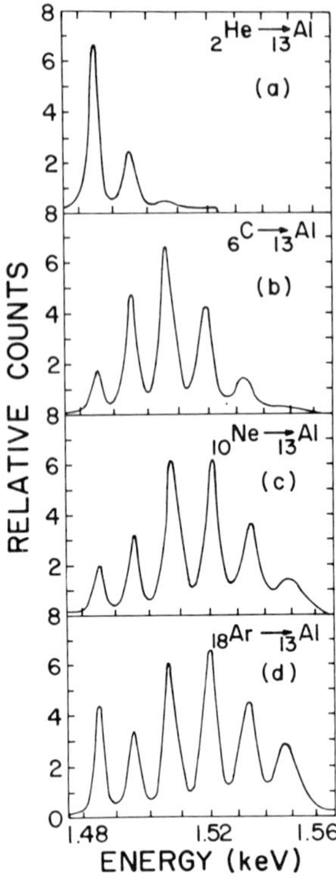

Figure 3.18. High-resolution crystal spectrometer spectra for Al K_α x rays from 1.7 MeV/amu He, C, Ne, and Ar ions (Wa 74).

In measurements where the various satellites cannot be resolved, determinations of the centroid of the K_α peak yield, upon comparison with Hartree–Fock calculations of the transition energies for different numbers of L-shell vacancies, the mean number of L-shell vacancies at the time of x-ray emission. Since the energy shift of the K_α line per additional L vacancy is roughly independent of the number of vacancies (see Figure 3.18), average L-vacancy populations can be extracted from the x-ray K_α centroids in a straightforward manner (Lu 77). The results of such a procedure are given in Figure 3.19 (Vi 77) for Ni + Ni collisions at various bombarding energies. At the highest projectile velocity (95 MeV beam energy), K_α vacancy production is accompanied by the excitation of, on the average, nearly four L-shell vacancies. Therefore, heavy-ion encounters result in substantial activity in the outer shells, and the role played by outer-shell phenomena in inner-shell excitation and decay processes is often quite significant. This, in turn, provides for a rich spectrum of interesting phenomena which can influence observations of inner-shell excitations.

3.3.1.2. Impact Parameter Dependences: Investigations of the differential properties of K-vacancy production cross sections have provided

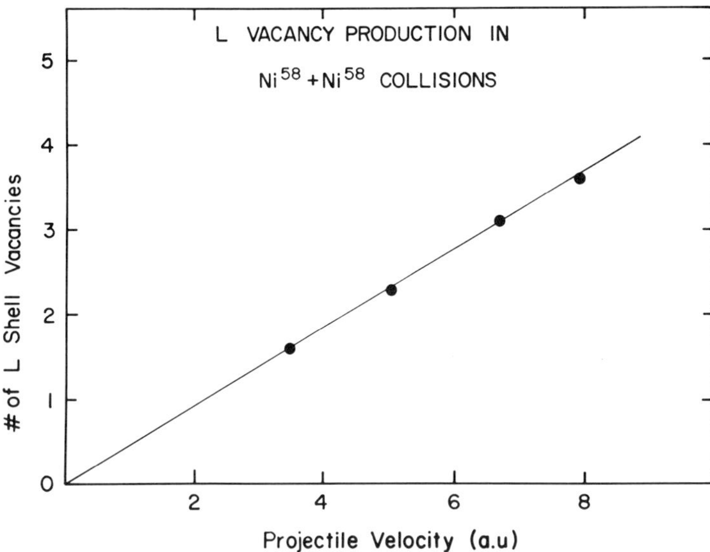

Figure 3.19. Number of L-shell vacancies in Ni + Ni collisions versus projectile velocity determined by Hartree–Fock calculations of observed K_α, K_β energy shifts (Vi 77).

critical checks of inner-shell excitation calculations. The scattering angle or impact parameter dependence of K x rays from asymmetric and symmetric collisions often displays distinctive features which can be utilized to more uniquely define the atomic reaction mechanisms active in the encounters. In general, the more detailed information provided by differential cross section measurements increases precision in defining the initial and final states of the atoms involved leading to a more refined understanding of the collisions. In this section, we will select examples of coincidence studies which have been particularly crucial to the definition of the range of validity of the molecular orbital model.

Figure 3.20 shows results of impact parameter studies of Cu K x-ray production in a variety of asymmetric collision systems (An 76). Total cross-section measurements for very asymmetric collision systems have (Figure 3.8) indicated that agreement between theory and experiment is achieved by calculations within an essentially "atomic" picture. In such calculations, the projectile is treated as a perturbation on the initial K-shell atomic wave function of the target atom. Within the context of such a simple perturbation theory approach, the target excitation probabilities should increase with the square of the projectile charge (Z_p) for projectiles of the same velocity (Ba 73). The ordinate of Figure 3.20 plots the ratio of the ionization probability to Z_p^2. Contrary to the expectations of simple perturbation theory, the scaled ionization probabilities fail to form a universal

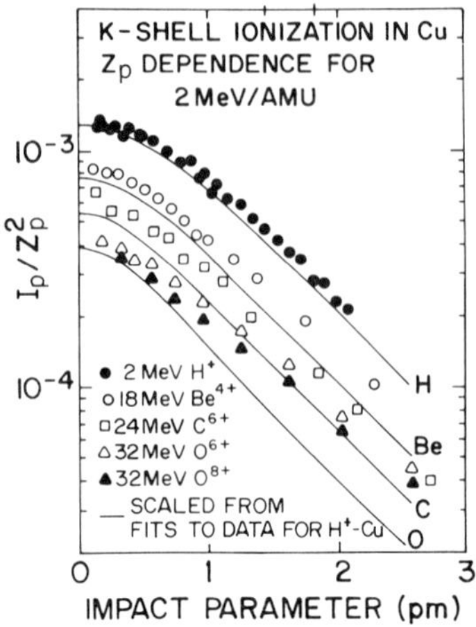

Figure 3.20. Impact parameter dependence of Cu K-shell ionization probabilities reduced by factor $1/Z^2_{\text{projectile}}$ for H^+, Be^{4+}, C^{6+}, O^{6+}, and O^{8+} beams at 2 MeV/amu. Solid curves are scaled from parametrized proton beam results (An 76).

curve, but exhibit decreasing scaled intensities with increasing Z_p. This tendency was observed in earlier total K-shell vacancy production studies and has been attributed to increases in the K-shell binding energy during the collision due to the presence of the projectile deep inside the target K-shell radius ("binding energy correction") (Br 66, Br 73). From a molecular standpoint, one would say that excitation takes place out of the $1s\sigma$ molecular orbital, and the binding energy of this state increases with increasing combined projectile–target charge. The solid lines in Figure 3.20 are predictions of the ionization probabilities based on a scaling procedure which utilizes experimental results from protons on Cu (Lu 74) at various bombarding energies to determine the binding energy correction. The scaling is successful at small impact parameters, but significant discrepancies occur for more distant collisions. Nevertheless, the results provide strong, relatively direct evidence for increases in K-shell binding during the course of these encounters, and, as such, illustrate the quasimolecular aspects of these collisions (Co 73, Ra 76b).

For symmetric or nearly symmetric adiabatic collisions, studies of K x-ray cross sections discussed in Section 3.3.1.1 have indicated that excitation proceeds principally by rotational coupling of the $2p\sigma$ and $2p\pi$ orbitals. These results can be further tested through measurements of the impact parameter dependence of the excitation probabilities. Figure 3.21 (Sa 74) shows results of one such study using Ne^+–Ne collisions at 235 and 363 keV. In such symmetric encounters one cannot distinguish between the projectile and target atoms. Therefore, in measurements at forward angles contributions from both forward scattered projectiles and recoiling target atoms will be present. As has been discussed in Section 3.2 above, the rapid decrease in Rutherford scattering cross section with increasing projectile

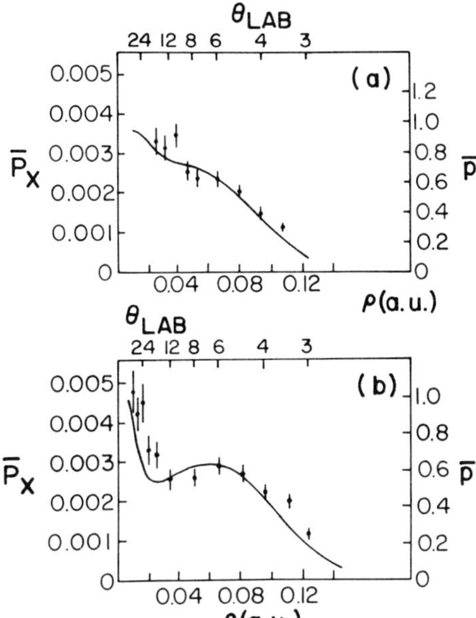

Figure 3.21. Experimental x-ray production (left-hand scale) and theoretical vacancy production probability (solid curve, right-hand scale) in Ne^+–Ne collisions versus impact parameter ρ of the distant collisions. (a) Ion energy of 235 keV; (b) ion energy of 363 keV (Sa 74).

scattering angle usually leads to strong attenuations of target recoil contributions. However, in cases where the excitation probability of the process being studied decreases rapidly with increasing impact parameter, recoil contributions are important and only symmetrized excitation probabilities can be determined.

The solid lines in Figure 3.21 are calculations (Br 72) of the symmetrized excitation probabilities due to simple two-state coupling of the initial $2p\pi$ vacancy to the $2p\sigma$ orbital. The striking change in the measured excitation probability shapes in going from 235 to 363 keV is very well reproduced by the calculations. The results provide strong support for the dominance of two-state $2p\pi$–$2p\sigma$ rotational coupling in the production of K-shell vacancies during these collisions.

In sufficiently asymmetric collisions, K vacancies in the heavy collision partner contain contributions from direct $1s\sigma$ excitation as well as from vacancy sharing of $2p\sigma$ holes. Each collision system must be examined on an individual basis in order to ascertain the relative contributions of these two excitation mechanisms. In our previous discussion of vacancy sharing, we presented the results of Meyerhof's (Me 73) formulation of the charge transfer theory of Demkov (De 64) [see Equations (3.3), (3.4) and associated discussion]. The vacancy sharing process was induced by radial coupling between the $2p\sigma$ and $1s\sigma$ orbitals. Demkov showed that this coupling takes place during the range of internuclear separations where the molecular wave functions change character, becoming essentially atomic in nature. Since these wave function changes are adiabatic, the vacancy sharing process should exhibit an impact parameter dependence. Briggs (Br 74) has derived an approximate solution to this two coupled state problem based on a method due to Crothers (Cr 71, Cr 73). Results for the vacancy sharing ration P_{vs} are given by the simple analytic formula

$$P_{vs}(b) = \exp[-2\Delta E y(b)/v] \tag{3.8}$$

where ΔE is the energy difference between the two separated atom $1s$ levels, $y(b)$ a function representing the impact parameter dependence (Br 74), and v the projectile velocity. The vacancy sharing ratio $P_{vs}(b)$ is defined as the ratio of K-shell ionization probabilities for the heavy (H) and light (L) collision partners: $P_{vs}(b) = P_K(H)/P_K(L)$.

In Figure 3.22, the results of experimental tests of Equation (3.8) using 35-MeV Cl on Ti and Ni targets are shown (Sc 77). The dashed and solid lines represent the predictions of Equation (3.8) for the Ti and Ni cases, respectively. Although discrepancies in absolute height between experiment and theory are indicated, the good agreement in shape supports the dominance of $2p\sigma$–$1s\sigma$ vacancy sharing in the generation of high-Z collision partner K-shell vacancies in these collisions and demonstrates the applicability of this particular calculation of the sharing probabilities (Sc 79b).

The success of the simple description of K-shell excitation in terms of only three molecular orbitals ($2p\pi$, $2p\sigma$, $1s\sigma$) has been one of the most convincing demonstrations of the validity of the molecular orbital picture. The range of validity of the three-state description has been explored in a variety of systematic studies of K x-ray impact parameter dependences. In symmetric or nearly symmetric collisions, the two-state description ($2p\pi$,

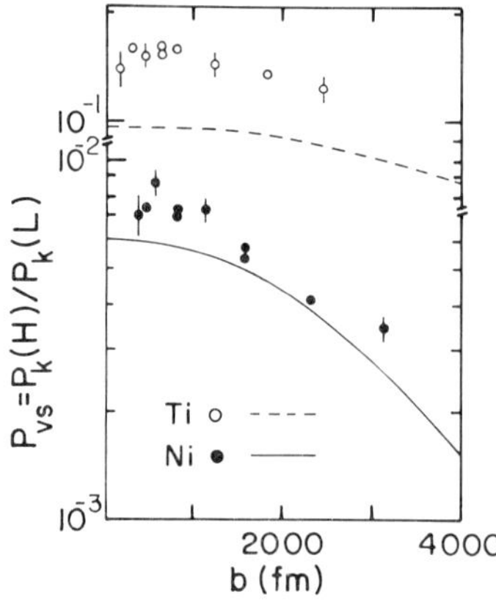

Figure 3.22. Measured vacancy-sharing ratios P_{vs} as a function of the impact parameter for 35 MeV Cl on Ti (open circles) and Cl on Ni (solid circles). Solid and dashed lines are vacancy-sharing calculations (Sc 77).

$2p\sigma$) has been successful in reproducing experimental results from lighter collision partners. However, for heavier systems, systematic deviations from the two-state calculations begin to occur. The source of these discrepancies is best examined by comparing systematic impact parameter dependent studies of K-shell excitation to the predictions of a theoretical scaling law for $2p\pi$–$2p\sigma$ rotational coupling (Ta 76, Br 76b).

Derivation of this scaling law is based on three simplifying assumptions (Ta 76): (1) straight line trajectories, (2) the proportionality of the $2p\pi$–$2p\sigma$ energy gap to R^2, and (3) the rotational coupling matrix element = 1. Under these conditions, the coupled equations assume a particularly simple form, yielding impact parameter probabilities which depend solely on the variable $b' = [\alpha(Z_p, Z_T)/\hbar v]^{1/3}\, b$, where Z_p and Z_T are the projectile and target nuclear charges, v the projectile velocity, and b the impact parameter. The function $\alpha(Z_p, Z_T)$ scales the energy splitting between the $2p\pi$–$2p\sigma$ molecular orbitals for various systems.

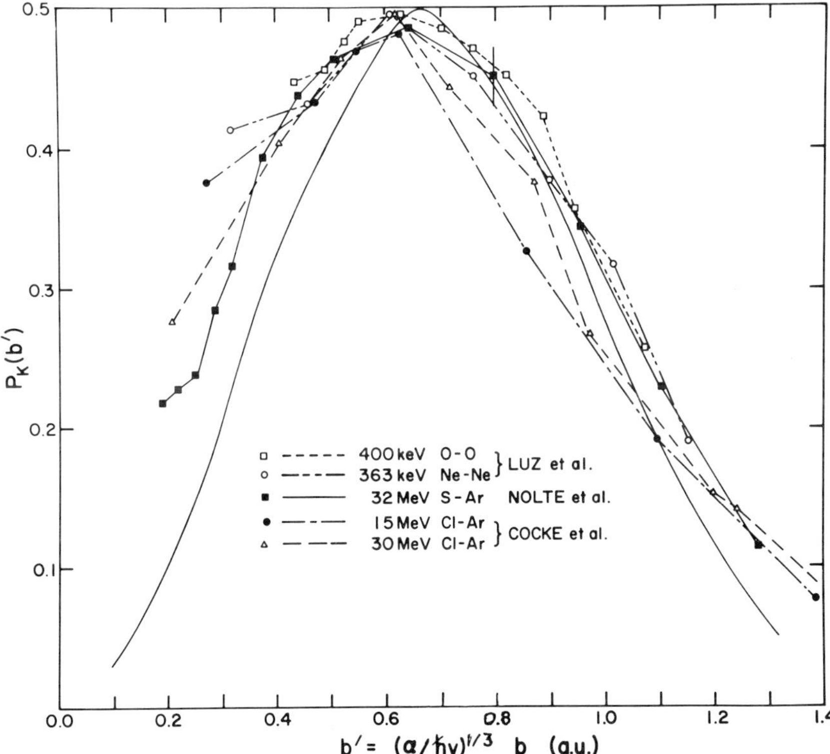

Figure 3.23. Normalized K-shell vacancy production probabilities for $Z_p \simeq Z_T \leqslant 20$ versus reduced impact parameter b'. Solid line represents universal $2p\pi$–$2p\sigma$ rotational coupling calculation (Sc 80).

Comparison between experiment and scaling theory is given in Figure 3.23 for a variety of collision systems (Sc 80). The solid line in the figure plots the prediction of the simplified $2p\pi$–$2p\sigma$ scaled rotational coupling calculation. The absolute heights of the data and calculation have been normalized to each other at the adiabatic peak maximum. The scaling of these data from this wide variety of collision systems (Co 76, Lu 79) according to the predictions of the simplified two-state description, and the good agreement in shape between calculation and scaled experimental results both justify the $2p\pi$–$2p\sigma$ molecular orbital description of these collisions. Collision systems up to a combined Z of 35 are presented in the figure.

A similar scaled plot for heavier collision systems is given in Figure 3.24 (Sc 80), where united atom Z values between 56 and 100 are shown. The data (Gu 78, An 79e, Jo 79, Sc 79c) display a common scaling in qualitative agreement with the two-state description, but the centroid of the reduced impact parameter distributions is shifted noticeably downward from the position of the adiabatic peak. Several authors (An 77b, Ja 80) have examined relativistic corrections to the assumed $2p\pi$–$2p\sigma$ energy gap, and have concluded that such corrections cannot account for the observed shift.

In addition to the adiabatic peak mentioned above, $2p\pi$–$2p\sigma$ rotational coupling exhibits a second characteristic feature at smaller impact

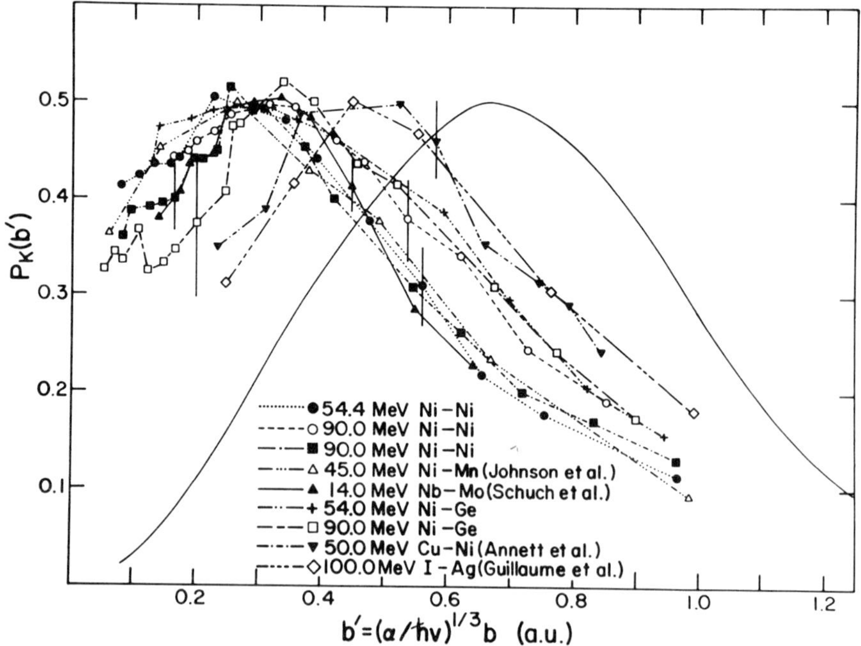

Figure 3.24. Same as Figure 3.23 except for $Z_p \simeq Z_T \geqslant 25$ (Sc 80).

parameters corresponding to the region of 90 degrees scattering in the center of mass. At this $\theta_{c.m.}$ angle, the molecular axis rotates through 90° during the course of the collision, thereby causing strong mixing of the $2p\pi$ and $2p\sigma$ states. Two-state rotational coupling calculations predict a second peak (referred to as the "kinematic" peak) in the impact parameter dependence centered at 90° scattering with a $\sin^2 \theta_{c.m.}$ shape. Subsequent measurements of K-shell excitation in 30-Mev Ca + Ti (Pf 80, Sc 81) and 60-MeV I + Ag (An 80) collisions in the "kinematic" peak region have been in good agreement with the rotational coupling prediction. Furthermore, investigations of the shape and size of K x-ray impact parameter dependences in nearly symmetric systems for various target thicknesses, τ (An 79e, St 81b) and incident projectile charge state (Ha 79, No 80, Sc 82b), q, have been carried out. The shape of the impact parameter dependences was found to be independent of target thickness and incident charge state although the absolute height increases with increasing τ and q.

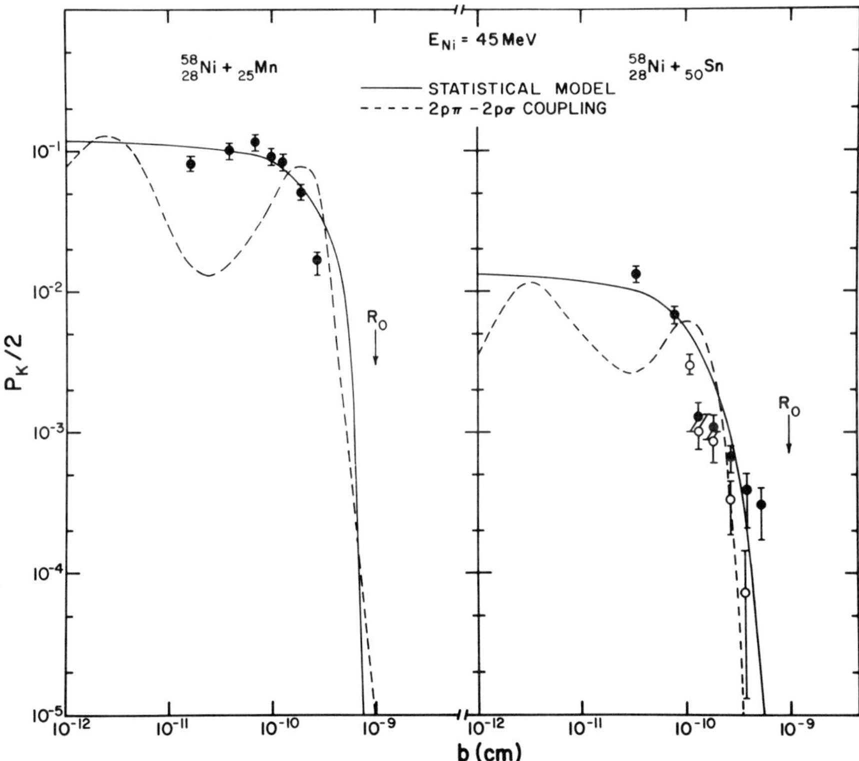

Figure 3.25. Comparison between experimental K-shell ionization probabilities P_K and the predictions of the statistical and rotational coupling models. R_0 is the interaction radius of statistical model used in the calculations (Jo 79).

Figure 3.25 (Jo 79) plots results for the impact parameter dependence of K x-ray excitation in 45-MeV Ni collisions with Mn and Sn targets. The dotted curve is the prediction of the $2p\pi$–$2p\sigma$ rotational coupling model (Ta 76) and the solid line the results of the statistical model (Mi 67, Br 76, Jo 76). Note the characteristic two peak structure exhibited by the rotational coupling calculations. The peak at low b is the "kinematic" peak, and that at large b the adiabatic peak. The increase in experiment over rotational coupling calculations in the region between the two peaks is typical for higher-Z systems. This "filling in" of the rotational coupling "valley" is well reproduced by the predictions of the statistical model.

The picture emerging from all these and other studies is that, although $2p\pi$-$2p\sigma$ rotational coupling dominates K-shell excitation in nearly symmetric collisions, other couplings of the $2p\sigma$ orbital to higher-lying bound and continuum states become increasingly important at higher Z. Figure 3.26 illustrates this point with measurements of K-shell excitation in the 12-Mev Ar + Ar system (Sc 82b). The ratio of the incident projectile $q = 9^+$ to $q = 4^+$ charge state yields is plotted versus impact parameter. The 9^+ yields are increased over the 4^+ yields in the kinematic ($b \cong 100$ fm) and adiabatic ($b \cong 1000$ fm) peak regions. The "valley" region shows no increased yield at higher q. The kinematic and adiabatic regions therefore increase with the number of $2p\pi$ vacancies brought into the collision, whereas the "valley" region does not. The authors attribute this behavior to contributions from direct excitation of the $2p\sigma$ (Br 75b, Fo 76b) state into vacant higher-lying molecular orbitals and the continuum. These additional excitations add to $2p\pi$–$2p\sigma$ rotational coupling to produce the measured results.

Therefore, although total cross sections for K-shell excitation in adiabatic nearly symmetric collision systems are dominated by $2p\pi$–$2p\sigma$ rotational coupling, coincidence studies of the differential cross sections demonstrate the presence of additional couplings to other states which become increasingly important at high Z. These results are typical of the important role played by impact parameter dependence measurements in refining our understanding of K-shell excitation processes (Lu 78, Ts 78,

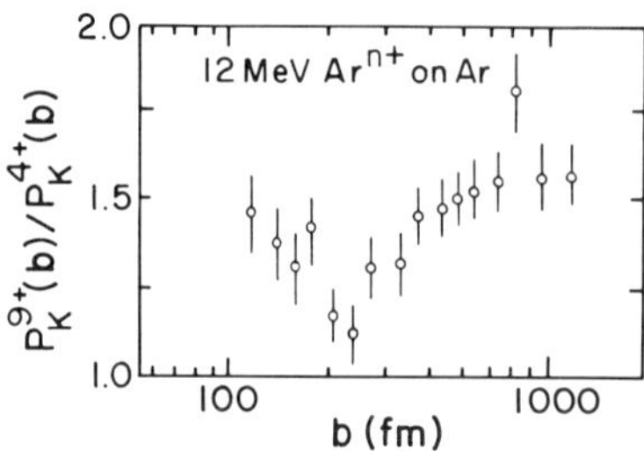

Figure 3.26. Ratio of experimental K-shell excitation probabilities $P_K^{q+}(b)$ for different charge states $q = 9^+$, 4^+ (Sc 82b).

Ha 82). The systematics of impact parameter measurements are particularly important in the extremely heavy systems, where the success of scaling laws provides striking evidence for the strong binding of $1s\sigma$ electrons in the heaviest possible quasiatoms. Discussion of these points follows in Section 3.4 below.

3.3.2. MO X Rays

We have seen significant amounts of evidence for the molecular orbital description of atomic collisions involving light systems based on studies of characteristic K x-ray emission. Although emission of the x rays takes place after the collision, vacancy production during the collision reflects the properties of the molecular orbitals. The evidence for molecular formation is therefore somewhat indirect. More direct evidence for the formation of these orbitals can be gained through studies of x-ray transitions which occur actually during the course of the collision itself.

Figure 3.27 illustrates schematically the source of MO x-ray transitions in heavy-ion collisions. Vacancies in the $1s\sigma$ and $2p\sigma$ molecular orbitals can be filled with electrons from higher-lying levels leading to the emission of x rays whose energies exceed those of the characteristic K lines. Typical spectra were illustrated in Figure 3.1. The essential features of these data are their continuum nature and extension to energies in excess of the united atom K x-ray transition frequency. Both of these features reflect the dynamic nature of transient molecule formation.

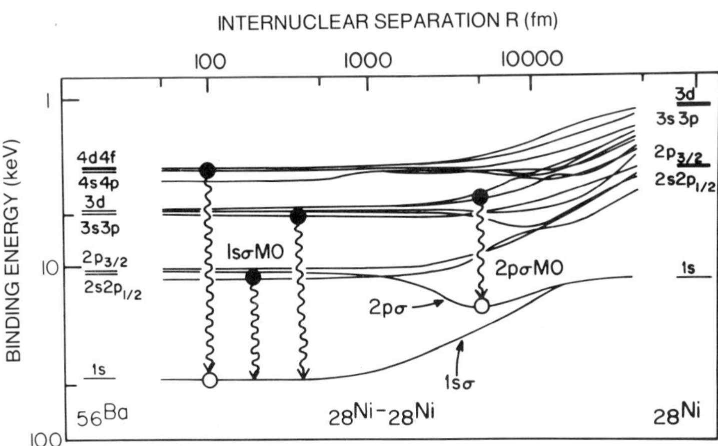

Figure 3.27. Molecular-orbital correlation diagram for the Ni + Ni quasimolecule showing the decay of $1s\sigma$ and $2p\sigma$ MO x rays (Mu 76b, Vi 81).

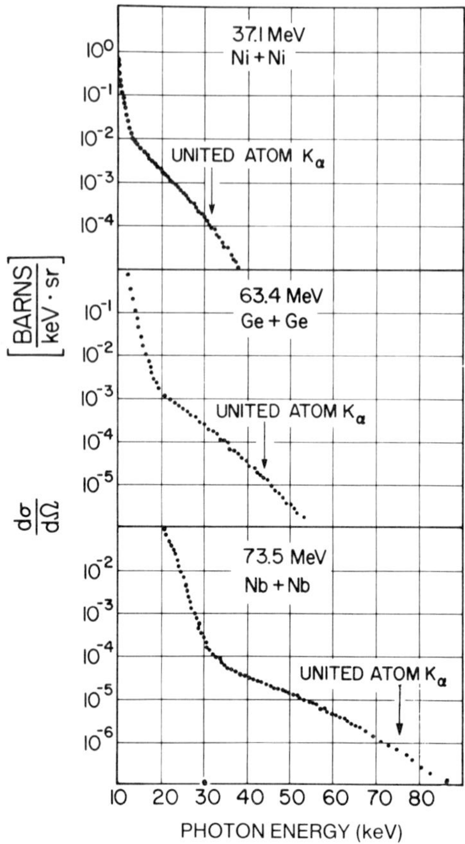

Figure 3.28. Cross sections for continuum x-ray production in three selected collision systems illustrating the two component features of the MO x-ray radiation (Vi 77, Vi 81).

Figure 3.28 (Vi 77, Vi 81) displays MO transitions from a series of symmetric collision systems after correcting the continuum spectra for external absorption and detector efficiency. The data display a two-component continuum structure which appears to scale with the combined projectile–target Z_u. The low-frequency component has been identified with transitions into the minimum in the $2p\sigma$ orbital (He 76, Ka 77) (labeled $2p\sigma$ MO's in Figure 3.27), and the high-frequency part with transitions into the $1s\sigma$ state (Me 73b, Da 74) ($1s\sigma$ MO's in Figure 3.27). These assignments have been aided by studies of the angular distributions of the x rays which display maxima corresponding to the energies where the $2p\sigma$ and $1s\sigma$ radiations are of comparable strength, and near the $1s\sigma$ united atom transition energy (Gr 74). The angular distribution studies will be discussed in detail below. Clearly, transitions to vacancies in higher-lying orbitals will also lead to the emission of lower-energy MO x rays (Kr 74, Wo 78, Mo 81b,

Mo 83). In keeping with our primary interest in high binding phenomena, we will restrict our discussion to $2p\sigma$ and $1s\sigma$ MO x rays only.

To produce continuum molecular orbital x-ray emission a vacancy is required in the $1s\sigma$ or $2p\sigma$ orbitals. We have noted previously that $2p\sigma$ vacancies are generated by rotational coupling to the $2p\pi$ level or by direct excitation to higher-lying molecular states and to the continuum. Processes for $1s\sigma$ vacancy production leading to MO x-ray emission are illustrated in Figure 3.29 for the case of symmetric collisions.

In general, $1s\sigma$ MO production proceeds by either a one- or a two-collision process. In the one-collision process, direct excitation of the $1s\sigma$ electron to vacant bound states and to the continuum is followed by MO x-ray emission in the same collision. In the two-collision process, $2p\sigma$ vacancies generated in a first collision are carried into the $1s\sigma$ level in the second collision. Although two collisions are required for the second production mechanism, the long lifetimes of K vacancies and the very small cross section for direct excitation of the $1s\sigma$ orbital actually make the two-collision mechanism more likely than the one-collision process in lower Z

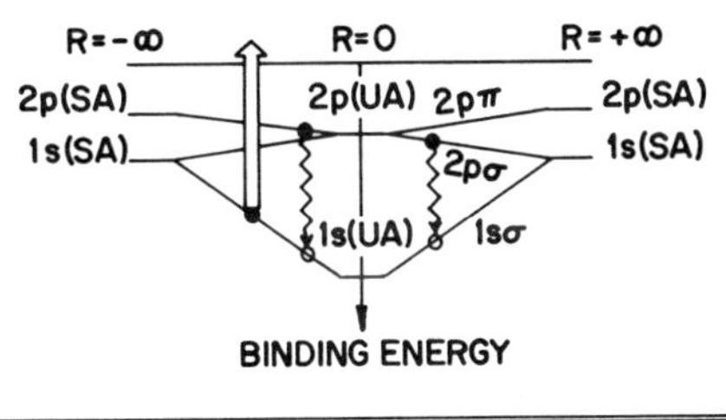

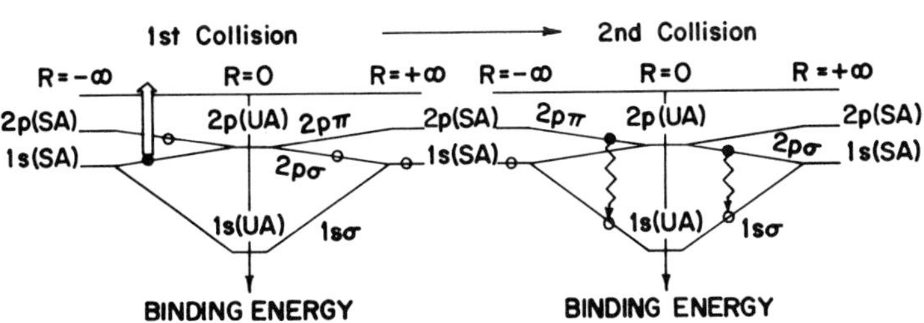

Figure 3.29. Schematic comparison of one- versus two-collision MO K x-ray production processes in symmetric systems (Vi 77).

systems and solid targets. At higher bombarding energies, higher Z and small impact parameters, the one-collision process grows in importance. It turns out that this dominance of two-collision excitation has been important in understanding the physics of MO x-ray production in the lower Z systems since calculations of one-collision $1s\sigma$ MO x-ray emission are difficult and have only been performed for the H^+–H system (Th 77).

There have been numerous studies of MO x-ray emission from light collision systems. Generally, the physics that has been extracted from such investigations has fallen short of goals initially envisioned. The reasons become evident from our discussion. We have selected for discussion those experiments that illustrate the main points which have emerged from studying the MO x rays.

In Section 3.3.2.1 salient features of MO x-ray cross sections in lighter collision systems are reviewed. Systematics of two-step production versus Z are treated, followed by discussion of gas target and charge-state dependence studies of the spectral yields. The distinctive directional anisotropy of the continuum K x rays is referred to in Section 3.3.2.2. Particularly the use of this anisotropy, in conjunction with the Doppler shift of the x rays, to extract information on the emitting source of the radiation is illustrated. Consideration of cross-section measurements and angular dependences in gas targets and results of polarization studies of the continuum x rays from selected encounters complete the discussion of the MO x-ray yields.

Section 3.3.2.3 considers particle–x-ray coincidence measurements. Experimental impact parameter dependences for MO x rays from the Ni + Ni system are presented as well as the MO x-ray–K x-ray cascade studies using the Nb + Nb quasimolecule. In a further discussion of MO x rays from the lighter collision systems we present measurements of the azimuthal anisotropy of the MO x-ray intensities followed by a discussion of interference effects in studies of impact parameter dependences.

3.3.2.1. Cross-Section Measurements. The results of systematic studies of $1s\sigma$ MO x-ray production are shown in Figure 3.30 and compared to a two-collision theory (An 78). In the two-collision picture, MO $1s\sigma$ x-ray cross sections are expected to scale with the asymmetry of the collision system, $\xi = Z_{\text{heavy}}/Z_{\text{light}}$, and the photon energy, $\nu = \omega/\omega_0$, $\hbar\omega$ being the energy of the x-ray and $\hbar\omega_0$ a characteristic frequency for the quasimolecule which can be taken to be the K_{α_1} transition energy of the united atom. In general, both the shapes and sizes of the $1s\sigma$ MO x-ray yields will also depend on the beam velocity, v, through the parameter $q = \omega_0 a_{KL}/v$, where $a_{KL} = Z_L^{-1}$ a.u. is the K-shell radius of the lighter collision partner. However, if one restricts study to intermediate x-ray energies ($0.4 \leqslant \nu \leqslant 1$), it can be shown that the shapes of the MO x-ray yields depend only on ξ and ν (An 78).

Figure 3.30 compares experimental results for continuum x-ray yields

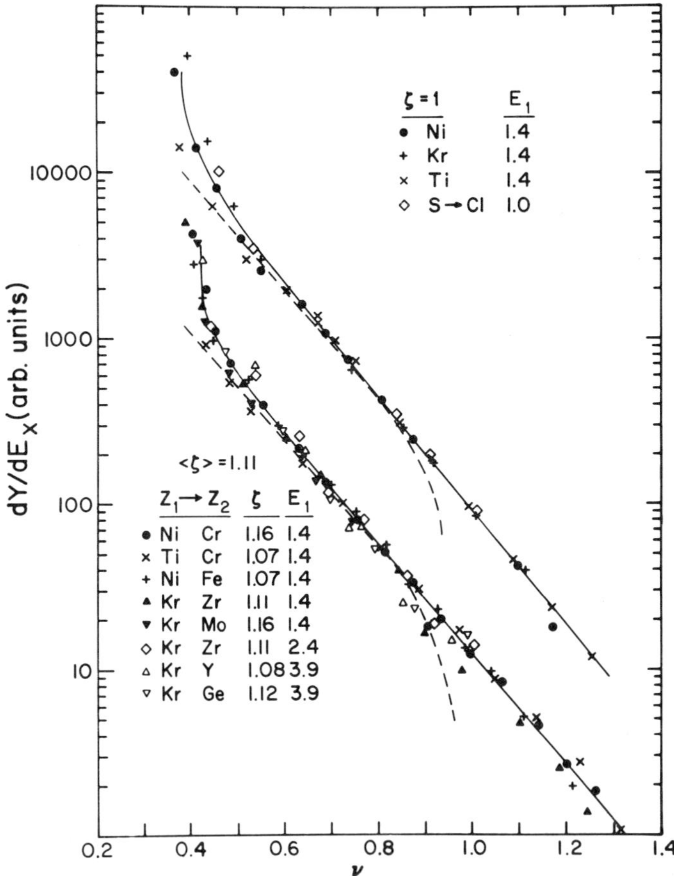

Figure 3.30. Thick target MO x-ray yields versus the reduced parameter ν. Solid line is fit to data. Dashed line, quasistatic two-step calculations. E_1 is the projectile energy in MeV/amu. See text (An 78).

with the two-collision theory for a variety of collision systems. Thick targets have been used in these studies primarily for reasons of experimental convenience. The data have been arbitrarily normalized to each other at $\nu = 0.7$. The scaled experimental results at equal ξ form universal curves in agreement with the predictions of the two-collision calculations. Furthermore, the agreement between these calculations (dashed lines in Figure 3.30) and experiment is good for $0.5 \leqslant \nu \leqslant 0.9$. These results confirm the molecular origins of the radiation and display the dominance of the two-collision production mechanism for lower Z collision systems when using solid targets.

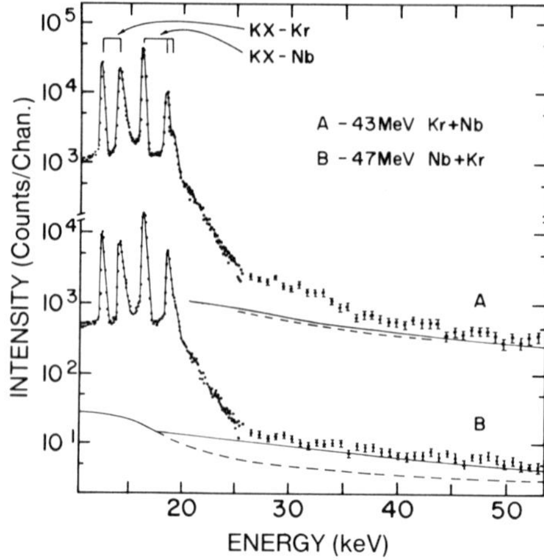

Figure 3.31. Continuum x-ray spectra from 43-MeV Kr → Nb (upper) and 47-MeV Nb → Kr (gas) target collisions (lower). Dashed line is measured background and solid line summed spectra of background and electron plus nuclear dipole bremsstrahlung (Fr 78).

In dilute gas targets, K vacancies generated in a first collision will nominally decay before entering a second collision, thereby resulting in the dominance of one-collision MO x-ray production. This feature is illustrated in Figure 3.31 (Fr 78) with measurements of $1s\sigma$ and $2p\sigma$ MO x rays from Kr + Nb (solid target) and Nb + Kr (gas target) collisions. The solid lines represent summed backgrounds from electron and nuclear bremsstrahlung (see Section 3.5 below). The two-component structure of the MO yields is evident in the figure. Only solid target data display an excess of experiment over background for the high-frequency ($1s\sigma$ MO) component. It is clear that in the solid target case there is a dominance of the two-collision mechanism. The low-frequency ($2p\sigma$) component, however, displays little change in intensity relative to the K x-ray lines in comparing the two targets. This result indicates that the one-collision mechanism is most likely responsible for the generation of $2p\sigma$ MO x rays (He 83).

An interesting experimental situation becomes available with targets of molecular structure where studies of the bombarding energy dependence of one- and two-collision MO x-ray production can be contrasted (Sa 77). Figure 3.32 (Le 81b) shows the spectra obtained from collisions of 300-keV O with O_2 and H_2O (gas) targets. For the diatomic O_2target, the close proximity of the two bound O atoms allows transfer of $2p\sigma$ vacancies to the $1s\sigma$ state in a second collision. For the H_2O target, the large separation between molecules suppresses two-step MO x-ray production. The full curve drawn in the H_2O target data is a superposition of the O_2 data which has been normalized to the O K_α peak. The large excess of O_2 to H_2O intensity between 1.3 and 2 keV is a beautiful illustration of the important role of the two-collision process in these light systems. The large peak at 1.1 keV represents two-electron one-photon x-ray transitions (see Section 3.5 below)

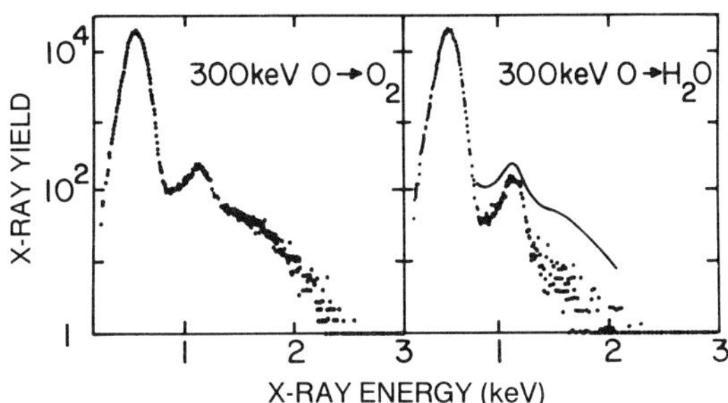

Figure 3.32. X-ray spectra from $O^+ \rightarrow O_2$ and H_2O (gas) collisions. The characteristic K_α lines are at 0.5 keV. Spectra are normalized to equal number of scattered particles at 45° to the beam. The solid curve in the right-hand picture is the superposition of the $O^+ \rightarrow O_2$ data (Le 81b).

in the O atom. Extension of these studies to higher beam energies (1 MeV) showed only slight differences between O_2 and H_2O targets. These latter results demonstrate the increasing contribution from the one-step process at higher projectile velocities due essentially to the rapid increase in direct $1s\sigma$ excitation with incident bombarding energy.

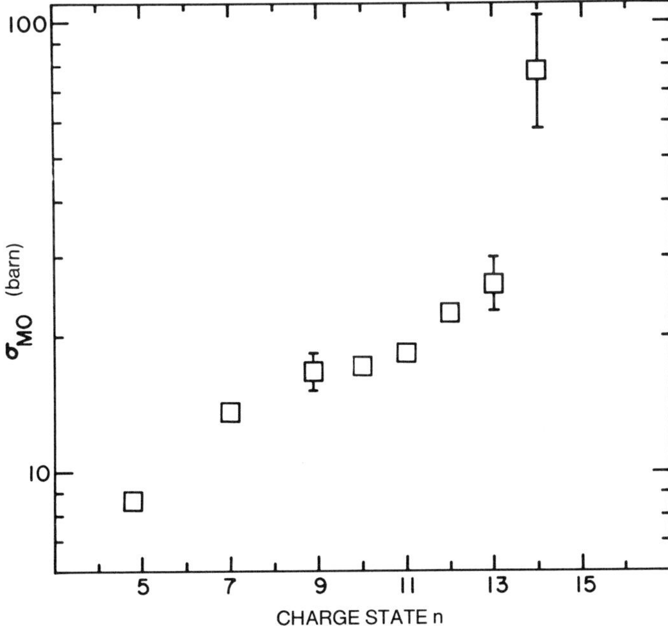

Figure 3.33. Total cross section for MO K x-ray production ($E_x \geqslant 4.2$ keV) versus projectile charge state n in 32-MeV S^{n+} + Ar collisions (Sc 78b).

More detailed information on MO x-ray production mechanisms can be gained from measurements of the charge state dependence of the cross sections. An example of one such study is given in Figure 3.33 (Sc 78b) where the charge state dependence of MO x-ray production in 32-MeV S^{n+} + Ar collisions is shown. Arguing that the use of a gas target limits production to one-collision processes only (Be 75, La 76), the authors interpret the increase in cross section with increasing incident charge state in Figure 3.33 as evidence for direct excitation of $1s\sigma$ electrons into empty bound orbitals. Such excitations are important in producing inner-shell vacancies in the heaviest collision systems as well (see discussion in Section 3.4 below).

The intrinsic dynamic nature of the quasimolecular system is directly reflected in the MO x-ray spectrum. Since a central role played by molecular orbital formation in very heavy systems is that of a vehicle for the generation of extremely strong binding during the course of these collisions, the extent to which these energies are well defined is an especially important question. The MO spectrum provides information on this question. Particularly, the extension of continuum x-ray emission to x-ray energies beyond the united atom static x-ray limit clearly illustrates collision broadening effects on the quasimolecular orbitals. For molecular orbital x rays, the continuum $1s\sigma$ radiation extends past the united atom kimit with a spectral energy dependence that can be described by the relation (Ma 74, Mu 75)

$$\frac{d\sigma}{d(\hbar\omega)} \alpha \hbar\omega [\exp(-\hbar\omega/2\Gamma)] \tag{3.9}$$

where $\hbar\omega$ is the radiated photon energy and Γ a characteristic falloff parameter. The experimental projectile velocity dependence of the half-width, H, of this experimental tail $[H = (2ln2)\Gamma]$ is given in Figure 3.34 (Vi 79) for

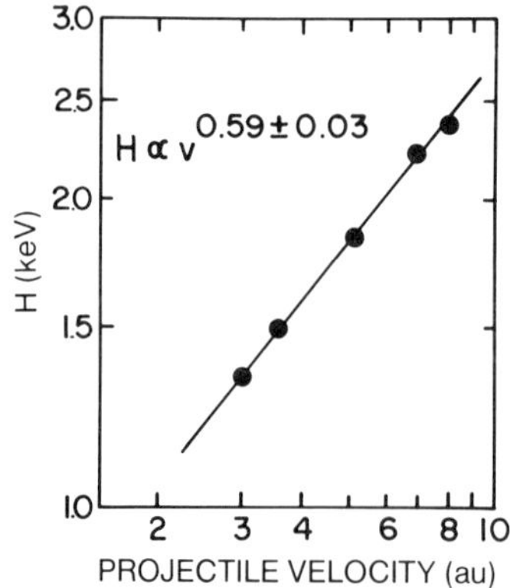

Figure 3.34. Half-width (H) values in keV as a function of projectile velocity in atomic units. The solid line is a least-squares fit to the data showing an $H \propto v^{0.6}$ dependence (Vi 79).

the case of Ni + Ni collisions. The half-width is found to increase with $v^{0.59 \pm .03}$, where v is the projectile velocity.

A number of theoretical studies have been directed towards the understanding of this velocity dependence (Ma 74, Mu 75, Be 75b, An 78b, Fr 79). In addition to its importance in the treatment of dynamic molecular effects, this tailing of MO emissions beyond the united atom limit has important consequences for the peaked angular dependence of the molecular orbital x rays which was alluded to earlier. Within the context of the two-collision picture of MO x-ray production, both the angular distribution and intensity of $1s\sigma$ x rays can be theoretically obtained by Fourier analyzing the dipole velocity matrix element between the initial and final molecular states (Br 77). Such a fully dynamical treatment of the collision broadening problem is necessary in order to describe the experimental angular dependences of the x rays (An 78b) (see discussion below). The results of such a dynamical study are shown in Figure 3.35 where measurements (Vi 77, Vi 79) of continuum x rays in Ni + Ni collisions (dashed lines) are

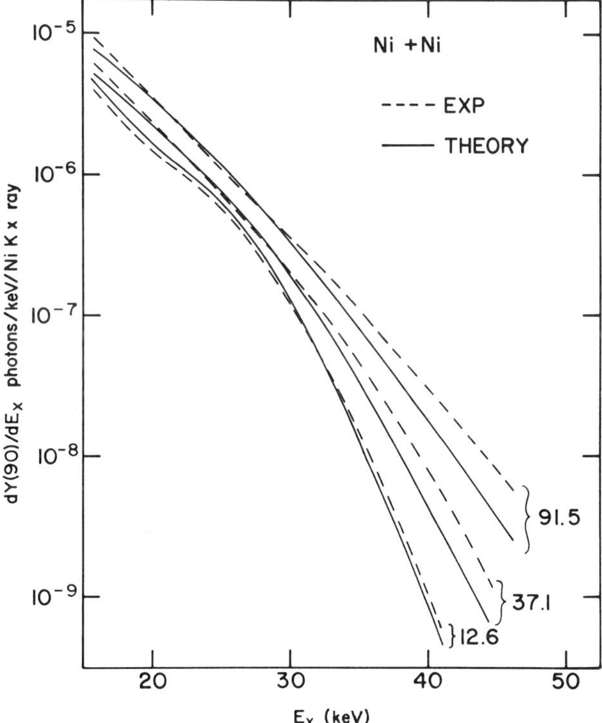

Figure 3.35. Experimental versus theoretical MO x-ray intensities in Ni + Ni collisions for 12.6, 37.1, and 91.5 MeV beam energies (Vi 77, Vi 79, An 78b).

compared to theory (An 78b) (solid lines) at various bombarding energies. The comparison is intitiated at 18 keV in order to exclude regions where $2p\sigma$ MO x rays contribute to the measured yields. Measurement and calculation agree very well at 12.6 MeV, but theory fails to predict the measured increase in yields beyond the united atom K_α x-ray energy of 32 keV at higher projectile velocities. These results may be interpreted as indicating additional contributions to the experiment from one-collision MO x-ray production at high x-ray and beam energies.

3.3.2.2. Angular Dependences. Except for the general features of scaling and the two-component structure, measurements of MO x-ray spectral yields display no additional signature or feature which can be exploited as a definitive indication of radiation from the combined projectile target system. This has been one of the limitations in utilizing MO radiation

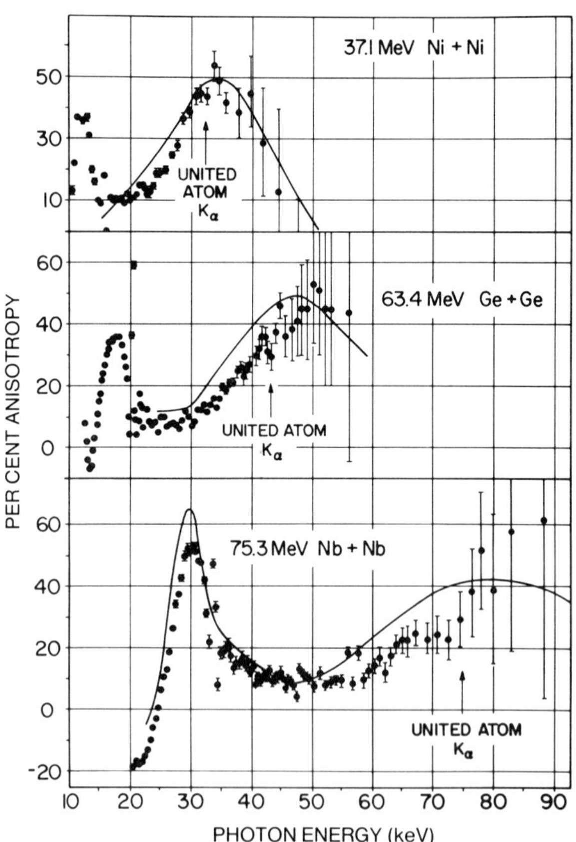

Figure 3.36. The directional anisotropy of MO x rays for three collision systems compared to theory (solid lines) (Vi 77, Vi 81, An 79, An 79b).

as a source of information on the structure of the quasimolecule. In this connection, the discovery (Gr 74) of the distinctive directional anisotropy of the $1s\sigma$ radiation first provided significant optimism for developing a spectroscopy of the combined atom using MO x rays. Figure 3.36 displays the percent anisotropy $= I(90°)/I(0°) - 1$ for selected collision systems (Vi 77, Vi 81). The striking peaking of these spectra at photon energies near the high-frequency ends of the $2p\sigma$ and $1s\sigma$ x-ray components (compare Figure 3.28) and the obvious scaling of these peaks with the combined projectile–target Z, generated enthusiasm for systematic exploration of this effect in lighter and heavier systems. The solid lines in Figure 3.36 are calculations of the directional anisotropies (An 78b, An 79, An 79b) which will be discussed further below (see Figures 3.39, 3.40, and associated discussion).

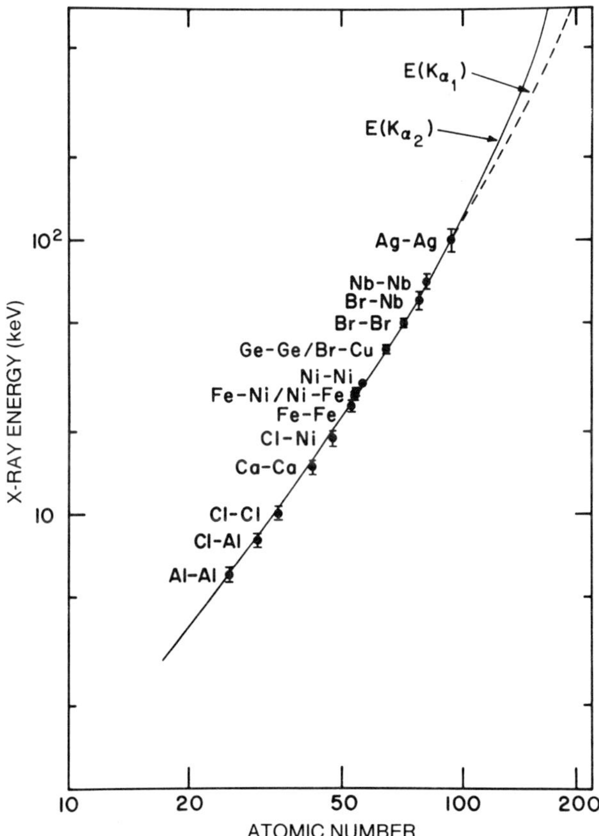

Figure 3.37. Plot of MO K x-ray anisotropy inflection point versus united atom number. The solid and broken curves indicate the united atom $K_{\alpha 1}$ and $K_{\alpha 2}$ transition energies (St 77).

Subsequent systematic studies of the $1s\sigma$ MO x-ray anisotropy peak led to a search for semiempirical correlations between the peak positions and the K x-ray energies of the united atomic system. The results of such a systematic investigation are shown in Figure 3.37 (St 77). Plotted are the x-ray energies at which the first derivative of the empirical $1s\sigma$ MO anisotropies reach their maxima for various collision systems in comparison to the K_α x-ray energies of the united atom. This striking correlation between the x-ray anisotropy points of inflection and the united atom K_α energies illustrates the degree to which scaling with the combined projectile–target Z is obeyed.

Similar results are true for the low-frequency peaks of Figure 3.36. These peaks are associated with $2p\sigma$ MO x rays and can also be found to obey a scaling law. Figure 3.38 (St 78b) plots the results of systematic studies of the $2p\sigma$ peak position in a variety of symmetric collision systems versus the atomic number of the separated atom. The peak position was found to equal approximately 1.5 times the characteristic K x-ray transition energy. The solid line in the figure is a fit to the data illustrating the Z^2 scaling of the results. Interpretation of these data in terms of molecular orbital correlation diagrams suggests that peaking occurs at transition energies corresponding to the minimum in the $2p\sigma$ molecular orbital (He 76, Ka 77) (see Figure 3.27; wiggly line labeled $2p\sigma$ MO).

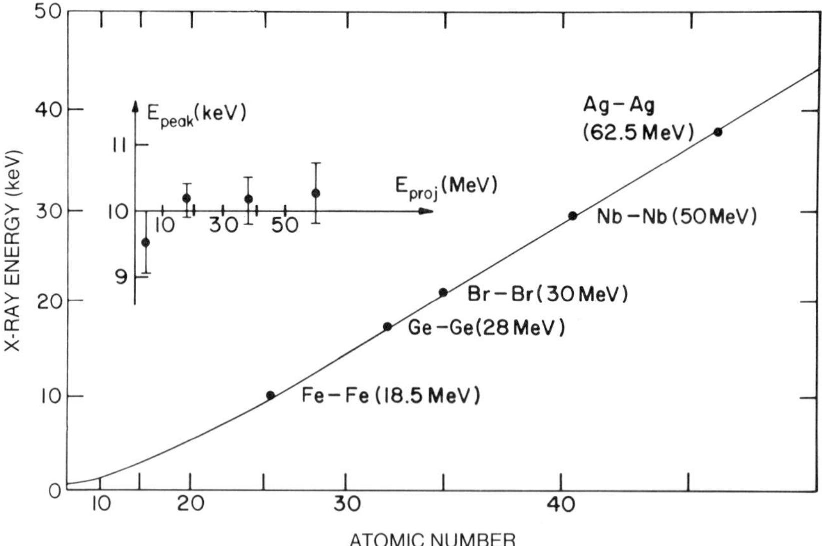

Figure 3.38. MO $2p\sigma$ anisotropy peak energy versus separated atom atomic number (quadratic abscissa scale). The solid curve is a fit to data using expression $E_x = a(Z - s)^2$. $a = 20.6 \pm 8$ ev and $s = 3.6 \pm 0.6$ from the fit (St 78b).

It turns out that this peaking of the directional anisotropy is a general feature of MO x-ray transitions not only to $1s\sigma$ and $2p\sigma$ final states, but also to higher-lying levels. Extensive studies of MO x-ray transitions to these higher-lying orbitals have been carried out, and have led to semiempirical scaling laws for the peaks (Wo 78, Mo 81b).

A number of theoretical explanations for the peaking of the directional anisotropy have been proposed (Wo 78b, Ha 78, An 78b). The most successful description (An 78b) entails a full dynamical treatment of the collision problem which includes a Fourier analysis of the radiative dipole matrix elements within a quasimolecular basis, and coupled-channel calculations of the time development of the strongly mixed $2p\pi$, $2p\sigma$ orbitals. The dynamic nature of these collisions leads to the inability of the $2p\pi$, σ molecular orbitals to follow the rapid rotation of the internuclear axis in close collisions. As a result, the electron clouds tend to remain fixed in space, while the molecular axis rotates within them. Since the π and σ states are defined by their angular momentum projections along this internuclear axis, complete "slippage" (Gr 76c) of the electron clouds for 90° scattering in the center of mass causes coupling between the $2p\pi$ and σ states with unit probability. This phenomenon is simply the kinematic peaking effect in $2p\pi$, $2p\sigma$ rotational coupling which was discussed earlier in Section 3.3.1.2.

It is, furthermore, necessary to include transitions from the $3p\pi$ orbital in order to explain the measured $1s\sigma$ directional anisotropies. Although these "K_β"-like transitions in the quasimolecule are weaker than the "K_α"-like x rays ($2p\pi$–$1s\sigma$), their higher energy leads to their eventual dominance at frequencies above the united atom K_β x-ray energy. The strong rotational or Coriolis mixing of the $3p\pi$ and $3p\sigma$ levels must be included in the calculations, although the rather weak coupling between the $3p$ and $2p$ MO orbitals can be neglected (An 78b).

The results of such calculations are shown in Figure 3.39 (An 78b) and compared to measurements of the Ni + Ni collision system (Vi 77, Vi 81). The methods described above do not, however, directly lead to a proper description of the empirical results. Additional alignment of a partially stripped $3p\pi$ orbital is an essential ingredient of the theoretical description. This level must be about half occupied at 12.6 MeV and about one-tenth full at 91.5 MeV in order to explain the measured anisotropies. Information on the occupation probabilities of the $3p\sigma$ and $2p\pi$ orbitals was derived from measurements of K x-ray line shifts in these collisions (Vi 77, Lu 77). These shifts reflect the L-shell occupation of the separated Ni atoms at the time of x-ray emission and can be used to estimate $3p\sigma$ and $2p\pi$ vacancy probabilities since both these molecular orbitals join the Ni L shell in the separated atom limit. Reasonable physical arguments can be made in support of the theoretical MO occupation amplitudes needed to provide agreement between theory and experiment. It is important to point out that,

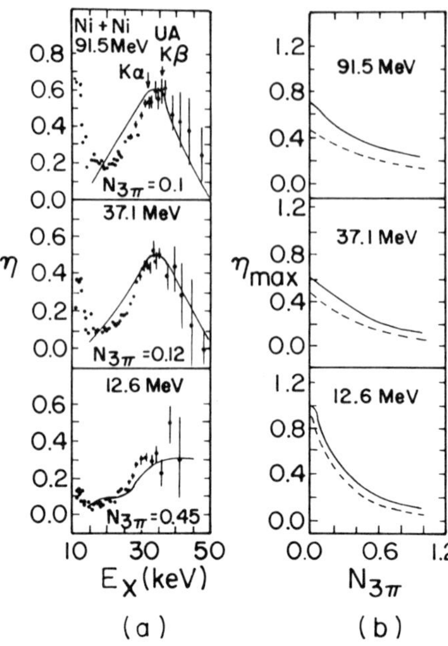

Figure 3.39. Theoretical versus experimental anisotropies calculated using best fit values for the 3π occupation probabilities ($N_{3\pi}$). Right-hand figure shows dependence of height of theoretical anisotropy peak η_{max} versus $N_{3\pi}$ for two different model assumptions (Vi 77, Vi 81, An 78b).

on the basis of these calculations, the author predicted that superheavy collision systems would not exhibit a $1s\sigma$ MO directional anisotropy (An 78b). This prediction was confirmed with subsequent measurements of the Pb + Pb system where no noticeable peaking of the $1s\sigma$ directional anisotropy was observed (St 80). Therefore, the dramatic scaling of the $1s\sigma$ peaks illustrated in Figure 3.37 does not extend into the relativistic regime, and, unfortunately, it cannot be used to provide a semiempirical spectroscopy of the $1s\sigma$ molecular orbital under very high binding conditions. Further discussion of these results follows in Section 3.4 below.

Similar studies may be carried out in order to describe results of $2p\sigma$ MO directional anisotropies. For the $2p\sigma$ case, contributions from the $3d\sigma$, π, δ and $4d\sigma$, π, δ orbitals must be considered. Analogous to the $1s\sigma$ MO case, Coriolis mixing between the $2p\pi$ and $2p\sigma$ orbitals and among the $3d\sigma$, π, and δ levels must be treated with coupled-channel calculations. Furthermore, the contribution of the $1s\sigma$ MO x rays to photon energy regions where $2p\sigma$ peaking occurs must first be subtracted from the data.

The results of such calculations (An 79) in comparison to measurements (Vi 77, Vi 81) of the anisotropies in the Nb + Nb collision system are shown in Figure 3.40. The solid and dashed lines are before and after correction for the $1s\sigma$ contribution, respectively. Theoretical examination of the projectile plus target Z_u dependence of the $2p\sigma$ anisotropy peaks explains the origins of the observed scaling (see Figure 3.38) while

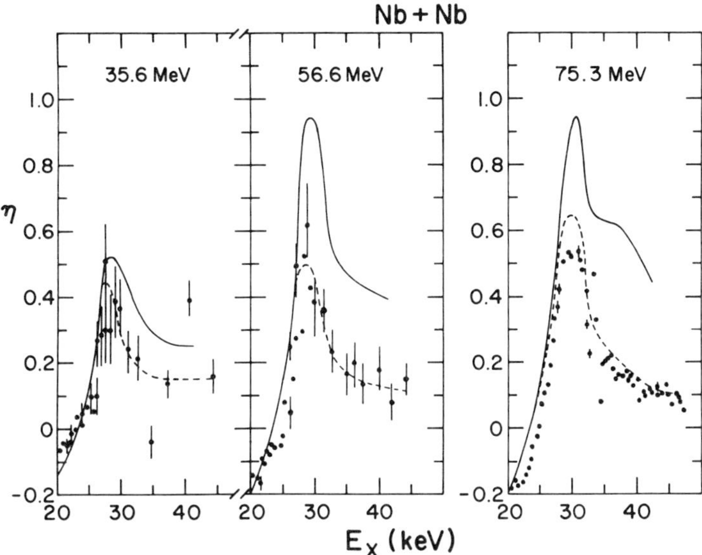

Figure 3.40. Measured versus calculated $2p\sigma$ MO anisotropies in Nb + Nb collisions. Solid line is theory; dashed line includes corrections for empirical $1s\sigma$ MO x-ray background contribution (Vi 77, Vi 81, An 79).

emphasizing that scaling occurs only when relativistic effects on the molecular orbitals involved are small. Thus, ironically, in collisions between very heavy atoms, where measurements of directional anisotropies could lead to spectroscopic information on the united system, scaling is expected to break down.

There is one important experimentally oriented point which has been utilized to establish an important conclusion. The continuum x rays discussed above are emitted from a quasimolecular system which moves with the velocity of the center of mass along the beam axis. Therefore, a correction has to be made to the observed laboratory directional anisotropies for the Doppler shift of the x rays. This correction increases the magnitude of the observed peaks in the directional anisotropy by factors of 2 and more and is, therefore, extremely important (Gr 74). Furthermore, the Doppler effect can be utilized to derive information on the velocity of the emitting source of the continuum x rays and, therefore, be used to identify them as quasimolecular in origin (Me 75, Vi 78).

Figure 3.41 shows measured spectra from the Ni + Ni system at 64.8 and 12.6 MeV bombarding energies after corrections for absorption and detector efficiency (Vi 78). The increased intensity at forward (45°) versus backward (−135°) laboratory angles is due primarily to the Doppler shift of the MO x rays. Assuming that the continuum x-ray intensities are

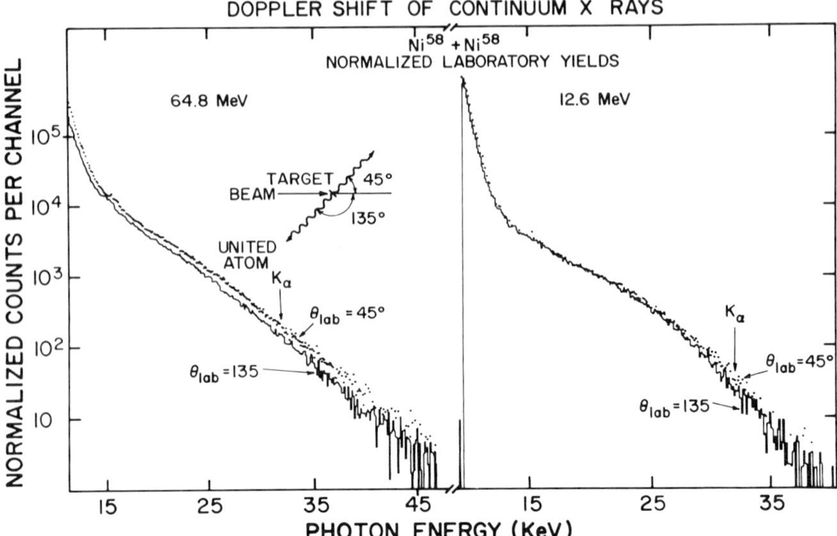

Figure 3.41. Absorber efficiency corrected, and normalized forward-backward continuum x ray Ni + Ni yields at 64.8 and 12.6 MeV (Vi 78).

forward–backward symmetric in their rest frame of emission, the observed shifts given in Figure 3.41 can be analyzed in order to extract the emitter velocity of the source of these x rays (Vi 78, Me 75, Fo 76, Fr 76). Corrections for relativistic solid angle effects, and angle shifts in going from the center of mass to laboratory frames, as well as for the x-ray Doppler shift must be included (Vi 78).

The results of such an analysis are given in Figure 3.42 (Vi 78). The top portions of the figure display the directional anisotropies, and the bottom portions the extracted emitter velocities of the x rays. For photon energies above 12 keV in the 64.8-MeV data, good agreement with a center-of-mass emitter velocity is obtained. The rise in both the directional anisotropy and its associated emitter velocity near 10 keV x-ray energy at 64.8 MeV is due to contributions from radiative electron capture (Sc 72, Kl 75, Sp 79). Above 22 keV, the 12.6-MeV data illustrate good agreement with the center-of-mass velocity. The origin of the very low emitter velocities in the 12–20 keV region at 12.6 MeV remains unexplained. Possible contributions from recoiling target atoms or secondary electron bremsstrahlung to the 12.6-MeV data do not appear to be the cause of the discrepancy. Further discussion of other sources of continuum x-ray emission in these collisions follows in Section 3.5 below.

An interesting result emerges from comparing anisotropies observed for MO x rays from gas and solid targets. Figure 3.43 shows results of studies of

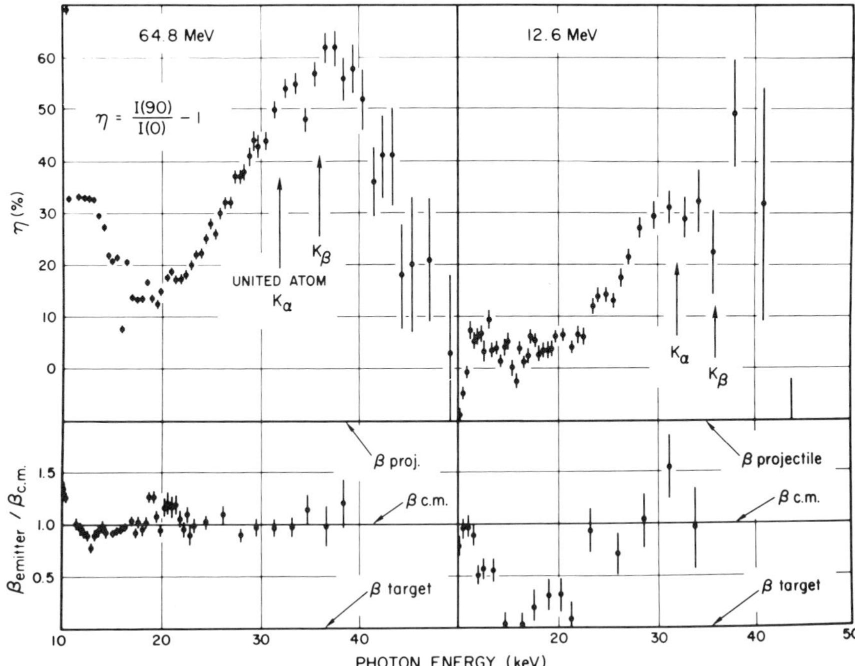

Figure 3.42. Anisotropies η (upper part of figure) and extracted emitter velocities in units of the center-of-mass velocity (bottom figure) for continuum Ni + Ni x rays using Doppler shift technique. See text (Vi 78).

directional anisotropies in gas versus solid targets at four bombarding energies (St 81). The solid lines guide the eye through the solid target data and the dashed lines through the gas target results. Qualitatively similar behavior in the gas versus solid target anisotropies is found except at low bombarding energies where the gas target result is small and shifted to the left. Adjustment of vacancy occupation numbers within the context of a two-step production mechanism described above cannot account for these low-energy gas target results. Apparently, a full one-collision calculation is necessary in order to explain these data.

One measurement of the linear polarization of MO x rays from light collision systems has been made. The results of this extremely difficult measurement are shown in Figure 3.44 (Mo 80) in comparison to calculation (Br 79) assuming different numbers of $2p\pi$ vacancies. The data are connected with dashed lines in order to guide the eye. Results agree well with calculation when two initial $2p\pi$ vacancies are assumed ($\frac{1}{4}N_{2p\pi} = 0.5$ in Figure 3.44). The MO x rays are found to exhibit a positive linear polarization parallel to the beam axis. A comparison of these polarization

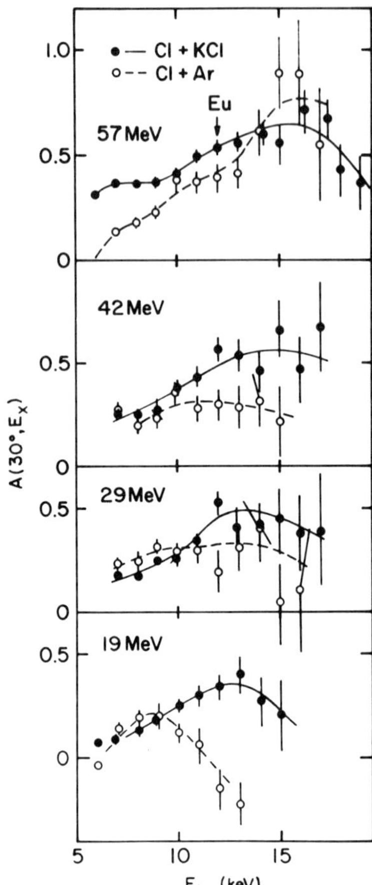

Figure 3.43. Gas (Cl + Ar) and solid (Cl + KCl) target anisotropies for 19–57-MeV Cl projectiles. Lines guide the eye (St 81).

results with determinations of the directional anisotropy show agreement if $E1$ radiation only is responsible for the MO x-ray yields.

3.3.2.3. Particle and X-Ray Coincidence Measurements. A small number of coincidence studies of MO x-ray production in lighter collision systems have been carried out. The additional detail provided by these studies has allowed simpler and more direct comparisons between experiment and calculation to be made. We present several selected examples of these experiments in this section.

The first example involves measurements of the impact parameter dependence (Ts 76, Sc 78c, Sc 80b, Sc 82) of MO x-ray radiation in 90-MeV Ni + Ni collisions. The results of these experiments are compared to two-collision mechanism calculations (An 82) of $1s\sigma$ MO x-ray production probabilities at 20 and 32 keV in Figure 3.45 (Sc 82). The top part of the

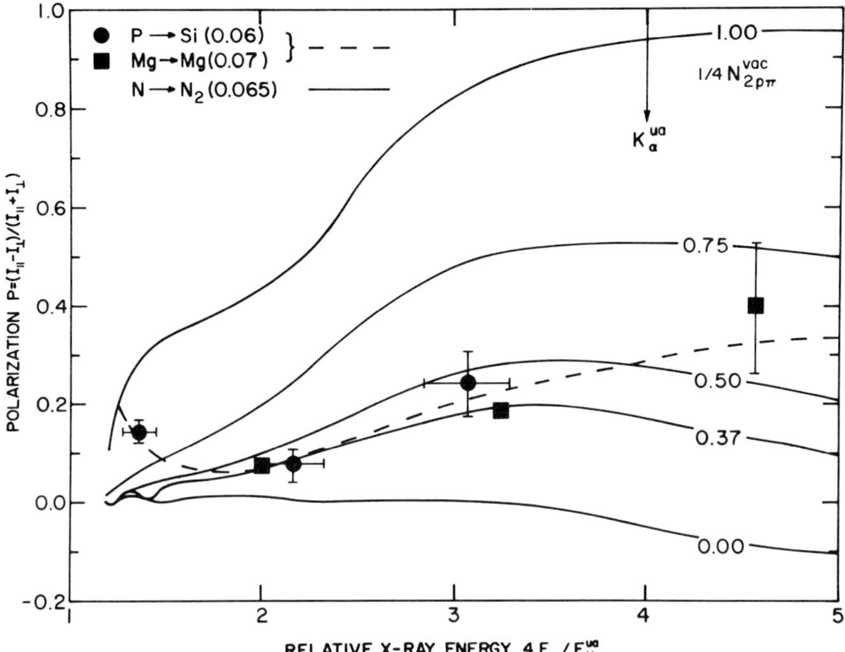

Figure 3.44. MO x-ray polarization versus scaled x-ray energy in 1.3-MeV Mg → Mg and 1.7-MeV P → Si collisions compared to theory (solid lines) for various values of the $2p\pi$ vacancy probability $N_{2p\pi}^{\text{vac}}/4$. The numbers in brackets following the collision systems are the adiabaticity ratios $v_{\text{proj}}/v_{\text{elec}}$. v_{proj} is the projectile velocity and v_{elec} the united atom K-shell electron velocity (Mo 80).

figure [labeled $P_K(b)$] plots the measured impact parameter dependence of the characteristic K x rays. The dashed lines in the bottom portion of the figure are the calculated two-step dependences assuming that the K vacancy generated in the first collision is carried into the second collision along the incident beam axis. This is equivalent to assuming that K vacancies are produced in the first collision at very large impact parameters only. The solid line represents the effect of folding the actual empirical projectile scattering angle dependence of K-vacancy production into the two-step MO x-ray calculations. Since this folding causes a mixing of laboratory azimuthal angles (ϕ) of the projectile in the first collision into various final observed projectile scattering angles with respect to the beam axis (θ) after the second collision, the calculations become somewhat complicated (An 82). Furthermore, it is clear that, under such conditions, the definition of an impact parameter no longer becomes meaningful. Instead, the abscissa in Figure 3.45 represents the pseudo-impact-parameter, b, defined as,

$$b \equiv \rho \cot \theta \tag{3.10}$$

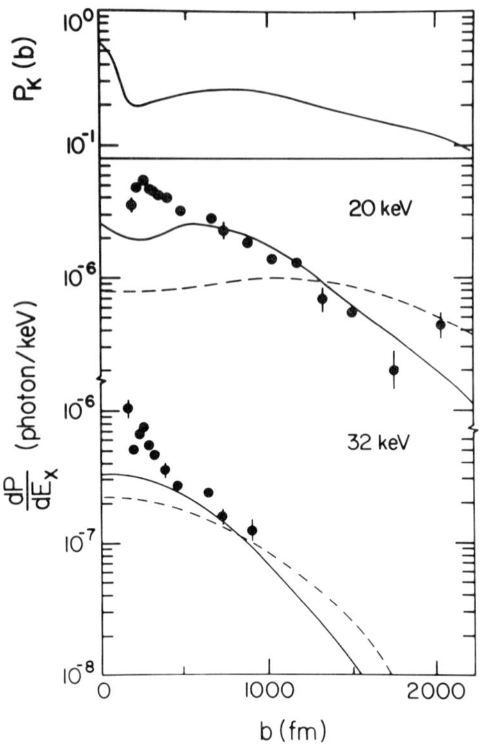

Figure 3.45. Impact parameter dependence for characteristic K x rays (top). Comparison of measurement to two-step collision calculations of the impact parameter dependence for MO x-ray production (bottom). Ni + Ni collisions at 90 MeV (Sc 82, An 82).

where ρ is one-half the distance of closest approach in a head-on collision and θ is the experimental scattering angle in the laboratory. The parameter b represents the actual impact parameter given by Rutherford scattering only when K-vacancy production in the first collision occurs at very large impact parameters and negligibly small θ. The results of Figure 3.45 show that this folding of the empirical impact parameter dependence in the first collision is necessary in order to correctly describe the measured results. The remaining excess of experiment over theory at small b can be interpreted as additional contributions from single collision MO x-ray production. One-step contributions are, therefore, important at high $1s\sigma$ MO x-ray energies and small impact parameters.

A second type of coincidence study has been performed between characteristic and continuum x-rays from the Nb + Nb collision system. The basic concept behind such an experiment is illustrated in Figure 3.46 (Ob 80), where the existence of a cascade relationship between certain MO x-ray transitions and separated atom K x rays is depicted. The upper part of the figure shows that $2p\sigma$ to $1s\sigma$ MO x-ray transitions result in the generation of a $2p\sigma$ vacancies which will, with high probability, survive the collision

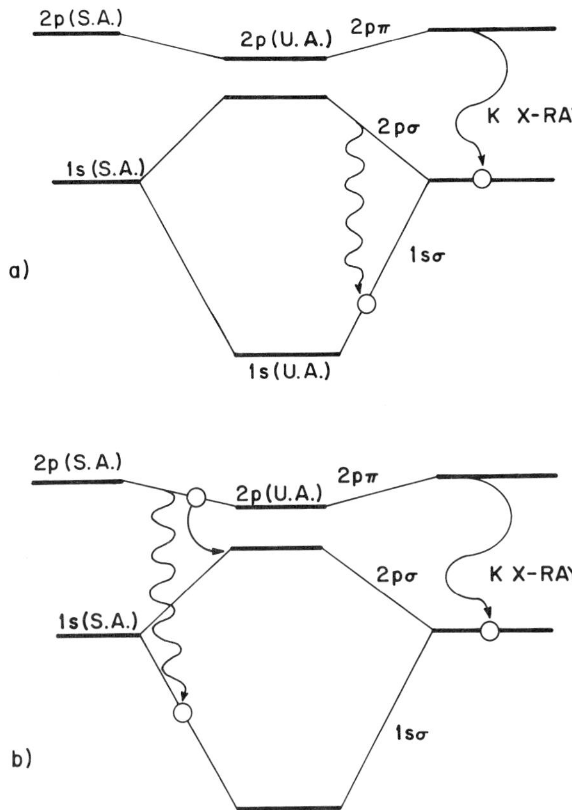

Figure 3.46. Schematic correlation diagrams showing (upper) the cascade relationship between $2p\sigma \rightarrow 1s\sigma$ transitions and separated atom K x rays. Also shown (lower figure) are the effects of rotational coupling leading to $2p\pi \rightarrow 1s\sigma$ cascades when $1s\sigma$ vacancies are present in incoming part of collision (Ob 80).

and decay into K x rays. In the bottom part of the figure, the contributions to K x-ray, MO x-ray coincidences from $2p\pi$ to $1s\sigma$ MO x-ray transitions with an associated rotational coupling of the resulting $2p\pi$ vacancy to the $2p\sigma$ orbital are illustrated. Coincidences between characteristic and continuum x-rays should therefore select $2p\pi$ and $2p\sigma$ to $1s\sigma$ MO transitions and suppress contributions from higher-lying MO orbitals to the $1s\sigma$ final state x-ray yields.

Figure 3.47 (Ob 80) shows the results of the first such coincidence measurements in the Nb + Nb system at three bombarding energies. The solid lines and their continuation into data points show spectra of continuum x rays in coincidence with characteristic K x rays. The dot-dashed curve in

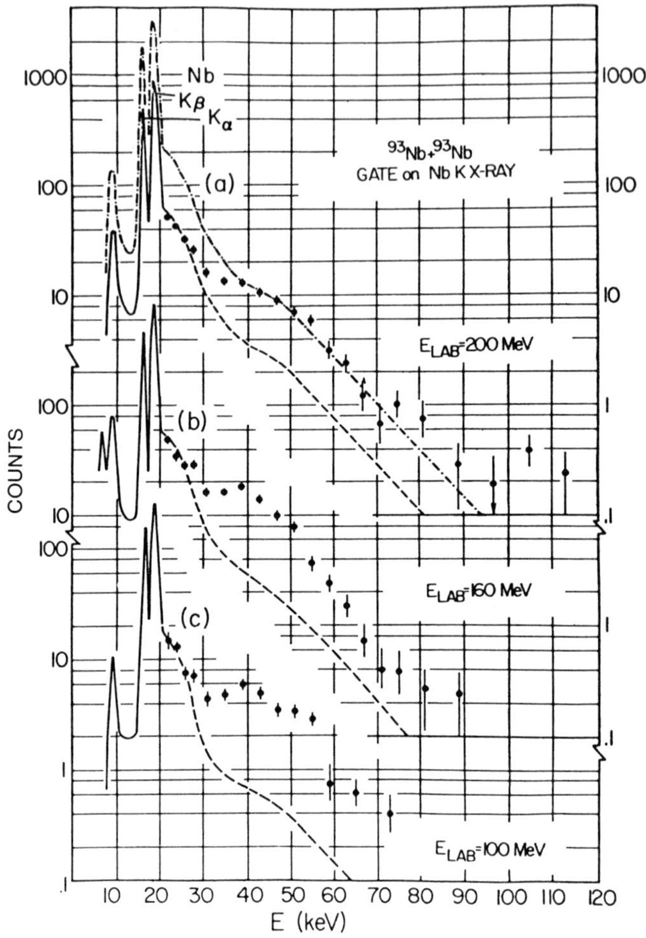

Figure 3.47. *K* x-ray, MO x-ray coincidence spectra in Nb + Nb collisions at 100, 160, and 200 MeV. See text for explanation (Ob 80).

part (a) of the figure shows the singles spectrum normalized to the coincidence data in the region 40–70 keV. This latter comparison implies that the intensity of continuum x rays below 40 keV is reduced relative to those above 40 keV when the coincidence condition is imposed. This result is consistent with the assignment of x rays below 40 keV to primarily $2p\sigma$ final state transitions.

The appearance of characteristic *K* x rays in the coincidence data of Figure 3.47 reflects the relatively large probability of double *K*-vacancy production in these collisions. Such double vacancy production in the quasimolecule implies the presence of a second source of *K* x-ray, MO x-ray coincidences due to multiple excitations which compete with the cascade events of interest. Solid lines and their continuation into dashed curves in Figure 3.47 represent estimates of these multiple-vacancy contributions to the coincidence yields. The authors conclude that multiple-vacancy processes cannot account for the measured coincidence intensities above 40 keV, and that significant contributions from $2p\sigma$ and $2p\pi$ to $1s\sigma$ cascade events have, in fact, been observed.

A third type of coincidence study has been performed in order to test various theoretical descriptions of the angular distribution of the molecular orbital x rays. Different model calculations of MO x-ray emission predict conflicting dependences for the x-ray intensity on the azimuthal angle of emission ϕ, where ϕ is the angle between the scattering plane and the plane formed by the beam axis and the x-ray detector. The experiment consists of detecting the continuum x-ray yields parallel and perpendicular to the scattering plane using particle, x-ray coincidence techniques (Bu 81, Sc 81b). Results of measurements on the Ni + Ni system at 90 MeV are shown in Figure 3.48 (Sc 81b). The percent azimuthal anisotropy ($I_{\text{perpendicular}}/I_{\text{parallel}} - 1$) is small at all x-ray energies and impact parameters. The solid lines (An 81), which are full dynamical predictions (An 78b) of the anisotropies, are in better agreement with experiment than other available model calculations (Ha 78, Wo 78b).

Recently, several authors have explored interference effects produced by transitions to vacant $1s\sigma$ states during the incoming and outgoing parts of the collision trajectory. By colliding hydrogenlike 10-MeV C1 ions with Ar

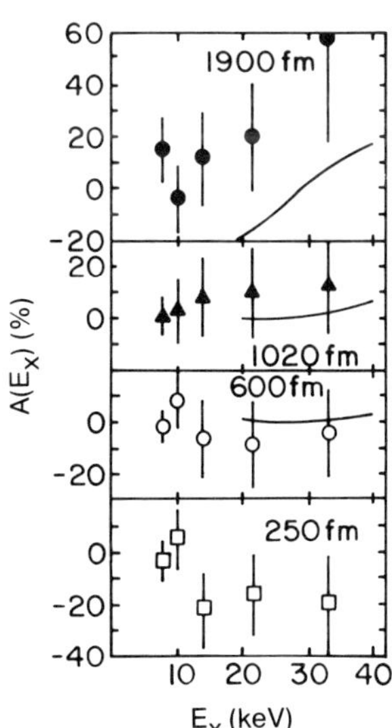

Figure 3.48. Azimuthal anisotropy $A(E_x)$ versus x-ray energy E_x for different impact parameters. Solid lines, calculation (Sc 81b, An 81).

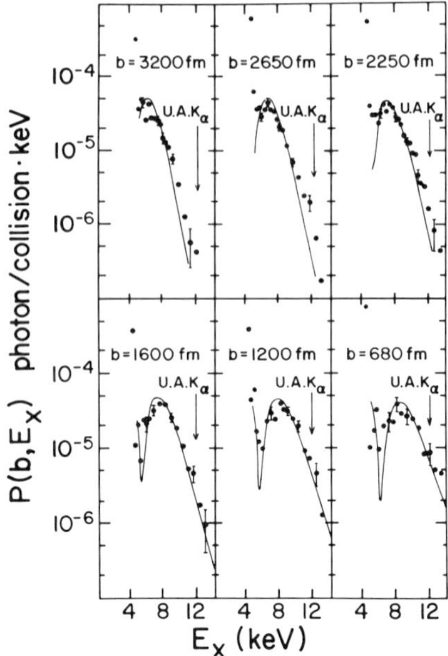

Figure 3.49. X-ray production probability versus photon energy for various impact parameters. Arrows indicate united atom (U.A.) K_α transition energies. Solid lines are dynamical calculations (Ts 83, An 83).

gas targets, an essentially 50% probability of bringing a $1s\sigma$ vacancy into the collision is established. Under such conditions, oscillations in the resulting MO x-ray spectra are expected if the impact parameter of the collision is defined by a particle–MO x-ray coincidence condition. Figure 3.49 (Ts 83) compares experimental MO x-ray production probabilities per keV for the MO x-ray intensities at various impact parameters with $1s\sigma$ MO x-ray theory (An 83). The agreement is quite good over the range of photon energies and impact parameters shown. A striking dip in MO x-ray intensity at about 6 keV is experimentally observed and theoretically predicted at small impact parameters. Corresponding dips in the singles and large impact parameter data are neither experimentally observed nor theoretically predicted.

3.4. X Rays from Heavy Collisions

In extending our discussion from lighter systems to the very heavy collision partners, it is important to recall the similarities and differences between the relativistic and nonrelativistic encounters. Clearly, in both cases, adiabatic molecular orbital levels are involved; for the heavy systems, the adiabatic condition is extremely well satisfied by the $1s_{1/2}\sigma$ and $2p_{1/2}\sigma$ orbitals. However, the large spin–orbit splitting of the molecular levels in the very heavy systems (compare Figures 2.5, and 3.27) leads to a different topography of the levels in energy space, resulting in the dominance of different couplings among orbitals in the two cases. In particular, the rapid increase in the inner-shell binding energies with decreasing internuclear separation, and the strong localization of the $1s_{1/2}\sigma$ wave function near the nuclei for small internuclear distances (Figure 2.6), are aspects of the relativistic quasimolecule which cause qualitatively different behavior of measured excitation and decay probabilities. The effect of these strong binding and relativistic phenomena on the *K* x-ray total and differential production cross sections will be the primary focus of this section.

In the following pages, discussion of x rays from the relativistic quasimolecule is begun with consideration of total *K* x-ray cross section measurements in Section 3.4.1. The *Z* dependence of the cross sections is contrasted with similar results from lighter systems, and $1s\sigma$ vacancy production cross sections extracted from these *K* x-ray studies are compared to theory. The scaling properties are then discussed, followed by review of vacancy sharing in these collisions and discussion of the *K* x-ray impact parameter dependences obtained from line-shape measurements in Section 3.4.2.

Section 3.4.3 treats coincidence measurements of impact parameter dependences for $1s\sigma$ excitation. The results are discussed within the context of the scaling law for $1s\sigma$ excitation presented in Section 2.3. The role of coupled-channel effects, excitations to empty bound states, and screening corrections in the $1s\sigma$-vacancy production process is then reviewed. The feasibility of quantitatively extracting binding energies from measurements of the scaling behavior of $1s\sigma$ excitation in such impact parameter measurements is examined.

In Section 3.4.4, results of MO x-ray measurements in the Pb + Pb quasimolecule are presented. The cross sections, directional anisotropies, and emitter velocities of the continuum x-ray yields are considered. Finally, impact parameter measurements of $1s\sigma$ and $2p\sigma$ MO x rays are treated, and x-ray, MO x-ray cascade experiments are presented.

3.4.1. K X-Ray Cross-Section Measurements

Important aspects of the very heavy quasimolecule are illustrated by measurements of the *Z* dependence of total *K* x-ray cross sections. Figure 3.50 (Be 79b) plots the results of such measurements using 3.6, 4.7, and 5.9 MeV/amu Pb projectiles on various targets. This graph can be compared directly to the results from lighter systems (Figure 3.7). The two plots are qualitatiavely similar, both illustrating the regions of direct $1s\sigma$ excitation (see 3.6 MeV/amu data of Figure 3.50 and labeled data points—this was region "i" in Figure 3.7), of *K–K* vacancy sharing (region "iii" in Figure 3.7), of $2p\sigma$ excitation (region "ii" in Figure 3.7), and of *K–L*

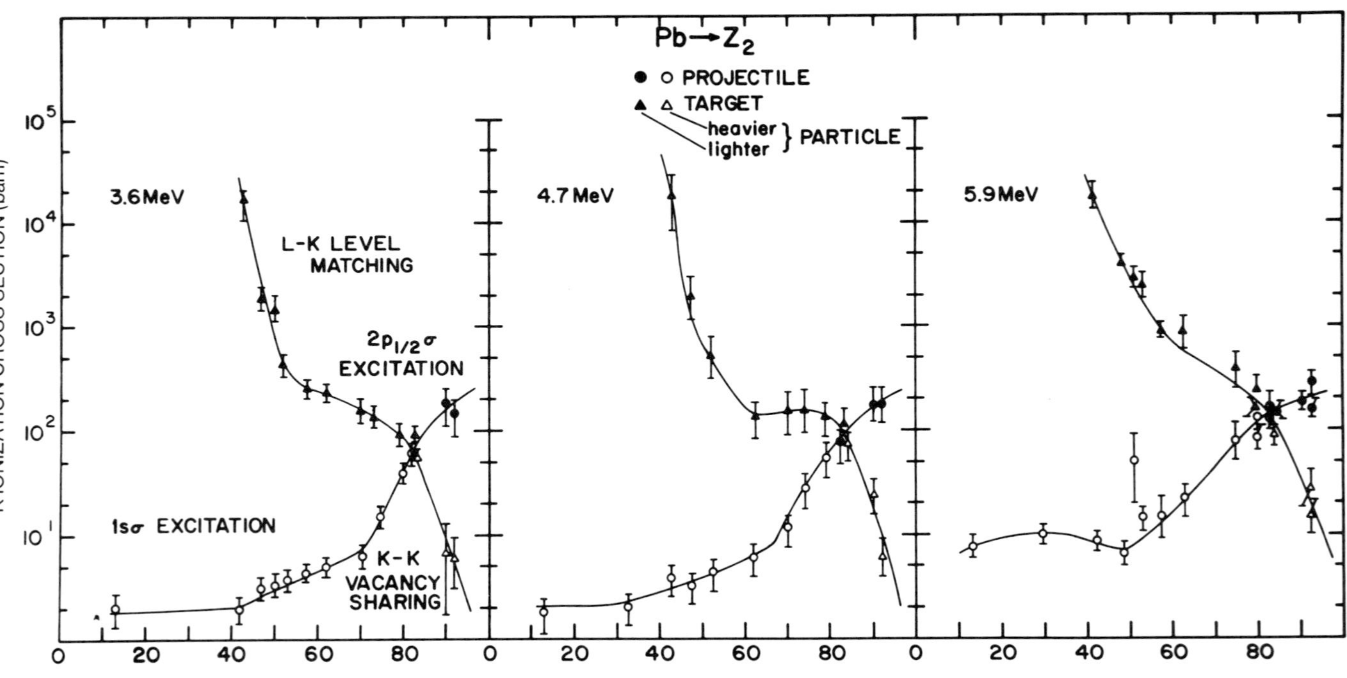

Figure 3.50. Projectile and target K-shell ionization cross sections for 3.6, 4.7, and 5.9 MeV/amu Pb ions. Solid lines guide the eye. Regions where $1s\sigma$ and $2p_{1/2}\sigma$ vacancy production and K–K, L–K vacancy sharing dominate are indicated (Be 79b).

level matching (region "iv" Figure 3.7). The major qualitative difference between Figures 3.50 and 3.7 is simply the displacement of the entire characteristic Z-dependent excitation profile from lower- to higher-target Z.

One particularly significant aspect of the total cross-section data of Figure 3.50 is their large absolute magnitude. We recall at this time earlier discussion concerning the role played by increased $1s\sigma$ orbital binding in decreasing the magnitude of vacancy production probabilities from lighter collision systems (recall Figure 3.20 and associated discussion). Indeed, this "binding energy" effect led to early predictions (Me 74, Fo 76b) that K x-ray production cross sections in collisions where $Z_{\text{projectile}} + Z_{\text{target}} > 100$ would be several orders of magnitude smaller than the actual measured results of Figure 3.50. Although the increased binding of the $1s\sigma$ orbital in very heavy collision systems does, in fact, make $1s\sigma$ ionization more difficult, the strong relativistic contraction of the $1s\sigma$ wave function accompanying this binding and the rapid increase in this localization with decreasing internuclear separation act to increase the ionization probabilities (Be 76, An 78d). The net effects are cross sections between 2 and 10,000 barns shown in Figure 3.50.

Figure 3.51 (Mo 83) illustrates these latter remarks more explicitly by comparing extracted $1s\sigma$ excitation cross sections (An 79c, Be 79b) to semiempirical (An 78d) and perturbed stationary state calculations (So 78) (solid and dashed lines, respectively). The perturbed stationary state calculations were done using the monopole approximation and include excitations from the $1s\sigma$ orbital to the continuum only. Excitations to empty bound states were neglected. The decreased magnitude of $1s\sigma$ cross sections predicted by perturbation theory is typical of calculations of this kind; coupled-channel calculations including excitations to both empty-bound and continuum states yield absolute excitation probability magnitudes which agree with experiment (see discussion in Section 3.4.3 below). The dotted-dashed lines in the ^{238}U projectile data of Figure 3.51 reflect calculations of contributions to U K-shell excitation due to vacancy sharing of lighter collision partner vacancies as symmetry is approached.

The solid lines in Figure 3.51 are semiempirical calculations of the $1s\sigma$ excitation cross sections (An 78d). These calculations are made within the context of the plane wave Born approximation (PWBA) (Me 58, Ja 57, Ba 73, Ba 78, An 77). Within this context one writes,

$$\sigma_{1s\sigma} = Z_L^2 \sigma_{\text{PWBA}} CBR \tag{3.11}$$

where Z_L is the atomic number of the lighter collision partner, and C, B, and R are factors for Coulomb deflection, binding, and relativistic effects. For 3.6 MeV/amu Xe + Pb, $C = 0.64$, $B = 6.3 \times 10^{-3}$, and $R = 3.4 \times 10^3$. Although the magnitude of these corrections sheds considerable doubt on the validity of the physical picture of excitation associated with the PWBA when

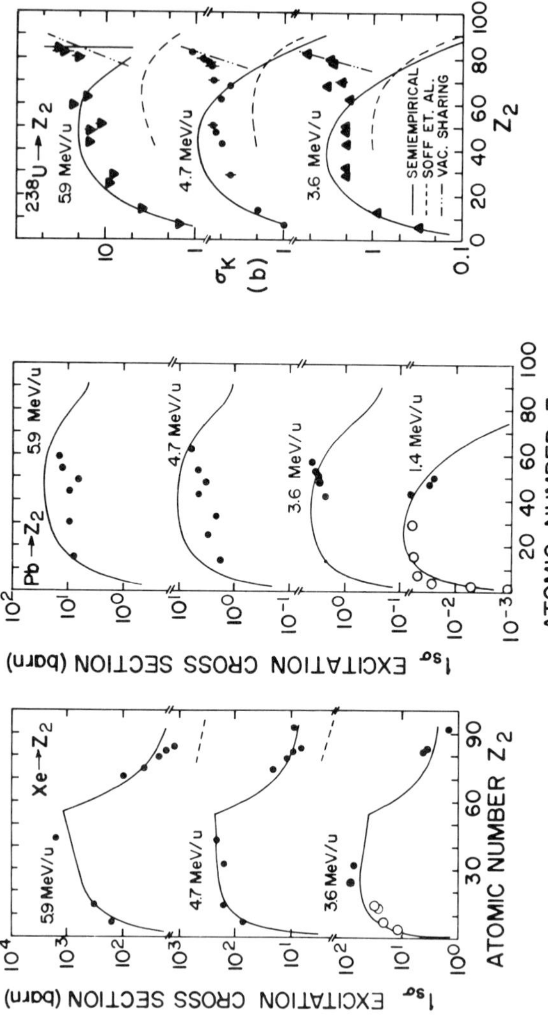

Figure 3.51. $1s\sigma$ excitation cross sections versus target Z_2 between 1.4 and 5.9 MeV/amu. Left graph Xe → Z_2; middle Pb → Z_2; right U → Z_2. Solid lines are semiempirical prescription for the $1s\sigma$ excitation. The dashed lines give monopole approximation results. Dashed-dotted lines are vacancy sharing for the U case (Mo 83).

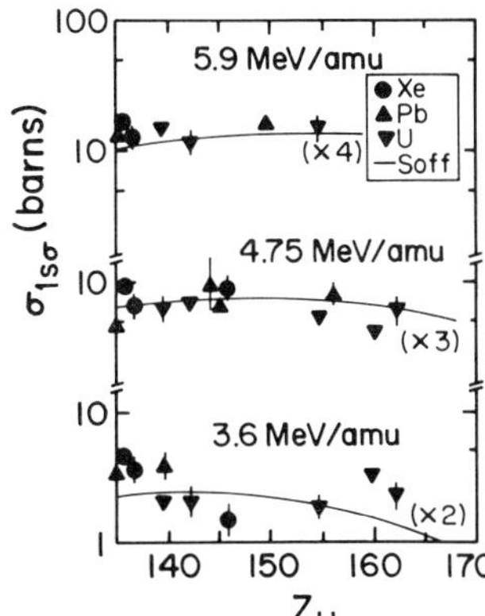

Figure 3.52. $1s\sigma$ excitation cross sections from 3.6, 4.75, and 5.9 MeV/amu Xe, Pb, and U ions versus $Z_u = Z_1 + Z_2$. Solid lines are monopole approximation times factors of 2,3, and 4 (bottom to top), respectively (An 79c).

applied to these collisions, it is useful to point out the fact that B is very small and R is very large. We, therefore, see that the large decrease in production cross sections caused by the increased binding of the $1s\sigma$ orbital is canceled, effectively, by the relativistic collapse of the $1s\sigma$ wave function, resulting in large production cross sections. Therefore, the very fact that these cross sections are large directly implies the dominant role played by relativistic effects on the molecular wave functions in these collisions.

Figure 3.52 (An 79c) illustrates the dependence of $1s\sigma$ excitation cross sections from various projectile–target combinations on the combined projectile + target Z_u. The solid lines compare measurement to perturbation

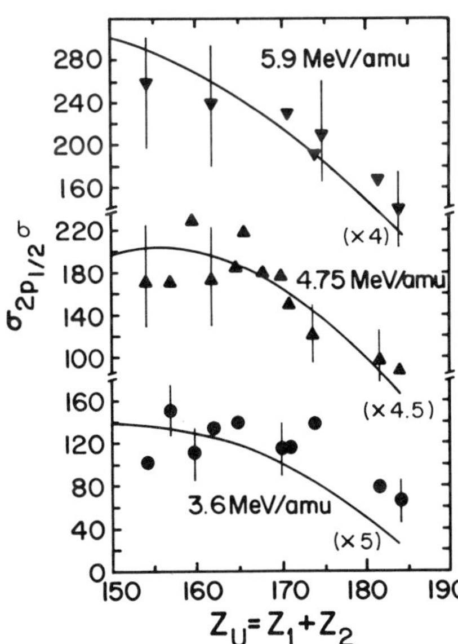

Figure 3.53. $2p\sigma$ vacancy cross sections for 3.6, 4.75, and 5.9 MeV/amu U ions compared to theory multiplied by factors of 4 to 5 as indicated adjacent to each curve (An 79c).

theory–monopole approximation calculations (Be 76, So 78b, Mu 78). These calculations have been arbitrarily multiplied by factors of 2–4 in order to bring them into agreement with experiment. Despite the poor agreement in absolute height, caused principally by neglected coupled-channel effects, the data are seen to form a universal curve which depends on Z_u. Furthermore, good agreement between measurement and the calculated shape of this Z_u dependence is seen. This scaling property is direct indication that excitation actually takes place from the combined projectile–target system ($1s\sigma$ molecular orbital), rather than from the heavier collision partner only. Similar results are shown for extracted $2p_{1/2}\sigma$ excitation cross sections in Figure 3.53 (An 79c).

Studies of vacancy sharing have been extended to these very heavy systems (Be 79b). Within the context of the vacancy sharing picture, one can write the ratio of K-vacancy production cross sections of the lighter partner σ_L to the heavier collision partner σ_H as a function of the vacancy-sharing probability, w [see Equation (3.5) and associated discussion]. If the cross section for $1s\sigma$ vacancy production ($\sigma_{1s\sigma}$) is small compared to that for $2p_{1/2}\sigma$ excitation ($\sigma_{2p_{1/2}\sigma}$) and if $w \ll 1$,

$$R = \frac{\sigma_H}{\sigma_L} = \frac{w}{1-w} + \left(1 - \frac{w}{1-w}\right)^2 \frac{\sigma_{1s\sigma}}{\sigma_{2p_{1/2}\sigma}} \tag{3.12}$$

$$\cong \frac{w}{1-w} + \sigma_{1s\sigma}/\sigma_{2p_{1/2}\sigma}$$

Combining Equations (3.3) and (3.12) then gives

$$R \cong e^{-2x} + \frac{\sigma_{1s\sigma}}{\sigma_{2p_{1/2}\sigma}} \tag{3.13}$$

where x is evaluated with Equation (3.4). Therefore, the usual expression for the vacancy sharing ratio ($R = e^{-2x}$) will, in general, have an additional term $\sigma_{1s\sigma}/\sigma_{2p_{1/2}\sigma}$. When $\sigma_{1s\sigma} \ll \sigma_{2p_{1/2}\sigma}$, as is the case in very light collision systems, the exponential term dominates. However, in very heavy systems, $\sigma_{1s\sigma}$ becomes large (compare, e.g., Figures 3.52 and 3.53) and the second term of Equation (3.13) becomes important. Figure 3.54 illustrates these remarks by plotting R (ordinate) versus $(Z_1 - Z_2)/v_i$ (abscissa in atomic units). The vacancy-sharing prescription including the exponential term only is plotted as a solid line in the figure. The dashed line calculates R including estimates of the second term of Equation (3.13). These results show that, in contrast to lighter systems, where R follows an exponential form over a range from 1 to 10^{-4} (Me 73) (Figure 3.12), for very heavy systems, vacancy sharing dominates only for values of $R \geqslant 10^{-2}$, i.e.,

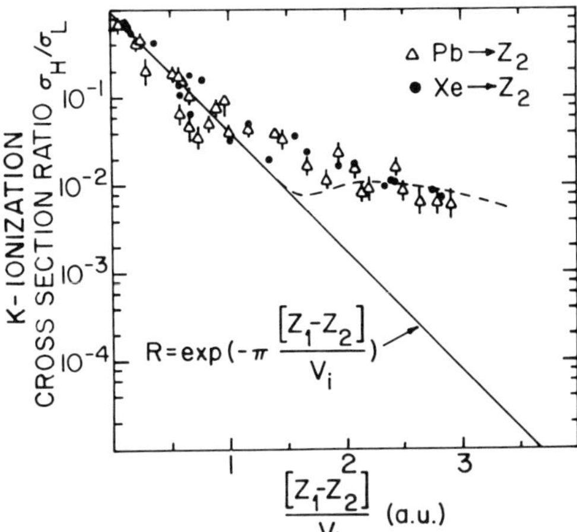

Figure 3.54. *K*-shell vacancy-sharing ratio versus $|\Delta Z|/v_i$ in atomic units. See text for explanation (Be 79b).

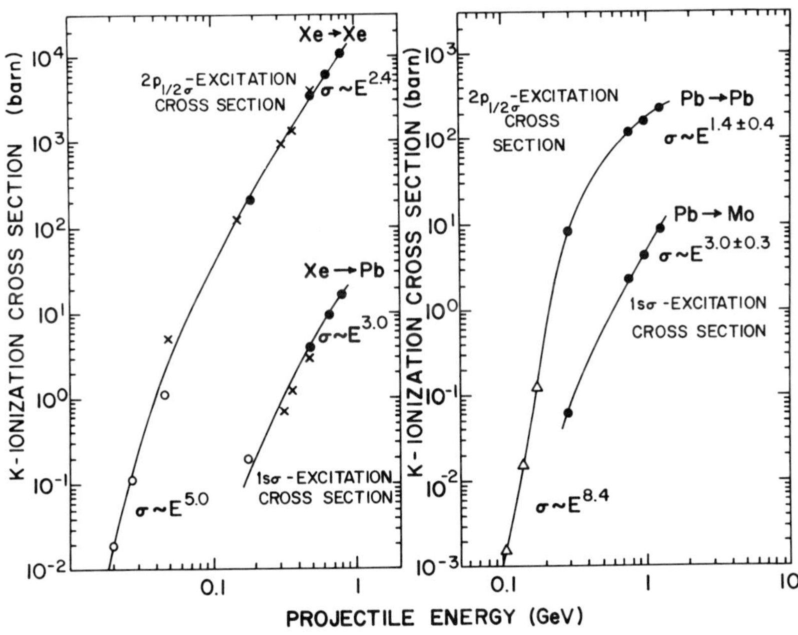

Figure 3.55. *K*-excitation cross sections versus incident impact energy. For symmetric systems, the summed *K* cross section is given representing $2p_{1/2}\sigma$ excitation. For asymmetric systems the heavier collision partner $(1s_{1/2}\sigma)$ excitation cross section is given. Curves guide the eye (Mo 83, Be 79b).

$|Z_1 - Z_2|/v_i \leqslant 1$. These deviations, in turn, give a direct indication of the relative strengths of the $1s\sigma$ and $2p\sigma$ cross sections as functions of the asymmetry and total Z of the collision systems.

The dependences of selected cross sections for $1s_{1/2}\sigma$ and $2p_{1/2}\sigma$ excitation on incident projectile energy are plotted in Figure 3.55 (Be 79b). The Xe + Xe and Pb + Pb data ($2p_{1/2}\sigma$ excitation) both display rapid increases in cross section at lower bombarding energies and slower increases at higher beam velocities. The large increase in $2p_{1/2}\sigma$ excitation in the Pb + Pb quasimolecule of nearly six orders of magnitude for a single order of magnitude increase in the projectile energy illustrates the difficulties experienced by earlier investigators in utilizing initial results at low energies to predict later results at high energies. The energy dependence of Xe + Pb $1s\sigma$ data ($E^{3.0}$) is well reproduced by perturbed stationary state calculations (So 78) in the energy range of 3.6 to 5.9 MeV/amu.

3.4.2. K X-Ray Differential Cross Sections

The scaling of both $1s\sigma$ and $2p_{1/2}\sigma$ cross sections with Z_u and their large absolute magnitudes, both provide evidence for the fact that inner-shell electrons are excited from states of the combined projectile–target system possessing relativistic binding energies. However, additional, and perhaps more direct, information on strong binding phenomena in quasimolecular encounters can be gleaned from studies of the impact-parameter dependence of K-shell excitation.

Recall our earlier discussion in Section 2.3 of the minimum momentum transfer ($\hbar q_{\min}$) necessary to ionize the K shell. This minimum momentum transfer is directly proportional to the binding energy and inversely proportional to the projectile velocity. For slow projectiles, the characteristic distance associated with this minimum momentum transfer $q_{\min}^{-1} \ll a_k$, where a_k is the K-shell radius. If excitation takes place out of the combined system ($1s\sigma$ or $2p_{1/2}\sigma$), the appropriate binding energy to use in calculating this minimum momentum transfer is the binding energy of the molecular orbital at the time of excitation. The higher this binding, the more localized the excitation must be at small internuclear separations. Therefore, excitation from highly bound orbitals is signified by impact parameter (b) dependent excitation probabilities which decrease rapidly with increasing b; excitation from loosely bound orbitals has impact parameter dependences which decrease slowly with increasing b. The slope of the impact parameter dependence of the excitation process is, therefore, directly related to the degree of binding of the electron at the time of excitation.

The first evidence that K-shell excitation in very heavy collision systems does, in fact, take place at extremely small impact parameters was determined by analysis of the Doppler broadened line shape of the K x-ray lines in Xe + Pb and U + Pb collisions (Gr 77). If an x-ray detector is placed at forward angles, Pb K x rays excited from, for example, Xe + Pb encounters will be Doppler shifted to higher x-ray energies due to the recoiling velocity of the target atom in the laboratory frame at the time of x-ray emission. If appreciable amounts of Pb target excitation take place at small impact parameters, the line shape of the Pb K x rays emitted at zero degrees to the beam direction will exhibit large contributions from Doppler-shifted x rays. Therefore, an analysis of the observed line shape can be utilized to extract the experimental impact parameter dependence of K-shell excitation.

When the detector is placed exactly at zero degrees, and the x rays are emitted isotropically in the laboratory, the relationship between the intensity of x rays as a function of Doppler-shifted energy and the impact parameter can be shown to be (for Rutherford scattering) (Gr 77)

$$\frac{d[N(\Delta E_x)]}{d(\Delta E_x)} = \frac{2\pi c}{E_{x0} v_{\text{c.m.}} K_i} P(\rho(\Delta E_x)) \frac{a^2}{4} \left\{ 1 + \left[\frac{\rho(\Delta E_x)}{a} \right]^2 \right\}^2 \left(\frac{E_x}{E_{x0}} \right)^2 \xi(E_x) \tag{3.14}$$

where ΔE_x is the laboratory Doppler shift of an x ray of center-of-mass energy E_{x0} given by

$$\Delta E_x = (E_{x0} v_{\text{c.m.}}/c)[1 + (-1)^{i+1} K_i \cos\theta_1]$$

to first order in the velocity. $v_{\text{c.m.}}$ is the center-of-mass velocity, and $i = 1,2$ refer to the projectile or target radiations, respectively, with $K_1 = A_p/A_T$ and $K_2 = 1$. $\xi(E_x)$ is the photon frequency response of the detector at $E_{x0} + \Delta E_x$, a is half the distance of closest approach in a head-on collision, $\rho(\Delta E_x)$ is the impact parameter, $P(\rho(\Delta E_x))$ is the impact parameter dependence of the excitation probability, and θ_1 is the scattering angle of the projectile in the c.m. system. Equation (3.14) is correct to first order in v/c.

Figure 3.56 shows results of such a line-shape study of $1s\sigma$ excitation in Xe + Pb collisions (Gr 77). The various dotted curves show (from top to bottom, respectively) the background-corrected K_α line shapes at 90° and 0°, the uncorrected shape at 0°, and the background contribution from the thick Ni backings employed in the measurement. The solid lines drawn alongside the 0° data are the K_{α_1} and K_{α_2} line shapes of Pb from a ^{207}Bi decay. The deviation of the in-beam line shapes from that of the radioactive source indicates significant amounts of ionization at small impact parameters.

Utilizing Equation (3.14), the authors extracted the impact parameter dependence given as an inset in Figure 3.56. The total intensity was normalized to independent measurements of the total cross section for Pb excitation ($\sigma_K = 5$ barns) (Be 79b). These results show that the $1s\sigma$ excitation probabilities climb rapidly for impact parameters below 100 fm, reaching values of 7% for central collisions. The decrease in $P(\rho)$ below 15 fm is not

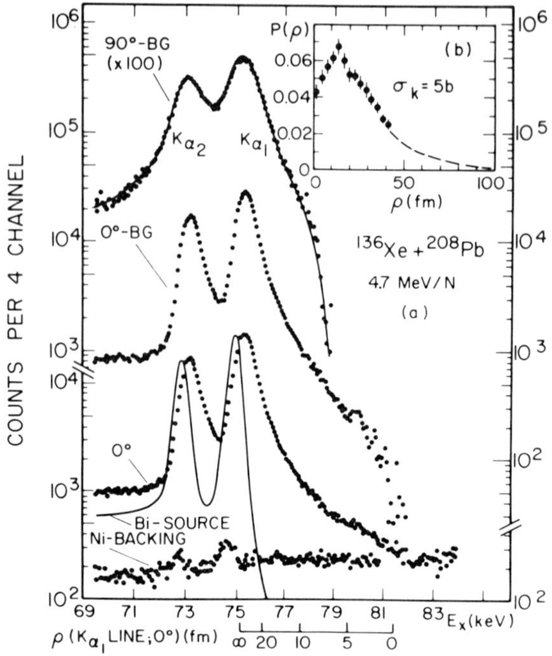

Figure 3.56. $K_{\alpha 1}$ and $K_{\alpha 2}$ line profiles of Pb from $^{136}Xe \rightarrow {}^{208}Pb$ collisions observed at 0° before and after background (BG) subtraction, and at 90° after BG subtraction. The solid line through the 90° spectrum represents the 90° line shape calculated with the $P(\rho)$ (inset) extracted from the 0° line shape (Gr 77).

well understood. Subsequent particle x-ray coincidence studies of $P(\rho)$ in the Xe + Pb system have yielded nearly constant excitation probabilities for $\rho < 20$ fm, which theory appears to favor (Am 78, Re 84). Further discussion of these latter results follows in Section 3.4.3 below.

Similar studies have been performed for the case of $2p_{1/2}\sigma$ excitation utilizing the Doppler-shifted line-shape analysis of target radiation from U + Pb collisions. The resulting impact parameter dependent excitation probabilities are shown in Figure 3.57 (Gr 77). The characteristic decreases at small impact parameters cannot be produced by direct excitation of electrons out of the $2p_{1/2}\sigma$ molecular orbital, but must reflect contributions from rotational coupling. The smooth curves drawn in Figure 3.57 are calculations (He 78) of the $P(b)$ ($b \equiv \rho =$ impact parameter) for Pb + Pb (upper) and U + U (lower) collisions. The dashed lines are the direct excitation contribution and the dashed-dotted lines the rotational coupling portion due to the mixing of the $2p_{1/2}\sigma$ with a (theoretically assumed) vacancy in the $2p_{3/2}\pi$ orbital. The solid lines are coupled-channel calculations including both effects. Note that, in this manner, a peaking of the $P(b)$ values can be theoretically generated. However, the position of the experimental peak is not precisely reproduced. The impression given by these results is that $2p_{1/2}\sigma$ excitation is considerably more complex than the $1s_{1/2}\sigma$ case and not nearly as well understood. This impression is further substantiated by coincidence measurements, which will be discussed in the following section.

In contrast to lighter quasimolecules where rotational coupling dominates $2p_{1/2}\sigma$ excitation, the strong spin–orbit splitting (see Figure 2.5) of the $2p\pi$ and σ quasimolecular orbitals exhibiting strong relativistic effects suppresses rotational coupling in very heavy collisions. The peak in the rotational coupling curves of Figure 3.57 is simply the characteristic kinematic peak at 90° scattering in the center of mass, which, for lighter systems, approaches unit probability in height.

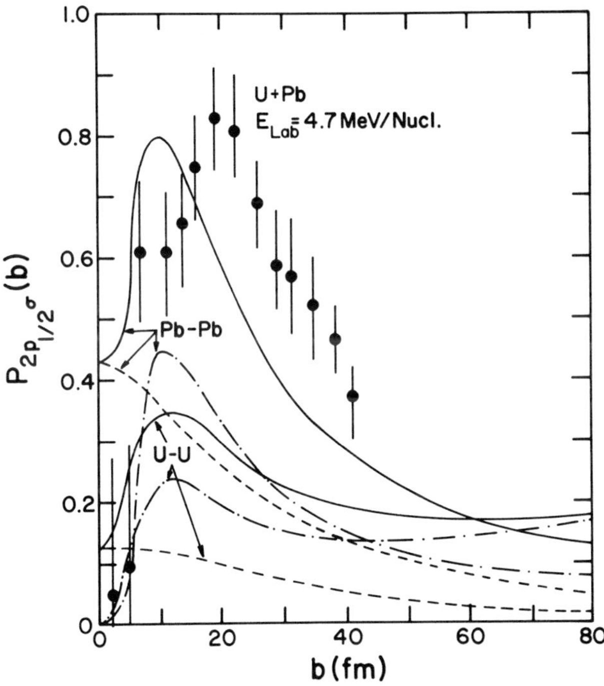

Figure 3.57. Excitation probability for U + Pb collisions at 4.7 MeV/amu compared to theory for Pb + Pb and U + U. Solid lines give summed theory for direct $2p_{1/2}\sigma$ ionization (dashed lines) plus $2p_{1/2}\sigma$ to $2p_{3/2}\pi$ coupling (dashed-dotted lines) (Gr 77, He 78).

3.4.3. *K X-Ray Impact Parameter Dependences from Coincidence Experiments*

The primary role played by the application of coincidence techniques to measurements of *K* x-ray production in very heavy collision systems has been the determination of impact parameter dependences. Studies of these dependences have been particularly interesting largely due to the discovery of scaling law for 1sσ excitation which might be utilized to extract spectroscopic information on the large transient binding energies generated during heavy-ion–atom encounters. Aspects of this scaling law were discussed in Section 2.3. The simplicity of the empirically observed scaling has been an especially attractive result of these experiments, and has generated considerable study of the theoretical origins of the scaling behavior.

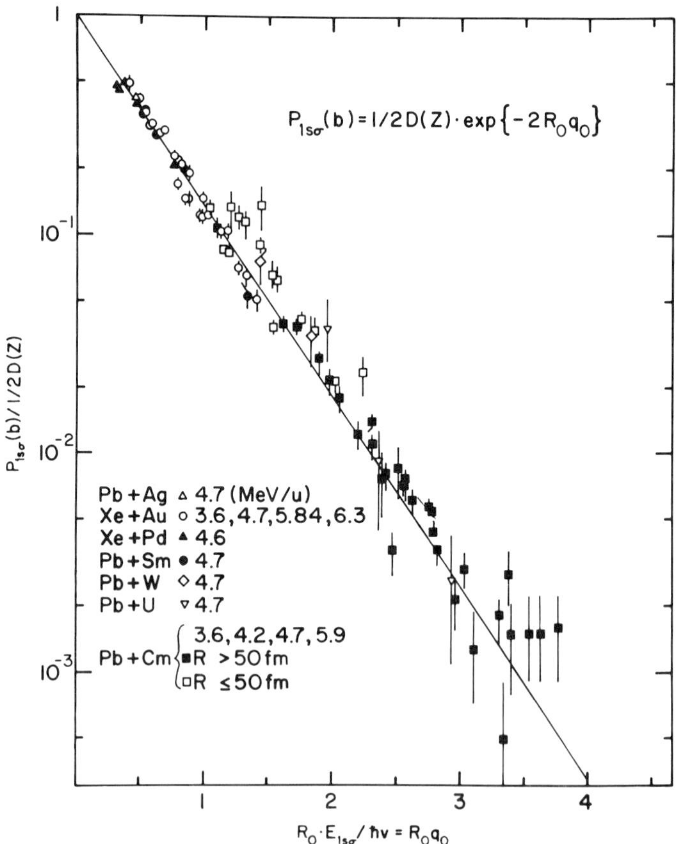

Figure 3.58. Reduced 1sσ excitation probabilities versus R_0q_0 for systems with $Z_{\text{united}} > 137$. The solid line is the result of Equation (3.15) (Li 80c, Bo 80).

Figure 3.58 (Bo 80, Li 80c) plots values of the reduced experimental $1s\sigma$ excitation probabilities $P_{1s\sigma}(b)$ (ordinate) versus the reduced impact parameter R_0 (abscissa) from a variety of heavy asymmetric collision systems. The impact parameter (b) expressed in terms of the distance of closest approach, R_0, at that particular value of b is given by Equations (2.20) and (2.21). $E_{1s\sigma}(R_0)$ in Figure 3.58 is simply the binding energy of the $1s\sigma$ molecular orbital at internuclear separation R_0, and v is the projectile velocity. As was discussed in Section 2.3, $\hbar q_{\text{min}} = E_{1s\sigma}/v$ is the minimum momentum transfer necessary to ionize the $1s\sigma$ level. The solid line in the figure plots the relationship

$$P_{1s\sigma}(b) = 1/2D(Z)\exp(-2R_0 q_{\text{min}}) \tag{3.15}$$

where $D(Z)$ is a universal, smoothly varying function of the united atom Z only, taken from theoretical calculations and given by Equation (2.19).

The striking result of Figure 3.58 is that all these $1s\sigma$ excitation data, which represent a wide variety of collision systems, lie on a universal curve which is extremely well described by a simple exponential function. The theoretical justification for such a scaling law was the subject of discussion in Section 2.3. The general result of this discussion was relation (2.23) for the impact parameter dependence of delta-electron emission. The corresponding relationship for $1s\sigma$ vacancy production is simply given by the integral over the entire electron spectrum. If one assumes that $1s\sigma$ electrons are ionized into continuum states just above the ionization threshold, and if one furthermore takes $\gamma(Z) \simeq 2$ [Equation (2.19)], as well as $\tau_0 \simeq R_0/v$ [Equation (2.22)], the result of Equation (3.15) is obtained.

The agreement between this approximate evaluation of first-order perturbation theory and experiment provides justification for such a semiempirical description of these excitations. However, it must be emphasized that the description given above is an approximate calculation of the result of first-order perturbation theory. Exact calculations within the perturbation theory picture yield results for the $P(b)$ values which are factors of 2–3 less in absolute magnitude than the approximate evaluation presented above (Mu 83).

Figure 3.59 illustrates these remarks by comparing exact calculations of first-order time-dependent perturbation theory with measurements of $1s\sigma$ excitation from 4.6 and 7.2 MeV/u Xe + Pb collisions (An 79f, So 81b). The exact perturbation theory results (dashed line) are well below measurement. The solid lines are coupled-channel calculations using the monopole approximation to the two-centered potential. Agreement between measurement and theory is only obtained when full coupled-channel calculations are performed.

The coupled-channel results illustrated in Figure 3.59 have two

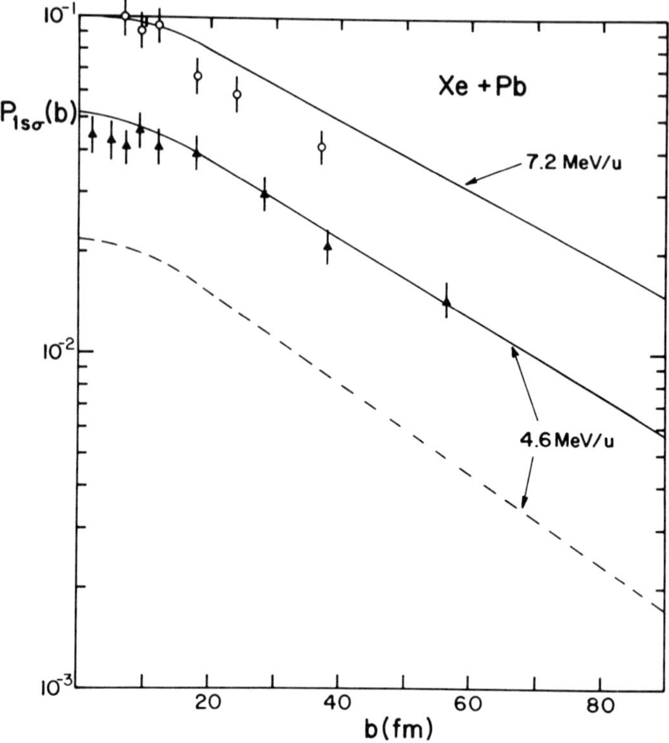

Figure 3.59. Experimental $1s\sigma$ vacancies in Xe + Pb collisions at 4.6 and 7.2 MeV/amu versus impact parameter b. Dashed line, first-order time-dependent perturbation theory. Solid lines, coupled channels calculations (So 81b, An 79f).

important aspects, both of which contribute to increased excitation probabilities. The first aspect is simply that the inclusion of multistep and coherent excitation processes in the coupled-channel calculations yield enhanced probability of $1s\sigma$ excitation to the continuum, by as much as a factor of 3. The second aspect is that excitation to empty bound states can further increase the excitation probabilities by another factor of 2. If one assumes a Fermi level for the electrons above which the quasimolecular levels are considered empty, resulting excitation probabilities can be calculated as a function of this Fermi level.

Figure 3.60 (Re 79) shows the results of such calculations for the Pb + Cm collision system at 4.7 MeV/amu. The dashed line is the exact perturbation theory result, and the solid lines the coupled-channel predictions for different assumed Fermi surfaces F. The labels in the figure indicate the last occupied molecular orbital. These calculations were done using eight bound states ($1s\sigma$, $2s\sigma$,..., $8s\sigma$). Due to the use of the monopole approx-

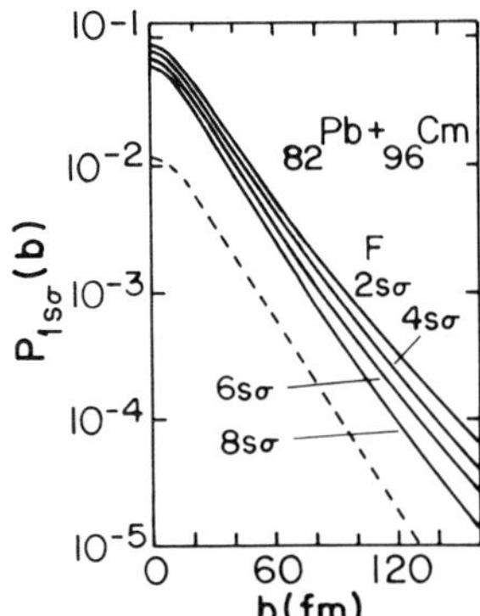

Figure 3.60. Probability of $1s\sigma$ excitation versus impact parameter for 4.7 MeV/amu Pb → Cm collisions. Solid lines, coupled-channels results belonging to different positions of the assumed Fermi surface F. Dashed line, perturbation theory (Re 79).

imation, only s states were considered. The primary cause of enhanced $P(b)$ values is the coupled-channel effects themselves; excitations to empty bound states have significant influence at large b, but introduce only modest enhancements at small impact parameters.

In view of the above discussion, some questions must be raised as to why the approximate perturbation theory was, in fact, so successful in describing the experimental results. This question has been recently explored (Mu 83), and the results of the study are shown in Figure 3.61. In the figure, the probability of creating $1s\sigma$ vacancies $P_{1s\sigma}(b)$ in 5.9 MeV/u Pb + Cm collisions is calculated. Graph (a) compares various approximate model calculations of first-order perturbation theory (dashed, dotted, and solid lines) with the predictions of Equation (3.15) (dot-dashed line). Part (b) compares an exact perturbation theory calculation (dashed line) with a coupled-channel calculation of the kind discussed in connection with

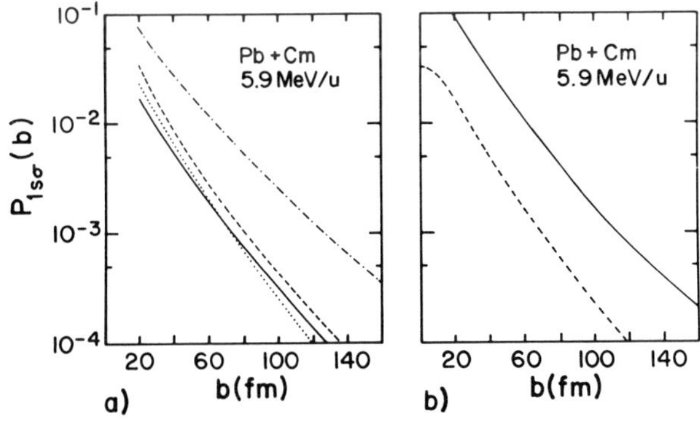

Figure 3.61. Theoretical calculations of $1s\sigma$ vacancy production versus impact parameter in 5.9 MeV/amu Pb + Cm collisions using various approximations. See text for explanation (Mu 83).

Figures 3.59 and 3.60 (solid line). Note that all the approximate perturbation results agree well with each other and with the exact perturbation theory results, except for the prediction of Equation (3.15). They all lie, furthermore, substantially below the coupled-channel result, again, except for the result of Equation (3.15), which displays much better agreement in absolute height with the coupled-channel calculations than the other approximate and the exact perturbation theory curves.

The principal cause of these discrepancies is the approximation made by the proponents of Equation (3.15) that excitation takes place only to very low-lying states in the continuum. This approximation was justified on the basis that there is excitation to empty bound states [Re 79] as well as to the continuum. However, proper integration over the continuum leads to the lower excitation probabilities illustrate in Figure 3.61. Since the scaling law given by Equation (3.15) does provide good agreement with experiment, it may be taken as a semi-empirical description of the excitation probabilities (Bo 80, Mu 83), and one possible approach is simply to utilize the success of this relationship over a wide range of collisions to attempt to extract information on the binding energy of the $1s\sigma$ orbital at the time of excitation, $E_{1s\sigma}(R_0)$. One measures $P(b)$ and utilizes relation (3.15) to extract $E_{1s\sigma}(R_0)$. The results of such an interpretation are given in Figure 3.62 for the

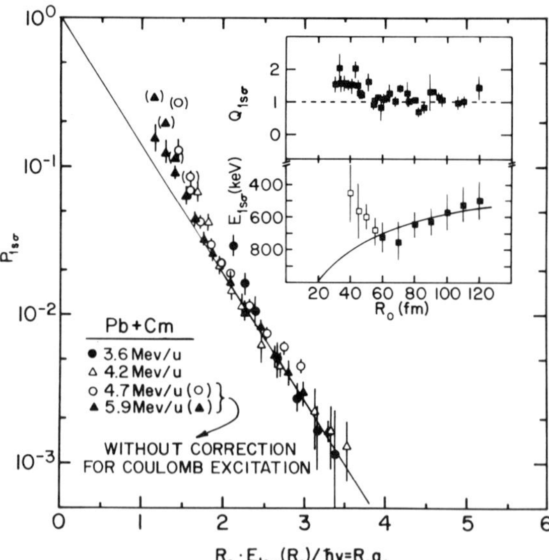

Figure 3.62. $1s\sigma$ excitation probabilities versus $R_0 q_0$ at different projectile energies. Data in brackets compared to those without brackets at the same value of $R_0 q_0$ give the contribution of nuclear Coulomb excitation to atomic vacancy production. The solid line is prediction of Equation (3.15). The inset figure is explained in text (Li 80c).

Pb + Cm system (Li 80c) The large picture shows measured $P_{1s\sigma}(b)$ values in the reduced representation of Equation (3.15) (solid line). Agreement between measurement and Equation (3.15) is very good for values of $R_0 q_{\min}$ above 2. Below these values, the measured excitation probabilities are in considerable excess of those predicted by Equation (3.15). The inset figure plots the ratio, $Q_{1s\sigma}$, of the measured $P_{1s\sigma}(b)$ to Equation (3.15) predictions versus R_0 (upper part of inset). The deviation below $R_0 = 50$ fm is quite noticeable. The lower part of the inset plots the extracted binding energies at the time of excitation versus R_0. The solid line in the lower inset figure is the result of the solution to the two-centered Dirac equation for the binding energies. The agreement between "measurement" and theory is good for $R_0 > 60$ fm. However, the increased $P_{1s\sigma}(b)$ values for $R_0 < 60$ fm lead to decreasing $E_{1s\sigma}(R_0)$ magnitudes, in disagreement with theory.

These results imply that the $1s\sigma$ electrons do not actually become bound by the values predicted by the solutions to the two-centered Dirac equation at small R_0 (Li 80c). However, before drawing such conclusions, careful examination of the approximations leading up to Equation (3.15) must be made. Deviations at low R_0 could just as easily be due to the breakdown of Equation (3.15) as to the occurrence of anomalous binding energies for the $1s\sigma$ orbital. Furthermore, proper subtraction of background x rays following the internal conversion of Coulomb excited nuclear states must be critically examined. Independent investigations of this latter background indicate that improper background subtraction is the most likely cause of the increased $P(b)$ values at small R_0, rather than the suggested anomalous binding energies (It 82). Therefore, considerable care and caution should be exercised in the extraction of $1s\sigma$ binding energies from such measurements (Li 80c).

Further refinements to the calculation of $P_{1s\sigma}(b)$ are illustrated in Figure 3.63 (So 80, So 81b). Here the effect of electron screening on the calculated $P_{1s\sigma}(b)$ values is shown. Screening lowers the binding energy of the $1s\sigma$ orbital, thereby leading to increased ionization probabilities which are in better agreement with experiment.

The picture of $1s\sigma$ excitation presented in this section was rather detailed. In fact, the size and shape of $P_{1s\sigma}(b)$ in very heavy collision systems is fairly well understood. The same cannot be said of $2p_{1/2}\sigma$ excitation. In Figure 3.64 measured values of $P_{2p_{1/2}\sigma}(b)$ from Pb + Au collisions are compared to calculation (So 78). The agreement at low bombarding energies is poor, and that at higher energies only fair. Unlike the $1s\sigma$ orbital, which is well isolated in energy from other molecular levels and which satisfies the adiabatic condition for quasimolecular orbitals very well, the $2p_{1/2}\sigma$ level is strongly rotationally and radially coupled to neighboring $2p$ and $3p$ molecular orbitals. Considerably more theoretical work must be done in order to completely understand the more complicated case of $2p_{1/2}\sigma$ excitation.

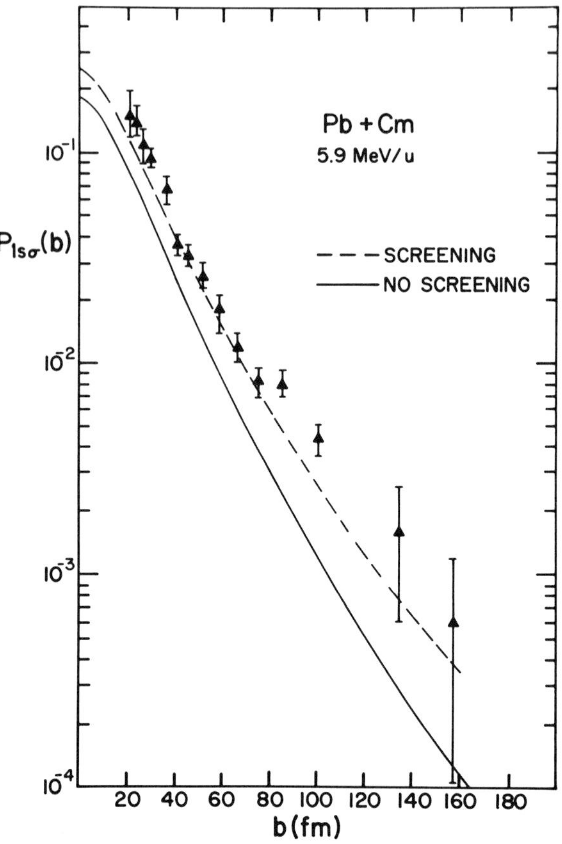

Figure 3.63. Comparison between experimental and theoretical results for $1s\sigma$ excitation probabilities in the Pb → Cm system at 5.9 MeV/amu. In the dashed-dotted curve, electron screening effects have been taken into account (So 81b, So 80).

In closing this section, we would like to emphasize the simple and direct statement made by measurements of total and differential x-ray production in these collisions about the behavior of the $1s\sigma$ orbital. The large magnitudes of the cross sections and their concentration at small impact parameters, taken together, provide direct evidence for strong binding and associated relativistic contraction of the $1s\sigma$ electron wave functions. Details concerning the precise applicability of scaling laws or the feasibility of quantitatively extracting a $1s\sigma$ binding energy from $P_{1s\sigma}(b)$ measurements should not detract our attention from the principal results of these studies. The presence of vacancies in the $1s\sigma$ orbital with large probabilities at small internuclear separations provides an essential initial condition for the observation of the spontaneous positron process. In general the establishment of these large,

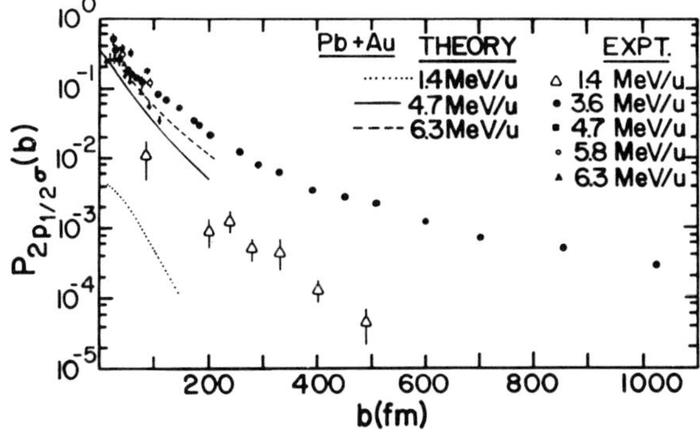

Figure 3.64. Summed Pb, Au K-vacancy probabilities versus impact parameter b at different projectile energies. Theoretical curves give $2p_{1/2}\sigma$ excitation probabilities for direct excitation to the continuum (Mo 83, So 78).

concentrated $1s\sigma$ excitation probabilities through the K x-ray measurements constitutes important results in the quasimolecular physics of the heaviest possible collision systems.

3.4.4. MO X Rays from Superheavy Systems

The properties of very heavy quasimolecules which determine the behavior of K x-ray total and differential cross sections in heavy collision systems are also relfected in studies of continuum x-ray emission. As has been emphasized throughout this article, MO x-ray studies provide complementary information about and insight into strong binding and relativistic phenomena in the superheavy quasiatom. In order to extract such information, differences between measurements of MO x-ray emission from quasimolecules exhibiting relativistic versus nonrelativistic features must be understood.

Perhaps the principal change in going from light to heavy systems is the increased importance of disturbing backgrounds from the Coulomb excitation of nuclear levels. Lighter systems are associated with continuum x-ray emission below 100 keV photon energies; in heavy systems, strongly bound inner molecular orbitals generate MO x rays with energies in excess of 1 MeV. Heavy nuclei have many levels in the 100 keV to 1 MeV region of excitation which are Coulomb excited readily in energetic heavy-ion encounters. Deexcitation of these states by γ-ray emission is a troublesome source of background in MO x-ray measurements. For close collisions, the strong relativistic contraction of the $1s\sigma$ wave function (see Figure 2.6) localizes this electron to the spatial vicinity of the colliding nuclei. As a result, there are large probabilities for excitation of the nucleus in collisions which excite the quasiatom and vice versa, resulting in troublesome background problems which are difficult to suppress experimentally. These backgrounds are so serious that up to this point, only investigation of the spherical nuclear $^{208}Pb + ^{208}Pb$ system has been possible.

A second major difference between MO x-ray production in lighter versus heavier systems is the growing dominance of the one-step production mechanism with increasing united atom Z. As has been discussed in Section 3.3.2, both the decreased lifetime of K vacancies, and the rapid rise of direct $1s\sigma$ excitation probabilities with increasing Z are responsible for this result. This dominance of the one-collision production process considerably simplifies the interpretation of MO x-ray impact parameter dependences, since distortion of the relationship between laboratory scattering angle and impact parameter caused by the two-collision mechanism does not occur (recall Figure 3.45 and associated discussion).

Figure 3.65 shows the measured pulse height distributions from continuum x-ray emissions in $^{208}Pb + ^{208}Pb$ collisions at 4.2 MeV/u (Me 79, St 80, Mo 81b). A 3-mm Sn plus 1-mm Cu absorber was used in these measurements to attenuate the K x-ray lines, thereby preventing dead time and pulse pile-up problems. The continuum band extending from the characteristic Pb K_α and K_β lines to 1.5 MeV photon energy is dominated by MO x rays. Also shown in the figure is the room background originating predominantly from trace radioactivity in the concrete walls and floors of the experimental target room. Thick (several inches) Pb shielding surrounded the counters in order to minimize this background. The data were taken with a thick, enriched ^{208}Pb target employing a large volume ($\sim 100 \text{ cm}^3$) coaxial Ge(Li) γ-ray detector.

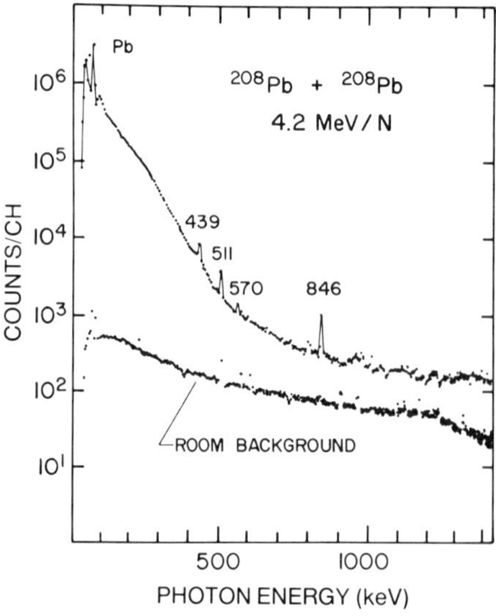

Figure 3.65. Pulse height distribution of x rays from 4.2 MeV/amu $^{208}Pb + ^{208}Pb$ collisions using a highly enriched ^{208}Pb target. Corresponding room background also shown (Mo 81b, St 80).

Various contaminant lines from ^{23}Na (439 keV), positron annihilation (511 keV), ^{207}Pb (570 keV), and ^{27}Al (846 keV) are evident in the figure. The Al contamination was caused by stray beam striking the target frame. Despite the use of highly enriched ^{208}Pb targets (0.7% ^{207}Pb, 0.2% ^{206}Pb, and <0.05% ^{204}Pb) and beams (<1% ^{207}Pb), small amounts of the 570-keV ^{207}Pb state excitation are still noticeable in the data. Clearly, use of a target with many levels in the 100 keV to 1 MeV region (^{238}U, for example) would yield a spectrum where the MO x-ray component would be dwarfed by large contributions from nuclear Coulomb excitation.

Choice of the incident bombarding energy for such a measurement can optimize the ratio of MO x-ray intensity to background. For the ^{208}Pb + ^{208}Pb collision system, the first nuclear level is at 2.6 MeV. For such a high-lying state, the increase in nuclear excitation probability with incident bombarding energy is much steeper than that for $1s\sigma$ vacancy production. The background from room radioactivity is clearly constant in time (and independent of incident bombarding energy). Therefore, at very low beam energies, $1s\sigma$ vacancy production is small and room background large; at high beam energies, Coulomb excitation of the 2.6-MeV state is large and the subsequent γ decay of this nuclear level produces Compton scattering in the detector and disturbs the measurement. The optimum beam energy will therefore depend on the rates of these backgrounds and on the incident beam intensity available. It turns out that 4.2 MeV/u was optimal for this experiment.

Figure 3.66 (St 80) (upper picture) plots the continuum x-ray data after correction for absorption and detector efficiency. A small (~10%) correction for the tails of the photon line shape (Compton scattering) was also subtracted (unfolded) from the data. The dashed line is the calculated contribution from nucleus–nucleus bremsstrahlung (NNB), and the solid line that of MO x rays (Ki 78, Li 79, Ki 80). The calculations of the MO x-ray yield include contributions to $1s\sigma$ final states only. The lower curve shows the results of measurements of the directional anisotropy of the continuum radiation $[I(90°)/(I(0°) - 1]$ in comparison to calculation (solid line) (Ki 80).

Calculations of the MO x-ray yields in these collisions are very difficult. In certain respects, the one-step process complicates matters considerably since coherent solutions for the vacancy production and radiative decay amplitudes as a function of time (internuclear separation) must be found. The effect of including rotational as well as radial coupling in the collision problem was examined theoretically; rotational coupling was found to strongly increase the MO x-ray intensities above 600 keV. The theoretical MO x-ray yields displayed in Figure 3.66 include an approximate treatment of both rotational and radial coupling (Ki 80).

The calculations suffer from a number of difficult theoretical problems.

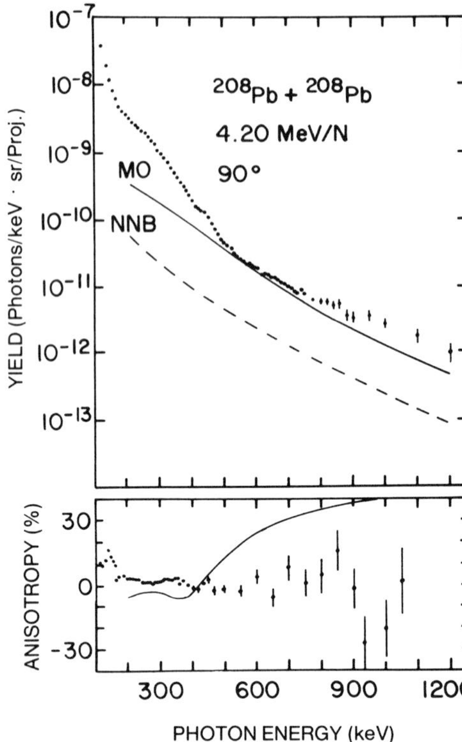

Figure 3.66. Thick target photon yield for 4.2 MeV/amu $^{208}Pb \rightarrow {}^{208}Pb$ collisions compared to nucleus–nucleus bremsstrahlung (dashed) and $1s\sigma$ MO x rays (solid) theory. The lower curve shows corresponding directional anisotropies (St 80, Ki 78, Ki 79, Ki 80).

The long-wavelength approximation was used to calculate the radiative matrix elements, and analysis of the validity of this approximation suggests that its accuracy is questionable. In addition, the matrix elements were attenuated smoothly at large internuclear separations (~2000 fm) in order to simulate the unknown influence of translation factors. Since the two-centered Dirac continuum wave functions are unknown, excitations to continuum *s* states only have been included (Ki 80).

The calculations suffer from inaccuracies in the long-wavelength approximation at high photon energies. At low photon energies, transitions at large internuclear separations occur and the precise behavior of the artificially simulated translation factors becomes important. A comparison between calculation and measurement in Figure 3.66 shows good agreement at high x-ray energies and poor agreement below 400 keV. At low photon frequencies, transitions into higher-lying molecular orbitals (i.e., $2p_{1/2}\sigma$) will also contribute. The results of coincidence studies (see below) indicate that $2p_{1/2}\sigma$ MO x rays dominate the spectrum below approximately 200 keV. The discrepancy between theory and measurement in the 200–400 keV region does not, therefore, appear to be attributable to additional contributions from

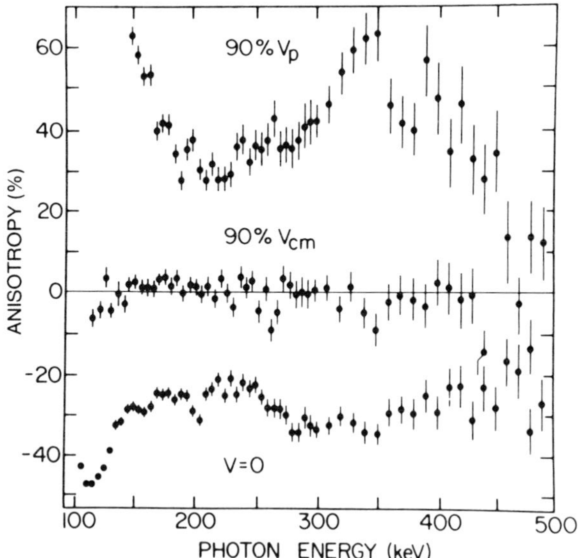

Figure 3.67. Forward-backward continuum x-ray anisotropies calculated for three different emitter velocities of the continuum x rays. Symmetry is obtained only for the case where the radiation is assumed to be emitted from the projectile–target center of mass (St 80).

higher-lying states. Nevertheless, rough overall agreement between experiment and theory is obtained, implying that the observed intensities are principally due to MO x-ray transitions into a $1s\sigma$ final state. The extension of the x-ray intensities to nearly 1.5 MeV photon energy is direct illustration of the achievement of high binding energies in these collisions.

The bottom portion of Figure 3.66 plots the results for the directional anisotropy. The small bump at 150 keV most likely reflects contributions to a $2p_{1/2}\sigma$ final state (Mo 81b). The $1s\sigma$ MO x rays exhibit no anisotropy, contrary to the expectations of the displayed calculations. This disappointing result precludes a straightforward extension of the semiempirical united atom K-shell spectroscopy discovered in lighter systems to the interesting very heavy encounters (recall discussion of Section 3.3.2.2).

In analogy to discussion associated with Figures 3.41 and 3.42, measurements of forward–backward continuum intensities can be utilized to extract an emitter velocity of the continuum radiation. Figure 3.67 (St 80) shows the results of such a Doppler shift measurement for the 4.8-MeV/u $^{209}Bi + {}^{209}Bi$ system. In the figure, the anisotropy ratios $I(137.5°)/I(42.5°) - 1$ are plotted versus photon energy for various assumed emitter velocities. The upper data (labeled v_p) are the ratios corrected for Doppler shifts and relativistic solid angle effects corresponding to the projectile velocity, the middle points for the velocity of the center-of-mass system, and the lower points assume no Doppler shift. Only when the measured intensities are corrected for Doppler effects and relativistic solid angle corrections corresponding to 90% of the center-of-mass velocity, is forward–backward symmetry achieved. The use of 90% $v_{\text{c.m.}}$ is due to the use of thick targets in these experiments, and simply reflects energy loss of the projectile ions in the thick target. These results provide further evidence that the observed emissions are indeed quasimolecular in origin.

Two kinds of coincidence experiments with MO x rays from the $^{208}\text{Pb} + {}^{208}\text{Pb}$ collision system have been carried out. The first experiment yielded the impact parameter dependence for the MO x rays shown in Figure 3.68 (Vi 81, St 82). The top portion of the figure plots results for the K x-ray $P_K(b)$ and the bottom portion those for the MO x rays at various photon energies. Close examination of the $P_K(b)$ data in Figure 3.68 indicates a slight rise in the probability below approximately 50 fm. At MO x-ray photon energies, this rise becomes increasingly apparent, and the two-component structure of the x-ray $P(b)$ is quite clear.

Earlier discussion of characteristic K x-ray impact parameter dependence studies in heavy collisions stated that these dependences provide infor-

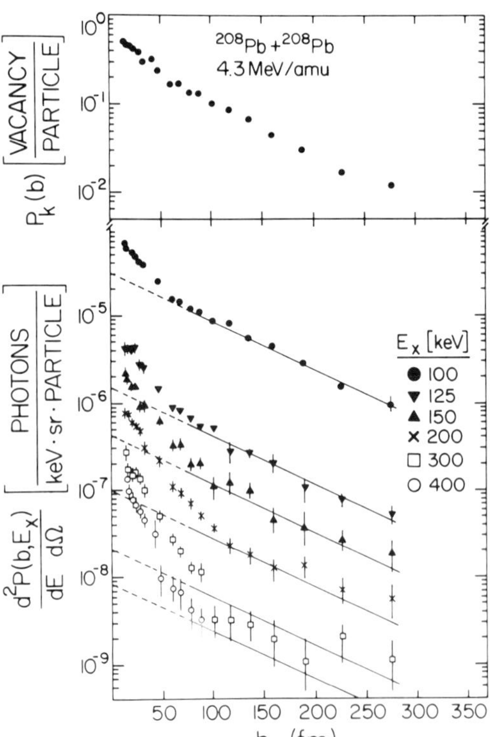

Figure 3.68. Impact-parameter dependence of characteristic (upper figure) and MO (lower figure) x rays from 4.3-MeV/amu $^{208}\text{Pb} \rightarrow {}^{208}\text{Pb}$ collisions showing two component structure of the $P(b)$ curves (Vi 81, St 82).

mation on the vacancy excitation process occurring during the collision. In those studies, x-ray emission took place after the collision was over, and was, therefore, well decoupled from the excitation process itself. However, for the case of MO x rays, the decay of the generated vacancies takes place during collision, and therefore cannot, in principle, be treated independently from vacancy excitation.

Within the context of the first-order perturbation theory scaling of the K x-ray $P(b)$ curves (recall discussion in Section 2.3), a relationship exists between the impact parameter dependence of the excitation process and the binding energy of the molecular orbital from which the vacancy originates. In fact, one can show that (Ba 79), under certain assumptions, $\langle b\rangle\langle q_{\min}\rangle \cong 1$ [see Equation (2.27) and associated discussion] where $\langle b\rangle$ is the average impact parameter associated with the observed $P(b)$,

$$\langle b\rangle \equiv \frac{\int_0^\infty b^2 P_K(b)\, db}{\int_0^\infty b P_K(b)\, db} \tag{3.16}$$

and $\hbar\langle q_{\min}\rangle = \langle E\rangle / v$ is the mean momentum transfer to the electron bound by mean energy $\langle E\rangle$ at the time of excitation.

If, in analogy to the K x-ray studies, one assumes that the slopes of the MO x-ray $P(b)$ curves reflect excitation of vacancies, rather than their decay, one can fit exponential functions to the data of Figure 3.68, extract an average impact parameter for each component, and interpret the results in terms of a minimum momentum transfer (St 82). For the steeper component $\langle b\rangle_1 \cong 50$ fm, and for the flatter component (solid lines in Figure 3.68), $\langle b\rangle_2 \cong 160$ fm. Using $\langle b\rangle\langle q_{\min}\rangle \cong 1$ then yields $\langle E\rangle_1 \cong 380$ keV for the steeper component and $\langle E\rangle_2 \cong 120$ keV for the flatter component. Examination of the correlation diagram for the Pb + Pb quasimolecule (Figure 2.5) shows these values to qualitatively agree with expectations for $1s\sigma$ and $2p_{1/2}\sigma$ excitation, respectively. The two-component structure of Figure 3.68 may therefore be interpreted as mixed, photon energy dependent contributions from $1s\sigma$ (steeper component) and $2p_{1/2}\sigma$ (flatter component) final-state MO x-ray transitions, respectively. Note that such an analysis is greatly facilitated by the one-step nature of the MO x-ray production mechanism (recall Figure 3.45 and associated discussion). A semiempirical separation of mixed contributions from $1s\sigma$ and $2p_{1/2}\sigma$ MO x-ray based on studies of their impact parameter dependences therefore appears quite feasible.

Within the context of such an interpretation, the $2p_{1/2}\sigma$ portion of the $P(b)$ curves may be subtracted out, leaving the $1s\sigma$ component only. This has been done in Figure 3.69 (St 82). The solid lines in the figure are fits through the $1s\sigma$ data using $\langle b\rangle_1 \cong 50$ fm, and the dotted lines are theoretical

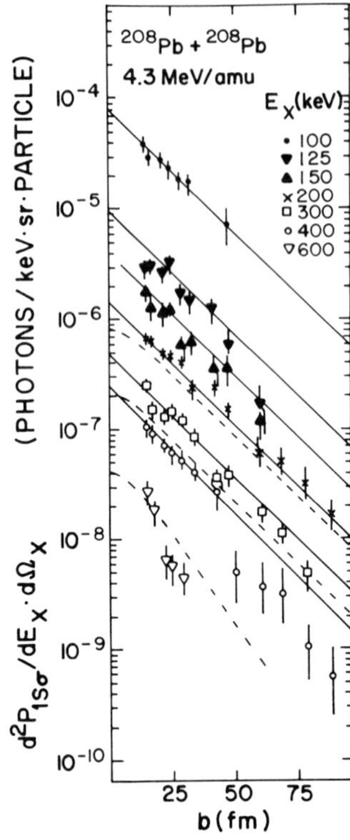

Figure 3.69. Doubly differential probability for the emission of $1s\sigma$ MO radiation in 4.3-MeV/amu ^{208}Pb → ^{208}Pb collisions for various x-ray energy regions. See text for further explanation (St 82).

calculations of the $1s\sigma$ MO $P(b)$ at 200, 400, and 600 keV (Ki 80). The agreement in slope between theory and experiment is good. However, if one normalizes theory to experiment at 400 keV, the absolute height of the theory is too low at 200 keV. This latter fact is further indication of problems in the theory in underestimating $1s\sigma$ MO x-ray yields at low photon energies, as was mentioned previously in connection with Figure 3.66.

A second type of coincidence experiment was performed between K x rays and MO x rays (recall Figure 3.47 and associated discussion). Although, the idea behind such a study was discussed in Section 3.3.2.3 in connection with Nb + Nb results (Ob 80), relativistic effects in the Pb + Pb quasimolecule lead to qualitatively different behavior of the coincidence yields in this high-Z system. We, therefore, repeat some of this earlier discussion here. As indicated in Figure 3.70 (Vi 81), a $2p_{1/2}\sigma$ to $1s_{1/2}\sigma$ MO x-ray transition produces a vacancy in the $2p_{1/2}\sigma$ orbital which survives the collision with large probability and decays into a K x ray. In addition, $2p_{3/2}\sigma$ to $1s_{1/2}\sigma$ transitions with the subsequent (highly probable) coupling of the $2p_{3/2}\sigma$ vacancy to the $2p_{1/2}\sigma$ orbital on the outgoing part of the collision, also result in the production of one continuum and one charac-

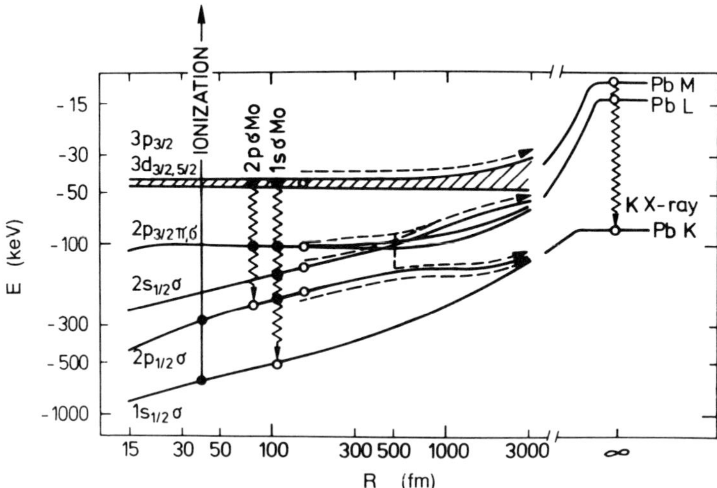

Figure 3.70. The Pb + Pb quasimolecular correlation diagram showing the excitation and decay of inner quasimolecular orbitals during and after the collision (Mu 76b, Vi 81).

teristic K x ray. Therefore, a coincidence experiment between characteristic and continuum K x rays might isolate these cascades from other transitions in the quasimolecule.

However, Figure 3.70 schematically indicates that independent excitation of two electrons in the same collision will also occur, with one vacancy decaying into an MO x ray during, and the other vacancy into a K x ray after, the collision. This latter process will also produce one continuum and one characteristic x ray. The coincidence requirement between MO and K x rays will therefore result in a continuum x-ray spectrum whose quantitative aspects reflect the competing strengths of various MO transitions into the $1s_{1/2}\sigma$ and $2p_{1/2}\sigma$ molecular orbitals, as well as the probability of single and double vacancy production in these orbitals.

Figure 3.71 shows the results of such MO x-ray, K x-ray coincidence measurements in 4.8-MeV/u ^{208}Pb + ^{208}Pb collisions (Vi 81). The top part of the figure displays extracted coincidence and singles MO x-ray cross sections in comparison to theory. The theory calculates "singles" radiation into the $1s\sigma$ molecular orbital only. The bottom part of the figure shows the ratio of measured coincidence to singles cross sections as a function of photon energy. The two parallel lines are calculated (Vi 81) coincidence to singles ratios from independent excitation of double vacancies, with one vacancy decaying into a $1s\sigma$ (top line) or a $2p\sigma$ (bottom line) MO x ray. The larger calculated ratio for $1s\sigma$ MO radiation compared to that for $2p_{1/2}\sigma$ MO x rays reflects the steeper impact parameter dependences of $1s\sigma$ excitation.

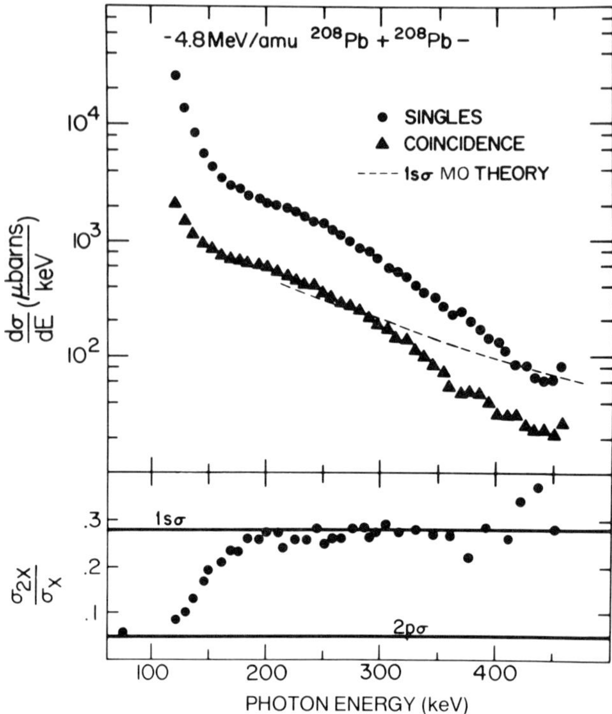

Figure 3.71. Singles and K x-ray, MO x-ray coincidence cross sections from 4.8-MeV/amu $^{208}\text{Pb} \rightarrow {}^{208}\text{Pb}$ collisions. Dashed line, theoretical $1s\sigma$ MO x-ray singles normalized to the data above 400 keV. See text for further explanation (Vi 81).

The bottom part of Figure 3.71 illustrates that the measured cross section ratios can, in principle, be reproduced assuming coincidences come from independently excited double vacancies only, simply by choosing a suitable mixture of $1s_{1/2}\sigma$ to $2p_{1/2}\sigma$ MO x-ray intensities as a function of photon energy. Moreover, the mixture necessary to explain the measured ratios of Figure 3.71 agrees with the mixture which one obtains from MO x-ray impact parameter dependence studies through analyses of their two component structure (St 82). Therefore, in contrast to results from lighter systems, where dominant contributions from cascades were obtained (Ob 80), a consistent description of the results of Figure 3.71 in terms of double vacancy production only can be made. This description, in turn, indicates a significantly higher cross section for $1s_{1/2}\sigma$ MO x rays than that calculated theoretically at photon energies below 400 keV (Ki 80).

In concluding this section, we reiterate mention of the unique role played by quasimolecular x-ray emissions in providing direct evidence for the formation of strongly bound molecular orbitals in these collisions. The straightforward spectroscopy of the united quasiatom based on directional anisotropy measurements in lighter systems cannot be extended to the superheavy quasimolecule. However, impact parameter dependence studies provide a means by which $1s_{1/2}\sigma$ and $2p_{1/2}\sigma$ MO x-ray yields can be separated and distinguished from each other, and K x-ray, MO x-ray coincidence measurements provide further complementary information on relative excitation and radiative decay probabilities in the innermost molecular orbitals. It is not clear whether or not such studies can be extended to still heavier collision systems due to the dominance of disturbing backgrounds from nuclear Coulomb excitation.

3.5. Other X-Ray Measurements

Energetic collisions between heavy atoms, although often adiabatic from the point of view of the velocities involved, are quite violent for the atomic electrons. Even quasimolecular orbitals generate electron binding energies which are only a small fraction of the total available energy in the center of mass of the collision system. As a result, large amounts of atomic ionization occur with the associated production of both discrete and continuum x rays of various, sometimes exotic, origins. In addition to characteristic K x rays and quasimolecular x rays, other x-ray yields from two-electron–one-photon atomic transitions (TET), radiative electron capture (REC), secondary electron bremsstrahlung (SEB), and nucleus–nucleus bremsstrahlung are observed.

3.5.1. Two-Electron–One-Photon Transitions (TET)

TET (Wo 75, St 77b, Gr 77b, Lu 77) occurs when an atom loses two electrons from the K shell in a heavy-ion collision. Usually, this excited atomic state will decay with the sequential emission of two K x rays, the first x ray being the hypersatellite atomic transition (K_α^h or K_β^h) and the second x ray, a "normal" satellite transition. The second transition is designated as satellite since, in heavy-ion collisions, double vacancy production in the K shell is nearly always accompanied by simultaneous ionization of the outer shells as well. Alternatively, double vacancy decay can proceed by the simultaneous jumping of two electrons from higher atomic shells into the K shell with the emission of a single photon. These latter events are designated as two-electron–one-photon transitions.

The first identification of TET events in the x-ray regime (Wo 75) was made during efforts to measure quasimolecular x-ray production in heavy-ion collisions. An example of typical experimental spectra is given in Figure 3.72 (Gr 77b) for the case of 17.9 to 91.5 MeV Ni + Ni collisions. The linear plots show the prominent, intense contributions from (strongly

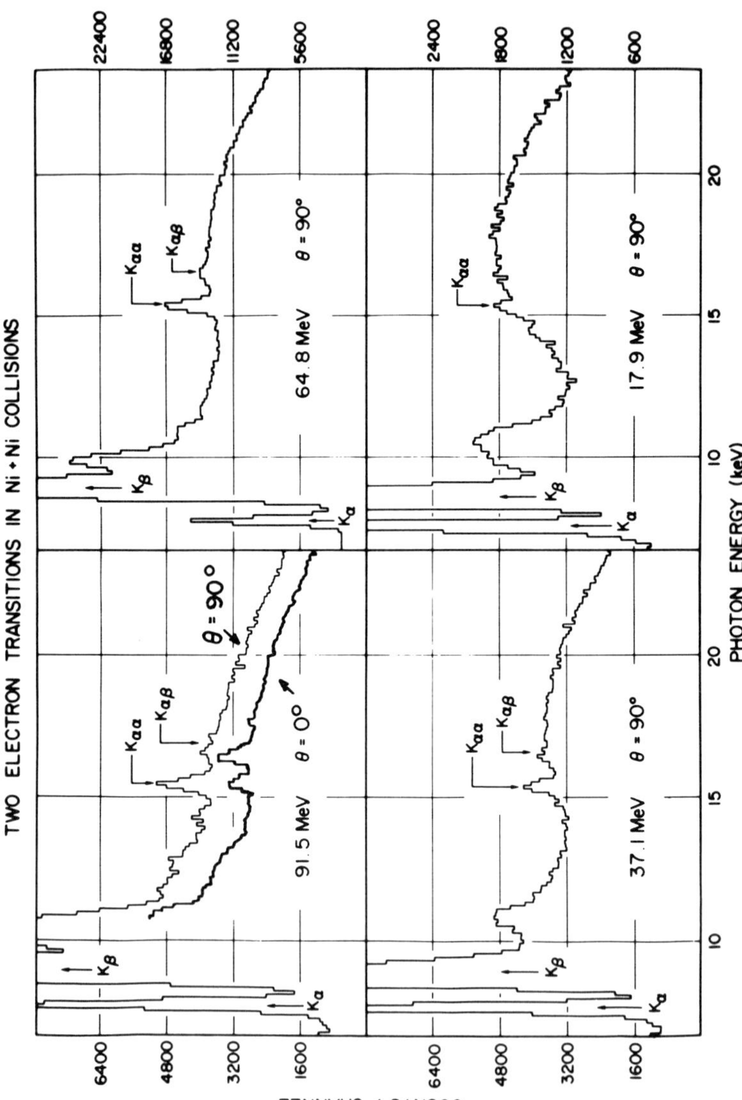

Figure 3.72. X-ray spectra from Ni + Ni collisions. Data are at 90° to the beam direction except at 91.5 MeV where both 90° and 0° distributions are shown. Peaks at ~15 keV are due to two-electron–one-photon transitions (Gr 77b).

absorbed) characteristic Ni K_{α} and K_{β} x-rays below 10 keV x ray energy, and the quasimolecular x-ray continuum extending to higher photon frequencies. The small bumps in the 15 to 16 keV region labeled $K_{\alpha\alpha}$ and $K_{\alpha\beta}$ are TET events. The $K_{\alpha\alpha}$ peak originates from the jumping of two L-shell electrons, and the $K_{\alpha\beta}$ from the jumping of one L-shell and one M-shell electron. The inset in the 91.5-MeV data is the spectrum at zero degrees to the beam direction where the prominent Doppler-shifted projectile $K_{\alpha\alpha}$ transition is superimposed upon the target $K_{\alpha\beta}$ peak. Note that the use of an absorber in these studies is essential not only to prevent distortion of the TET lines from the high-energy detector line-shape tails of the K x rays, but also to reduce pulse pile-up to negligible proportions.

Through attempts to theoretically calculate the branching ratio for double vacancy decay by TET, it has become clear that the transitions are fundamentally many-electron processes in which electron–electron correlations play an important role in both the initial and final states (Lu 77). A number of authors have carried out theoretical studies of the TET decay rates and transition energies (Vi 32, Ab 76, Ga 76, Ke 76). Results show that the experimentally observed transition energies can be adequately explained by calculations which include the state of L-shell ionization at the time of TET emission.

Figure 3.73 (St 77b) plots the results of measurements of the branching ratio for TET versus hypersatellite decay as a function of atomic number Z in comparison to various theoretical predictions (Vi 32, Ab 76, Ga 76,

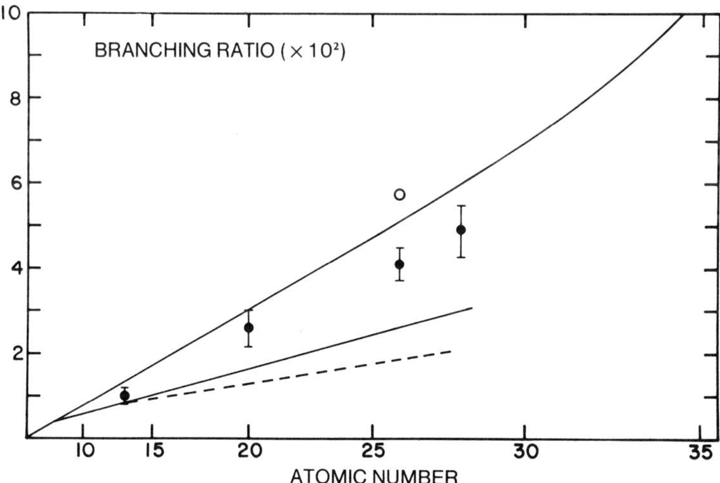

Figure 3.73. Comparison of measured TET branching ratio to calculations (St 77b). The atomic number Z is displayed in a quadratic scale. Lower dashed line, theory from (Ga 76). Lower solid line, theory from (Ab 76). Upper solid line, theory from (Vi 32). The hollow data point is from (Ke 76).

Ke 76). The experimental hypersatellite to TET ratios show the characteristic Z^2 increase predicted by theory, although certain discrepancies in absolute magnitude still seem to exist. Both the $K_{\alpha\alpha}$ and K_α^h transition rates depend on the state of the atomic L shell. These dependences, as well as possible experimental errors, are somewhat reduced by utilizing the $K_\alpha^h/K_{\alpha\alpha}$ ratio as the quantity of interest.

Discussion of TET events up to this point has touched only on the atomic structure aspects of the problem. However, the fact that these transitions are observable at all in heavy-ion encounters stems from the large double K-shell excitation probabilities in nearly symmetric and symmetric adiabatic collisions. The mechanism of excitation of the double vacancies can be studied within the context of the molecular orbital model of electron promotion in atomic collisions. TET events therefore provide for additional tests of the applicability of the quasimolecular description of these encounters (Gr 77b).

For nearly-symmetric and symmetric collision systems, the K vacancy production cross section is given by

$$\sigma(K) = \frac{R}{6} 2\pi \int_0^\infty P(b) b \, db \tag{3.17}$$

where b is the impact parameter, $P(b)$ the probability of a $2p\pi$ to $2p\sigma$ rotational coupling, and R the mean number of $2p$ atomic vacancies prior to the collision. It is important to note that R represents dynamically induced vacancies produced during the earlier stages of the collision, or possibly in a prior collision. No static L-shell vacancies are present on the incident projectile atom.

The cross section for double K-shell excitation can be shown to be (Gr 77b)

$$\sigma(KK) = \frac{R^2}{144} 2\pi \int_0^\infty P^2(b) b \, db \tag{3.18}$$

This double vacancy cross section depends on statistical factors and the squares of the vacancy factor R and single vacancy rotational coupling probability $P(b)$. These relations, conjoined with theoretical calculations of $\sigma(K)$ for $R = 1$, yield for the $K_{\alpha\alpha}$ TET (Gr 77b)

$$\sigma_{K_{\alpha\alpha}} = \sigma_{\text{rot}} \frac{1}{48} \eta S R^2 \omega_{\text{khs}} \tag{3.19}$$

where σ_{rot} is the theoretical $\sigma(K)$ for $R = 1$, η is the hypersatellite to $K_{\alpha\alpha}$ branching ratio, S the probability that double vacancies will both wind up on the same atom, and ω_{khs} the hypersatellite fluorescent yield. The only quantity in Equation (3.19) which cannot be calculated is the vacancy factor R. However, using experimental K_α cross sections, $R = \sigma(K)/\sigma_{\text{rot}}$ can be empirically determined, and the result used in Equation (3.19) to predict $\sigma_{K_{\alpha\alpha}}$ (Gr 77b).

The results of such an analysis, performed for the Ni + Ni collision system, are shown in Figure 3.74 (Gr 77b). The upper portion of the figure compares experimental cross sections to theoretical predictions for $R = 1$. The comparison shows that, for these collisions, $R = 1$ at about 65 MeV bombarding energy, falling to values below 1 at lower beam energies and rising above 1 at higher projectile velocities. The bottom portion of the figure compares the measured $K_{\alpha\alpha}$ cross sections with cross sections given by Equation (3.19). The excellent agreement between theory and measurement furnishes independent confirmation of the essential validity of the σ_{rot} calculations, when modified to include the concept of dynamically induced vacancies (the factor R).

Studies of TET x rays in asymmetric collision systems provide an opportunity to extend understanding of the vacancy sharing phenomena to

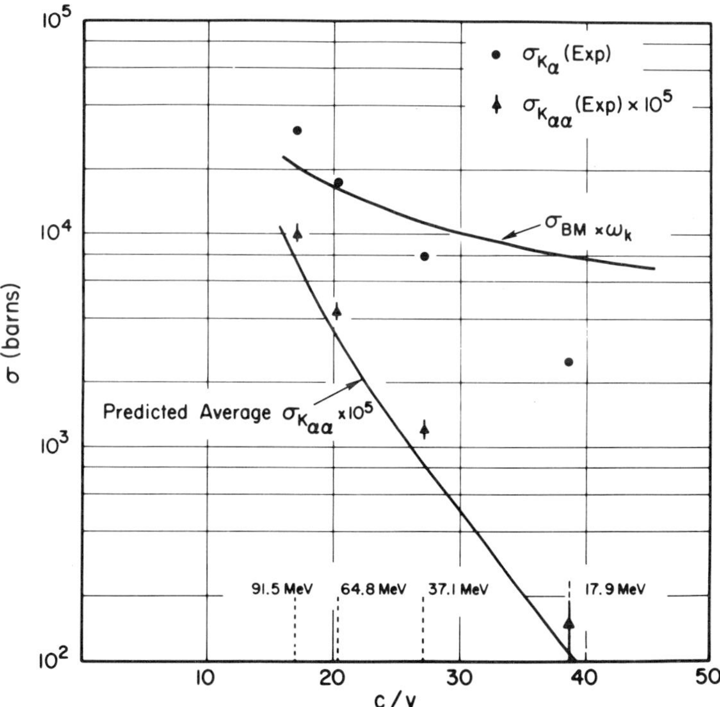

Figure 3.74. Experimental K x-ray (σ_{K_α}) and TET ($\sigma_{K_{\alpha\alpha}}$) cross sections in Ni + Ni collisions compared to theory (solid lines) as a function of projectile velocity (Gr 77b).

the case of double vacancy transfer. Our discussion in Section 3.41 showed that the ratio of heavy- to light-collision partner vacancies in asymmetric collisions is given by [Equation (3.13)]

$$R_k = e^{-2x} + \sigma_{1s\sigma}/\sigma_{2p\sigma} \tag{3.20}$$

where x is given by Equation (3.4) in Section 3.3.1.1 and $\sigma_{1s\sigma}$, $\sigma_{2p\sigma}$ are the excitation cross sections for the $1s\sigma$ and $2p\sigma$ molecular orbitals, respectively. For lighter collision systems, we can neglect contributions from term two in Equation (3.20) since $\sigma_{1s\sigma} \ll \sigma_{2p\sigma}$.

If one assumes that, for the case of double-vacancy sharing, the two vacancies are each shared independently, then the probability of double sharing is simply the square of the single vacancy sharing probability. Within the context of such a description, the ratio of heavy- to light-collision partner double vacancies is given by (Le 78)

$$R_{kk} \simeq e^{-4x} \tag{3.21}$$

TET data can be utilized to test the predictions of Equation (3.21). By measuring the ratio of heavy- to light-collision partner TET intensities over a range of collision systems, one can determine the fractional population of double K-shell vacancies in the two atoms involved. Clearly, corrections for fluorescent yields must be considered. However, use of a ratio rather than an

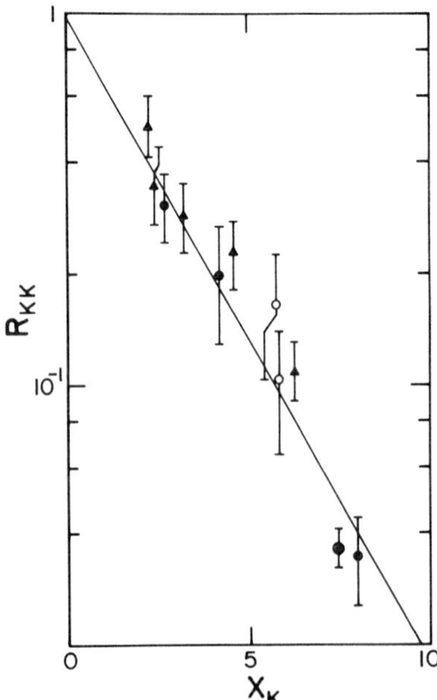

Figure 3.75. Comparison between theoretical (solid line) and experimental double K vacancy sharing ratios versus the reduced sharing parameter x as determined by TET measurements (Le 78).

absolute comparison minimizes error from fluorescent yield variations, particular in nearly symmetric collision systems.

The results of such a study are given in Figure 3.75 (Le 78). The solid line is the prediction of Equation (3.21). The agreement between experiment and Equation (3.21) is excellent over nearly two orders of magnitude in R_{kk}, and thereby provides strong evidence for the uncorrelated nature of the double-vacancy transfer process.

3.5.2. Radiative Electron Capture (REC)

The continuum character of the MO emissions has led to studies of other possible competing sources of x-ray emission in heavy-ion collisions. SEB and REC are two candidates for continuum x-ray production processes which could generate intensities comparable to or greater than those expected from MO x rays in some collision systems. Therefore, both as background processes to MO emission, as well as for reasons of their intrinsic interest, these competing sources of continuum x-ray production in heavy-ion–atom encounters have been explored.

REC is a process by which a weakly bound target electron is captured into a projectile atom with an inner-shell vacancy. The resulting decrease in total energy of the collision complex is compensated for by the emission of an x ray. Conservation of energy for this process gives

$$\hbar\omega = E_f - E_i + \frac{p_0^2}{2m} + \mathbf{v} \cdot \mathbf{p} \tag{3.22}$$

where $\hbar\omega$ is the energy of the radiated photon, E_f the binding energy of the electron in its final state, E_i the initial electron binding energy, m the mass of the electron, $\mathbf{v}$ the projectile velocity, $p_0 = mv$ the momentum of the target electron viewed in the frame of the projectile due to the motion of the projectile, and $\mathbf{p}$ the momentum of the electron due to its presence in a target atomic orbital. Thus, the photon energy distribution for REC will exhibit a peak at $\hbar\omega = E_f - E_i + p_0^2/2m$ with a width reflecting the projection of the target electron momentum distribution (Compton profile) along the projectile direction ($\mathbf{v} \cdot \mathbf{p}$). The cross section for REC will be related to the capture cross section and the probability of having projectile vacancies.

Figure 3.76 shows the results of REC spectra from studies of F^{n+} + He (gas) collisions at 15 and 30 MeV (Ta 82). The peaks at low x-ray energies are F K x-rays, and those at higher photon energies are REC x rays due to capture into the projectile K shell. Note that, at 15 MeV, REC produces a shoulder which is not resolved from the K x rays. At 30 MeV the peak has moved out to higher energies and is well resolved from the K x-ray intensity. The centroids of REC peaks have been found to be in good agreement with Equation (3.22). However, studies show that the REC peak

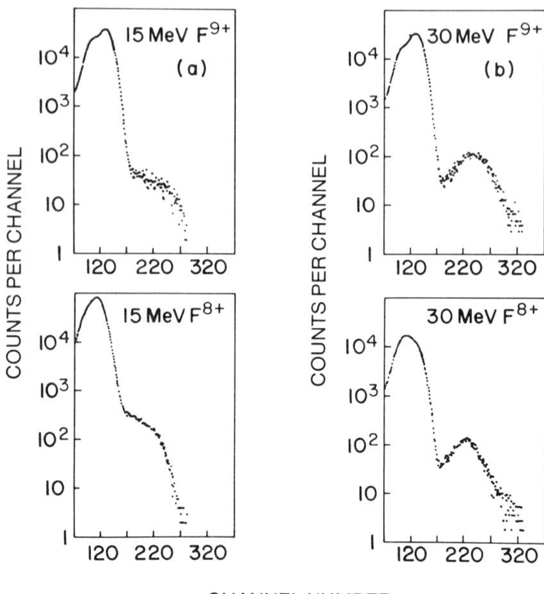

Figure 3.76. Typical x-ray spectra from F^{8+} and F^{9+} ions incident on a He gas target at 15 and 30 MeV under single collision conditions. Peak at left due to F *K* x rays, and that at right due to REC x rays (Ta 82).

widths increase somewhat for projectiles with two as opposed to one *K* vacancy in contrast to simple theoretical expectations.

The magnitude of REC cross sections for *K*-shell capture will reflect the *K*-vacancy probability of the projectile. In experiments where thin solid targets are used, this vacancy probability will change with target thickness reflecting competition between excitation and quenching processes in the foil (recall discussion in Section 3.3.1.1). The relevant cross section for REC is the cross section per projectile *K* vacancy, which can be determined by experiments as a function of target thickness when solid targets are used (Ta 78, Ta 81).

Figure 3.77 plots (Ta 81) the results of experimental REC cross sections per projectile vacancy for 20–80 MeV Cl beams incident on thin solid C and Cu targets. The dashed line is the free electron theory for REC (Be 77), and the two solid lines multiply the dashed curve by factors of 4, and 11, respectively. The authors interpret the excellent agreement between experiment and the solid curves as evidence that only the most weakly bound target electrons with velocities

$$\frac{v_e}{v} = \left(\frac{E_B/m_e}{E/M}\right)^{1/2} \leqslant 0.1 \tag{3.23}$$

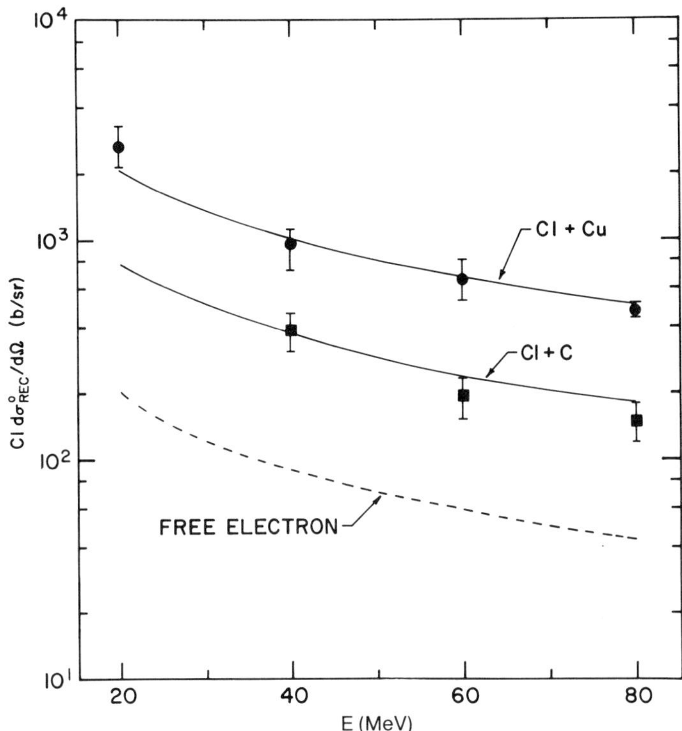

Figure 3.77. Cl REC differential cross sections at 90° per K vacancy versus incident Cl ion energy. Squares, Cl + C. Circles, Cl + Cu. Dashed curve is free electron Bethe and Salpeter theory assuming $\sin^2\theta$ REC angular distribution. Solid curves are dashed curve times 4 and 11 for C and Cu, respectively, representing the effective number of free electrons in these target atoms (Ta 81).

contribute to the REC cross section (Ta 81). In Equation (3.23), v_e is the target electron velocity, E_B the target electron binding energy, m_e the mass of the electron, E the projectile energy, M the projectile mass, and v the projectile velocity. This criterion is satisfied by the four C L-shell electrons and the 11 outermost Cu electrons. The plausibility of this interpretation was confirmed by subsequent, more precise gas target studies (data of Figure 3.76), which also found good agreement between cross-section measurements and REC calculations (Ta 82).

3.5.3. Secondary Electron Bremsstrahlung (SEB)

Another source of continuum x-ray emission in heavy-ion collisions is secondary electron bremmstrahlung (SEB). This is a two-step production mechanism; the electron is ejected in the first collision, and bremsstrahlung

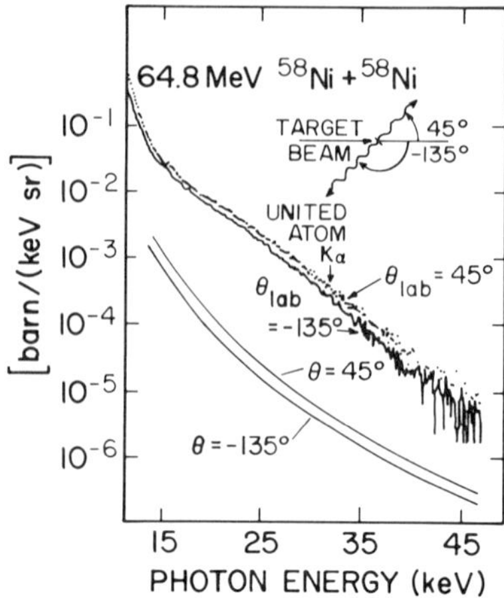

Figure 3.78. Comparison between experimental continuum x rays (upper curves) and extracted SEB yields (lower curves). Both cases exhibit forward–backward asymmetry (Vi 78, Be 82).

is produced in a subsequent collision between the ionized electron and a target atom. Problems in estimating the size of SEB in heavy-ion collisions stem principally from theoretical uncertainties in the ionized delta-electron intensities and energy distributions. In addition, electron bremsstrahlung can be produced from collisions with materials surrounding experimental targets (frames, chamber walls, etc.) leading to SEB yields which depend on experimental geometry. It has been generally assumed that SEB yields are small in studies of MO x-ray production (Vi 78, St 80). However, only recently has experimental evidence in support of this assumption been obtained.

Figure 3.78 plots results of cross sections for continuum x-ray emission in Ni + Ni collisions at 45° and 135° with respect to the beam direction (Vi 77, Vi 79, Vi 78). The solid lines in the figure show results of calculated x-ray intensities from SEB based on measured delta-electron spectra from these collisions (Be 82). Clearly, SEB intensities are too small to explain the observed data thereby providing strong evidence that backgrounds from electron bremsstrahlung can be safely neglected in these studies of MO x-ray emission from Ni + Ni collisions.

3.5.4. Nucleus–Nucleus Bremsstrahlung (NNB)

Nucleus–nucleus bremsstrahlung (Re 76, Tr 76) can be experimentally studied in very asymmetric systems where MO x-ray yields are suppressed by the small probability of producing $1s\sigma$ vacancies. Examination of the angular distribution of NNB in selective collision systems has illustrated the importance of dipole–quadrupole interference effects on the measured intensities, and demonstrated the validity of a semiclassical theoretical approach to the calculation of NNB yields. NNB can constitute a large source of background in MO x-ray measurements, particularly at high photon frequencies in asymmetric collision systems.

The cross section for NNB can be expressed as a multipole expansion of the prominent dipole term plus the weaker quadrupole contribution as follows (Re 76):

$$\frac{d\sigma}{dE_x\, d\Omega} \propto \frac{Z_p^2 Z_T^2}{E_x \beta^2} [D^2 W_{E1}(E, E_x, \theta) + \mu^2\beta^2 Q^2 W_{E2}(E, E_x, \theta) + 2\mu\beta DQ W_{E1,E2}(E, E_x, \theta)] \tag{3.24}$$

$D = Z_p/A_p - Z_T/A_T$ comes from the dipole moment $\mathbf{d} = \sum_v \varepsilon_v \mathbf{r}_v$ transformed to the center-of-mass system, and $Q = Z_p/A_p^2 + Z_T/A_T^2$ is associated with the quadrupole moment. W_{E1}, W_{E2}, and $W_{E1,E2}$ represent the dipole, quadrupole, and interference parts of the angular distribution at bombarding energy E, photon energy E_x, and emission angle θ. μ is the reduced mass, and $\beta = v_{\text{proj}}/c$.

Examination of Equation (3.24) shows that when Z_p (projectile) = Z_T (target), the most prominent contribution from dipole radiation vanishes leaving only weaker quadrupole contributions. Therefore, MO x-ray experiments should be performed with isotopically pure, symmetric collision systems. Furthermore, although the quadrupole and dipole terms are each separately symmetric about 90°, the interference term results in a forward–backward asymmetry for the NNB angular distributions.

Figure 3.79 (Tr 76) plots measured photon spectra at 0° and 135° for the 13-MeV $^{18}O + ^{58}Ni$ collision system. A 0.86-mm Al absorber has been used in the measurements. Extending above the x-ray contributions from Ni K x rays and a continuum bump at low energies are contributions from NNB. At zero degrees, the dipole ($E1$, dashed lines) and quadrupole $E2$ (dotted lines) interfere destructively to produce the total (T) (solid line) intensity. At backward angles dipole and quadrupole contributions interfere constructively ($E1$, $E2$ interference is dashed-dotted line) to produce the total. At both angles, absolute agreement between experiment and theory (Re 76) is very good.

By changing the sign of the D term in Equation (3.24), the sign of the interference term can be changed. Illustration of this effect is given in Figure 3.80 (Tr 76), where comparison between the $^{18}O + ^{58}Ni$ ($D < 0$) and $^{12}C + ^{58}Ni$ ($D > 0$) systems shows a dramatic reversal of the forward–backward angular distributions. Alternatively, simple interchange of the projectile with the target ($^{58}Ni + ^{12}C$ versus $^{12}C + ^{58}Ni$) changes the sign of D and reverses the angular distribution. Agreement between experiment and theory in absolute magnitude is excellent.

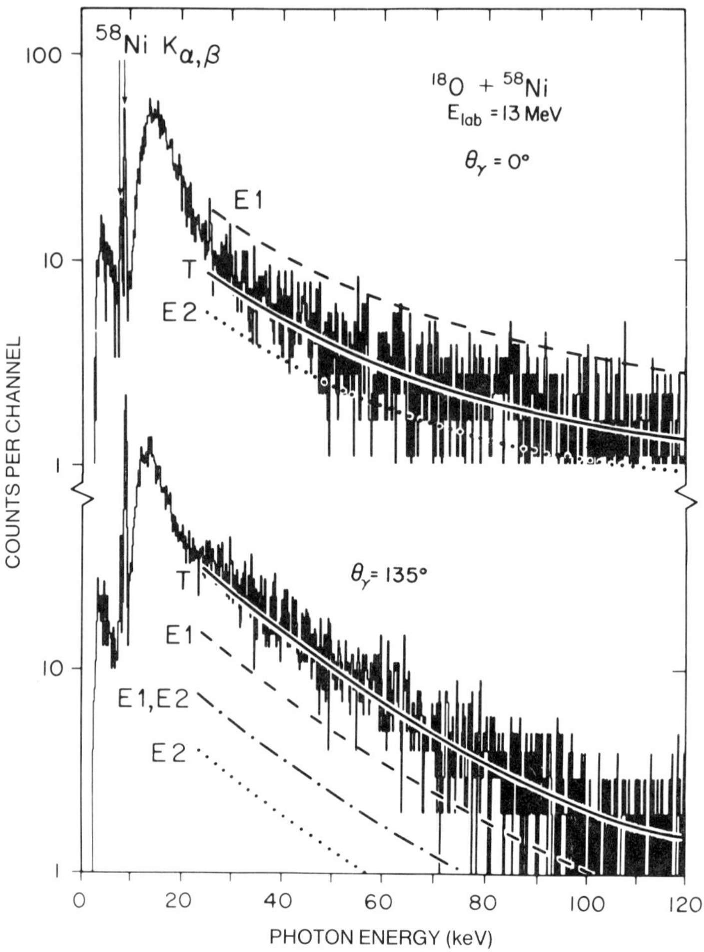

Figure 3.79. X-ray spectra from $^{18}O \rightarrow {}^{58}Ni$ at forward and backward angles obtained using an intrinsic Ge x-ray detector and 0.86-mm Al absorber. Alsolute intensity calculations for the $E1$ (dashed), $E2$ (dotted), and $E1$-$E2$ (dashed-dotted) interference components are shown (Tr 76).

For the very heavy collision systems, contributions to continuum x-ray yields from REC, SEB, and NNB have been studied. An example of a typical experimental spectrum is given in Figure 3.81 (Vi 81), where data taken with a 3×3 in. NaI crystal in 4.8-MeV/amu $^{208}Pb + {}^{208}Pb$ collisions are shown. Contributions from NNB (dashed line), and an upper bound to SEB yields (solid line with arrows) are shown in the figure. The SEB calculations were performed using measured delta-electron cross sections

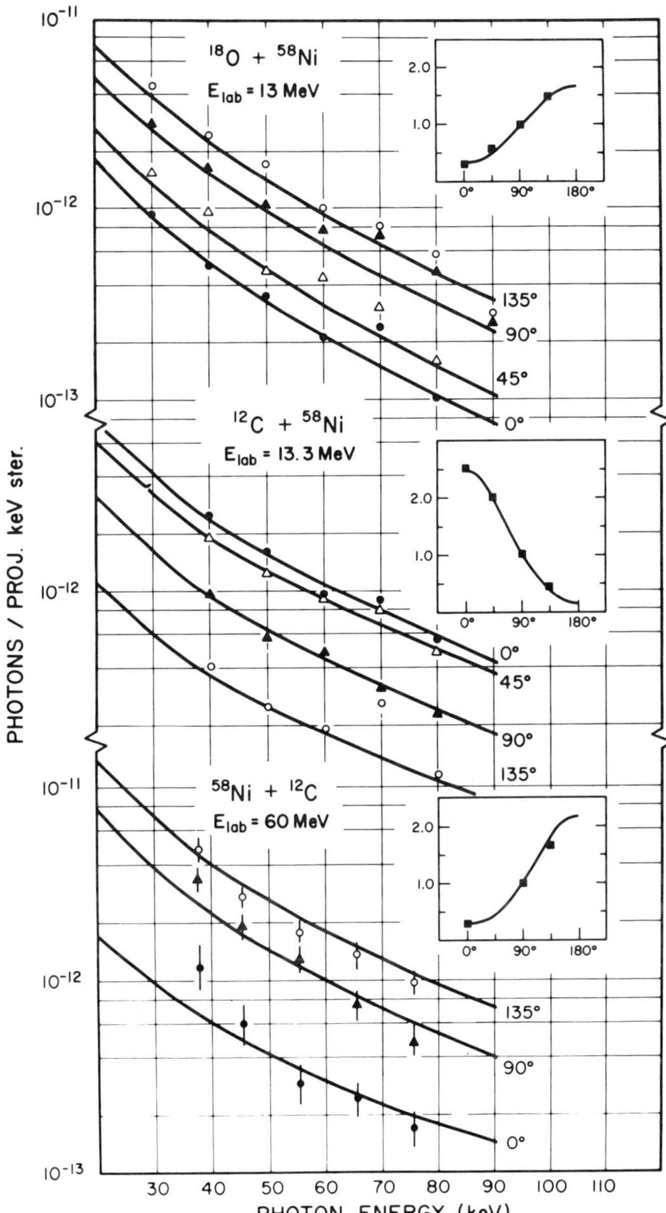

Figure 3.80. Comparisons of calculated NNB absolute thick target yields with experimental x-ray intensities versus photon energy at four angles of observation. Each datum point represents an integration over a 10-keV photon energy interval. The insets (vertical scale relative) show a comparison of calculated angular distribution (solid lines) with the data (squares) for the photon energy interval 40–60 keV (Tr 76).

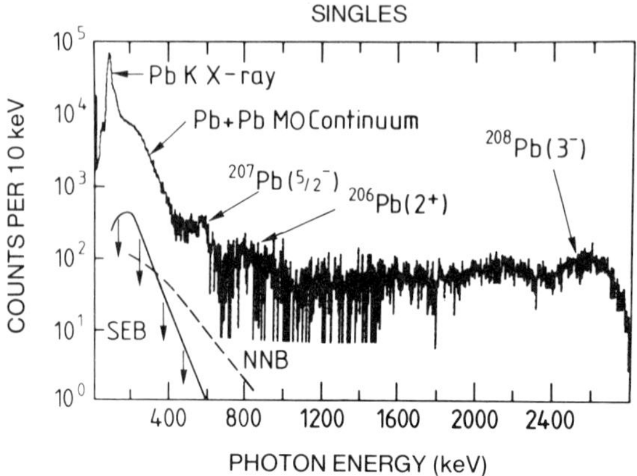

Figure 3.81. Singles spectrum from 4.8-MeV/amu $^{208}Pb \rightarrow {}^{208}Pb$ collisions showing SEB (solid line) and NNB (dashed line) backgrounds. Coulomb excited target and projectile contaminant nuclear states are indicated (Vi 81).

(Be 79). Large contributions from Coulomb excitation of the 3^{-} ^{208}Pb line at 2.6 MeV and the resulting Compton pulse height distribution provide the largest source of background to MO x-ray intensities above 800 keV photon energy. Despite the use of isotopically enriched targets and beams, small contaminations (<1%) of ^{207,206}Pb yield large contributions due to Coulomb excitation of their low-lying nuclear levels. Clearly, excitation of the nucleus is the largest source of background in measurements of continuum x-ray emission from very heavy collision systems. Backgrounds due to internal conversion of nuclear states cause similar disturbances in K-vacancy production studies.

4. Delta-Electron Spectroscopy

4.1. Introduction

Continuous spectra of delta electrons from atomic collisions have been under investigation since the early days of nuclear physics (Me 58). A number of recent reviews discuss topics related to electron production in ion–atom collisions (Og 72, Ru 72, Se 72, Bu 72, St 78). In many electron spectroscopy studies the primary interest is in electron line spectra such as internal conversion electrons in nuclear physics or Auger electrons in atomic physics. Of course, in such investigations the continuous spectra are merely an annoying background. The intrinsic potential of delta electrons as a source of information on atomic structure has not escaped recognition, however, and a number of investigations of these continuous spectra have been carried out with light projectiles (Zu 54, Me 58, Hu 53, Ch 69, Ch 70, Fo 75) and, most recently, with heavy ions (Ko 82, Ko 77, Bo 78, Gu 82). Following the emphasis throughout this chapter, our discussion will deal principally with the more recent heavy-ion experiments.

Some general reasons for interest in delta-electron spectroscopy are readily apparent. It was pointed out in the Introduction that information on inner shells, and particularly on the quasiatom, begins with the ionization process. Delta-electron emission is clearly the primary product of ionization and therefore provides information both on the excitation mechanism and on the wave function of the quasiatom. Moreover, delta electrons constitute a principal source of such information undiluted by the deexcitation processes. In Section 3 we saw that studying vacancy formation with the deexcitation characteristic x rays omits vital information on the final state of the excited electron. Particularly, detecting the presence of the vacancy by x-ray deexcitation alone excludes knowing whether the electron was excited into bound states or ionized into the continuum. For example, this information is required to determine the energy transfer, which is an important parameter in the ionizing process and appears in the scaling law discussed in Section 2. With the additional information obtained by detecting the electron's energy and direction of emission as well as the nuclear scattering states, however, the final state of the ionization process becomes more clearly defined.

Of course, determination of the binding energy associated with the initial state of the ejected electron is one of the goals of ionization studies. Obviously, the inclusive detection of the electron spectrum is not sufficient to define this state since the spectrum is then a superposition of ionization from several inner-shell states. But in some collisions the initial state can be selected by identifying the characteristic x rays that follow the excitation (Ko 77). Even for the heaviest atoms these x-rays are emitted in $\sim 10^{-17}$ s, which is long after the collision time of $\sim 10^{-20}$ s when the atoms are again well separated. Therefore, irrespective of whether the initial states involved in the excitation are atomic or molecular in character, the vacancy decays predominantly while associated with the separated atom. For collisions with $Z_1 \ll Z_2$ and $v_{\text{ion}} > v_e$, where atomic states are excited, the deexcitation x ray readily identifies the initial atomic shell excited. Where the quasimolecular picture is valid, it is also possible to select the individual contributions from particular quasiatomic shells from the total delta-electron spectrum in sufficiently asymmetric collision systems where vacancy sharing is small. We have noted in Section 3 that in such cases the K x rays from the heavier collision partner are associated principally with the excitation of the $1s\sigma$ state of the quasiatom while the K x rays of the lighter separated atom follow after $2p_{1/2}\sigma$ vacancy formation. Thus, for example, delta electrons measured in coincidence with the characteristic K x rays of the heavier atom probe the innermost $1s\sigma$ quasiatomic shell.

The most complete type of experiment, therefore, consists of detecting the delta electrons simultaneously with the scattered nuclear particles and the deexcitation characteristic x rays. In such a measurement the initial and final electronic states are defined as well as the ion trajectory which determines essential parameters such as the distance of minimum approach between nuclei. Obviously such triple coincidence measurements are very demanding on achieving large efficiencies for all detectors, and, therefore, only very few of these measurements have been reported to date. Simpler experiments which measure the delta-electron spectrum alone or in coincidence with either the scattered ions or x rays also yield significant information, although obviously not as complete as potentially available from the latter type measurements. But even these simpler experiments are presently limited in number, with the field just beginning to be exploited by groups at Munich, Heidelberg, and at GSI Darmstadt (Ko 77, Ko 82, Bo 78, Gu 82). We will consider these new results discussed recently by Kozhuharov (Ko 82).

4.2. Qualitative Features of Delta-Electron Spectra

A comprehensive theory of delta-electron emission in heavy-ion collisions has been developed as part of the general attempt to understand the ionization processes. In large part the experimental observations have found an explanation within this theoretical framework developed by G. Soff, B. Müller, J. Reinhardt, W. Greiner, and co-workers (Re 85). We return to these detailed comparisons later, but let us begin with a qualitative examination of some delta-electron data together with simple kinematic considerations which reveal many of the central points emerging from the delta-electron studies. In general, the recurrent theme is the one previously emphasized by the vacancy production investigations via detection of characteristic x rays. Mainly it was found in those studies that the $1s\sigma$ vacancy production was governed by two principal ingredients, the kinematics of the collision and the coupling of the quasiatomic wave functions to the demands of the kinematics. These set the conditions for the spatial confinement of the ionizing process and the magnitude of the excitation probability, respectively. Of course the delta-electron spectra also reflect these two ingredients as well as providing additional detailed information on the electronic states involved in ionization.

For example, Figures 4.1–4.4 illustrate some of the main features of delta-electron emission in heavy-ion collisions.

(i) Figure 4.1 shows measured cross sections as a function of the energy of the ejected electron for two collision systems (Ko 82). In both

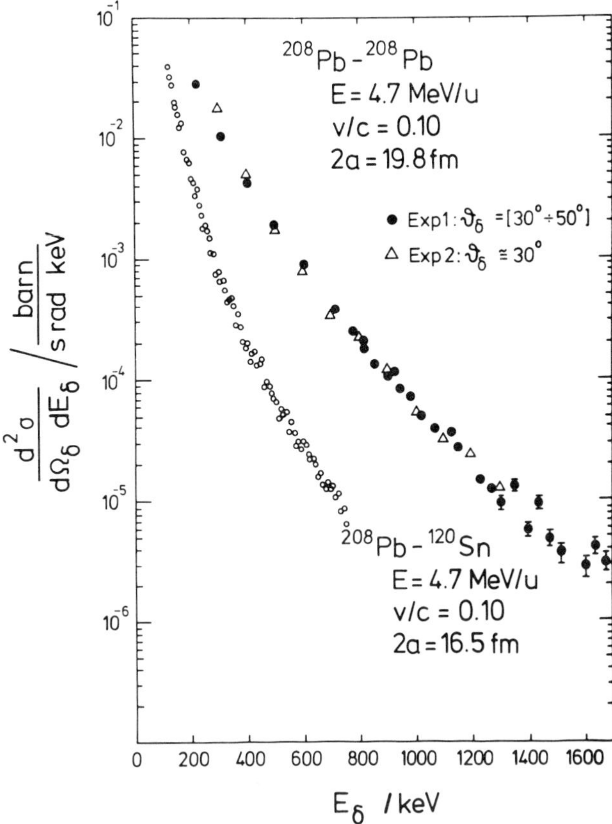

Figure 4.1. Delta-electron spectra from Pb + Pb ($Z_u = 164$) and Pb + Sn ($Z_u = 132$) collision systems measured at the same relative velocity $v/c = 0.10$. Also indicated are distances of closest approach for a head-on collision, $2a$ (Ko 82).

cases the bombarding energies exclude nuclear contact between collision partners. An interesting feature of the spectrum from Pb + Pb collisions is the presence of electrons with energies of ~1.5 MeV at an intensity level corresponding to ~10 μb/keV. This energy is approximately 150 times the energy that can be transferred to a free electron by the projectile (~10 keV), and approximately 20 times larger than the binding energy of the *K* shell for the Pb atom. Both the strong dependence of the intensity on electron energy and the growing divergence with increasing electron energy between the heavier and lighter collision systems are also striking properties. Almost equal intensities at low electron energies become differentials of more than an order of magnitude for electron energies of a few hundred keV. Clearly, since the projectile is the same in both cases, this behavior resides with a property of the combined system.

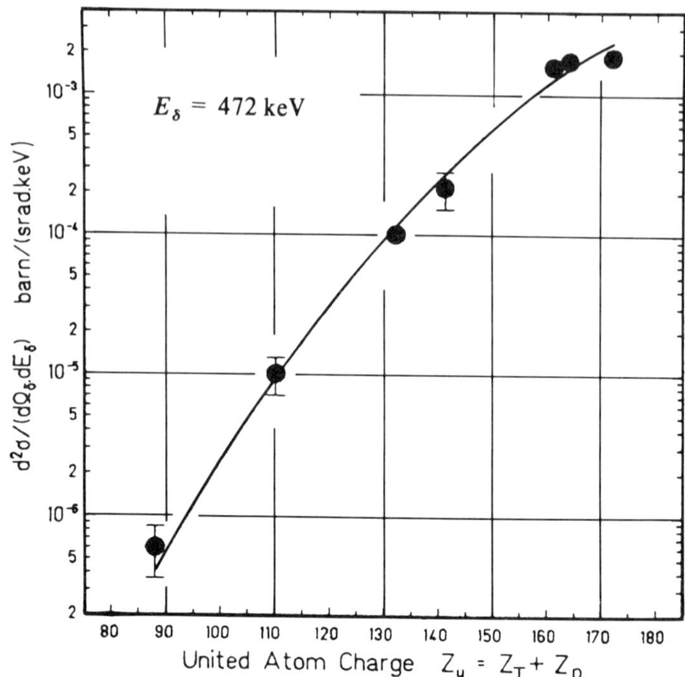

Figure 4.2. Double-differential cross section for emission of delta electrons of 472-keV kinetic energy as a function of the united charge Z_u of the collision system for a relative collision velocity $v/c = 0.10$ (Ni + Pb extrapolated from $v/c = 0.091$) (Ko 82).

(ii) Figure 4.2 presents results of a measurement of the double differential cross section $d^2\sigma/d\Omega_\delta\, dE_\delta$ at a given delta-electron kinetic energy of 472 keV, as a function of the combined charge Z_u of the collision systems C + Pb, Ni + Pb, Pb + Sn, Pb + Pr, Pb + Au, Pb + Pb, and Pb + Th (Ko 82). All measurements were carried out with the same relative velocity, 0.10c. There is a remarkable initial increase of the cross section with $(Z_u)^{12}$ for $Z_u\alpha < 1$. It levels off at higher Z_u where the binding energy of the quasiatoms becomes comparable to or exceeds the ejected electron kinetic energy.

(iii) Results from an investigation of the projectile velocity dependence of the cross section are depicted in Figure 4.3 (Gu 82). There is a strong variation of $d^2\sigma/d\Omega_\delta\, dE_\delta$ with projectile velocity; it increases by a factor of 20 for a change in projectile velocity of only 37%. This increase seems to be independent of the kinetic energy of the electrons.

(iv) The spatial definition of the ionization process is illustrated by data such as shown in Figure 4.4 (Ko 82). The delta-electron emission probability is displayed as a function of the distance of minimum approach,

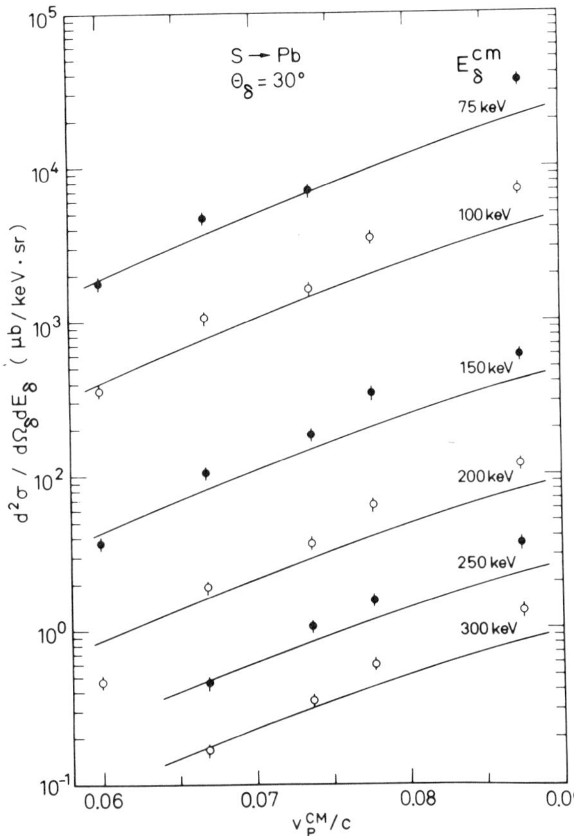

Figure 4.3. Excitation function for the system S + Pb ($Z_u = 98$): Double differential cross sections $d^2\sigma/d\Omega_\delta\, dE_\delta$ as a function of the center-of-mass projectile velocity, $v_p^{\rm cm}$, plotted for various center-of-mass kinetic energies, $E_\delta^{\rm cm}$. The solid lines are Born approximation calculations with hydrogenlike wave functions of the united atom (Gu 82).

R_0, for fixed electron kinetic energy in the center of mass. Generally for this fairly heavy system it drops off rapidly with increasing R_0 well within the mean radius of the united atom K shell; $\langle r_{1s}\rangle \simeq 350$ fm for $Z = Z_p + Z_T = 135$. But it is important to notice also that the slope of the exponential fall-off of $P(R_0)$ depends on the electron kinetic energy, especially when the kinetic energy is greater than the binding energy of the inner shells.

As mentioned above, the qualitative explanation for all these observations can be found in the kinematics of the collision and in the role played by the wave function of the ionized state. As was the case when characteristic x-ray production was considered, the effects of the kinematic

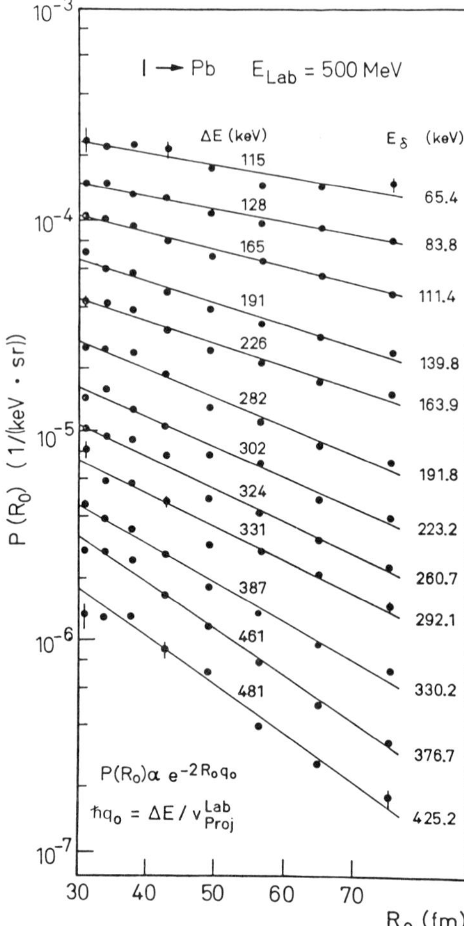

Figure 4.4. Delta-electron emission probability for I + Pb collisions as a function of the distance of closest approach, R_0, for fixed delta-electron kinetic energy, E_δ, in the center of mass. The solid lines represent results of the fit to the slope of the data with an exponential function $P \propto \exp(-2R_0 \Delta E/\hbar v)$. Also indicated in the figure are the values obtained from the fit for the energy transfer required, ΔE (Ko 82).

constraints on delta-electron emission can be readily demonstrated. We need only modify the previous discussion by including explicitly the freedom to choose the electron kinetic energy which can be comparable to or exceed the binding energy of the initial state. We had noted previously that for the major part of the vacancy production cross section little kinetic energy is transferred to the ejected electron so that E_δ is neglected with respect to $|E_B|$. Using Equation (2.25), the energy transfer now consists of the sum of the binding energy and the kinetic energy parts, $\Delta E = E_\delta + |E_B|$, and the momentum transfer takes on its minimum value of $q_{\min} = \Delta E/(\hbar v) = (E_\delta + |E_B|)/(\hbar v)$. The adiabatic nature of the collision ($v/c \ll 1$) and the associated large magnitude of the momentum transfer directly and indirectly determine the main features of the delta-electron spectrum as they did the vacancy production probability.

We see, for instance from Equation (2.28), that the emission of high-energy electrons associated with large momentum transfer can only come from close encounters. This correlation also follows from the more general semiclassical consideration that the electromagnetic pulse generated by the time changing Coulomb field contains in its spectrum Fourier frequencies $\tilde{\omega}$ sufficient to supply the energy transferred:

$$\Delta E = \hbar\tilde{\omega} \tag{4.1}$$

Encounters with average impact parameter $\tilde{b}$ and velocity v are characterized by Fourier frequencies $\tilde{\omega} \sim (v/\tilde{b})$. To produce ionization we, therefore, require that (Ar 78, Ba 59)

$$\Delta E = E_\delta + |E_B| \approx \hbar\tilde{\omega} = \hbar(v/\tilde{b}) \quad \text{or} \quad \tilde{b} \approx \hbar v/(E_\delta + |E_B|) \tag{4.2}$$

Substitution of parameters of interest, such as those associated with Pb + Pb collisions shown in Figure 4.1, $E_\delta \sim 1$ MeV and $(v/c) \sim 0.1$, yields $\tilde{b} < 20$ fm. Hence high-energy electrons with $E_\delta > 1$ MeV originate from collisions involving internuclear separations approaching nuclear dimensions. In fact, for the range of delta-electron energies that are usually measured (>150 keV), the localization of the ionization is restricted to dimensions which are well within the radii of the atomic electron shells involved. Therefore, kinematic arguments alone suggest that the high-energy part of the delta-electron spectrum can be used to investigate the properties of the wave functions and energies of the united atom system. More specific evidence for this important opportunity is presented below.

It is also important to point out that the degree of spatial confinement of the ionization probability depends both on $|E_B|$ and E_δ. Only for $E_\delta \ll |E_B|$ is a rapid falloff of the ionization probability with increasing impact parameter a signature for strong potential binding. Since, as we mentioned earlier, total vacancy production is indeed dominated by very small E_δ, the impact parameter dependence of vacancy production detected by x-ray emission provides a measure of $|E_B|$.

Thus these simple kinematic considerations indicate that, indeed, it is interesting to study high-energy delta electrons. However, to address the questions regarding the accessibility of this part of the delta-electron spectrum and, more generally, the composition of this spectrum, we have to turn to the structure of the wave functions involved. In doing so, we will see that kinematic considerations will again elucidate some relevant points.

It is characteristic of slow ion–atom collisions that, in large part, the momentum transfer also specifies the initial momentum $\hbar\mathbf{k}_i$ of the ejected electron. Momentum conservation gives $\mathbf{k}_i = (\mathbf{k}_f - \mathbf{q})$, where $\hbar\mathbf{k}_f$ is the final electron momentum. But with $v \ll c$, we find that $q_{\min} \gg k_f$. For example, for a 1-MeV delta electron and $v/c = 0.1$, $\hbar q_{\min} > 10$ MeV/c and $\hbar k_f \simeq$ 1.4 MeV/c. Therefore, since the ionization cross section is dominated by the smallest momentum transfers allowable, then $k_i = |\mathbf{k}_f - \mathbf{q}| \simeq q_{\min}$. Hence for ionization to occur with the emission of a delta electron with energy E_δ, electrons with the appropriate momentum $\hbar k_i \approx (E_\delta + |E_B|)/v$ have to be accessible in the initial state; high-energy components of the delta-electron spectrum must reflect high-momentum components of the bound-state wave function.

Using for illustration our previous example of 1-MeV delta electrons from Pb + Pb collisions, we find that in this case $\hbar k_i \approx 9$ MeV/c. This momentum is 30 times larger than the mean momentum of $\langle \hbar k_i \rangle \simeq 0.3$ MeV/c for the 1s electron in Pb obtained from its average kinetic energy through the virial theorem. The related mean orbiting radius in Pb is $\langle r_{1s} \rangle \approx 750$ fm. Therefore, the large initial momentum components required to emit a 1-MeV delta electron clearly have to be found in a wave function much more strongly condensed in space than the inner shells of the colliding atoms if there is to be an appreciable probability for ionization. The inverse square growth with decreasing internuclear separation R of the electron density around the two nuclear centers of a superheavy collision system, due to the relativistic contraction of the wave function, produces the necessary enhancement of the high-momentum components of the wave function. We noted previously that in the monopole approximation (Mu 78, So 79) the matrix element for a direct transition to the continuum by radial coupling, $\langle \psi_f | \partial/\partial R \, | \psi_i \rangle$, grows with $1/R$. This relativistic property ensures the crucial presence of the high-momentum components at the small internuclear separations where the large momentum transfers dictate that they are needed. Thus high-energy delta electrons are emitted from a restricted region well within the K-shell radius of the quasiatom. In fact, their detection with large cross sections is one of the definitive signatures for the formation of the quasiatom. It bears emphasis that the connection between large E_δ and spatial confinement is largely kinematic and that the enhancement of the cross section for ionization of high-energy delta electrons by relativistic effects is the new physics found in superheavy collision systems. Thus the role played by the formation of the quasiatom is the essential ingredient in the superheavy systems since the extreme relativistic properties are effectively only associated with the quasiatom and not with the properties of the individual collision partners.

All these qualitative features of delta-electron emission outlined above are explicitly displayed in the scaling law which was referred to in Section 2 (Ba 59, Ba 79, Ba 81b, Mu 78, Mu 83, Bo 80). We recall the conditions under which it is derived: it is based on first order perturbation theory, uses the monopole approximation for binding energies of and transition matrix elements between quasimolecular states, assumes the dominance of radial coupling to the continuum and bound states, and exploits the concentration of the ionization near the turning point of the collision to associate the location of the ionization with the distance of minimum approach R_0 between nuclei. We have also noted previously that although multistep processes (Re 79, So 80, So 81b) play an essential role in ionization and cannot be neglected, nevertheless, the scaling law seems to anticipate the behavior of Coulomb ionization of superheavy systems on the energy and momentum transferred and on the electron density at the nucleus: there seems to be little dependence on how the energy and momentum are transferred. Some theoretical justification is given for this interesting feature where two-step processes are involved (Ba 81b).

In Section 2 we noted that before integration over the electron energy distribution, the differential probability for excitation of the $1s\sigma$ state with binding energy $E_{1s\sigma}$ into a continuum state with kinetic energy E_δ and solid angle $d\Omega_\delta$ is given by (Mu 83)

$$\frac{dP(E_\delta, b)}{dE_\delta\, d\Omega_\delta} = \frac{\pi}{2} d_0^2(Z) \left(\frac{mc^2}{E_\delta + mc^2}\right)^\gamma \exp\left\{-\frac{2\tau_0[E_\delta + |E_{1s\sigma}(R_0)|]}{\hbar}\right\} \tag{4.3}$$

Recalling our previous discussion, the particular coefficient in the exponential is associated with the $1/R^2$ scaling of the electron density $|\psi(0, R)|^2$ at the origin of very heavy electronic systems and the subsequent parametrization of the matrix element by

$$\langle E_\delta s\sigma| \frac{\partial}{\partial R} |1s\sigma\rangle \equiv \frac{d(E_\delta, Z)}{R} \simeq \frac{d_0(Z)}{R} \left(\frac{mc^2}{E_\delta + mc^2}\right)^{\gamma/2} \tag{4.4}$$

for small and intermediate distances R and $E_\delta < 1$ MeV. We see that the energy spectrum of ionized electrons falls off exponentially apart from a weak dependence of the matrix element on the electron energy. It should be emphasized that the particular factor given above which determines the slope of the exponential reflects systems with $Z\alpha > 1$ and that, in principle, the slope can be different for lighter systems or $2p_{1/2}\sigma$ excitation. Generally, the slope will decrease if the electron density grows less rapidly with decreasing R than $|\psi(0, R)|^2 \sim 1/R^2$ (Mu 78, Bo 80). However, the exponential decrease with the product $R_0 q_{\min}$ seems to be a general feature of Coulomb ionization (Ba 59, Ba 79, Ba 81b).

Thus, this scaling law exhibits in a transparent manner the competing features that govern the delta-electron spectra from superheavy collision systems. The coupling strength to the continuum, represented by $d_0^2(Z)$, increases very rapidly with Z, while the binding energy term in the exponential has the opposite effect, reflecting the larger momentum transfer required to liberate a very deeply bound electron as Z becomes very large.

The influence of the binding energy becomes most apparent for high-Z systems and where the binding energy becomes comparable to E_δ. For example, explicit calculations (So 79) for the sum of $1s\sigma$ and $2p_{1/2}\sigma$ ionization in the superheavy systems $Z = 140$, 170, and 190 show that for $E_\delta = 1000$ keV and $R_0 = 20$ fm $d\sigma/dE_\delta$ increases by almost one order of magnitude from $Z = 140$ to $Z = 170$ and decreases then by a factor of 2 between $Z = 170$ and $Z = 190$.

Another interesting qualitative aspect of the delta-electron spectrum from heavy collision systems is found in the relative ionization probability from the $1s\sigma$, $2p_{1/2}\sigma$, and less bound states. For heavy-ion collisions with $Z_p \ll Z_T$, the high-energy component of the delta-electron spectrum is dominated by emissions from the K shell (Ch 69, Ch 70, Ko 77). This behavior merely reflects the stronger spatial localization and much higher prevalence of high-momentum components in the $1s$ wave function compared to the higher shells. The binding energy has little consequence here since $E_\delta \gg |E_B|$ so that $\Delta E \simeq E_\delta$. However, in superheavy collision systems the higher shells contribute significantly to the high-energy delta-electron spectrum and may even dominate. The radial density distributions shown in Figure 4.5, calculated by Soff *et al.* (So 79) for an internuclear separation of 20 fm for the Pb + Pb quasimolecule, illustrate why. The radial densities for the $1s\sigma$ and $2p_{1/2}\sigma$ states are very similar, leading to transition matrix elements of almost equal magnitude (So 79). Moreover, with E_δ and $|E_B|$ comparable in magnitude, the Fourier frequencies required to ionize are lower for the $2p_{1/2}\sigma$ state than for the $1s\sigma$ state. Therefore, as shown in Figure 4.6, the delta electron spectrum from the ionization of the $2p_{1/2}\sigma$ state is predicted to constitute the largest part of the total spectrum from Pb + Pb at kinetic energies above 1 MeV.

Thus in terms of the effects of kinematics and of the coupling of the wave functions to the demands of the kinematics, we can summarize our understanding of the qualitative features of the data selected in Figures 4.1–4.4 as follows.

The exponential falloff of the spectra in Figure 4.1 reflects the

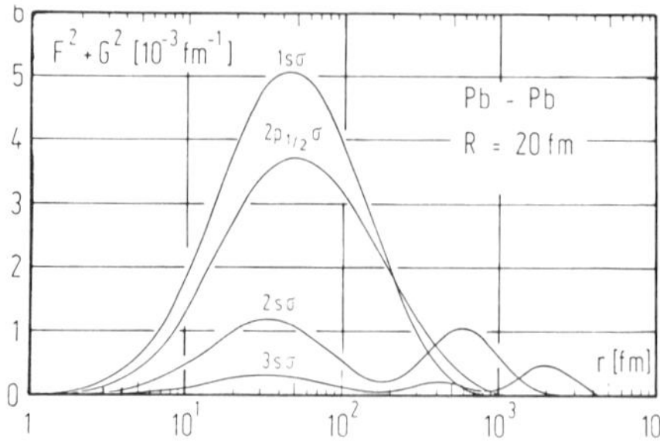

Figure 4.5. Radial density distribution $|\psi r^2| = F^2 + G^2$ for the $1s\sigma$, $2p_{1/2}\sigma$, $2s\sigma$, and $3s\sigma$ states in the Pb + Pb quasimolecule at a separation $R = 20$ fm (So 79).

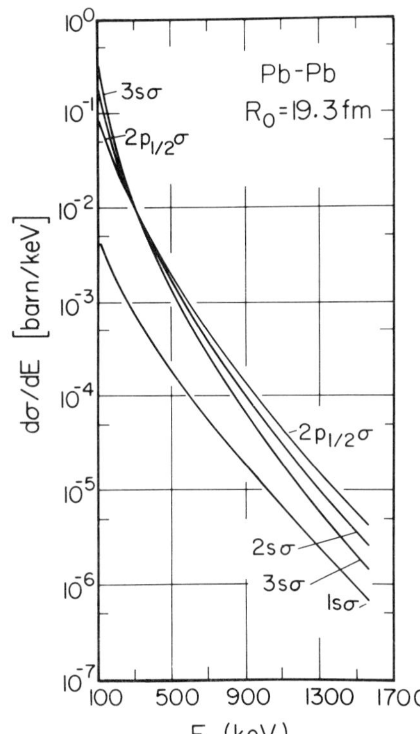

Figure 4.6. Delta-electron spectrum in Pb + Pb collisions with $R_0 = 19.3$ fm. The initial bound states $1s\sigma$, $2s\sigma$, $3s\sigma$, and $2p_{1/2}\sigma$ are considered (So 79).

increasing momentum transfer required with increasing electron energy. The observation of electrons with energies above 1 MeV in the Pb–Pb system is testimony to the fact that the electrons are emitted from bound quasiatomic states where the unusual relativistic contraction of the inner-shell wave functions makes it possible to transfer the required large momentum to localized initially bound electrons. A much steeper falloff found for the Pb–Sn spectrum relative to the Pb–Pb spectrum only demonstrates the relative lack of high-momentum components in the quasiatomic bound states of the former collision system compared to the latter. The strikingly rapid increase of high-energy delta-electron emission with Z_u illustrated in Figure 4.2 further demonstrates the essential role played by the enormous increase in the electron density of the quasiatomic wave functions for high Z_u. It should be noted that for the low Z_u data plotted in Figure 4.2 $\Delta E \simeq E_\delta$, since the ionized electrons originate mainly from the L shell (Kr 78). For the heaviest systems the contribution from the $1s\sigma$ state becomes comparable with emission from the L shell. The leveling-off of the cross section is thus related to the increase in ΔE contributed by the increasing binding energy which is no longer negligible with respect to E_δ.

Similarly the velocity dependence observed in Figure 4.3 reflects the decreasing momentum transfer with increasing v, while the localization of the ionization exhibited in Figure 4.4 directly derives from the energy transfer $\Delta E = E_\delta + |E_B|$ which is involved in the scaling relation $\tilde{b} = \hbar v/\Delta E$.

4.3. Delta-Electron Spectrometers

Our discussion has pointed out the central role played by the momentum and energy transferred in the ionization process. However, in contrast to electron scattering experiments, these important quantities cannot be measured directly in the high-energy heavy-ion scattering experiments under discussion. The energy resolution required is far beyond that which can be achieved, and the Coulomb interaction between the nuclear centers dominates the kinematics. We can only deduce these quantities indirectly.

However, the kinematics does define important experimental parameters. Among these are the impact parameter, b, and the distance of minimum approach, R_0, in a sub-Coulomb barrier scattering. As noted before, it is clear that a classical description of the heavy-ion trajectories through these quantities is meaningful in our heavy-ion collisions since the Sommerfeld parameter for the typical collisions we have been considering is very large, $\eta = a/\lambda \simeq 500$. We have also noted the importance of specifying experimentally these spatial aspects of the collision since they are connected directly with the physical consequences produced by the relativistic effects, and, thus, with the essential features of electron excitation and ionization. The kinematic parameters b and R_0 have been defined previously in terms of the measured scattering angles.

We pointed out in Section 4.1 that the most comprehensive measurements of delta-electron emission involve the detection of the delta electrons simultaneously with the scattered nuclear particles and with the deexcitation x rays. The x ray is used to define the initial state of the electron, under favorable conditions. Such extensive measurements include the specification in multiparameter format of a number of quantities depending on the differential information to be obtained.

A complete description of the electron parameters includes the electron energy, E_δ, the emission angle relative to the incident beam, θ_δ, the emission angle relative to the scattering plane, ϕ_δ, and the polarization state of the electron. The large range of electron energies of interest dictates a tunable spectrometer, preferably with broad-band characteristics for rapid data collection capabilities. Although the focus is on continuous spectra and good energy resolution is not a primary requirement, backgrounds from line spectra, such as internal conversion of nuclear γ-ray transitions, impose the need for fair energy resolution. In this connection it is important to consider the effects of Doppler shift and broadening on the electron spectra. These effects can be exploited to some advantage both in electron and positron spectroscopy with geometries where the Doppler line profiles are readily determined. The need for large solid angles in coincidence experiments and the simultaneous desire to specify the electron emission angle also emphasizes broad-band capabilities in emission angle as well as in energy. The spin polarization state of the delta electron is an interesting quantity (So 81); but since its determination is not in the mainstream of the delta electron studies and involves very specialized measurement techniques, it effectively has to be treated as a separate problem.

The heavy-ion scattering information is contained in the projectile scattering angle, θ_p, for sub-Coulomb barrier scattering. In terms of the above experimental variables, the most detailed measurement is provided by the triple differential cross section $d^3\sigma/(dE_\delta\, d\Omega_\delta\, d\Omega_p)$ as a function of E_δ, θ_δ, θ_p, and ϕ_δ for a given collision system (Z_pM_p), (Z_TM_T) colliding at a laboratory energy E_p. The detection of the characteristic x ray, to identify the initial electron state, involves a further correlation with the x-ray angle of emission and its solid angle. Because of the obvious problems with normalization of cross sections, the quantity which is most convenient to measure directly is the emission probability leading to a particle scattering angle θ_p,

$$\frac{d^2P(E_\delta, \theta_\delta, \phi_\delta, \theta_p)}{dE_\delta\, d\Omega_\delta} = \frac{d^3\sigma/(dE_\delta\, d\Omega_\delta\, d\Omega_p)}{(d\sigma_R/d\Omega_p)}$$

which is derived from the above cross section by dividing by the Rutherford cross section. Experimentally this probability is obtained directly from the ratio of (electron)–(scattered particle) coincidence rate and the particle rate at the angle θ_p, taking due account of the electron energy acceptance, solid angles, and efficiencies (Ko 82). Measuring the excitation probability therefore eliminates uncertainties associated with monitoring the parameters of the beam intensity and target density. However, the latter quantities are introduced again in the measurement if the scattered particles are not observed and the inclusive double differential cross section $d^2\sigma/(dE_\delta\, d\Omega_\delta)$ is measured:

$$d^2\sigma/(dE_\delta\, d\Omega_\delta) = \int d\Omega_p\, d^3\sigma/(dE_\delta\, d\Omega_\delta\, d\Omega_p)$$

In the latter experiments the product of beam intensity and target density can be obtained from a monitor counter with a well-defined detection angle and solid angle viewing Rutherford scattering from the target simultaneously with the electron spectrometer (Ko 82).

Thus delta-electron studies call for a flexible spectrometer which can operate over an energy range from keV to several MeV with an efficiency conducive both to detecting small cross sections and carrying out coincidence measurements while discriminating against copious backgrounds found in the vicinity of a target bombarded by heavy projectiles. In addition good kinematic definition is required for the heavy-ion scattering. Two experimental arrangements have been used to carry out most of the recent measurements with heavy projectiles (Ko 77, Ko 82, Bo 78, Gu 82). They possess somewhat complimentary features, and we describe their principal functions below briefly.

One experimental arrangement is illustrated schematically in Figure 4.7 (Bo 78, Gu 82). The electron spectrometer is based on an achromatic magnetic channel with a momentum acceptance interval $\Delta p/p = \pm 12\%$. The efficiency of this channel within this momentum interval is constant so that a distortion of continuous spectra is avoided. After transmission by the magnetic channel the electrons are detected by a Si(Li) counter with a resolution of 2.5 keV at 1 MeV. The counter subtends a solid angle of 4.1 msr at the target.

This type of arrangement, where a wide-band magnetic channel is used to transport the electrons away from the target environment of large gamma-ray and heavy particle backgrounds while employing an auxiliary detector to provide the energy resolution, is typical of both experiments studying delta-electron and positron emission. In all such measurements the magnetic filter is necessary to suppress the low-energy electrons whose fluxes are several orders of magnitude larger than the electron intensities in the energy regions of interest. Using the magnetic field as a transport device and relying on the detector for energy information, however, does introduce possible distortion of spectra by outscattering from the detector which can be a function of the energy. For this spectrometer this effect was investigated

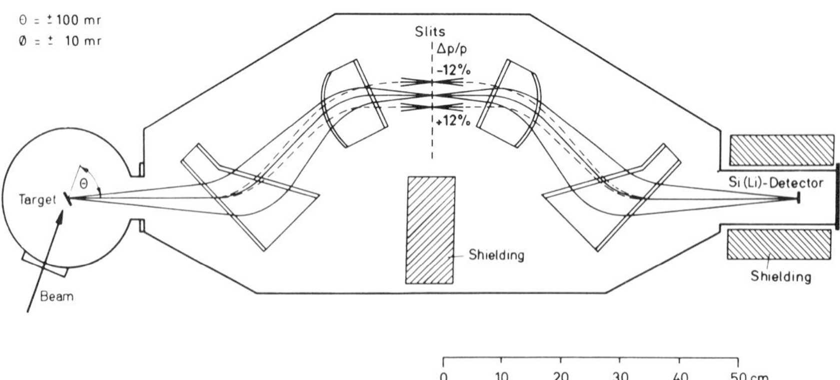

Figure 4.7. Electron spectrometer consisting of an achromatic electron channel with four dipole magnets which focuses electrons from the target onto a Si(Li) detector. The momentum acceptance $\Delta p/p = 24\%$ and the solid angle $\Delta\Omega = 4$ msr (Gu 82).

using a beta emitter and an electron channel with narrow momentum acceptance so as to simulate a monoenergetic source with variable electron energy. Such studies show that, to a good approximation, outscattering is independent of energy and occurs for $23.3 \pm 0.9\%$ of the electrons incident on the Si(Li) detector (Bo 78).

Parallel-plate avalanche counters are utilized in this apparatus to correlate delta-electron emission with the scattered ion trajectories (Ko 82). An annular geometry provides θ_p sensitive anode and ϕ_p sensitive cathode readouts. Coincidences between electrons and characteristic x rays are measured using two $2 \times 1/4$ in. NaI(Tl) detectors mounted at 120° relative to the beam direction in the plane of the beam and electron, and at 45° out of this plane. The energy resolution of this detector is 12% at 80 keV. The magnetic channel can be rotated so that it samples electron emission between 10° and 120° relative to the beam direction.

The other experimental arrangement which has provided a major part of the data available on heavy collision systems employs an iron-free "Orange"-type β spectrometer (Mo 65) at its central function of electron detection. A schematic representation of the spectrometer and auxiliary detectors is given in Figure 4.8 (Ko 82). The spectrometer utilizes a toroidal magnetic field

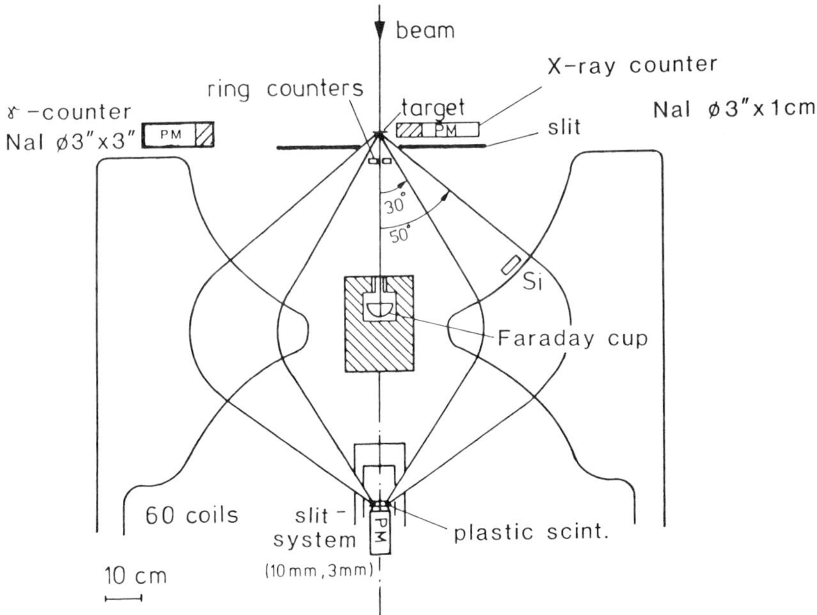

Figure 4.8. Schematic view of iron-free, "Orange"-type beta spectrometer used for delta-electron studies. Momentum resolution $\Delta p/p = 0.08$, transmission $T = 1$ srad. Coincidence measurements with scattered particles or x rays can be performed (Ko 82).

produced by 60 current coils to momentum analyze and focus electrons from a target onto a cone-shaped scintillation detector. In one configuration used, electrons are accepted with θ_δ between 30° and 50° in an annular geometry relative to the beam direction and with $\Delta p/p = 0.018$. Another configuration employs $50° \leqslant \theta_\delta \leqslant 70°$ with $\Delta p/p = 0.014$. Unlike the achromatic channel described above, in the arrangement shown in Figure 4.8 the energy resolution is dictated by the momentum bandpass, with the plastic scintillator providing only a clean signal. However, the detector at the focal point can be modified to further subdivide the momentum channel and thus provide smaller subdivisions in energy. Typically electrons varying between 150 keV and 2.4 MeV have been studied by selecting the magnetic field.

One of the features of this spectrometer is a large transmission efficiency of 0.08 of 4π, which makes it especially useful for detecting small intensities and for performing coincidence experiments with x rays and scattered particles. The x rays are detected in a 3-in.-diam by 1-cm-thick NaI(*Tl*) counter mounted 4.5 cm from the target to obtain high efficiency. The scattered ions are detected in an annular geometry about the beam axis covering two alternative angular regions of 9.5° to 27° and 16° to 48°, corresponding to the two configurations for detecting the electrons noted above. This experimental arrangement also utilizes parallel-plate avalanche counters whose active area is subdivided into eight concentric rings with individual readouts. It is claimed that each output ring is capable of operating up to counting rates of 1 MHz (Ko 82). Over a small angular region kinematic particle coincidences between rings are possible for asymmetric collision systems, allowing in these cases the unambiguous definition of particle trajectories. Both spectrometers utilize elastically scattered particles, detected by a surface-barrier Si detector, for normalization.

With the large spatial separation of target and detector volumes achieved by the magnetic transport systems in the two experimental arrangements, the principal backgrounds in either apparatus originate from internal conversion and internal pair conversion of nuclear gamma-ray transitions following Coulomb excitation or other nuclear reactions. For example, direct interactions of gamma rays from the target with the electron detector can be effectively shielded, while Compton electrons or electrons from external pair creation can be suppressed sufficiently by limiting the amount of material and its Z composition in the immediate target area.

The most troublesome background is produced by internal conversion, particularly if the collision partners consist of nuclei with a well-developed collective energy level structure below a few MeV in excitation energy. As was noted earlier in the discussion, dealing with these line spectra is complicated by Doppler broadening. High velocity of the emitting system and high electron velocity can sufficiently broaden the line so that it may be difficult to differentiate the contribution from internal conversion from the bulk of the continuous delta-electron spectrum, particularly if closely spaced lines are involved. In principle, kinematic reconstruction can be utilized to reproduce the original line spectra if the emitter velocity as well as the emission angle between the electron and emitter are known (Ba 78). Internal conversion presents a serious limitation to delta-electron studies in the heaviest collision systems such as $U+U$ where densely packed, easily excited, low-lying nuclear energy levels exist. Of course, performing coincidences with x rays does not eliminate this background.

The background from internal pair conversion of nuclear transitions is generally smaller than that from internal conversion and in many situations can be neglected. However, where the magnitude of this background is significant relative to the delta-ray intensity, it is difficult to evaluate due to the intrinsic broad energy distribution associated with its electron spectrum (So 81c, Sc 78, Sc 79). Fortunately, a coincidence between electrons and characteristic x rays can filter out the pair conversion background.

In the sections that follow we select some major aspects of the results which have emerged from experiments carried out with the two experimental arrangements shown in Figures 4.7 and 4.8. Unfortunately, the literature on this newly developing subject is very limited in scope, consisting mainly of a few research papers and some reviews in conference proceedings (Ko 77, Ko 82, Bo 78, Gu 82). Our review, therefore, centers mainly on these papers.

4.4. Selected Results

The experiments on delta-electron emission from slow heavy-ion collisions divide somewhat naturally into two general groups. One of these has been concerned primarily with demonstrating the feasibility of experiments and with establishing the scope of the information that can be gleaned from studying delta-electron emission. Included in the latter category are investigations of the important roles played by kinematics and the relativistic properties of the quasiatomic wave functions which we have already sampled, to some extent, in Section 4.2. These types of studies constitute the bulk of the data available. The other group of investigations is essentially just beginning. They are pursuing the questions associated with extracting specific information on the quasiatomic structure and are, therefore, part of the more general investigations, referred to in the Introduction to this article, concerned with the spectroscopy of such a dynamic system. We have already mentioned in connection with both the vacancy formation studies and delta-electron emission that scaling laws may be able to provide such a spectroscopic tool.

We organize our discussion of these selected data generally along the lines of these two categories of investigation. Chronologically the experiments have evolved from light to heavy collision systems. This is also a good progression to follow for discussion purposes since some of the main ideas are most easily demonstrated with the light collision systems and new physics is introduced in the heavy systems.

4.4.1. Light Collision Systems

Two recurring considerations enter into all our descriptions of delta-electron emission from heavy-ion collisions below the Coulomb barrier: (1) The adiabatic adjustment of the inner-shell electrons to the slowly changing nuclear Coulomb fields justifies a molecular basis for describing the ionization process. (2) For high-energy ejected delta electrons, the internuclear separations $R = 1/q_{\min}$ (Ba 59, Ar 78), which contribute predominantly to electron emission, are well within the K-shell radius of the united atom. Therefore, it is appropriate to consider the quasimolecular wave functions and the binding energies associated with internuclear separations close to the united atom limit as good representations of the initial state of the electron. Particularly for the lighter collision systems, $Z_u < 130$, the two-center potential does not vary appreciably at small R, e.g., $R < 150$ fm for $Z_u = 99$ (Mu 76b). Therefore, especially for the lighter collision systems, the binding energy and wave function of the initial state can be treated as time independent and they can be identified with the eigenenergies and eigenfunctions of the united atom (Ko 77, Bo 78).

This united atom picture, suggested by Briggs (Br 75b), leads, through a first-order perturbation treatment of the ejection of electrons from a molecular orbital, to a simple physical description of ionization of an initial united atom state $|i\rangle$ by the slow motion of both nuclei moving coherently with respect to the center of charge to which $|i\rangle$ is referenced (Br 75b, Me 78b). For an asymmetric system $Z_p \ll Z_T$, this is equivalent to the Born amplitude for ionization of the united atom by the projectile, if we neglect the effect of the target nucleus moving much slower than the projectile relative to the center of charge. The emergence of the excitation amplitude in this form for the lighter asymmetric systems makes them particularly useful for studying the effects of kinematics and the wave function structure since these appear in an especially transparent manner in the Born approximation. For this reason the earlier experiments have used the light asymmetric collision systems to establish some of the basic concepts of delta-electron emission in heavy-ion collisions (Ko 77, Bo 78). However, it is important to keep in mind that as we move from the particular consideration of light projectiles colliding with heavy targets to descriptions of more symmetric and heavier systems, the first-order theory and Born approximation lose their validity, the combined action of the projectile and target charges becomes increasingly important, and multistep excitations assume an important role. Including the effects of the molecular behavior then becomes a prime consideration, as has been shown in the detailed calculations by Soff and co-workers (So 79, So 80, So 81b, So 82b, Re 79).

More specifically, it was presented in Section 2 that the first-order amplitude for the excitation of an electron from an initial one-electron state $|i\rangle$ to a final state $|f\rangle$ is given by

$$a_{fi} = -\int_{-\infty}^{\infty} dt \, \langle f| \frac{\partial}{\partial t} |i\rangle \exp\left[\frac{i}{\hbar}\int_{-\infty}^{t} dt' \, (E_f - E_i)\right] \tag{4.5}$$

Within our approximation $|i\rangle$ and $|f\rangle$ are time-independent states of the quasiatom and E_f and E_i are the unperturbed energies (Br 75b). The use of the identity

$$\langle f| \frac{\partial}{\partial t} |i\rangle = -(E_f - E_i)^{-1} \langle f| \frac{\partial V}{\partial t} |i\rangle \tag{4.6}$$

leads to

$$a_{fi} = \int_{-\infty}^{\infty} dt \, (E_f - E_i)^{-1} \langle f| \frac{\partial V}{\partial t} |i\rangle \exp\left[\frac{i}{\hbar}\int_{-\infty}^{t} (E_f - E_i) \, dt'\right] \tag{4.7}$$

where V is, in general, the full molecular potential. But, as mentioned above for light asymmetric systems, only the emission due to the Coulomb field of the projectile is taken into account (Ko 77, Bo 78, Gu 82) with the contribution from the motion of the target nucleus neglected since its velocity with respect to the center of charge, to which the electron cloud of the united atom is anchored, is small. Then $\partial V/\partial t$ takes the form (Ko 77)

$$\frac{\partial V}{\partial t} = \dot{\mathbf{R}} \nabla_R \frac{Z_p e^2}{|\mathbf{r} - \chi \mathbf{R}|} \tag{4.8}$$

where $\mathbf{r}$ is the electron coordinate of the one-electron wave function with respect to the center of charge, and $\chi = Z_T/(Z_p + Z_T)$ is a correction factor reflecting the correlation of the atomic wave function to the center of charge. Assuming a straight line ion–atom collision then yields the cross section for the ejection of an electron from the K and L shells into a continuum state with kinetic energy E_δ, solid angle $d\Omega_\delta$, and final state density dN/dE_δ (Ko 77, Bo 78, Gu 82),

$$\frac{d^2\sigma}{dE_\delta \, d\Omega_\delta} = \sum_{K,L} 8\pi \left(\frac{Z_p e^2}{\hbar v \chi}\right)^2 \frac{dN}{dE_\delta} \int_{q_0}^{\infty} \frac{q \, dq}{q^4} |F(\mathbf{k}_i)|^2 \tag{4.9}$$

where

$$q_0 = q_{\min}/\chi$$

For $\chi = 1$ this reduces to the well-known Born approximation (Ko 77, Bo 78, Gu 82, Mc 70). The term in the integral, $1/q^4$, reflects the difficulty of transferring large momentum. The relativistic form factor $|F(\mathbf{k}_i)|^2$, given by (Bo 78, Gu 82)

$$|F(\mathbf{k}_i)|^2 = \sum_s \left| \int \bar{\psi}_f \gamma^0 \psi_i e^{i\mathbf{q}\cdot\mathbf{r}} d^3r \right|^2 \tag{4.10}$$

where the summation sign indicates averaging over initial states and summation over final states, provides a measure of the properties of the initial state. For example, with the simple choice of a plane wave for the final state, the form factor reduces, in essence, to the Fourier transform of the united atom state $\psi_i(Z_u)$, giving the momentum distribution of the bound electron. The differential cross section given above contains only a sum over electrons from the K and L shells. This restriction has been made in analyzing experiments on the basis that the contributions from the higher M, N,... shells are small for electrons of high kinetic energy, $E_e \gg E_B(M, N,...)$, since for the corresponding high momenta k_i

$$|F_{M,N}(\mathbf{k}_i)|^2 \ll |F_L(\mathbf{k}_i)|^2$$

A number of asymmetric collision systems have now been studied and analyzed within the framework of these ideas (Ko 77, Bo 78, Gu 82). Plane waves for final states and united atom bound states in the limit of pointlike central charges have been used in these analyses. Although it becomes readily apparent that this first-order description deteriorates with approach to symmetry in the collision partners and to high Z_u, the essential role played by the united atom is revealed. To a good approximation it has been found that the high-energy tail of the delta-electron spectrum is reproduced by the form factor of the quasiatom at the point of closest approach of the colliding nuclei.

The first of these studies was carried out by Kozhuharov *et al.* (Ko 77) using 83-MeV ^{16}O ions colliding with ^{197}Au. Experimentally they established that high-energy delta electrons from heavy-ion collisions could be effectively detected above background such as internal conversion, and that delta electrons from the K shell could be isolated from the rest of the spectrum by employing coincidences between electrons and K x rays. This measurement was first to suggest that the measured spectrum is a reflection of the K-shell momentum distribution of the quasimolecule.

As part of an effort to follow up this type of investigation systematically, additional detailed information emerged from a study of ^{32}S + ^{208}Pb collisions performed at the Heidelberg MP-Tandem accelerator (Bo 78). The double differential cross section, $d^2\sigma/d\Omega_\delta\, dE_\delta$, was examined over 5 orders of magnitude with good statistical accuracy for electron energies ranging from ~50 to ~400 keV. The bombarding energy was 120 MeV, corresponding to $v/c = 0.09$ and a minimum internuclear separation of 18.2 fm. Figure 4.9 displays the data and an analysis following the Born approximation outlined above using plane waves for the outgoing electron and a point charge nuclear Coulomb field for the bound state wave functions. For the reason noted previously only K and L electrons are included in the calculation. The high-energy electrons are found to be associated predominately with the 1s and 2s shells; the contribution from the 2p electrons is small for this system, but it should be noted that it grows in importance with increasing Z_u, finally contributing a majority of the delta electrons for very high Z_u. The authors emphasize that in the calculation no normalization or fitting parameter is used to adjust the calculated cross section to the measurements.

In Figure 4.9 three interesting comparisons are made between calculations and data. One curve (solid line) shows the sum of the cross sections from the 1s, 2s, and 2p states of the *united atom* ($Z_u = 98$) using relativistic wave functions, while the dash-dotted line displays results from similar calculations using nonrelativistic wave functions. A third curve (dashed line) represents a relativistic Born calculation where the electron states are identified with the *target atom*. The calculations using nonrelativistic wave functions or target atom related wave functions, clearly, predict significantly smaller cross sections than those measured, with the deviations becoming larger with increasing electron kinetic energy. It is

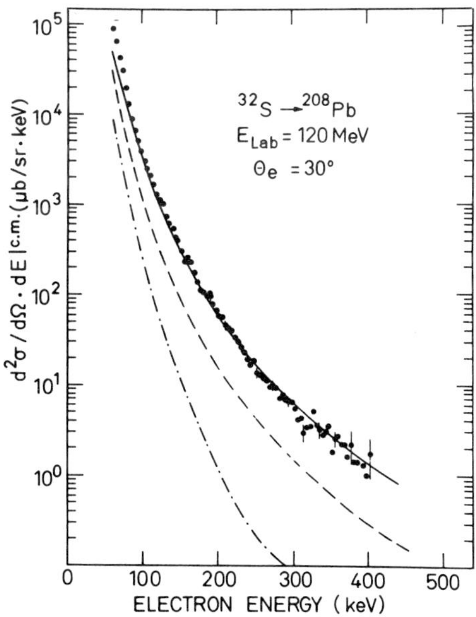

Figure 4.9. Center-of-mass differential cross sections for the emission of delta electrons from 120-MeV $^{32}S + ^{208}Pb$ collisions. The solid line represents cross sections calculated in the relativistic Born approximation for emission from the 1s, 2s, and 2p states of the united atom with $Z_u = 98$. The dashed line is a similar calculation using target atom wave functions, while the dashed-dotted line denotes the nonrelativistic Born calculation for the united atom wave function (Bo 78).

evident from these comparisons that the formation of the quasiatom is not only the crucial aspect of the high-energy delta-electron emission mechanism, but that the relativistic effects, enhanced by the increased charge of the combined atom, are essential to explain the high-energy tail of the electron spectrum. This basic result, demonstrated so simply within the framework of the Born approximation where we can directly relate the form factor of the inner-shell electrons of the quasiatom to the spectral distribution of energetic delta electrons, in fact, extends beyond the range of validity of the Born approximation, as the more detailed calculations of Soff *et al.* (So 79) show. As we have stated on several occasions, the relativistic enhancement of the high-momentum components of the electrons in the united atom is the dominant effect.

The excellent agreement of the Born calculation with the data in Figure 4.9 may be better than expected, given that plane waves and a point nucleus were used in the calculations. Indeed, using a variety of data presently available (Gu 82) spanning combinations of projectile and targets with $Z_u \leqslant 117$ and $0.06 \leqslant v/c \leqslant 0.09$, it is found that, although these types of Born approximation calculations reproduce the slopes of the spectral distributions quite faithfully, they underestimate the intensities by factors of 1.5 to 2. Figures 4.10–4.12 illustrate several examples of such data. The

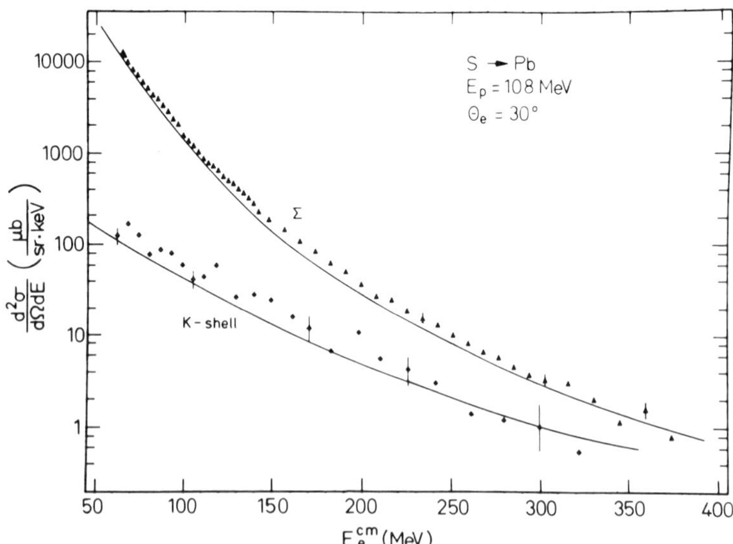

Figure 4.10. Delta-electron spectra from S + Pb collisions at 108 MeV. The upper spectrum reflects all electrons emitted at 30° to the beam direction, while the lower spectrum was obtained by requiring a coincidence with Pb K x rays. The solid curves represent calculation using the Born approximation with the united atom wave function (Gu 82).

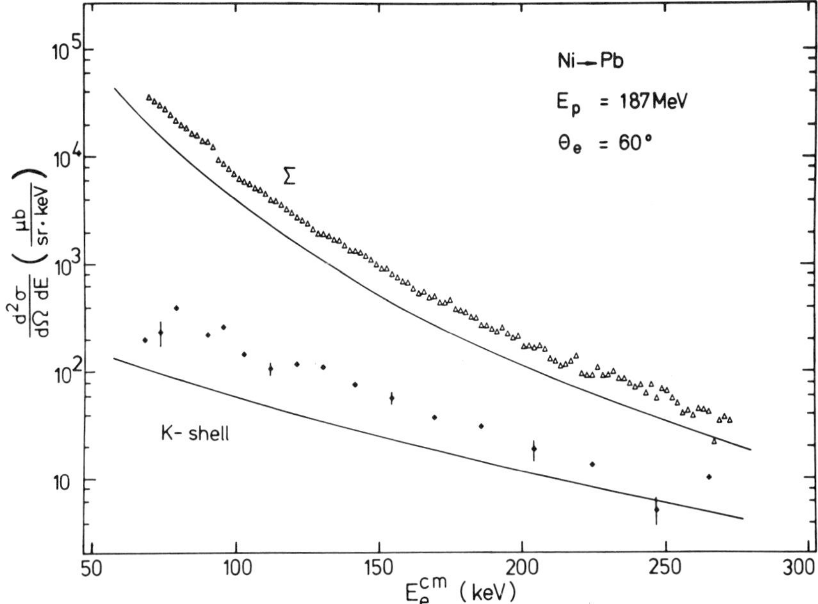

Figure 4.11. Similar study as in Figure 4.10 for Ni + Pb system (Gu 82).

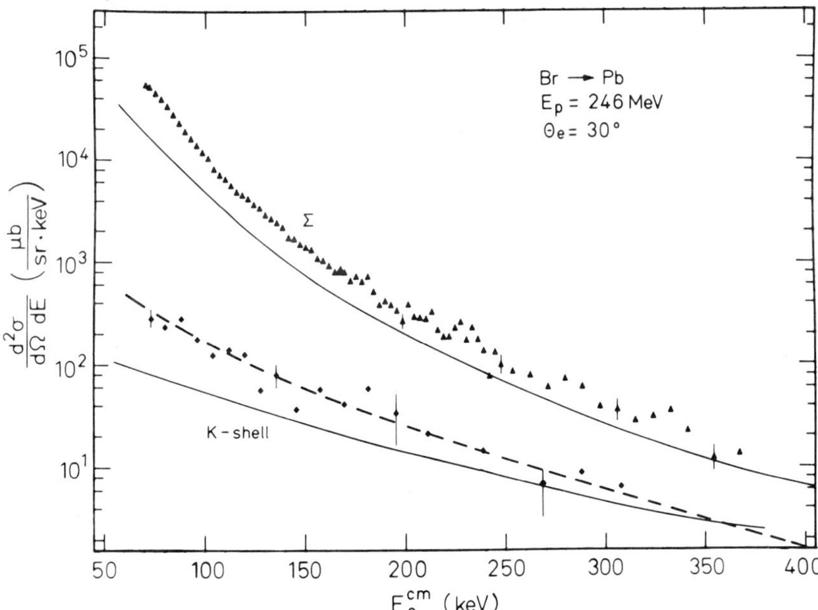

Figure 4.12. Similar study as in Figure 4.10 for Br + Pb system (Gu 82). In addition the dashed line compares a coupled-channel calculation by Soff *et al.* (So 82b) to the coincidence data.

disagreement between calculation and experiment seems to increase with increasing Z_u.

Of course, one of the sources of uncertainty in attempting to understand these data, where the total electron spectrum is detected, is the undefined initial state of the electron. The electron can originate from any shell. The analyses quoted above chose to ignore all contributions from shells less bound than the L shells for the reasons specified in our discussion of this approximation. We noted previously that this ambiguity can be avoided and the contribution from a particular quasiatomic shell can be isolated if the delta electrons are detected in coincidence with a characteristic x ray from one of the collision partners (Ko 77). We also pointed out in Section 4.1 that particularly the K x rays of the heavier collision partner in an asymmetric collision system are correlated with the K shell of the quasiatom. Of course, the K-shell electrons have a special relevance to our studies of strong potential binding and, therefore, it is fortunate that they can be separated from the rest of the spectrum. However, several considerations limit the degree to which this isolation can be carried out. The experimental problems encountered reflect the large ratio of total to K-shell related delta electrons which lead to additional coincidence events that especially can contribute at the lower electron kinetic energies.

One of these results from vacancy sharing (Me 73) from the L shell of the united atom into the K shell of the heavier collision partner. Güttner *et al.* (Gu 82) have calculated this contribution for the collision systems in Figures 4.10–4.12. They estimated the fraction from vacancy sharing by

$$\sigma_{\text{shar}}/\sigma_{\text{coinc}} = (\sigma_{\text{tot}} - \sigma_{\text{coinc}})/\sigma_{\text{coinc}} \cdot P_{\text{vac}} \tag{4.11}$$

where P_{vac} is the probability for vacancy sharing given by

$$P_{\text{vac}} \simeq \exp[-\pi\alpha c \Delta Z/v] \tag{4.12}$$

with $\Delta Z = [Z_T - Z_p]$ (Me 73, Mo 78). It depends exponentially on the projectile–target Z asymmetry for asymmetries not too large. For the systems in Figures 4.10–4.12, the vacancy sharing contribution to electron K x-ray coincidences is less than 1%. However, it should be kept in mind that it can grow rapidly with decreasing ΔZ when $\Delta Z < 20$.

The other spurious source of electron K x-ray coincidences occurs in situations where the probabilities for both the total delta-electron and K x-ray emissions are large. For example, typically in heavy systems colliding with energies close to the Coulomb barrier several shells can be ionized simultaneously resulting in a multiplicity of delta electrons at the low kinetic energies greater than unity. We, therefore, can have a situation due to the finite detection efficiencies for both radiations that an electron is detected from the L or higher shells of the united atom in coincidence with a K x ray produced by K-vacancy production within the same collision. Güttner *et al.* (Gu 82) have estimated this background as given by the probability $P_K(\bar{b})$ for K vacancy production at a mean impact parameter $\bar{b}$ where L-shell electrons are emitted with the measured kinetic energy. As we have done previously, we can estimate $\bar{b}$ from $\bar{b} = 1/q_{\min}$. The probability $P_K(\bar{b})$ is taken from the literature, when available, assuming that it is not affected by the multiple vacancy production. Güttner *et al.* have calculated that such multiple vacancy production contributes to the electron–K x-ray coincidence cross section less than 3.5% in the S + Pb and less than 5% in the Ni + Pb coincidence spectra shown in Figures 4.10 and 4.11. No data are available to make this estimate for the Br + Pb system shown in Figure 4.12.

A comparison of the total and coincidence electron spectra in Figures 4.10–4.12 shows that for these collision systems the quasiatomic K-shell contribution to the total electron emission is small over the range of electron energies studied. However, the K-shell fraction does increase by approximately an order of magnitude over the kinetic energy range sampled. These two observations again demonstrate the basic qualitative features of delta-electron emission emphasized in Section 4.2, e.g., the competing roles played by the energy transfer and the localization of the inner-shell wave functions. Obviously, for a fixed electron kinetic energy the energy transfer $\Delta E = E_\delta + |E_B|$ is larger for the K shell than for the other more weakly bound electrons. Because of the strong dependence of the cross section on the energy transfer, the stronger binding leads to a suppressed K-shell excitation. The degree of suppression depends on competition between the larger energy transfer needed and the enhancement of the excitation probability caused by the availability of higher momentum components in the highly localized K-shell wave functions. As the kinetic energy of the electron increases, the binding energy becomes a smaller fraction of ΔE and therefore is less of a factor in determining the excitation probability. The much shallower rate of decrease of the cross section observed for the K-shell spectrum than for the total spectrum with increasing electron kinetic energy is due to this feature and to the presence of the high-momentum components in the K-shell wave function which are needed to produce the higher-energy delta electrons. Alternatively stated, the different slopes found for K-shell and total spectra imply that, for fixed kinetic energy of the electron, the impact parameter dependence of delta-electron emission varies with the parent shell involved, decreasing more rapidly with increasing impact parameter for the most tightly bound electrons. Moreover, the impact parameter dependence becomes steeper as the kinetic energy is increased for emission from a particular shell. These properties are displayed in some data presented below.

Born approximation calculations for the coincidence measurements are shown also in Figures 4.10–4.12 (Gu 82). Again we find that they underestimate the data and possess a somewhat shallower slope. Therefore, we have seen that although the Born approximation with united atom wave functions reveals some of the main features of delta-electron emission for the systems we have considered, a full explanation of the data only comes with the more detailed theoretical consideration presented by Soff and co-workers (So 79). As was the case for K x-ray emission discussed in Section 3, multistep processes play an important role and the ionization of the united atom electron cloud by the motion of the target nuclear charge cannot be neglected, particularly as symmetry in the collision is approached.

As an example of calculational success achieved in the region of $Z_u \simeq 100$, in Figure 4.12 we compare coupled-channel calculations (So 82b) for delta electrons ejected from the $1s\sigma$ state of the Br + Pb system colliding at 246 MeV to the coincidence data. In the calculations the exact two-center wave functions are replaced by solutions of the Dirac equation containing the monopole part of the potential only. This is justified on the basis that the ionization of the $1s\sigma$ state in relativistic systems is predominantly induced by radial coupling which acts at the small internuclear separations (Be 76). The authors emphasize that the theory yields absolute results and that the open bound states available in the multiple excitation process were not used as a fitting parameter but were chosen according to the Massey criterion that electronic states with an orbital velocity below the impact velocity are highly ionized. The agreement of the calculation with experiment is excellent. We note again that this agreement can only be achieved by including multistep excitations.

Thus the full coupled-channel calculations point out the limitations of the united atom Born approach as described above. One of these is that the coherent motion of both target and projectile charge centers have to be taken into account. Nevertheless, we have seen that the Born approximation is a useful way to systematize a great deal of data providing some of the principal features, as we have already noted. In this connection it is of interest to note that a so-called spectral function has been introduced (Ko 82, Gu 82) which displays the main features of the cross section exclusive of the dominant kinematic effects. It is related to the differential cross section $d^2\sigma/dE_\delta\, d\Omega_\delta$ by

$$S(q_0, Z_u) \equiv \left(\frac{d^2\sigma}{dE_\delta\, d\Omega_\delta}\right)\left(\frac{\partial E_\delta}{\hbar\, \partial q_0} v_p^{\text{c.m.}} \chi^2 \frac{1}{\hbar k_f} \frac{1}{Z_p^2}\right) \tag{4.13}$$

$$\chi = \left(\frac{Z_T}{Z_T + Z_p}\right), \qquad \left[\frac{\partial E_\delta}{\hbar\, \partial q_0}\right] = v_p^{\text{c.m.}} \tag{4.14}$$

where $v_p^{\text{c.m.}}$ is the projectile velocity in the center-of-mass system. The terms multiplying the cross section, respectively, convert the energy scale into a momentum scale, correct for the velocity dependence of the flux, transform the ion trajectory $R(t)$ into the united atom system to account for the dominance of electron excitation by the projectile, take account of the final-state density for the case when the final states are plane waves, and factor out the dependence of the emission cross section on the projectile charge Z_p. Thus the spectral function represents the united atom form factor $|F(\mathbf{k}_i)|$ weighted by $1/q^4$:

$$S(q_0, Z_u) \propto \sum_{K,L} \int_{q_0}^{\infty} |F(\mathbf{k}_i)| \frac{1}{q^4} q \, dq \tag{4.15}$$

The integral is dominated by the contribution from $q = q_{\min}$. Moreover, since $k_f \ll k_i$ and $k_f \ll q_{\min}$ so that $k_i = |\mathbf{k}_f - \mathbf{q}| \simeq q_{\min}$, the experimental data can be described principally with the variables Z_u and $q_{\min}$. We had noted previously the strong dependence of the cross section on the velocity. Güttner *et al.* (Gu 82) point out that some residual velocity dependence remains in $S(q_0, Z_u)$ which originates from the neglect of the Coulomb deflection in the analysis.

Through the spectral function $S(q_0, Z_u)$ it is particularly apparent how the variables q_0 and Z_u reflect some of the interesting aspects of the physics associated with delta-electron emission. Figure 4.13 compares the spectral

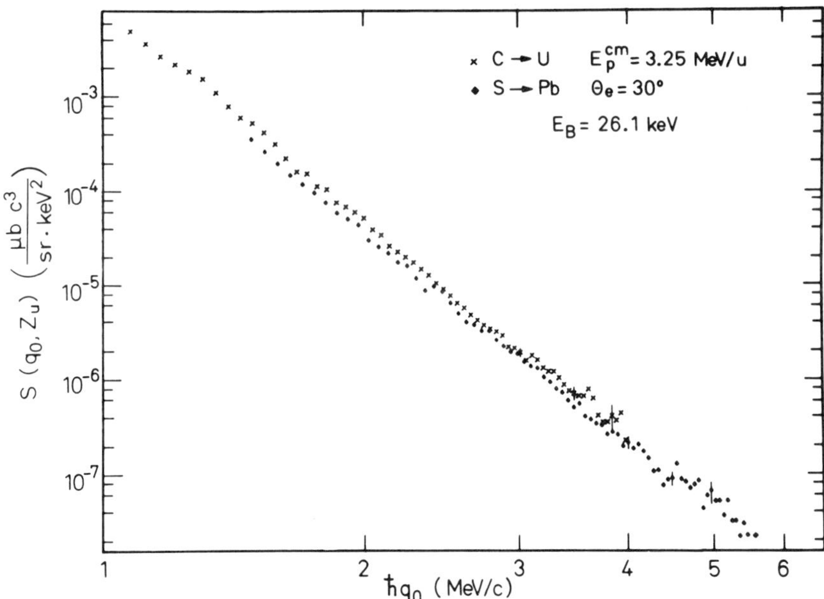

Figure 4.13. Spectral function $S(q_0, Z_u)$ as a function of the minimum momentum transfer $\hbar q_0$ for S + Pb and C + U (Gu 82). E_B is the 2*s* binding energy of the united atom and $E_p^{\text{c.m.}}$ specifies energies of the projectile in the c.m. system.

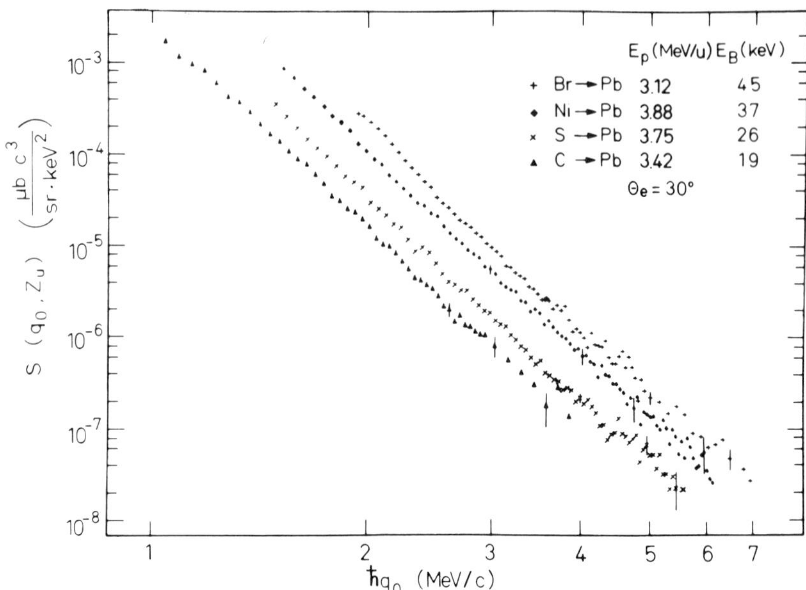

Figure 4.14. The spectral function $S(q_0, Z_u)$ for various systems plotted as a function of $\hbar q_0 = (|E_B| + E_\delta^{\text{c.m.}})/v_p^{\text{c.m.}}$. Here E_B is the $2s$ binding energy of the united atom (Gu 82).

functions for the two systems S + Pb and C + U with the same Z_u, while Figure 4.14 studies the effect on $S(q_0, Z_u)$ of varying Z_u by bombarding Pb with C, S, Ni, and Br projectiles. These studies were carried out by Güttner *et al.* (Gu 82). In all cases q_0 values are calculated using the binding energies for the L shell of the united atom taken from experiments (Le 78b) and from Fricke *et al.* (Fr 77) for $Z_u > 100$. As was mentioned previously, for systems with $Z_u < 130$ where the two-center potentials are independent of the internuclear distances for $R \lesssim 150$ fm, the united atom limit is a reasonable approximation for the binding energy since the internuclear distance $R \simeq 1/q_0$ contributes most to the electron emission.

Figure 4.13 shows that for the two systems with equal $Z_u = 98$, the spectral functions agree within systematic normalization errors of ~20% if a small contribution from electron internal conversion from transitions in excited U nuclei is taken into account. On the other hand Figure 4.14 demonstrates the strong increase of $S(q_0, Z_u)$ with increasing Z_u; for an increase in Z_u of ~30% there is a 12-fold increase in $S(q_0, Z_u)$. This behavior of $S(q_0, Z_u)$ exhibits in an especially transparent fashion the observations established previously regarding the quasiatomic origin of the delta-electrons from these collision systems and the dominant role played by relativity through the high Z of the united atom.

The effects of relativity on $S(q_0, Z_u)$ are readily seen. The parallel lines in the double logarithmic plots of Figures 4.13 and 4.14 suggest that the spectral function can be reproduced with a power law in q_0 (Gu 82):

$$S(q_0, Z_u) \simeq S(Z_u) q_0^{-m} \tag{4.16}$$

For collision systems with $80 < Z_u < 137$, in fact, the slope constant m is found to lie within the values 6.8 to 7.5 decreasing slightly with increasing Z_u. Within the Born approximation description of the process, m should reflect the slope of the momentum distribution of the bound electrons. Indeed, expanding the form factor in a power series in $1/k_i$, and using relativistic hydrogenlike wave functions for a point central charge (Be 77) yields

$$|F(k_i)|^2 \sim 1/k_i^{4+2\eta} + \cdots \qquad \text{with} \quad \eta = [1 - (Z_u \alpha)^2]^{1/2} \tag{4.17}$$

Letting $k_i \simeq q$ and performing the integration over momentum transfer, which, as mentioned previously, is dominated by q_0, leads to (Ko 82, Gu 82)

$$S(q_0, Z_u) \sim 1/q_0^{6+2\eta} \tag{4.18}$$

The slope constant m is obviously underestimated here since assuming a pointlike nucleus overestimates the high-momentum components in the momentum distribution. However, numerical treatment of this question (Gu 82) substantiates the principal point obtained with this analytical treatment: the slope of the spectral distribution is dominated by relativistic effects. The nonrelativistic limit leads to $m = 10$ for $1s$ or $2s$ states and $m = 12$ for $2p_{1/2}$ states and, obviously, both these values of m are in large disagreement with the experimentally determined value of ~ 7. More generally, it is clear that a momentum transfer of several MeV/c inherently implies the presence of relativistic effects so that the high-energy region of the delta-electron spectrum is always dominated by such effects irrespective of the charge of the united atom.

The role of the united atom in delta-electron emission is exhibited by the magnitude of the spectral function $S(q_0, Z_u)$. For example, Figure 4.15, again taken from the work of Güttner *et al.* (Gu 82), illustrates the exponential dependence of $S(q_0, Z_u)$ on Z_u. We recall that the principal kinematic effects have been factored from this plot. The rapid increase in $S(q_0, Z_u)$ is not correlated to the projectile or target, but, as previously considered, to the behavior of the electron density of the united atom wave function at small distances on the order of $1/q_0$. Quantitatively these data indicate that, for large $k_i \simeq q_0$, the momentum distribution for the united atom $I + U$ ($Z_u = 145$) exceeds that for the combined system C + Pb ($Z_u = 88$) by three orders of magnitude.

The spectral function is, thus, a sensitive probe of the small distance behavior of the wave function where the united atom Z and relativity exert

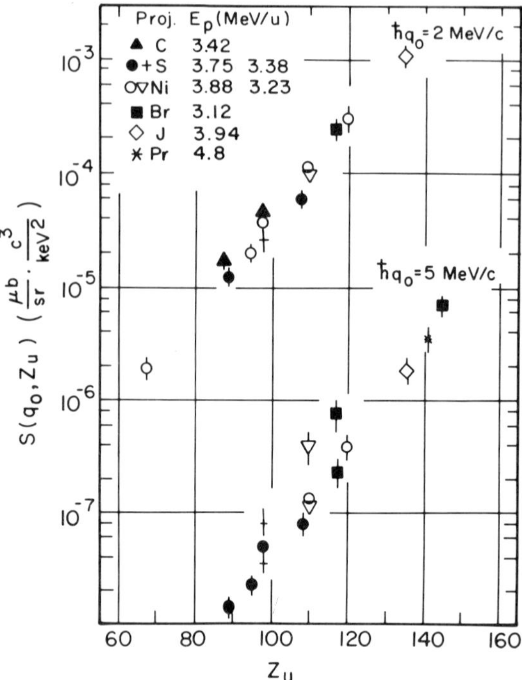

Figure 4.15. Magnitude of spectral function $S(q_0, Z_u)$ for various projectile–target combinations as a function of the united atom Z_u for $\hbar q_0 = 2$ MeV/c and 5 MeV/c (Gu 82).

pivotal influences. As emphasized before in this context, the experiments pursued with the lighter collision systems have provided important insight into some of the main features of delta-electron emission in heavy-ion encounters. However, again we point out that the work is only on the threshold of detailed quantitative understanding. More detailed theoretical calculations are particularly needed for the heavier systems where the perturbation considerations are inadequate. We deal with the available data on heavy collision systems in the section that follows.

4.4.2. *Heavy Collision Systems*

In large part the recent work in delta-electron emission has been directed towards understanding the very heavy collision systems in connection with the questions associated with supercritical binding and spontaneous positron emission. Paralleling the investigations on inner-shell vacancy production via x-ray emission measurements, the experiments have focused on defining the role played by the quasiatom in delta-electron emission and on the question of extracting spectroscopic information on the inner-shell structure of quasiatoms where binding energies can approach and exceed $2mc^2$. As implied by the x-ray experiments and already established for the lower united charge systems discussed in Section 4.4.1, the source of the delta electrons from the very heavy collision systems, clearly, is found to reside with the quasiatom. Some of the evidence for this conclusion has already been presented in Section 4.2. We discuss more such relevant data below.

On the other hand, the question of using delta electrons as a source of spectroscopic information for determining the binding energy of quasiatomic states of superheavy systems has been much less well defined. As in the case of vacancy formation studies, such information is derived from inverting the scaling law discussed in Section 4.2, using a measurement of the impact parameter dependence to determine the average energy transfer in the ionization process, $\Delta E = E_\delta + |E_B|$. We noted earlier that the advantage of using delta-electron over x-ray detection in such a measurement resides with the determination of the ejected electron energy in the former. However, the inherent features that one is measuring the average energy transfer in applying the scaling law and that collision times $\sim\hbar/mcv < 10^{-20}$ s imply an uncertainty of at least 50 keV in the energy determination still remain.

There are two other general points which influence and, in some ways, limit the experiments on very heavy systems. First, comparing theory with experiments is not tractable unless the initial state is specified, e.g., via coincidence measurements with K x rays. For the present, theoretical treatment of only K-shell excitation appears to be meaningful quantitatively so that useful tests of the theory are confined to sufficiently asymmetric systems where vacancy sharing can be neglected. Second, there are only a few truly superheavy systems which are accessible experimentally without interference from excessive backgrounds from nuclear excitation and following deexcitation via internal electron conversion processes. Indeed, there exists only one very heavy collision system, $^{208}Pb + {}^{208}Pb$, where there is an unique opportunity to study the delta-electron spectrum free of distortion from nuclear processes. For bombarding energies below the Coulomb barrier, the singular source of background originates from the decay of the 3^- state at 2.614 MeV by internal pair conversion, after it is excited weakly by the Coulomb excitation process. Unfortunately, in this symmetric system $1s\sigma$ excitation cannot be singled out with K x-ray coincidence studies. Moreover, backward and forward scattering cannot be distinguished and the ion trajectory can only be defined unambiguously for the scattering angle $\theta_{\text{c.m.}} = \pi/2$.

In the very heavy collision system regime we are, thus, mostly dealing with a situation in which at least one of the collision partners is a nucleus which is soft to nuclear excitations. A glance at the chart of nuclei shows that, except for the closed shell nuclei like ^{208}Pb, we are confined to the rare earths below lead and actinide nuclei above Pb. These nuclei are usually deformed with well-developed rotational and vibrational bands which are readily excited by Coulomb excitation. Low-energy transitions, high Z, and $E2$ and higher multipolarity all lead to strong internal conversion amplitudes for deexcitation. Clearly uranium is desirable both as a projectile and target to achieve some of the highest-Z systems available. Unfortunately, this nucleus is strongly excited by Coulomb excitation. In principle, the internal conversion electron background can be specified by identifying line structures in the electron spectrum if the Doppler broadening is not excessive, or by measuring the gamma ray spectrum and from it reconstructing the electron spectrum from internal conversion using measured or calculated internal conversion coefficients. Generally, unknown multipolarities lead to uncertainties in applying this procedure. Measurements with uranium projectiles or targets have been made (Ko 81) but reported only in a preliminary manner (Ko 81b).

Generally, investigations pursued into the heavy collision domain, $Z_u\alpha \gtrsim 1$, have demonstrated that the good understanding of delta-electron emission achieved with the lower Z_u systems discussed in the last section can be carried over into the heavier systems. Figures 4.16 and 4.17 illustrate data

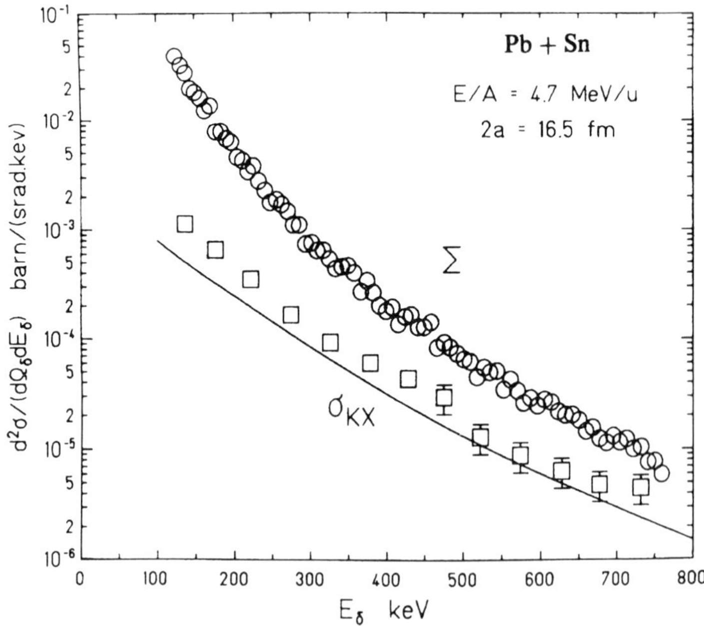

Figure 4.16. Delta-electron spectra from Pb + Sn at 4.7 MeV/u with $v/c = 0.10$. The singles (Σ) spectrum and the spectrum taken in coincidence with Pb K x rays are shown (Ko 82). The solid line is the coupled-channel calculation by Soff *et al.* (So 80, So 82c) for electrons from the $1s\sigma$ orbital including multistep processes.

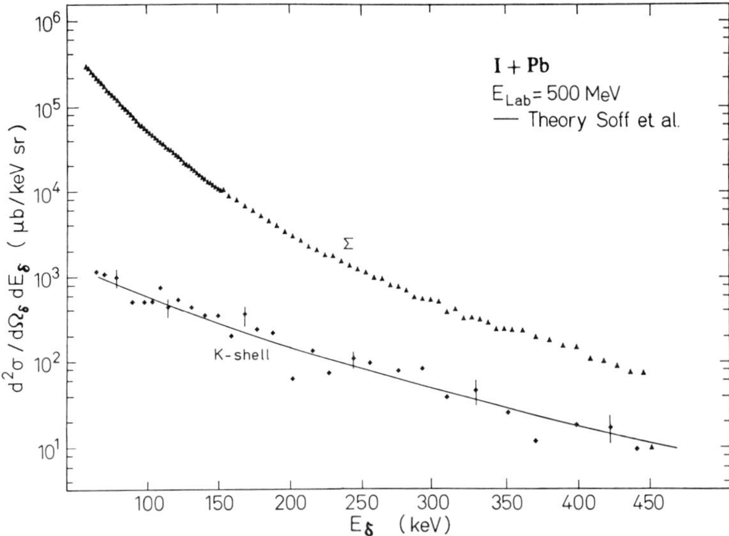

Figure 4.17. Delta-electron spectra from I + Pb collisions at $v/c = 0.09$, similar to those described in Figure 4.16. Solid line represents coupled-channel calculation by Soff *et al.* (So 80, So 82c).

for examples where two collision systems with similar Z_u have been studied (Ko 82). In each case both the total delta-electron spectra as well as delta electrons in coincidence with K x rays from the heavier collision partners were measured. Also in each case the collision is sufficiently asymmetric so that this coincidence selection identifies $1s\sigma$ excitation according to our quasimolecular scheme.

The measurements on the ^{208}Pb + ^{120}Sn system were carried out with the "Orange"-type β spectrometer situated at the GSI laboratory in Darmstadt at a relative collision velocity of $v/c = 0.10$. The corresponding experiments on the ^{127}I + ^{208}Pb system was performed with the achromatic electron spectrometer at the MPI in Heidelberg at a relative velocity of $v/c = 0.09$. We note that the two cases invert the role of projectile and target for the heavier and lighter collision partners. Clearly, the results for the two collision systems are very similar. On a quasimolecular basis, theory (So 80, So 82c) provides an excellent account of the data if $1s\sigma$ excitation is assumed to be represented by coincidence selection specified above. The calculations reproduce both the energy dependence and the absolute magnitude of the cross sections. As was found in the attempt to describe the Br + Pb collision data shown in Figure 4.12 (So 82b), a key aspect of the calculation is the inclusion of multi step processes which also take into account the possibility that the ejected electron can be rescattered during the collision to higher energy (Re 79, So 80, So 81b). The contribution of K-shell excitation relative to the total cross section, again, is understood in terms of the higher momentum transfer required by the latter excitations relative to excitation of the weaker bound L-shell and higher-shell electrons, and the decreasing contribution to the energy transfer required by the binding energy as the kinetic energy of the electrons increases. As mentioned above, the calculations are well defined and the theory is successful when the initial state selection is simple. The necessarily complex description required to deal with the total spectrum prohibits its comparison to theory for the present.

Data such as shown in Figures 4.16, 4.17 and other such data to be published (Ko 81), thus, indicate that the delta-electron emission process from the innermost shells of systems with $Z_u \alpha > 1$ is reasonably well understood. A particularly striking example of the success achieved in describing the details of the differential cross section is illustrated in Figure 4.18 (Ko 82). This is the type of experiment referred to earlier which is capable of providing a maximum amount of information. Here triple coincidence measurements have been performed which detect the electrons associated with particles scattered at a specified angle and with K x rays from the Pb atom. Figure 4.18 displays the probability of delta-electron emission from I + Pb collisions at $E_{\text{lab}} = 500$ MeV, tagged by K x rays of Pb, as a function of closest approach between collision partners for four delta-electron energies specified in the center of mass. The solid lines represent calculations by Soff *et al.* (So 82c). The excellent agreement with experiment, especially in absolute magnitude, again, particularly emphasizes the multiple excitation mechanisms which necessarily have to be included in the calculations in order to achieve the impressive description of the data shown.

However, the importance of the multistep processes in accounting for the absolute magnitude of the cross section is not reflected in a necessity to include these processes to describe the dependence of the emission probability on the distance of closest approach. In fact, the scaling law discussed in Section 4.2, which is based on first-order perturbation theory, reproduces the exponential dependence on R_0 and on the minimum energy transfer for a specified electron energy. As was found in the case of total ionization investigations and noted again in Section 4.2, the detailed modes of transfer of the total momentum and energy seem not to be relevant to this aspect of the emission probability. Only the total energy and momentum transfer at fixed energy affect the scaling behavior.

With this in mind, a number of measurements have been carried out (Ko 82) to pursue the question of scaling and its application to determining the energy transfer and, through the latter quantity, to obtaining some measure of the binding energy within the context already discussed. An example of such data using the I + Pb collision system has been presented in

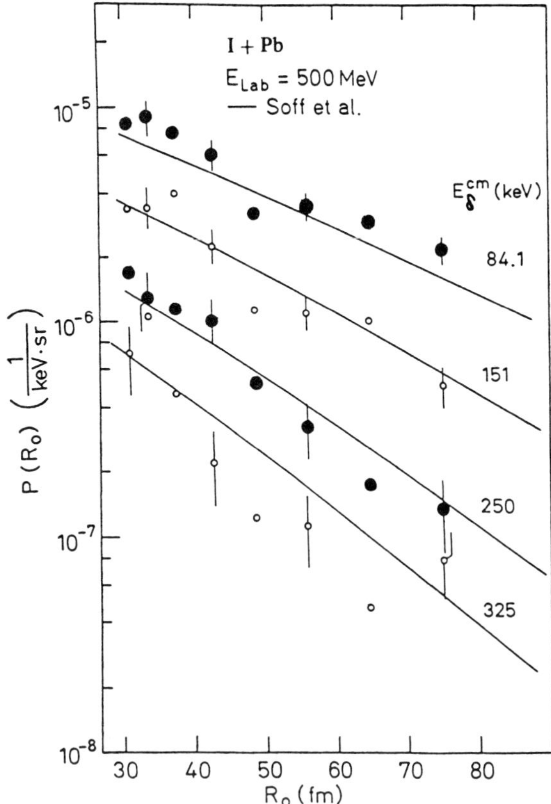

Figure 4.18. Emission probability for delta electrons associated with Pb K x rays ($1s\sigma$ excitation) for I + Pb collisions as a function of the distance of closest approach, R_0, for specific delta-electron kinetic energies ($E_\delta^{\text{c.m.}}$) as indicated (Ko 82). The solid lines show the results of calculations by Soff *et al.* (So 82c).

Figure 4.4. The results of another such recent study with the Pb + Sn system at $v/c = 0.10$ are given in Figure 4.19 (Ko 82). The general feature of the exponential falloff with R_0 is clearly displayed. The solid lines in the two figures represent least-squares fits to the data (arbitrary normalization) using

$$P \propto \exp(-2R_0 q_{\min}) \tag{4.19}$$

from which $q_{\min}$ is extracted. We had previously noted that the two-center potential energy diagrams for these two systems show that the binding energies of the inner-shell quasimolecular states can be considered as varying little over the region of R where the ionization is concentrated. Assuming the

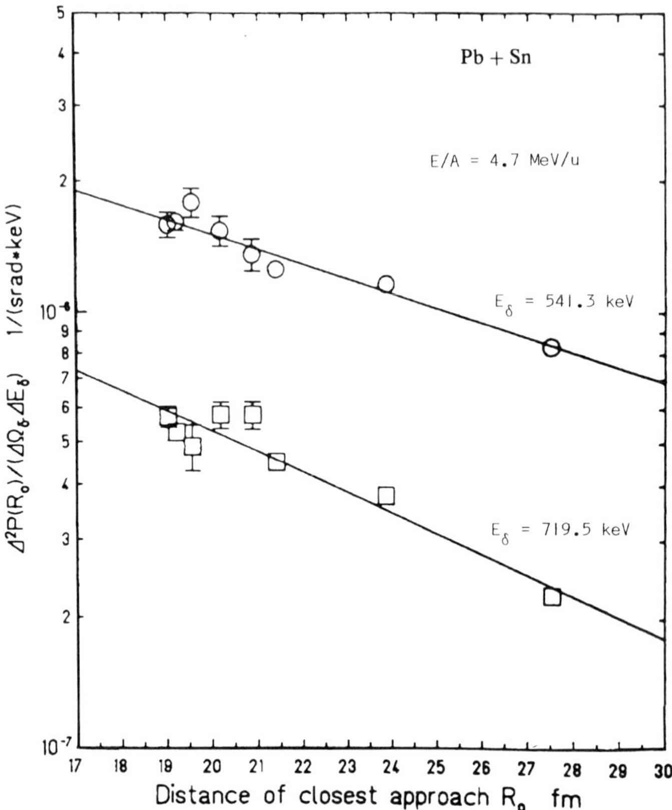

Figure 4.19. Emission probability for delta electrons from Pb + Sn collisions as a function of distance of closest approach, R_0, for two delta-electron kinetic energies. The solid lines represent fits to the data with $P \propto \exp[-2R_0(E_\delta + |E_B|)/\hbar v]$ and $v/c = 0.10$ (Ko 82).

binding energy is constant during the ionization process, $E_B(R_0)$ is deduced from

$$q_{\min} = [E_\delta + |E_B(R_0)|]/\hbar v \tag{4.20}$$

For Pb + Sn, selecting $E_\delta = 720$ keV for which the K-shell contributions are significant leads to $|E_B| \equiv 310$ keV. There seems to be good agreement between this binding energy and the value of 320 keV calculated by Fricke and Soff (Fr 77). The choice of $E_\delta = 540$ keV, however, leads to $|E_B| =$ 250 keV. It is suggested (Ko 82) that a larger contribution from the L shell in this case lowers the effective value of the binding energy extracted.

The data on the I + Pb system in Figure 4.4, again, show an increase in the slope of $P(R_0)$ with increasing E_δ, as expected from the scaling law. The authors have extracted an average binding energy of 60 keV in this case with the comment that the low values for E_δ chosen favor the emission of delta electrons from the L shell and thus account for the low average E_B deduced.

Thus these and other data (Ko 81) indicate that the delta-electron excitation probability scales with the product of $R_0 q_{\min}$ as did the total ionization probability. Indeed, it is somewhat surprising that L-shell ejection, which dominates Figures 4.4 and 4.19, is so well described by the law derived for $1s\sigma$ ionization. It appears that this part of the scaling is generally characteristic of the Coulomb ionization process as is also indicated by the observation, referred to above, that it is the total energy transfer which is the relevant quantity irrespective of the established multistep character of delta-electron emission.

But more detailed tests of scaling are provided by detecting the ionized electrons which are not available in studying vacancy production via x-ray emission. For example, closer scrutiny reveals that the scaling law in Section 4.2 does not account for the delta-electron energy dependence measured for a fixed impact parameter or R_0 (Ko 82). One question that has to be answered in this connection is how the contribution from the L shell affects this test. It is clear that more data are needed which isolate the contribution from $1s\sigma$ excitation. It is also evident that the further use of this scaling law to interpret delta-electron emission for spectroscopic analysis of binding energies requires further theoretical understanding of the influence of multistep processes on scaling laws since it has been established that delta-electron emission can only be understood within the framework of coupled excitations. For the present this type of analysis is providing a qualitative measure of the binding energies involved. As has been evident from our discussion, this information, although crude, is of primary interest since it provides one of the more direct indications that the deeply bound states of the quasiatom are involved in the ionization process.

Of course, such information is sought particularly for the heaviest collision systems that can be assembled. Unfortunately, as noted above, the data available are presently very sparse. Experiments have been carried out with ^{208}Pb projectiles bombarding targets of Au, Pb, Th, and U (Ko 81). Only the results on the $^{208}\mathrm{Pb} + {}^{208}\mathrm{Pb}$ system have been reported in any detail (Ko 82, So 81b). Actually the first measurements of high-energy delta electrons from very heavy systems were carried out on this system at a bombarding energy of 4.7 MeV/u, considerably below the Coulomb barrier, to take advantage of the low background conditions referred to previously.

The measured double differential cross sections are displayed in Figures 4.20 (Ko 82) and compared with the results of coupled-channel calculations by Soff *et al.* (So 81b) which include multistep processes. Here data have been combined both from the "Orange" spectrometer at GSI and from the achromatic electron channel at Heidelberg referred to above. As discussed in Section 4.2, the slope of the spectrum implies that the highest-

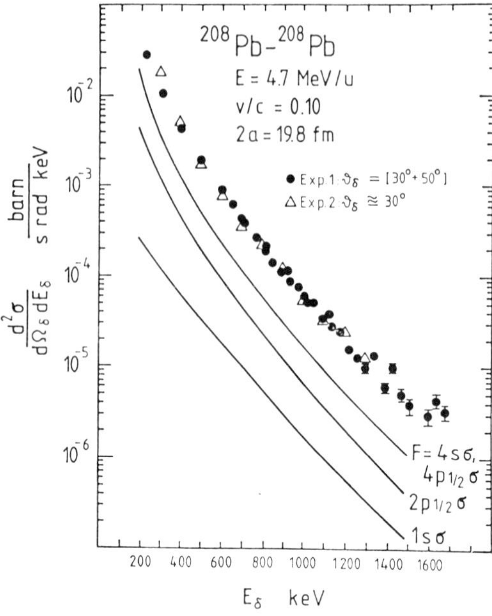

Figure 4.20. Delta-electron spectra from $^{208}Pb + ^{208}Pb$ collisions at a bombarding energy of 4.7 MeV/u. Exp 1 refers to data from "Orange" spectrometer, and Exp 2 is data obtained from the achromatic electron channel (Ko 82). The solid lines represent coupled-channel calculation by Soff *et al.* (So 81b) for the total spectrum ($F = 4s\sigma$, $4p_{1/2}\sigma$ denotes the Fermi surface) and for the contributions from the $1s\sigma$ and $2p_{1/2}\sigma$ states individually.

energy electrons can only originate from states of the quasiatom and implies the important conclusions regarding the relativistic effects already emphasized.

The theoretical calculations in Figure 4.20 are for a total delta-electron distribution which represents a sum of all initial σ states up to $4s\sigma$ and $4p_{1/2}\sigma$ states. Emission from the $1s\sigma$ state is predicted to constitute only a small fraction of the total spectrum. The fact that the slopes of the spectra from the $1s\sigma$ and $2p_{1/2}\sigma$ states are very similar reflects the property, noted earlier in Figure 4.5, that the $2p_{1/2}\sigma$ wave function closely resembles the $1s\sigma$ wave function in this very relativistic domain. Even at very high kinetic energies of ~2.0 MeV the binding energy still plays an important role in the superheavy systems.

The slope of the data is reproduced very well by the calculations, but the calculations fall short by approximately a factor of 3 in predicting the magnitude of the cross sections. It should be noted that perturbation theory calculations underestimate the data by more than a factor of 5 (Ja 79, So 78). The source of the discrepancy between theory and experiment is not understood presently. This problem may be associated with the one encountered previously of not being able to account in detail for the contribution from many higher bound states which together can add considerably to the total spectrum. It bears emphasis that the theory only considers radial coupling, neglects rotational coupling, and employs the monopole approximation. The effect of these approximations is presently not completely explored.

The results of triple coincidence experiments, in which delta electrons associated with Pb K x rays and scattered particles are measured in the attempt to isolate electron emission exclusively from the $1s\sigma$ and $2p_{1/2}\sigma$ states, do not clarify this situation very much. Within relatively large statistical errors of ~25%, the energy distributions and absolute magnitudes of the total and K x-ray coincident probabilities at a fixed particle scattering angle are similar (Ko 82). Of course, it is expected that the energy distributions are not changed by the K x-ray coincidence requirement. Presumably one explanation for the similar intensities observed with and without K x-ray coincidences is that the K x rays are not associated entirely with electrons excited directly from the $1s\sigma$ and $2p_{1/2}\sigma$ states, but originate from events where a vacancy originally is produced in a higher shell and subsequently couples to these inner-shell orbitals in the outgoing phase of the collision. Another possibility is that the $2p_{1/2}\sigma$ ionization probability is of the order of unity.

These and other questions regarding this and other collision systems are presently being resolved with experiments. As noted in the introduction to this section, investigations into the very heavy systems and their spectroscopy is only beginning. But the experiments on delta-electron emission already have established, to a large extent, the validity of the quasimolecule formed in collisions near the Coulomb barrier as a vehicle for exploring superheavy atomic systems and the dominance of relativistic effects in such systems.

5. Positron Spectroscopy with Heavy-Ion Collisions; Search for the Decay of the Vacuum

5.1. Introduction

The focus on the physics associated with superheavy atomic collision systems has in turn led to the emergence of a probe unusual for atomic collision studies, positron emission. Positron creation becomes an important aspect of these investigations on two levels. Similar in some effects to x-ray and delta-electron emission, dynamic mechanisms of positron production in superheavy collision systems reveal in an even more dramatic fashion the growth of the electron density around the nuclear centers arising from relativistic effects. With a minimum energy transfer of $2mc^2$ required to create an electron–positron pair, dynamic positron creation is particularly sensitive to the electronic configurations at very small internuclear separations for the reasons outlined in our discussion on vacancy formation and delta-electron emission. Especially because of this property, the dynamic aspects of positron emission test our theoretical understanding of the superheavy quasiatomic systems. In fact, of all the electron excitation and deexcitations mechanisms we have discussed, pair creation reflects the relativistic properties most directly since they are responsible for the occurrence of pairs at a level which is at all observable in superheavy systems. As we shall further discuss in more detail, the rapid increase of the dynamic pair production probability with $\sim Z_u^{20}$ accounts for the absence of positron emission for low Z_u collision systems, reflects the important features characterizing the interaction of electrons with very strong electromagnetic fields and, thus, provides very convincing evidence for a quasiatomic mechanism in which the two nuclear charge centers act together to generate the necessary strong electric field.

But more importantly, as we noted in the Introduction to this chapter, positron emission from a transiently assembled supercritical charge of $Z > 173$ may be able to provide the most direct signature of a fundamental new phenomenon, the rearrangement in the structure of the QED vacuum in supercritical static electric fields (Pi 69, Ge 69, Ge 70, Ra 71). This process of spontaneously emitting positrons is, clearly, the most striking observable consequence expected from supercritical fields. As we have previously noted, the search for this process has largely motivated the investigation of superheavy collision systems. The vacuum structure and its change by external conditions, of course, is of great general interest in quantum field theory. Because it is singularly accessible to experimental observation, the change in the vacuum state of QED in the vicinity of a nucleus with supercritical charge is particularly interesting as the prototype of such phase transitions.

Therefore, a central theoretical question associated with the spectroscopy of electronic states of superheavy atomic systems is whether their binding energy can exceed $2mc^2$ as predicted by theory (Mu 72, Mu 72b, Ze 72). The parallel central experimental task is to assemble a supercritical charge and to identify a viable signature for spontaneous positron emission. In the Introduction to this chapter we briefly outlined how superheavy collision systems may provide the conditions for supercritical binding for a short period of time. This scenario is illustrated schematically in Figure 5.1 for a U + U system colliding with an energy sufficient for the nuclei to touch. To penetrate well within the quasiatomic limit, where the innermost electrons experience the full impact of the combined nuclear Coulomb charges, requires a relative velocity of $\sim 0.1c$. As shown more quantitatively in Figure 1.3, if the quasimolecular picture represents a valid description of the collision under these conditions so that the electronic levels follow the nuclear motion adiabatically to the quasiatomic limit, then binding energies greater than $2mc^2$ are expected for the $1s\sigma$ state. For such

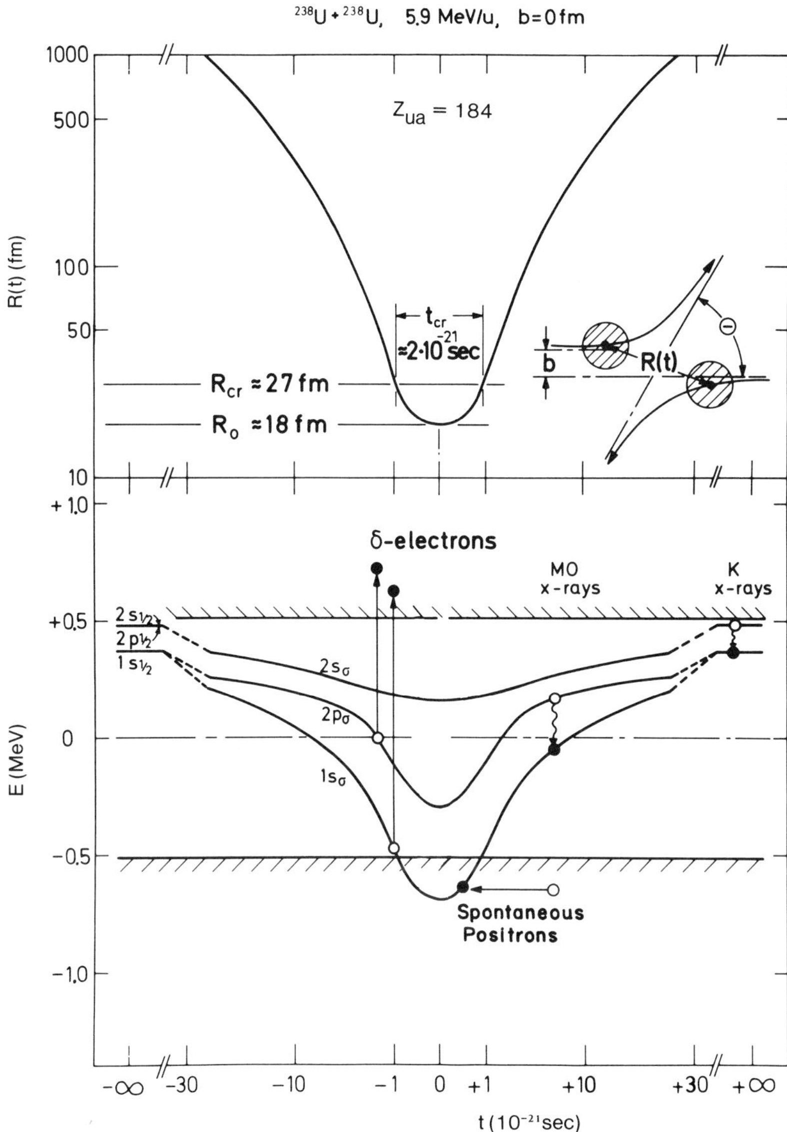

Figure 5.1. Schematic presentation of the time evolution of quasimolecular orbitals in U + U collisions showing the atomic excitation and deexcitation processes. Rutherford scattering is assumed. Spontaneous positron production can occur during the period of supercritical binding if a vacancy is present from prior ionization.

collisions at energies close to the Coulomb barrier Rutherford scattering is the dominant scattering process. Figure 5.1 illustrates how the internuclear separation R and the quasimolecular binding energies E_B vary with time during a Rutherford trajectory. In this case supercritical binding is expected to be achieved within a critical internuclear separation, R_{cr}, of ~30 fm (Wi 79, Li 80b, Mu 73, Ma 75b, Ra 76, Li 77, So 78b) corresponding to a time interval of $\sim 2 \times 10^{-21}$ s. [As will be discussed below, this period can be prolonged with nuclear reactions (Ra 78b).] Therefore, in principle, within this short time interval there is an opportunity to observe spontaneous positron emission if, as described in Section 1.5, a vacancy in the $1s\sigma$ state is created prior to or during the time of supercritical binding.

The studies of x-ray and delta-electron emission described in Sections 3 and 4 indeed suggest that some of the essential requirements for this scenario may, in fact, be met. We shall see that further evidence is supplied by dynamic positron production. All these data suggest that superheavy quasimolecules (quasiatoms) are being formed in collisions at bombarding energies close to the Coulomb barrier (~5.9 MeV/u for U + U). Although the question of supercritical binding has not been directly addressed by these experiments, nevertheless, they do demonstrate that the binding energies of the inner-shell electrons exceed the single-atom binding energies by a considerable amount. In addition, in Section 3 evidence has also been presented which indicates that the characteristics of $1s\sigma$ vacancy formation are favorable to spontaneous positron emission; the experiments on x-ray emission establish that the $1s\sigma$ ionization probability can be as large as 10%, and, moreover, it is found to be concentrated at small internuclear separations so that the vacancies are available when supercritical binding can also exist.

Therefore, the presence of conditions essential to utilizing the superheavy collision system for observing spontaneous positron emission appears to be founded in experimental observations. Unfortunately, the scenario for exploiting this situation described above is incomplete because of fundamental limitations associated with the dynamic aspects of the quasiatomic system. Similar to the situation that is found in attempting to establish spectroscopic information using x rays, MO x rays, and delta electrons, there is a basic difference between the dynamic mock-up of a superheavy atom and a stable atom which can modify the simple ideas represented in Figure 5.1 considerably. A key aspect of the experimental task is to find an unambiguous signature for spontaneous positron emission.

We have seen that apparently there is a very clear signal for the spontaneous decay of the vacuum in the vicinity of a bare supercritically charged stable nucleus. The spontaneous filling of the supercritically bound resonance states leads to the emission of narrow line spectra centered at an energy $E_B - 2mc^2$ (see Figure 1.4). The width of the line reflects the lifetime for spontaneous decay. For example, for $Z = 184$ (U + U) the mean positron energy is ~200 keV (Re 81), and the natural time scale for the unstable vacuum state is $\sim 10^{-19}$ s, corresponding to an intrinsic resonance width of <10 keV. This is, certainly, a very characteristic signal, particularly since other positron emissions predominantly appear in nature as continuum distributions. This distinction can be lost, however, with the collision formed supercritical charge, depending on how this charge evolves with time. With the loss of line structure, the signal for the spontaneous process is further obscured by additional contributions from dynamic amplitudes for positron production which, again depending on the time evolution of the collision, can overwhelm the spontaneous process (Re 81b, Sm 74, So 77, Ja 76, To 82, To 82b).

The source of these difficulties is apparent from some qualitative considerations. The broadening of the spontaneous spectrum can occur due to several intrinsic features of the quasimolecule. Since the two-center field varies with time, the adiabatic binding energy and, thus, the energy of the spontaneously emitted positrons is ill defined. We have previously noted this reason for spectral broadening in connection with MO spectra. In the case of superheavy collision systems with $Z\alpha > 1$, where the dominance of relativistic effects produces an enhanced dependence of the energy eigenvalues on the nuclear charge distribution ("the collapse to the center"), the adiabatic energies can change rapidly with internuclear separation R for $R < R_{cr}$ (see Figure 1.3, for example). This aspect of collision broadening, therefore, depends in a sensitive way on the details of how the two-center distance evolves with time (Re 81, Ra 78b) and of how the spontaneous emission amplitude develops with time for times less than the intrinsic lifetime (Re 81, Re 83, To 83, To 83b, Mu 83b). Moreover, it is clear that any shortening of the time during which supercritical binding is achieved introduces an uncertainty in the energy level structure and thus also broadens the positron spectrum. In fact, for lifetimes of the quasiatom corresponding to Rutherford collision times $t_{cr} \simeq 2 \times 10^{-21}$, during which, as Figure 5.1 shows, the $1s\sigma$ state should be supercritically bound in the adiabatic limit, the broadening from this source, $\hbar/t_{cr} \sim 300$ keV, exceeds the mean positron energy of ~200 keV anticipated for spontaneous emission from a charge with $Z = 184$. Of course, the short time scale associated with the quasiatom not only broadens the spectrum, but also diminishes the probability for observing the transition of the vacuum from the neutral to the charged phase. The typical time of 2×10^{-21} s noted above competes with the natural time scale for the unstable vacuum of $\sim 10^{-19}$ s so that the vacancy in the $1s\sigma$ state has little chance to decay.

The short collision times lead to another important consequence. They imply high Fourier frequencies associated with the collision and thus rapidly varying electromagnetic fields due to the moving nuclear charges. As was considered in connection with inner-shell x-ray and high-energy delta-electron emission, the high frequencies available reflect the ability to transfer large energies which is augmented by the strong fields made available by the combined action of the two nuclear charges of the quasimolecule. In a mechanism similar to the one in which the time-varying quasimolecular potential excites the needed $1s\sigma$ vacancies and gives rise to delta electrons, an ionization of the antiparticle continuum either into vacant bound states or into the positive energy continuum leads to positron emission. These are the dynamic processes of positron production referred to in the beginning of this section. In fact, in this respect, there is little to differentiate the states of the negative energy continuum from the bound states with strong binding. The same couplings and matrix elements responsible for ionization of bound states also ionize the Dirac sea. Like delta-electron emission from the inner shells, the spectrum is continuous except the low-energy part of the positron spectrum is inhibited by the Coulomb field. On the one hand these emissions are an unwanted background to spontaneous positron emission, but, as described earlier, on the other hand they are also a sensitive probe of the quasiatom and its extreme relativistic structure. Therefore, exclusive of the search for spontaneous positron emission, the dynamic positron production processes are interesting in their own right as representing a situation involving strong electric fields where perturbation theory is not applicable (So 77). They provide an additional approach to those already cited for exploring both bound and continuum electronic states in a highly relativistic domain.

Another consideration is that the rapidly varying electric fields not only excite electronic states that lead to pair creation, but they also excite nuclear states by multiple Coulomb excitation. Moreover, nuclear states can also be excited by nuclear contact. Deexcitation of these states by internal pair conversion is an additional source of positrons in the collisions, if the transition energy exceeds $2mc^2$. The time scale associated with nuclear decay is of the order of 10^{-12} to 10^{-16} s so that these events cannot be separated from the atomic component by the available timing techniques. The spectral distribution of these positrons reflects the spectrum of nuclear transition energies. Of course, this source of positrons represents additional background which has to be differentiated from the atomic sources of positrons. We will consider such a procedure based on the measurement of the γ-ray spectrum (Me 77b).

Therefore, it is apparent that although the quasiatom may provide the vehicle for assembling a supercritical charge, dynamic effects have to be a major consideration in the experiments to detect spontaneous positron emission. Relying on a threshold detection of positrons is clearly too simplistic. As a basis for discussion of the experimental results and to assess in more detail the experimental task of identifying spontaneous positron emission in the presence of the dynamic competing processes, in the following section we bring together the characteristic features expected for the different sources of positrons from heavy-ion collisions. For a more complete discussion of the theoretical background we refer to (Re 85) Chapter 1 in this volume. Here we will deal principally with those aspects which are particularly relevant to the practical experiments.

5.2. Positron Emission Mechanisms

Pair creation in charged particle collisions is clearly not a newly studied phenomenon. However, in the past, it has been considered mainly in connection with high-energy physics where, for example, electron–positron creation by high-energy electrons (trident production) is used to probe the small distance behavior of QED. The early unsuccessful attempts to detect pairs in slow proton–nucleus collisions (St 58) are now understood as due to the reduction of the cross section by Coulomb repulsion (Ba 59). More recently it was the clarification of the physics associated with the realignment of the vacuum in overcritical fields and the suggestion that this phenomenon could be observable in superheavy collision systems which led to a renewed interest in the subject of pair creation during heavy-ion collisions. Of course, in this connection, the principal points of immediate interest were centered on the level of positron production by the moving nuclei, on the shape of the positron spectrum, and on the relationship of the dynamic processes to the inherent static process of spontaneous positron emission. It speaks for the degree of interest in the experiments attempting to observe the spontaneous positron emission process that they were begun with only partial answers to these questions available and that they were pursued later even in the face of some discouraging prognosis that the dynamic effects could dominate the positron spectrum (Sm 74).

In fact, a detailed understanding of how spontaneous positron emission emerges under the dynamic conditions encountered in superheavy collision systems has only been available recently (Re 81, Re 81b, Re 83, To 82, To 82b, To 83, To 83b, Mu 83b). The first predictions of cross sections for spontaneous positron emission were obtained in the so-called "quasistatic approximation" where the resonance is allowed to decay at every internuclear distance R as if the system lives at each location of the trajectory for an infinitely long time (Pe 73, Ge 73, Po 73, Po 74, Ma 75b). The time dependence of the collision enters only in the parametric relation between R and the collision time, and the vacancy is assumed to be brought into the collision. Although these calculations led to the prediction of peaked spectra (Pe 73), which, of course, are appealing to experimental verification, they had the obvious shortcomings of ignoring collision broadening and the coherent effects of vacancy formation and positron emission. They, thus, only provided an approximate measure of the positron emission probability based on the assumption of an average $1s\sigma$ vacancy availability during the supercritical binding phase of the collision, and gave no insight into how important positron production would be due to the changing Coulomb field. Indeed, it was argued in one of the early reviews on this subject by Zel'dovich and Popov (Ze 72) and also in later papers by Popov (Po 74, Po 74b) that the dynamic mechanisms could be neglected, citing the reasons for this conclusion that they would be suppressed by a small frequency compared to $2mc^2$ associated with the collision and by the adiabatic nature of the collision which allows the electronic states to adjust to the changing Coulomb potential. However, this reasoning was clearly erroneous since we have already noted that $t_{cr} \simeq 10^{-21}$ s $(\hbar/t_{cr} \sim mc^2)$ and, as was demonstrated for the case of inner-shell ionization, energy transfer of $\sim 2mc^2$ is quite probable for collisions with impact parameters less than $R_{cr} \simeq 30$ fm and collision velocities $v/c \simeq 0.1$. Of course, inner-shell ionization and dynamic positron emission are closely related so that the understanding of the role played by the essential contribution of relativistic effects in superheavy systems to both processes evolved together, and the recognition that the $1s\sigma$ ionization probability is large and that dynamic positron emission could not be neglected occurred at about the same time.

The importance of elaborating on the quasistatic description and including the effects of the changing Coulomb field for positron production was first pointed out by Smith *et al.* (Sm 74). They particularly emphasized that the energy transfer required could be considerably reduced by filling vacancies in the $1s\sigma$ level which can lie close to the negative energy continuum. It was found that this so-called "induced" pair creation can be two orders of magnitude larger than spontaneous emission as then calculated and, furthermore, it occurs for collisions where supercritical binding is not achieved. In this calculation it was still assumed that an inner-shell vacancy was brought into the collision so that the coherence between vacancy formation and positron emission was not treated. However, although incomplete, these first calculations of dynamic positron production emphasized that the dynamic aspects of the collision could not be ignored in the search for spontaneous positron emission, and that a comprehensive theory is required which has to include the special status of the supercritically bound resonance state. Other later calculations also supported these initial results (Ja 76, Ja 76b).

We consider in the next sections some of the main results relevant to the experiments that emerge from the recent theoretical descriptions of positron emission from superheavy collision systems under the general heading of atomic processes. Under a separate section we discuss positron production originating from nuclear processes since, in this case, the approach is mainly empirical. The atomic and nuclear processes add incoherently to produce the total positron spectrum.

5.2.1. Atomic Positron Production Processes

When the total charge of the colliding nuclei $(Z_p + Z_T)$ is larger than the inverse of the fine structure constant, the large effective coupling, $(Z_p + Z_T)\alpha > 1$ makes it unwieldy to describe the dynamic pair creation processes in a Feynman graph approach. The lowest-order Feynman diagram, which yields a $Z_p^2 Z_T^2$ dependence, is clearly insufficient to account for the much stronger observed Z_u^{20} dependence. Moreover, while the induced vacuum decay may, in principle, be handled by the coherent action of many photon exchange diagrams (So 77), the spontaneous decay process cannot, even in principle, be treated by a perturbation expansion that somehow does not describe the diving of the 1*s* state. The Feynman graph approach is obviously not suited to this problem, which is inherently nonperturbative. Instead, it has been found (Re 81b) that with the proper choice of basis set (Th 79) we can visualize the excitation of the vacuum by the nuclear motion as a Coulomb excitation process analogous to the treatment of ionization in an adiabatic picture described in the earlier sections. Figure 5.2 illustrates the three basic transitions involving the vacuum and specific final states. We refer to them as direct, spontaneous, and induced emission although this is an artificial designation, as demonstrated below.

The nuclear motion plays a role in all three processes. However, for spontaneous positron emission this role is secondary, serving only to create a vacancy in the supercritically bound state. Of course, as previously described, the spontaneous vacuum transition that fills the vacancy occurs without additional energy needed and is basically a static mechanism. But the motion is still required to produce the essential vacancy. On the other hand, the direct transition to the positive energy continuum does not involve bound states at all and thus can be described in first-order perturbation theory. The "induced" emission only differs from spontaneous emission in that energy transfer is involved in the vacuum transition. Of course, the later process can also occur for noncritically bound states like the $p_{1/2}$ state shown. In lowest order it is a two-step process and, therefore, requires at least second-order perturbation theory for its description. Beginning and ending in identical configurations the amplitudes for all three processes have to be added coherently. It is also clear that an appropriate description of the bound states is essential to this approach, and that the theory has to treat these bound states and positron states in a uniform manner. As the critical charge is approached for Z_u, the lowest bound states of the quasimolecule approach the negative energy continuum so that they begin to assume a dominant role in the excitation of the Dirac sea electrons as the energy transfer required decreases. They may also serve as an intermediate state in multistate excitation which we saw was characteristic of atomic excitations in superheavy collision systems.

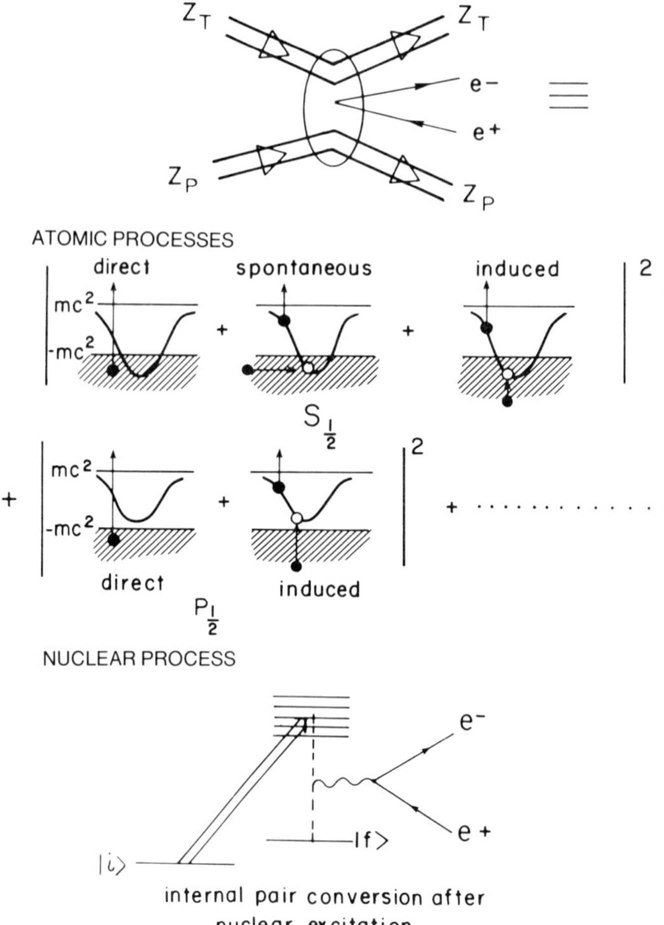

Figure 5.2. Coherent atomic positron production mechanisms in superheavy collision systems where the $1s\sigma(s_{1/2})$ state can be supercritically bound transiently. Internal pair conversion after nuclear excitation also contributes incoherently to the total positron emission intensity (Gr 80b).

Like the atomic excitation processes discussed in Sections 3 and 4, positron creation is described through the time-dependent Dirac equation with the excitation amplitudes $a_{ki}(t)$ given in Section 2,

$$\dot{a}_{ki} = -\sum_{j \neq i} a_{kj}[\langle\phi_i| \partial/\partial t |\phi_j\rangle + (i/\hbar)\langle\phi_i| H_{\text{TCD}} |\phi_j\rangle]$$
$$\times \exp\left[(-i/\hbar)\int_{-\infty}^{t} dt' \, [E_j(t') - E_i(t')]\right] \tag{5.1}$$

with the initial condition $a_{ki}(-\infty) = \delta_{ki}$.

Analogous to the other electronic excitations, pairs can be produced within this formalism by the time variation of the adiabatic quasimolecular states ϕ_i of the two-center Hamiltonian, H_{TCD}, due to the radial variation and rotation of the internuclear axis ($\partial/\partial t \to \dot{R}\,\partial/\partial R - i\boldsymbol{\omega}\cdot\mathbf{j}$). Moreover, in pair production, as in inner-shell excitation, we also get considerable simplification in handling the time-dependent part of the coupling in Equation (5.1) since the rotational coupling is of minor importance and the strong peaking of the matrix elements of $\partial/\partial R$, produced by the relativistic shrinking of the wave function, makes it possible to utilize the monopole approximation (Be 76, So 79). However, the descriptions of positron production and the other inner-shell excitation processes differ in an essential way. In the latter, the couplings are due entirely to the $\partial/\partial t$ operator unless there is electron–electron coupling since the potential coupling term in Equation (5.1) is nonzero only if the basis contains states which are not exact eigenstates of the Dirac Hamiltonian (Re 81b) (see Section 2.2). On the other hand, this potential coupling assumes a particular significance for collisions where supercritical binding can occur and the $1s$ state joins the continuum as a resonance. In this case the adiabatic $1s\sigma$ state is unstable even in the absence of any time dependence of the collision. It was first shown by Reinhardt, Müller, and Greiner (Re 81b) and later by Tomoda and Weidenmüller (To 82) that this instability is implicitly contained in the potential coupling term. Thus the formalism which describes the dynamic excitations and pair creation processes accounts for the resonance character of the $1s\sigma$ state and thus includes spontaneous positron emission.

In the method developed in (Re 81b) a normalized quasibound state Φ_R is defined at the resonance energy E_R which is degenerate with the positron continuum Φ_p. A new negative energy continuum $\tilde{\Phi}_p$ is constructed which includes a subspace orthogonal to Φ_R and substitutes for the old continuum Φ_p. The resonance state Φ_R and the modified continuum $\tilde{\Phi}_p$ are determined for all values of $R < R_{\text{cr}}$. They are included in the basis used to expand the time-dependent wave function $\Phi_k(t)$ so that it contains contributions from the positive energy continuum, the bound states, the resonances when there is supercritical binding, and the modified states in the negative energy continuum. In the new basis the states Φ_R and $\tilde{\Phi}_p$ are not eigenstates of the two-center Dirac Hamiltonian H_{TCD}. This leads to the important modification that the resonance couples to the positron continuum now by the two operators noted above:

$$\langle\tilde{\Phi}_p|\frac{\partial}{\partial t}+\frac{i}{\hbar}H_{\text{TCD}}|\Phi_R\rangle \to \dot{R}\langle\tilde{\Phi}_p|\frac{\partial}{\partial R}|\Phi_R\rangle + (i/\hbar)\langle\tilde{\Phi}_p|\,H_{\text{TCD}}\,|\Phi_R\rangle \tag{5.2}$$

The second term thus couples a vacancy in Φ_R to the positron continuum and leads to its spontaneous decay. The time development of the resonance and $\tilde{\Phi}_p$ only determine the instantaneous decay width which in the limit of a fixed $R(t)$ is given by (Re 81b);

$$\Gamma = 2\pi \, |\langle \tilde{\Phi}_{E_p} | \, H_{\mathrm{TCD}} \, | \Phi_R \rangle|^2 \tag{5.3}$$

with E_p at the resonance energy. The vacancy decays exponentially in time with the equivalent time constant $t = \hbar/\Gamma$. Thus, as noted above, the dynamic and spontaneous emission appear coherently together whenever an unstable resonance state Φ_R is formed in the lower continuum. There is a natural continuous transition from the subcritical to supercritical expansion basis which, with coherence, leads to important consequences discussed below. The construction of the resonance state Φ_R forms an important part of this description of positron creation in heavy-ion collisions. This part of the problem is treated in some detail in (Re 81b) and (To 82), where two calculational approaches are adopted and lead to similar results.

As noted previously, and as is clear from this description of positron production, the breaking up of the total positron amplitude into the three components illustrated in Figure 5.2 is somewhat artificial since for any given collision a measurement cannot distinguish among them. Nevertheless, it is convenient for purposes of discussing the experiments in the context of the theory to maintain this distinction. It is also useful to consider separately experimental conditions with and without supercritical binding present, as well as to distinguish collisions with pure Coulomb trajectories from collisions that can lead to reactions which can prolong the state of supercritical binding.

5.2.1a. Spontaneous Positron Emission. Some very unusual properties are associated with spontaneous positron emission if we ask how measurable parameters such as the energy, the energy distribution, and the emission probability of these positrons vary with controllable variables such as the total charge of the collision system and the nuclear configuration. Indeed, it is this exceptional behavior which makes it at all feasible to venture into these experiments. In the following discussion we consider first what to expect in a quasistatic-type situation where the nuclear motion is frozen or prolonged at one configuration for a long time, and subsequently discuss how the collision-type situation is expected to modify this outlook.

We should first note that there is one particularly important parameter associated with spontaneous positron emission from collision systems which is indigenous to the process and has a dominating impact on the experimental considerations. This is the critical distance at which the binding energy of the $1s\sigma$ state reaches the lower continuum. It would be expected, naively, that the spontaneous positron production cross section should scale with R_{cr}^2. However, the actual dependence on R_{cr} is much more dramatic. As discussed below, the degree of penetration of the $1s\sigma$ state into the negative energy continuum overwhelmingly determines the positron emission probability. The crucial importance of this parameter has, therefore, motivated a considerable number of calculations of R_{cr}. Some of these involve the solutions to the two-center Dirac equation with various approximations for the two-center potential and screening effects, and others utilize variational methods (Mu 73, Mu 76b, Ma 75b, Ra 76, So 74, So 78b, Li 77, Li 80b, Wi 79, De 81, as well as references therein). Results from one of the most recent calculations based on the monopole approximation, with the electron–electron interaction included within the relativistic Hartree–Fock–Slater model, are shown in Figure 5.3. Finite nuclear size is taken into

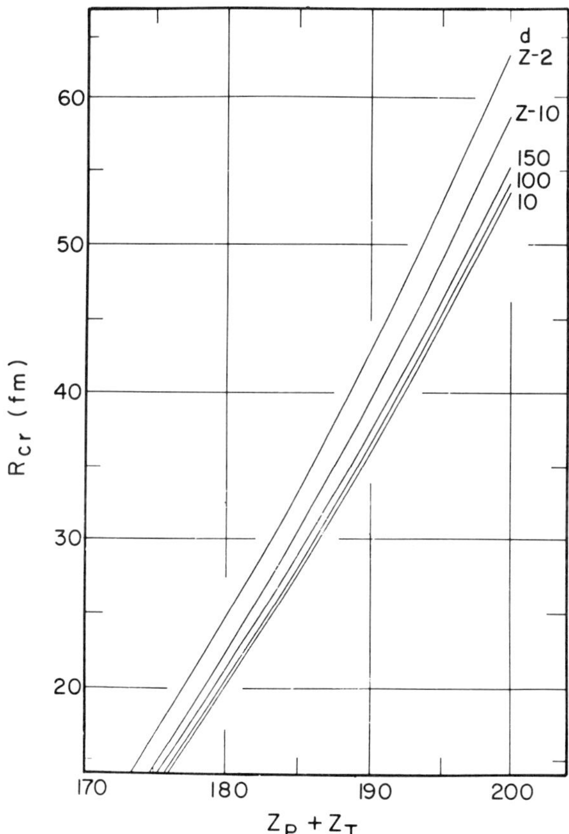

Figure 5.3. Critical internuclear separation required to produce supercritical binding for the $1s\sigma$ state as a function of the combined nuclear charge $Z_p + Z_T$. The electron–electron interaction has been calculated within the relativistic Hartree–Fock–Slater model. The individual curves are for different degrees of ionization (Wi 79).

account because the wave functions are highly localized. It is evident that the two-center distance required to achieve supercritical binding for accessible collision systems places the bombarding energy near the Coulomb barrier and essentially dictates the characteristics of the accelerator needed. The dynamic range available between producing supercritical binding and nuclear contact is very limited. It is also apparent that the choice of collision systems that are suitable and also accessible is even more confining. The lower limit for Z_u near threshold is ~180, while the upper limit is dictated by the heaviest projectiles and targets that can be practically considered. The highest Z projectiles presently available are ^{90}Th and ^{92}U. When combined with target elements of ^{90}Th, ^{92}U, ^{94}Pu, ^{96}Cm, and ^{98}Cf, which can be obtained and handled in sufficient quantities to make targets, a range up to $Z_u = 190$ is possible in principle. The interesting, and somewhat unprecedented, aspect of the spontaneous emission process is that this ~5% change in Z_u and equally small variations in the internuclear separation with R_{cr} produces very large measurable effects.

For example, Figure 5.4 shows the $1s\sigma$ decay widths $\Gamma = 2\pi |\langle \tilde{\Phi}_{E_R} | H_{\text{TCD}} | \Phi_R \rangle|^2$ calculated by Reinhardt *et al.* (Re 81, Re 81b) for U–Th, U–U, and U–Cm collisions as a function of internuclear separation. These calculations are based on the monopole approximation. They solve the radial Dirac equation with the monopole part of the potential created by two homogeneously charged spheres with sharp surfaces. When the spheres overlap normal nuclear charge density is maintained. Relativistic self-consistent field calculations in the Hartree approximation are used here assuming 50-fold ionization (Fr 77, Li 71). The authors note that these

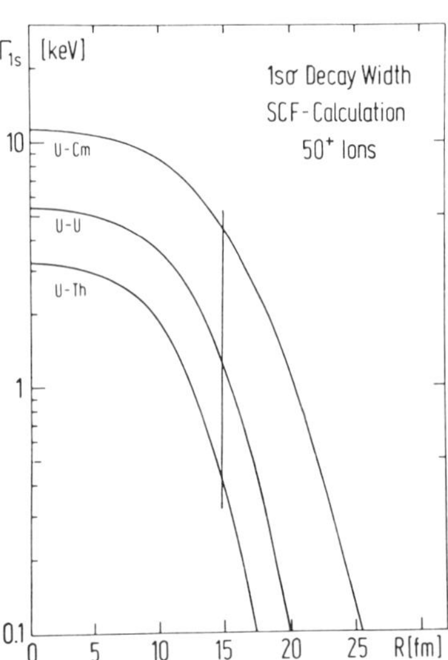

Figure 5.4. Decay width of the $1s\sigma$ resonance as a function of internuclear separation R. It is calculated in the monopole approximation for two charge spheres with radii given by 1.2 $A^{1/3}$ fm. In the region of nuclear contact, left of the vertical line, a geometry of two overlapping spheres with homogenous normal nuclear density has been assumed (Re 81).

calculations are supported by the results of direct phase shift type analysis of the continuum Φ_p (Mu 72b, Mu 73b). The spontaneous decay width is a very steep function of R, and it also varies very rapidly with Z_u at a fixed R value. In fact, in changing the collision system from U + U to U + Cm, a difference in Z_u of 4 units of charge, the spontaneous decay width changes by more than a factor of 4 when the nuclei are just touching. Within the range of available collision systems this translates to an unprecedented Z_u^{70} dependence on the quasiatomic charge. There is almost an equally sharp dependence of the kinetic energy of the positrons on total nuclear charge and on internuclear separation which is illustrated in Figure 5.5. Again, going from the $Z_u = 184$ to the $Z_u = 188$ system at nuclear contact results in an increase in the kinetic energy from ~200 to ~300 keV ($\sim Z_u^{20}$ dependence). We shall see that this increase in kinetic energy is an important factor in the detection schemes utilized as well as a significant signature for this process.

It should be noted that with the exaggerated dependence on Z_u, screening must have significant effects both on the positron energy and on the decay width. Thus the absolute values of these quantities have uncertainties associated with them since the ionization state of the system is a dynamic quantity. Müller *et al.* (Mu 83b) and de Reus *et al.* (De 83) discuss this question in a treatment of time-dependent screening based on an adiabatic approximation to the time-dependent Hartree–Fock formalism. A measure of the importance of screening is illustrated by their results: at an internuclear separation of ~17 fm for U + U collisions, there is a decrease of ~100 keV in the binding energy of the $1s\sigma$ state in going from the unshielded to the shielded configurations. The spontaneous decay width, Γ, also decreased by ~60%, but this is partially counterbalanced by an increase in the K-vacancy production probability which results in a net decrease in spontaneous positron emission of ~35%. Screening effects, therefore, are significant considerations here as they are in the other excitation processes in

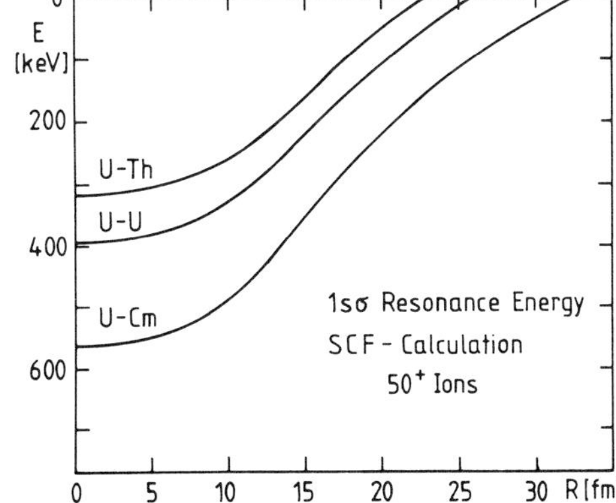

Figure 5.5. The position of the $1s\sigma$ resonance in the positron continuum as a function of internuclear separation R. The model for the nuclear charge distribution is the same as for Figure 5.4 (Re 81).

superheavy collision systems (De 83). Furthermore, the screening effects are exaggerated by the threshold character of spontaneous positron emission around $Z_u = Z_{cr}$. Nevertheless, screening does not change the unusual properties of spontaneous positron emission illustrated in Figures 5.4 and 5.5, although it is clear that to predict absolute quantities requires careful consideration.

Some physical insight into the exceptional behavior of spontaneous positron emission as a function of Z_u and R is obtained by considering several of the descriptions developed to view the formation of the quasistatic resonance and its decay. Of course the central point is the appearance of this resonance as the supercritical conditions are established. Figure 5.6 illustrates how the resonance reflects in the behavior of the radial probability density of the continuum Φ_p as a function of R for a supercritical system like U + U ($Z_u = 184$) but not for the subcritical system Pb + Pb ($Z_u = 164$) (Ba 82). The emergence of the resonance behavior is coincidental with the vanishing of the $1s\sigma$ state at an energy of $-mc^2$. One of the salient features

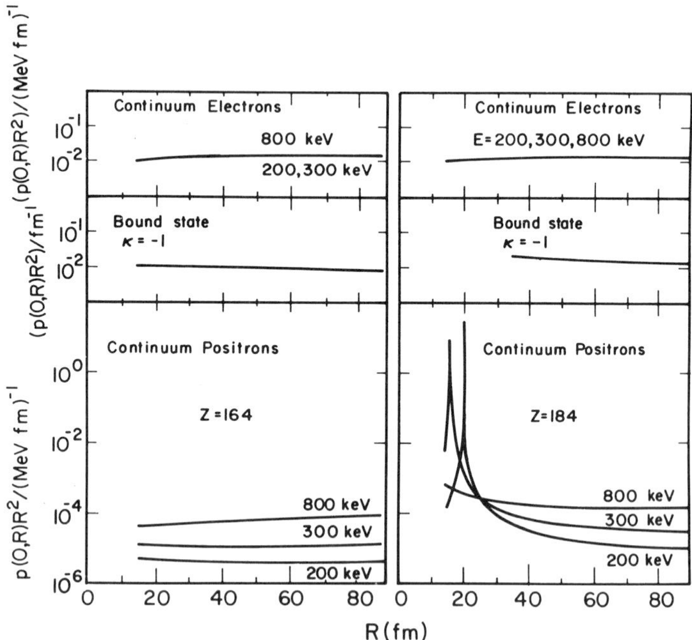

Figure 5.6. Radial probability distributions at the origin, $\rho(r = 0, R) R^2$ for electron and positron continuum and bound states as a function of internuclear separation of two Pb and two U nuclei. The monopole approximation is used and only $s_{1/2}$ ($\kappa = -1$) states are considered: upper panel for continuum electrons, middle panel for $n = 1$ bound state, and lower panel for continuum positrons [after (Ba 82)].

of the continuation of the bound $1s$ state into the lower continuum is that although it is not possible to ascribe its charge distribution to a stationary one-electron state, it remains localized very much like that of the $1s$ state with an energy close to $-mc^2$ (Ra 74). The quasibound character of this state is the essential aspect, and its decay characteristics can be understood in the following, qualitative terms developed quantitatively in a number of papers (Pi 69, Po 71b, Ze 72, Fu 73, Fu 74, Ra 74).

We can view a deeply bound K-hole with total negative energy as a bound positron with total positive energy $|E_{1s}|$. This viewpoint becomes a natural one if the vacuum is redefined as the state in which all solutions to the Dirac equation with negative total energy bound or unbound are occupied. The new vacuum has a lower energy than the usual vacuum and has the virtue of assuring the continuity of a stable vacuum state for all Z including $Z > Z_{cr}$. The old vacuum then becomes an excited state of the system with two bound positrons if the two atomic K orbitals have negative total energies. As the K-hole energy approaches $-mc^2$, the positron binding energy approaches zero, and at $Z = Z_{cr}$ the wavefunction ultimately becomes unbound and the positrons can escape. The binding of the positron and the weakening of this binding with increasing Z is a purely relativistic effect, as can be demonstrated by writing the radial equation in Schrödinger form and singling out an effective potential (Re 85). As discussed by Zel'dovitch and Popov (Ze 72), for negative total energy electron orbits, the bound positron moves in an effective potential which becomes shallower as Z increases. However, as Z reaches and exceeds Z_{cr} the freedom to escape the nucleus is restricted by the fact that the potential is attractive at small distances and repulsive at larger distances so that the positron, even with effective energy $>mc^2$, has to penetrate through a barrier. This characteristic of the relativistic Coulomb problem can be viewed as leading to the strong localization of the supercritically bounded state referred to above, and to its typical resonance character described earlier. In this description of spontaneous positron emission, the decay width, Γ, is governed by the barrier penetrability. Typical of barrier penetration phenomena such as alpha particle decay in nuclei, the decay width is a strong function of the positron energy. Furthermore, as noted by Popov (Po 71b) and Müller *et al.* (Mu 72b), the threshold behavior usually associated with penetrating the barrier leads to an exponentially decreasing positron production as $Z \to Z_{cr}$ and $E_{e+} \to mc^2$. The resonance width varies as $\exp(-\pi Z\alpha M_e/p)$ near threshold (Mu 76), reflecting the small probability for finding low-energy positrons near the nucleus. Thus the barrier penetration picture provides a physical interpretation for the strong dependence of Γ on Z_u and R which emerges from the results of more complete calculations depicted in Figure 5.4. The rapid variation of the positron energy as a function of R and Z_u, shown in Figure 5.5, can be understood as being produced by the sensitivity of the $1s\sigma$ binding energy to nuclear dimensions and Z noted in Section 2.

The Frankfurt group has made extensive studies of the resonance behavior of the supercritical state by examining the exact solution of the one-center Dirac equation (Mu 72b, Mu 73b). Their autoionization model (Mu 72, Mu 72b) which describes how the bound state mixes into the continuum and becomes a resonance provides a unique insight into how the resonance forms and establishes a useful form for how Γ and the energy of the resonance, E_R, vary with Z. This approach is valid over a broader range in Z than the barrier penetration picture. They find (Re 85) that $E_R \simeq \delta(Z - Z_{cr})$, and $\Gamma \simeq \gamma(Z - Z_{cr})^2$ with $\delta \simeq 29$ keV and $\gamma \simeq 0.04$ keV. As noted in (Re 85), due to the exponential damping of Γ near $Z \simeq Z_{cr}$, this expression for Γ is applicable only if $Z > (Z_{cr} + 3)$. Thus, for example, for U + Cm with $Z_u = 188$, and with $Z_{cr} \simeq 173$, $\Gamma \simeq 9$ keV. For a Breit–Wigner shape, this leads to lifetime for this state of $\tau = \hbar/\Gamma \simeq 7 \times 10^{-20}$ s, which was the basis of the lifetime quoted in the discussions previously. Therefore, if superheavy composite systems can be formed with lifetimes of this duration, indeed, we can expect to observe very narrow line positron spectra from the spontaneous emission process.

In this connection, it bears emphasis to point out that the sensitivity of the decay width and energy of the $1s\sigma$ resonance to changes in the nuclear charge distribution is a potential tool for obtaining information on the shape and extension of nuclear molecular charge distributions during the time of formation of such systems. As an example of what may be expected, Figure 5.7 (Re 81) shows the dependence of the $1s\sigma$ resonance energy on the quadrupole deformation parameter β. The nuclear charge distribution has been assumed to have a sharp surface, $R = C[1 + \beta P_2(\cos\theta)]$, where $C = 1.2(A_1 + A_2)^{1/3}$ at $\beta = 0$ and is renormalized to conserve volume. Of course, this sensitivity is a two-edged sword, since it is then difficult to predict the binding energy and decay width without prior knowledge of the

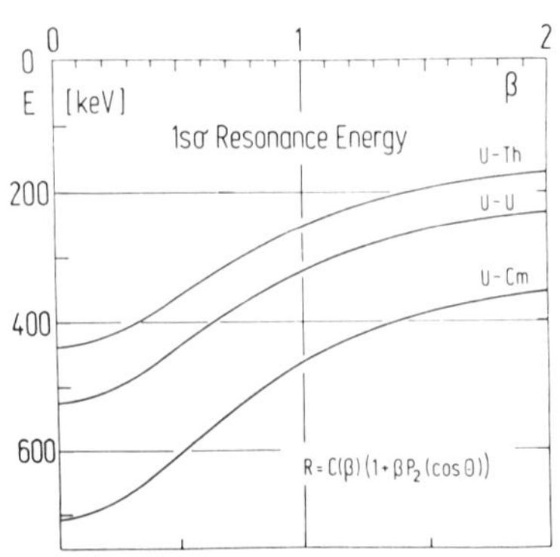

Figure 5.7. The energy of the $1s\sigma$ resonance for a deformed compound nucleus with a surface $R(\theta) = C(\beta)\ [1 + \beta P_2(\cos\theta)]$ as a function of the deformation parameter β. The radius parameter $C(\beta)$ is renormalized to conserve volume. No screening effects are included (Re 81).

nuclear physics. Reinhardt *et al.* (Re 81) also enumerate a number of other effects that can smear out the position and width of the positron line. These include (1) the variation of the shape and internuclear separation during contact, (2) different reactions, leading to different configurations, which may appear at the same scattering angles, (3) Zeeman splitting by the magnetic field created by the nuclear current, (4) fluctuation in the degree of electron screening, and (5) isomeric shape changes in the nuclear compound system. It is clear that this interplay of spontaneous positron production with nuclear phenomena can be of great interest to both fields, and we pursue this question further in the discussion of recent experimental results.

It is apparent, therefore, from the preceding discussion that, in the limit

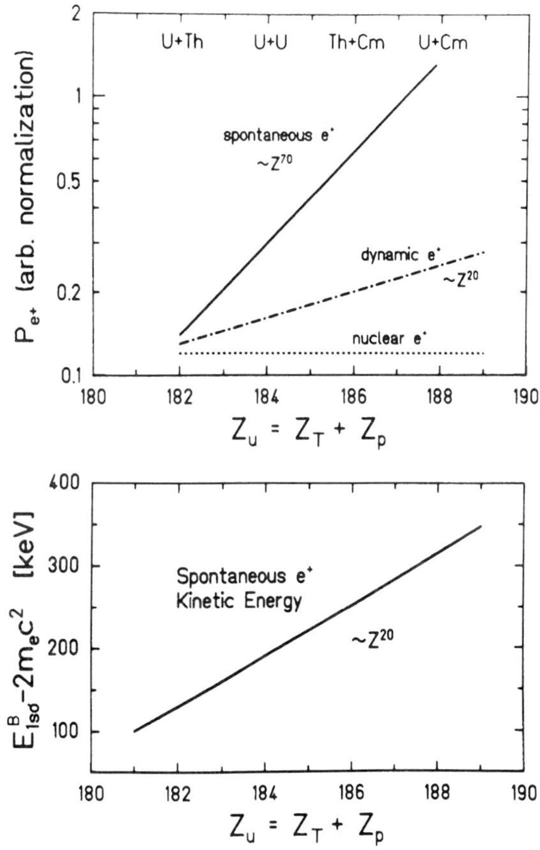

Figure 5.8. The variation of the relative intensity of positron emission and the positron energies for spontaneous emission as a function of Z_u for plausible collision systems at nuclear contact. Numbers are approximated from (Re 81) and (Re 81b) and measurements described herein.

of a long-lived supercritically bound system, there is not much ambiguity in the signature for spontaneous positron emission. Figure 5.8 summarizes the variation of the predicted decay widths and positron energies as a function of Z_u for plausible collision systems at nuclear contact. In this figure we have anticipated the backgrounds from continuous spectra taken from the discussions that follow. On the other hand, we have already emphasized that the signal for spontaneous positron emission can be modified considerably in quasiatoms due to the short time scales that can be associated with supercritical binding. The consequences of dealing with a dynamic system are explored in more detail in the following section.

5.2.1b. Dynamic Positron Creation. The similarities in the descriptions of pair production and electron excitation in superheavy collision systems were pointed out in Section 5.2.1. The principal difference is the introduction of the potential coupling term for pair creation when the resonance state appears in the negative energy continuum. If this complication is excluded by considering, for the time being, only collision systems that do not achieve supercritical binding, all the essential features that were noted previously for electron excitation and δ-ray emission can be carried over to pair production.

The initial state now becomes the antiparticle continuum which behaves, close to the nucleus, very much like the bound states except for the Coulomb repulsion factor for positrons. As before, there is a strong increase in the matrix element of $\partial/\partial R$ at small internuclear distances due to the contraction of the electron and positron wave functions as the nuclear charge distribution collapses. This increase is even steeper for pair emission than for the ionization process, although the absolute values of the pair production matrix elements are suppressed by about one order of magnitude by the Coulomb repulsion of the positron wave functions. A strong dependence of the matrix element on Z_u also exhibits the relativistic collapse of the wave functions. Using the monopole contribution alone remains a good approximation since, as for the corresponding ionization process, the strength of pair creation is concentrated at small internuclear separations. Multistep processes play an important role here as they do in vacancy production since there is a strong coupling between the lower continuum and deeply bound discrete states which, in turn, can be emptied by strong coupling to other bound states or to the positive energy continuum. Electron shielding affects the hole formation as well as the positron excitation. All these features have been described in a number of publications (So 77, Re 78, Re 80, Re 81b, Ba 81b, To 82) where the required amplitudes for positron emission have been calculated in first- and second-order perturbation theory and also obtained as numerical solutions of the coupled-channels differential equation (5.1). The assumptions adopted in the theory and its limitations are also discussed in these articles.

Coupled-channels calculations are required to give reliable results, but, as found in the case of vacancy formation and delta-electron emission, a great deal of the physics is exhibited by the results obtained from first-order time-dependent perturbation theory. The first-order amplitude can only couple the positron continuum to continuum electrons and bound electron states that are not occupied. Inner-shell states, originally occupied, are not involved in the first-order amplitude. It, therefore, represents the direct pair production process in Figure 5.2 which was first calculated by Soff *et al.* (So 77). From paralleling the discussion on ionization and delta-electron emission in Sections 3 and 4, it becomes apparent that the first-order amplitude is basically the Fourier transform of the matrix element which reflects the matching of the typical collision frequencies to the energy transferred in the pair production process. The same scaling behavior as found in electron ejection results in this case by using the monopole approximation and taking advantage of the $1/R$ dependence of $\langle \Phi_{e^-} | \partial/\partial R | \Phi_{e^+} \rangle$ for small R (Mu 78, Mu 83, Ka 80). The principal difference in comparison to delta-electron emission is that the Coulomb repulsion attenuates the wave functions associated with low energy positrons. The differential emission probability for positrons of kinetic energy E_{e^+} can be written as

$$\frac{dP_{e^+}}{dE_{e^+}} = g(E_{e^+}, \tau_0, Z_u) \exp\left[-2\frac{\tau_0}{\hbar}(E_{e^+} + 2mc^2)\right] \tag{5.4}$$

The main qualitative features of dynamic positron emission are displayed by this relation. The spectrum falls off exponentially for high energies due to the increasing energy $(E_{e^+} + 2mc^2)$ that must be transferred. The intensity drop-off for low-energy positrons, due to the repulsion by the Coulomb field of the nucleus, is contained in the function $g(E_{e^+}, \tau_0, Z_u)$, which also reflects the coupling strength. As shown in the Soff *et al.* (So 77) calculations, the energy distribution of the positrons from 5.8 MeV/u Pb + Pb collisions is bell-shaped, peaking at about 400 keV where the energy transfer matches a typical collision frequency given by $1/\tau_0 \simeq v/R_0$, where R_0 is the distance of closest approach in the collision. Unlike the ionization process the minimum energy transferred in pair creation is $2mc^2$. This sets a very narrow spatial confinement for the process to the vicinity where the colliding nuclei are very close together. Thus typically R_0 is less than 20 fm, as can be seen from the $\exp(-2mc^2R_0/\hbar v)$ term.

Two additional qualitative features stand out. One is the large cross section and the other is its previously noted strong dependence on the nuclear charge. Above $Z_u = 160$, the cross section increases approximately as Z_u^{20}. Both properties are due to the unusually rapid growth of distortion of the continuum states with increasing Z_u at small internuclear distances where the electric fields produced by the combined action of the two charges are very strong. These properties can only be obtained in a quasimolecular picture of the collision with the superheavy quasiatom as its limiting case. As such, this experimental verification, therefore, constitutes an important element in establishing the possibility of producing supercritical binding in such collisions, and points out the interest even in subcritical systems as a check on the underlying theory.

The essence of the results obtained with the first-order amplitude is substantially unaltered when higher-order excitations are considered (Re 78, Re 80, Ba 81b). Second-order calculations include the other process illustrated in Figure 5.2 where a vacancy is first created in an inner shell and it is subsequently filled by a transition from the positron continuum. As before, an electron and positron are found in the final state. Pair creation in subcritical systems has been calculated with direct and two-step transitions via the three innermost bound states involving the $s_{1/2}$ and $p_{1/2}$ angular momentum channels (Re 78, Re 80). As noted in these papers, these are expected to be the dominant contributing channels since they are the most severely distorted by the strong quasimolecular potentials. Using the monopole approximation, as was done in these calculations, decouples the s and p channels since they possess different parities. Clearly in this approximation no information emerges on a possible anisotropy in the emission of the positrons. It is of interest to note that in the $1s_{1/2}$ state the direct and two-step processes interfere destructively; the phase difference increases with increasing electron energy, with decreasing positron kinetic energy, and with increasing binding energy of the intermediate bound state (Re 80). Other principal features can be summarized as follows (Re 81b).

The $s_{1/2}$ and $p_{1/2}$ channels yield approximately equal contributions. It is also interesting that, although of higher order, the two-step induced excitations via the $s_{1/2}$ and innermost bound states are comparable in magnitude to the direct excitation process. Figure 5.9 (Re 80) illustrates the variation of the two amplitudes as a function of the total charge $(Z_T + Z_p)$ for 180° scattering and a minimum distance of approach of 20.9 fm. Since in the strongly relativistic domain the binding energies of the inner shells approach the lower continuum as Z_u is increased, a steep rise in positron production is expected for the induced process as Z_u approaches Z_{cr}. The induced process has the steeper rise with Z_u and constitutes the major fraction of positron production for $Z_u > 175$. Of course, in any experiment the one- and two-step processes are not distinguishable since we only measure the total coherent sum plotted in Figure 5.9.

Probably one of the more interesting results obtained from studying the second-order amplitudes is that the kinematic characteristics of the direct and two-step excitations are very similar (Ba 81b, Re 81b). It may be recalled that a similar conclusion was reached for electron excitations. As in that case, the important physical parameters in pair production seem to be

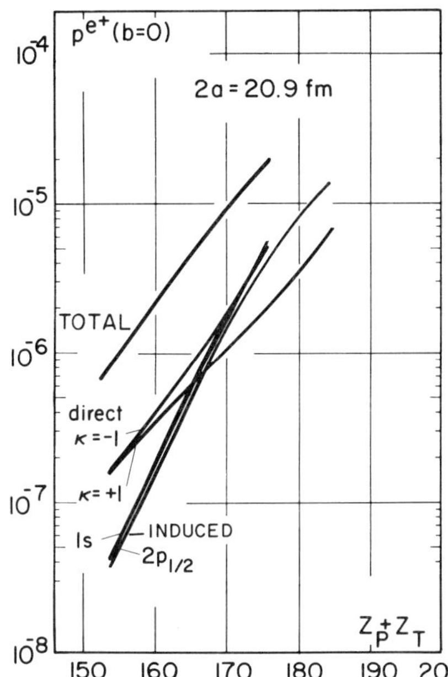

Figure 5.9. The contributions of the direct one-step and induced two-step dynamic positron production processes to the total positron emission as a function of the total charge Z_u (Re 80). The calculations use the monopole approximation and only $s(\kappa = -1)$ and $p(\kappa = +1)$ states are considered.

the total energy transferred and the scattering time which determine the exponential falloff of the spectrum. The fact that second-order and coupled-channel calculations (Re 81b, To 82, To 82b) only differ from the first-order theory by a constant intensity factor suggests that higher-order terms in the perturbation expansion contain the same dependence on these two parameters as the first-order term and that intermediate steps are of little consequence except to offer additional excitation routes.

Some sample coupled-channel calculations of energy spectra and the positron emission probability as a function of the minimum distance of approach for 180° scattering are included in Figure 5.10 for the subcritically bound systems Pb + Pb and Pb + U (Re 81b). Other such calculations are available in (To 82b). These results do not take into account the effect of electron screening which typically shifts the maximum to lower energies by about 50 keV and reduces the intensity by (5–15)% (Mu 83b, De 83), but does not otherwise alter the main features. Clearly, there is little to distinguish one system from another except the absolute emission probability. The shapes do not differ qualitatively from the perturbation results or from each other. The emission probabilities exhibit the almost pure exponential dependence on the minimum internuclear separation predicted by Equation (5.4) with the steep slope determined by the large energy transfer.

Therefore, for scattering systems that do not achieve supercritical binding, the positron spectra are characteristically uneventful. Except for the

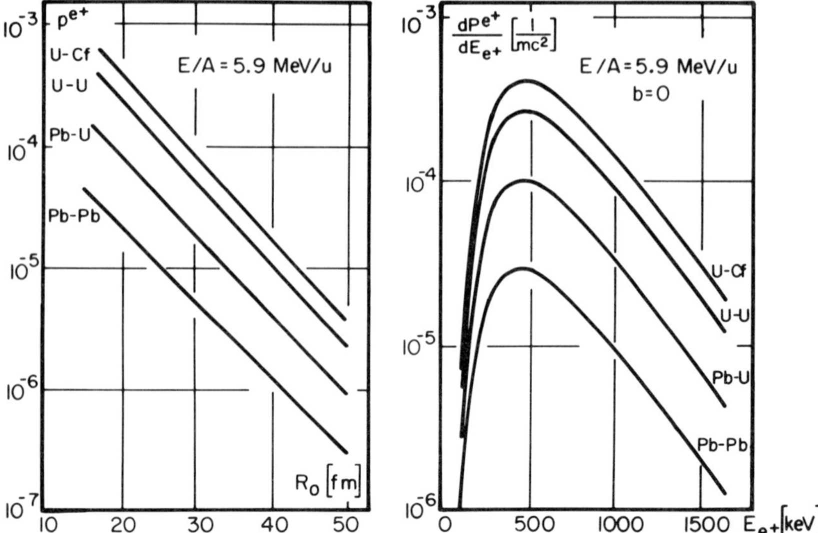

Figure 5.10. Calculated energy spectra and the total emission probability as a function of R_0 for positrons created in 5.9-MeV/u head-on collisions (Re 81b).

absolute intensity which reflects the Z_u of the system and collision kinematics, there is only a minor shift in the position of the maximum which is associated with the matching of the Fourier frequency components in the collision to corresponding components in the wave function as Z_u and the distance of closest approach change. However, there is some virtue in the fact that this continuum has a very constant shape, since as a background to spontaneous positron emission it can be readily identified. The question that naturally follows is, what detectable effects are expected from spontaneous positron emission as we pass from a subcritical to a supercritical quasiatomic system?

5.2.1c. Supercritical Collision Systems. It is apparent from the discussion associated with the Equation (5.1) that two considerations are particularly relevant in addressing the latter question. One concerns the relative strengths of the time derivative and potential couplings. The other is associated with the time development of the collision. The first governs the size of the signal above background that can be expected ultimately. However, the second has assumed a special importance in the current search for a believable signature for spontaneous positron emission for reasons which become evident from examining expectations for possible experimental scenarios.

If we confine ourselves to experiments involving scattering below the Coulomb barrier, the theory predicts a difficult experimental situation. (Without the benefit of the recent quantitative results, such experiments were first envisioned in order to minimize the background from nuclear origin.) We have previously discussed the qualitative effects of collision broadening which diffuses the spontaneous positron line. We have also noted the suppression of spontaneous positron emission by the short time available in the supercritical state relative to the natural time scale for the decay of the $1s\sigma$ state of $\sim 10^{-19}$ s. Indeed, the quantitative calculations confirm that the positron yields are dominated by the dynamic mechanisms if the nuclei follow Coulomb scattering trajectories. There is little to distinguish the contribution that spontaneous emission makes to the shape of a typical dynamic spectrum, except to alter the total intensity. As illustrated in Figure 5.10, the theory predicts that the features of positron emission evolve from subcritical to supercritical conditions without any major signature to mark a transition in crossing Z_{cr} or R_{cr}. There is no appreciable change in the spectra and there is no break or change in the slope of dP_{e^+}/dE_{e^+} with R. No qualitative effects can be expected. This lack of a threshold merely reflects the dynamical widths associated with both the induced and spontaneous amplitudes due to the short collision time and the small contribution from spontaneous emission. Thus, although nuclear background conditions may be favorable, an experiment carried out under conditions of sub-Coulomb barrier scattering has to rely on accurate absolute measurements and quantitative comparisons with theory at the several percent level. With the uncertainties encountered in calculating absolute probabilities when ionization states are uncertain, such comparisons are presently difficult even if the desired experimental accuracy can be obtained. It should be noted in this connection that when the experiments were first initiated the quantitative results presented in Figure 5.10 were not available so that the difficulty of the task to detect spontaneous positron emission in sub-Coulomb scattering was not appreciated fully.

Since the subordinate role played by the spontaneous component in sub-Coulomb collisions is an outgrowth of the short collision time which does not allow the spontaneous amplitude to develop, one way of altering this situation is to change the time development of the collision. More specifically, both coupling matrix elements enter in the positron production cross section via their Fourier transforms. Their relative contributions and their coherent sum, therefore, depend on how the collision evolves in time since the spontaneous and dynamic couplings exhibit characteristic, but different, functional dependences on the nuclear motion. The matrix element for spontaneous emission depends on static properties, e.g., the internuclear separation R. On the other hand, the dynamic coupling is only present if R is changing, i.e., it depends on the velocity $\dot{R}$. It can, therefore, then be expected that if the collision can be interrupted for some delay time during which the quasimolecular $1s\sigma$ orbital is in a state of overcritical binding, spontaneous positron emission will be enhanced since, during the delay interval, the radial velocity is zero and only the potential coupling between the $1s\sigma$ state and the positron continuum is effective. But equally important, a prolonged time, T, spent in overcritical binding not only changes the ratio of spontaneous to dynamic components, but, for sufficiently long delay times approaching the natural lifetime for spontaneous decay, the spontaneous emission component evolves a narrow line spectrum reflecting $\hbar/T$; this line can provide the clear signal sought above the dynamic continuum spectrum. For this reason Rafelski, Müller, and Greiner (Ra 78b) suggested using the finite delay times in nuclear reactions to explore this possibility.

Of course, there are several questions that arise in connection with this suggestion: (1) What time delays can be achieved in superheavy collision systems? (2) What is the form and distribution of the time delay? (3) What reactions should be explored? (4) What are the quantitative implications of time delay? (5) How does time delay affect the dynamically produced spectrum? The first three questions can only be answered by experiment. The latter two have been investigated using several models for a nuclear reaction.

Many of the salient features introduced by prolonging the time for supercritical binding are demonstrated in a simple modeling of the deviations from Coulomb trajectories examined by Reinhardt *et al.* (Re 81, Mu 83b). As illustrated in Figure 5.11, between the two hyperbolic Rutherford trajectories, followed by the incoming and outgoing ions, these authors introduce a well-defined delay time T at the distance of minimum approach R_0 during which the nuclei are in contact with each other. It is assumed, furthermore, that the nuclear configuration is fixed during this nuclear reaction period so that the binding energy of the $1s\sigma$ does not vary for the time T. Some of the interesting results obtained with this schematic model for the collision are displayed in Figure 5.11 (Mu 83b).

In a subcritical collision system the delay time T produces oscillations in the positron energy spectra due to the interference between the dynamic excitation amplitudes from the incoming and outgoing parts of the Rutherford trajectories. Their relative phases depend on the positron kinetic energy and the delay time T. The interference pattern is characterized by peak separation $\Delta E \simeq 2\pi\hbar/T$. If the system is supercritical we obtain, in addition to the oscillating pattern, a general enhancement of the positron production which funnels into a distinct peak located at the resonance energy position. The source of this enhancement is spontaneous positron emission. For values of T that are larger than the Coulomb scattering time but smaller than the spontaneous decay time, Reinhardt *et al.* (Re 81) show that positrons produced exclusively during the contact period T have an energy spectrum described in first-order perturbation theory by

$$\frac{dP_{e^+}}{dE_{e^+}} \simeq P_{1s\sigma} \frac{\Gamma}{2\pi} T^2 \left[\frac{\sin[(E_{e^+} - E_{1s\sigma})T/2\hbar]}{(E_{e^+} - E_{1s\sigma})T/2\hbar} \right]^2 \tag{5.5}$$

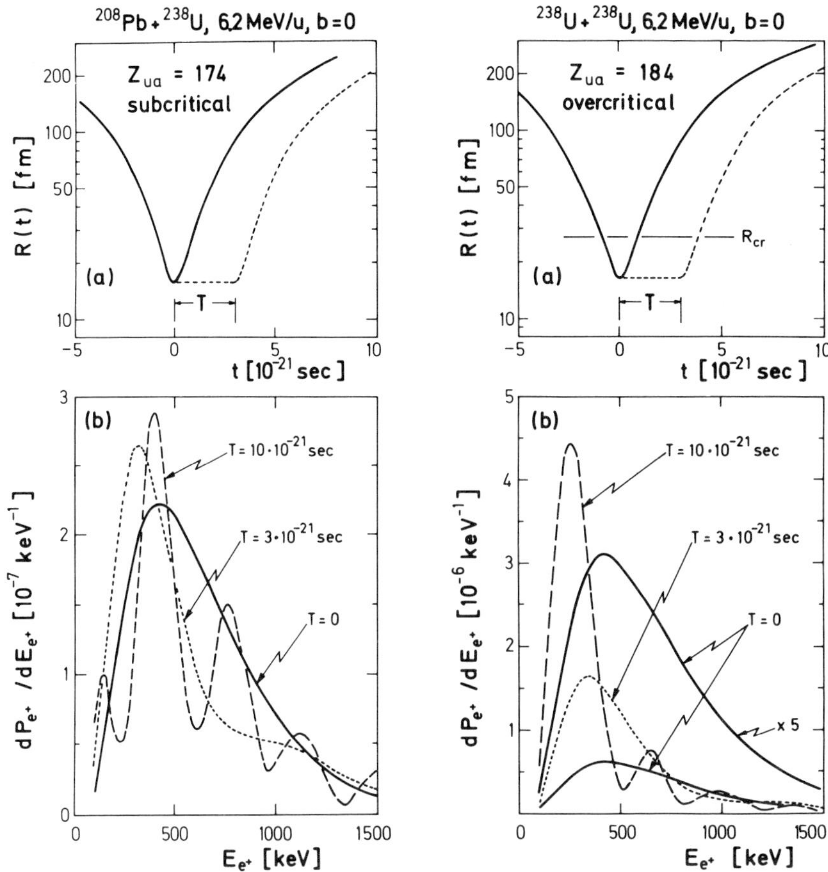

Figure 5.11. The influence of a fixed nuclear time delay T on positron spectra is displayed for systems with subcritical (U + Pb) and supercritical (U + U) binding (Mu 83b). In the supercritical system, with the growing time delay a narrow line appears due to the spontaneous decay of the vacuum which gradually approaches the position expected for the $1s\sigma$ resonance.

if a $1s\sigma$ vacancy is available with probability $P_{1s\sigma}$. Thus, due to potential coupling alone the spectrum of positrons becomes increasingly narrow as the delay time increases, and, in the limit of very long times for the reaction, the width approaches the natural width Γ discussed in Section 5.2.1a. For a well-defined delay $T \ll \hbar/\Gamma$, the full width half-mass (FWHM) is given by $\Delta E \simeq 5.56\hbar/T$ (Re 81), while the total intensity grows linearly with T, $P_{e^+} \simeq P_{1s\sigma}\Gamma T/\hbar$. As illustrated in Figure 5.11, a full coupled-channel calculation shows that with increasing delay time the position of the peak shifts from the first kinematic maximum to the energy of the spontaneous peak located at

$[E^B_{1s\sigma}(R_0) - 2mc^2]$ corresponding to the distance of minimum internuclear separation R_0.

The appearance of the peak due to spontaneous positron emission with the introduction of a fixed time delay is interesting but intuitive. But more interesting and less intuitive is that the peak can survive under more reasonable scenarios of modeling the nuclear reaction. It is not expected that the fixed time delay reflects real nuclear reaction times. More likely, the reaction time will be represented by a distribution $f(T)$. The effect of a distribution of delay times has been examined by introducing various forms of $f(T)$ in a further development of the calculations just considered (Re 83, Mu 83b), and by another approach using the autocorrelation function of the s matrix for the heavy-ion-induced reaction (To 83, To 83b). [Reinhardt *et al.* discussed the equivalence of these two approaches under specific conditions (Re 83).] Generally, it is found that even for a small spread in T, represented for example by a Gaussian distribution, the oscillating structure due to interference is damped out (Mu 83b). Indeed for an exponential dependence $f(T) = 1/\tau e^{-T/\tau}$, as would be obtained for a single resonance or a random distribution of resonances, this damping is complete and leads to a smooth spectrum. In contrast, in superheavy collision systems the first peak, associated with spontaneous positron emission, strikingly survives irrespective of the distribution in T so long as the mean value of T is sufficiently large. In fact, for an exponentially decaying state, i.e., for a Breit–Wigner line shape, the width of the line, given by $\Delta E \simeq 2\hbar/\tau$, is much narrower than that for a fixed delay $T = \tau(\Delta E \simeq 5.56\hbar/T)$. It also bears emphasis that the survival of the spontaneous peak is independent of the assumptions made regarding $R(t)$ on the incoming and outgoing path, and on the dissipation of nuclear kinetic energy and angular momentum in the reaction (Re 81). However, the peak is subject to dissipation due to effects already enumerated in the discussion on the resonance in Section 5.2.1a. It was also noted that the position of the peak is a sensitive mirror of the nuclear separation and the nuclear charge distribution.

Of course, the nuclear reaction part of the scattering does not determine the entire positron spectrum. Rutherford scattering also always contributes to the spectrum so that the total spectrum consists of an incoherent sum of positron emission from Rutherford scattering and nuclear reactions. If it is assumed that only the latter two mechanisms make up the positron spectrum, then the differential probability for emitting a positron of energy E_{e^+} with a scattering into an angle θ is defined as

$$\frac{dP_{e^+}(E_{e^+}, \theta)}{dE_{e^+}} = \frac{\dfrac{d\sigma^N(\theta)}{d\Omega}\dfrac{dP^N_{e^+}(\theta)}{dE_{e^+}} + \dfrac{d\sigma^R(\theta)}{d\Omega}\dfrac{dP^R_{e^+}(\theta)}{dE_{e^+}}}{\dfrac{d\sigma^N(\theta)}{d\Omega} + \dfrac{d\sigma^R(\theta)}{d\Omega}} \tag{5.6}$$

The Rutherford cross section usually overwhelms the nuclear cross section, i.e., $d\sigma^R/d\Omega \gg d\sigma^N/d\Omega$; then

$$\frac{dP_{e^+}(E_{e^+}, \theta)}{dE_{e^+}} \simeq q \left[\frac{dP^N_{e^+}(\theta)}{dE_{e^+}}\right] + \left[\frac{dP^R_{e^+}(\theta)}{dE_{e^+}}\right] \qquad (5.7)$$

where

$$q \equiv \frac{d\sigma^N}{d\Omega} \bigg/ \frac{d\sigma^R}{d\Omega}$$

Obviously, whether a spontaneous line spectrum is detectable above the continuum background depends on q, on the lifetime of the composite system which allows the decay to develop and the peak to build up into a sharp line signal, and on the probability of a $1s\sigma$ vacancy being present. For long nuclear lifetimes $>10^{-19}$ s, the spontaneous emission probability, $P^N_{e^+}$, can be as large as 10^{-2} with a peak width $\ll 100$ keV (Mu 83b). Under such circumstances, the fraction of collisions that lead to desirable nuclear reactions can be very small and still lead to a prominant signal above the underlying background from Rutherford scattering.

Therefore, in principle, nuclear reactions may be a filter for identifying spontaneous positron emission. At least these speculations provide interesting new prospects to be explored. On a historical point it should be noted that the original suggestion visualized reactions above the barrier such as deep inelastic scattering. Delay times in such reactions have in the meantime proved to be too short, so that it has come as a gratifying surprise to find that appropriate nuclear reaction conditions may be met near the Coulomb barrier where background conditions are more advantageous. These considerations will be discussed further in the sections that follow.

5.2.2. Positrons from Nuclear Transitions

In considering positron emission from atomic related sources the nucleus is merely viewed as the seat for the Coulomb field. On the other hand, it is not an inert object and, as mentioned briefly previously, it can be a major source of positrons in heavy-ion collisions. Even for sub-Coulomb barrier collisions multiple-Coulomb excitation can excite either the projectile or target. If we enter the realm of nuclear reactions as suggested by the discussion in the preceding section, then it can be expected that nuclear excitation will become even more prominent. The major source of positrons from nuclear transitions is internal pair conversion of gamma-ray transitions and beta decay. Unlike the atomic processes of positron emission in heavy-ion collisions which are just currently being examined, internal pair conversion and, of course, beta decay are well documented and understood.

It is a little surprising, however, to find that calculations on internal pair conversion were fragmentary until very recently when the experiments under discussion here motivated some comprehensive calculations (Sc 78, Sc 79, So 81c).

Beta decay can be discounted as a relevant background to these experiments. Because of the range of lifetimes associated with this weak decay process, experimental geometries and timing requirements eliminate it in any in-beam measurements. Internal pair conversion occurs whenever the deexcitation energy, E_x, in nuclear transitions taking place in the final state nuclei exceeds $2mc^2$; its intensity reflects the multipolarity of the gamma-ray transition. Because the lifetimes associated with the internal pair conversion process vary from 10^{-12} to 10^{-14} s, the positrons are emitted in the immediate vicinity of the target and cannot be differentiated spatially or by time selection from the atomic processes. The experiments, therefore, acquire the following task: if the measured spectra are continuous, the question arises as to what fraction of these spectra is contributed by internal pair creation. A continuous spectrum can be produced by a continuum of transitions which would, for example, be mirrored in continuous gamma-ray spectra emitted concurrently from the target. The accuracy to which such a background subtraction can be carried out then becomes an essential point. But the more compelling situation occurs if peak structure is observed in the positron spectra, and the question then arises whether nuclear transitions can account for this structure. We assemble here some of the elements that are involved in such considerations of the data.

The positron spectrum from internal pair conversion is also reflected in two other measurable quantities, (1) the gamma ray spectrum and (2) the electrons from internal conversion of the inner atomic shells. The central point which is utilized experimentally is that all three processes are related by branching ratios which, to a very good approximation, can be calculated without reference to the nuclear process since the nuclear transition matrix element is common to all. Thus, in principle, measured gamma-ray and electron spectra can be used to construct the positron spectra if the multipolarity is known (Me 77b). Except for electric monopole transitions (E0) for which gamma-ray emission is forbidden, electric multipole (EL) and magnetic multipole (ML) transitions provide two independent pieces of experimental information which can be cross-checked.

More specifically the branching ratios for pair emission and K-electron conversion can be defined if the number of K vacancies are known after the collision. In this it is assumed implicitly that the number of vacancies does not change by other means than by the conversion process, e.g., electron capture or radiative decay during the lifetime of the excited state. In addition to gamma-ray emission with probability P_γ, we consider the three principal conversion processes:

a. In the ordinary internal conversion of the K shell with probability P_{Ke^-}, a monoenergetic electron is ejected with energy $E_x - |E_B^K|$, where E_B^K is the binding energy of the K shell. Of course, other shells can also be converted but with a probability which is usually smaller by more than an order of magnitude. Because of the long nuclear lifetime, the linewidth is dominated by instrumental effects.

b. One pair conversion process, designated herewith probability $P_{e^+e^-}$, involves the transfer of the energy of the transition to the positron continuum resulting in the production of a continuum electron and a positron. The energy E_x is shared by the electron, positron, and recoiling nucleus so that, neglecting the energy carried off by the nucleus, the positron energy spectrum extends from zero energy to the maximum available energy, $E_x - 2mc^2$. In particular, for high-Z nuclei like uranium the Coulomb field repulsion leads to the triangular shaped energy spectra illustrated in Figure 5.12 (Sc 83), with a sharp cutoff at the maximum energy. The half-width of the distribution is $\sim$150 keV and it varies little with multipolarity. This characteristic feature should be readily recognizable.

c. Internal pair conversion can also lead to monoenergetic positron emission. In this process internal pair conversion is followed by the capture

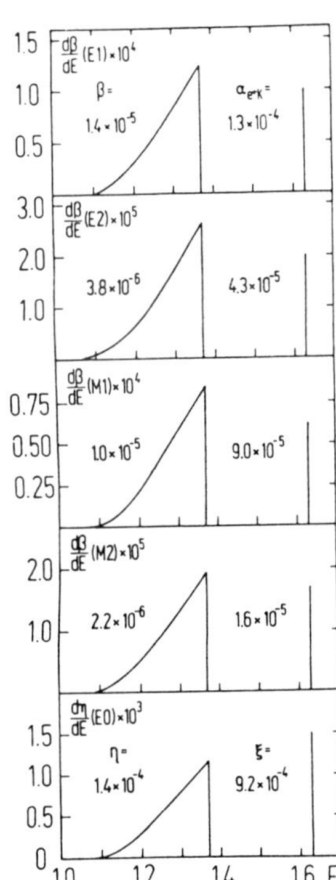

Figure 5.12. Positron spectra $d\beta/dE$ and $d\eta/dE$ from internal pair conversion of a 1.34-MeV transition in the uranium nucleus. The multipolarities $E1$, $E2$, $M1$, $M2$, and $E0$ are shown. Moreover, the position of the monoenergetic positron line is illustrated. The internal conversion coefficients β, η, ξ, and α_{e+K} are defined in the text (Sc 83).

of the electron into empty atomic K or L orbits instead of the electron being emitted into the continuum (Sl 49, Pe 62, Lo 62, Wi 63, Sh 64, Sc 83). Each atomic shell produces monoenergetic positrons with an energy $E_x - 2mc^2 + |E_B^{K,L}|$. The probability $P_{K,L_{e+}}$ of emitting such line spectra, of course, depends on the presence of a vacant bound state whose wave function has a large overlap with the nuclear volume. Capture into a K-shell orbital is most probable. Not only must such a vacancy be produced in the collision, but it must be available for a time comparable to the lifetime, τ_N, of the nuclear state since the probability of emitting the monoenergetic positron is proportional to $\tau_K/(\tau_K + \tau_N)$, where τ_K is the lifetime of the K vacancy. For a neutral atom, such as uranium, $\tau_K \simeq 10^{-17}$ is several orders of magnitude shorter than that of the competing nuclear transition so that this process, normally, is highly suppressed relative to ordinary internal pair conversion. Obviously, a mechanism which lengthens the vacancy lifetime so that τ_K approaches τ_N would enhance the emission of these nuclear monoenergetic positrons. It should also be noted that ordinary pair conversion always accompanies the line spectrum and internal electron conversion is also present unless the inner shells are completely vacant. As shown in Figure 5.12, the most favorable conditions for observing monochromatic positrons occur for E0 transitions. In this case for $\tau_K > \tau_N$ the ratio of line to continuum intensities is expected to be approximately 4:1. However, for this E0 transition $\tau_N \simeq 10^{-12}$ s so that, clearly, monochromatic positron emission has to be considered an improbable process which can occur only under exceptional circumstances when the vacancy lifetime is prolonged by several orders of magnitude.

Following Schlüter *et al.* (Sc 83), we can relate the three processes through the branching ratios and the number of K vacancies x $(0 < x \leqslant 2)$. For EL and ML transitions the branching ratios are defined by

$$\begin{aligned} P_{e^+e^-} &= \beta P_\gamma \\ P_{Ke^+} &= (x/2)\,\alpha_{e^+K} P_\gamma \\ P_{Ke^-} &= (1 - x/2)\,\alpha_{Ke^-} P_\gamma \end{aligned} \tag{5.8}$$

and for E0 where $P_\gamma = 0$, we redefine the last two relationships by

$$\begin{aligned} P_{e^+e^-} &= 1/(1 - x/2)\,\eta P_{Ke^-} \\ P_{Ke^+} &= (x/2)(1 - x/2)\,\xi P_{Ke^-} \end{aligned} \tag{5.9}$$

As an example, Figure 5.12 compares the coefficients for ordinary and monoenergetic pair conversion for a transition energy near one of experimental interest (Sc 83).

Generally we see that the positron spectra from individual transitions in high-Z nuclei like uranium have structure. The structure is broad for ordinary pair conversion, and it can be narrow for the rare process of monoenergetic pair conversion. If there are strong individual transitions well separated in energy, even the broad structure may be visible in the positron spectra. However, in superheavy collision systems we mainly deal with both target and projectile nuclei that are soft to excitation so that the spectrum of gamma-ray transition energies is principally continuous, as is demonstrated in the data to be discussed. Then ignoring for simplicity the monoenergetic conversion, the positron spectrum can be formed from the gamma-ray spectrum with the following folding integral:

$$\frac{dN_{e^+}(E_{e^+})}{dE_{e^+}} = \int_{2mc^2}^{E_c} dE_\gamma \frac{dN_\gamma(E_\gamma)}{dE_\gamma} \frac{d\beta(E_{e^+}, E_\gamma)}{dE_{e^+}} \tag{5.10}$$

where $d\beta/dE_{e^+}$ depends on the multipolarity, and E_c is a convenient cutoff energy for the gamma-ray spectrum.

As an example Figure 5.13 shows a typical measured gamma-ray spectrum and the calculated positron spectra assuming pure $E1$ or $E2$ multipolarity for the transitions. The main feature of this positron background is that it is continuous and smooth. The spectral distribution

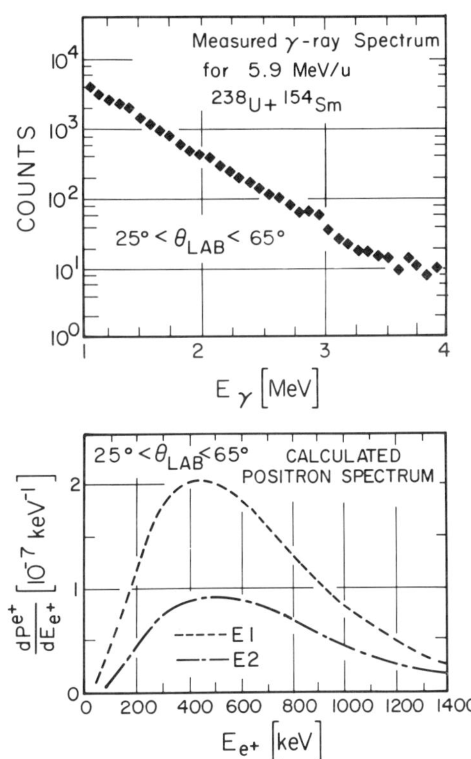

Figure 5.13. A typical measured γ-ray spectrum from $^{238}U + X$ collisions, and a deduced positron spectrum for $E1$ and $E2$ multipolarity for the transition (Bo 81, Sc 83b).

spans an energy region similar to that associated with the dynamic positron emission process, up to a few MeV. The reason for the similarity between the two spectra is that, because of the steep decrease of the gamma-ray intensity with E_γ, the main contribution to the integral derives from an energy interval between ~1 and 2.5 MeV in the gamma-ray spectrum and it peaks near the gamma-ray energy of approximately 1.5–1.6 MeV. We note also that there is a considerable difference in the intensity of the deduced positron spectrum depending on the choice of either the $E1$ or $E2$ multipolarity, which suggests that a knowledge of the multipolarity mix in the gamma-ray spectrum can be an important factor in evaluating the nuclear positron background from the measured gamma rays. However, it will be shown in the discussion of the experimental results that this turns out not to be a major problem. The similarity in the nuclear structure of the collision systems studied, which leads to very similar gamma-ray spectra, makes it possible to deduce an effective multipolarity mix with some confidence.

The continuous spectra make up the major fraction of the gamma-ray spectra and, as we shall see, also the electron spectra. They are a consideration in determining the absolute positron yield, but have little effect on prejudicing the identification of peak structure in the positron spectrum, except, of course, in that they determine the absolute background level. The question of discrete structure in the positron spectrum is addressed more specifically through the connection of the positron yield to the gamma-ray and electron yields implied by Equations (5.8) and (5.9). Therefore, good measurements of the gamma-ray and electron spectra form an integral part of the search for spontaneous positron emission. Particularly, the question regarding the contribution of strong isolated E0 transitions to the positron spectrum can only be examined through the electron spectrum. These points are further discussed in the presentation of the experimental results.

This discussion has only been concerned with nuclear transitions in the final-state nuclei. However, it is conceivable that nuclear transitions can also occur in a composite superheavy system during the short time that it stays together. The high Z of the composite system speeds up the internal conversion processes so that they can become competitive with the lifetime of the composite system. For example, an electron internal conversion decay of a nuclear transition which involves an electron in a supercritically bound $1s$ resonance state leads eventually to spontaneous positron emission when this hole is filled by an electron from the lower continuum, if, of course, the nuclear state survives for a time on the order of the spontaneous positron decay time. Müller *et al.* (Mu 83b) discuss other such exotic processes that can lead to multiple peak structure. Nothing is known about the nuclear structure of the giant nuclei in question so that, at the present time, these processes are clearly very speculative and can only be considered as possible explanations for structure that can appear in the positron spectrum. It is also interesting to note that, alternatively, positron emission can be used as a probe of nuclear structure in the giant nuclear systems.

5.3. Positron Spectrometers

Because there had been very little previous experience in in-beam positron spectroscopy measurements before these investigations began, the instrumentation to study positron emission in heavy-ion collisions underwent a gradual development as some of the experimental problems were identified with the first measurements and also as some of the physics being sought was clarified. At the time the first measurements were planned and initiated in 1973–1976 at the GSI laboratory in Darmstadt, West Germany, the theory on atomic positron production was only partially developed and very little information was available regarding the gamma-ray and delta electron backgrounds that would be generated in a collision between two uranium atoms near the Coulomb barrier. Particularly, since the $1s$ vacancy formation was uncertain, the first experiments, therefore, were especially concerned with determining in a general way the level of signal to background. Since those earliest measurements the experimental arrangements have evolved to address more specific goals. In this section the principal characteristics of some of the currently employed positron spectrometers and their auxiliary equipment will be outlined.

The essential features that have to be incorporated into an experimental arrangement to study positron emission from superheavy quasiatoms, and particularly to search for spontaneous positron emission, are evident from our discussion of the properties of the three principal contributions to the positron spectrum. Of primary concern is the expected small overall production probability, which is typically 10^{-6} to 10^{-7} per keV. Experience with the delta-electron and x-ray measurements has shown that this low flux of positrons has to be detected in an overwhelming radiation field of gamma rays, electrons, and nuclear constituents in the vicinity of the target. Therefore, as a general problem, these conditions dictate an apparatus with a large efficiency for detecting the primary positrons, an unambiguous identification of positrons in the presence of the large electron background, and a very effective suppression of secondary positron production from external pair creation by gamma rays and high-energy electrons. Optimizing the efficiency for all detectors is an especially important consideration since not only the small cross section but also the requirement for differential information, emphasized by the theory, leads to small data accumulation rates, long irradiation times, and slow assimilation of information.

More specifically the theory identifies a number of features and measurement parameters which should be particularly included in the experimental arrangement. Some of the considerations here parallel those already mentioned in association with experiments on delta electrons. This is not unexpected since at least the dynamic effects of positron production are closely related to electron ejection and, indeed, involve the same physics.

For instance, common to both is the emphasis on a good definition of the heavy-ion trajectory, and, in the case of nuclear reactions, of the reaction kinematics. In both cases it is important to be able to select the elastic scattering events among background reaction channels and to specify the key parameters R_0 and τ_0 associated with the behavior of the wave functions and principal dynamic effects. But more importantly, good kinematic resolution may be especially rudimentary to pursuing the search for the spontaneous positron emission process. If a clear signal for spontaneous emission relies on a reaction which introduces a significant time delay when supercritical binding occurs, then a discriminating selection of these reaction products is required. Such an arrangement can, of course, become quite complex, involving multiparticle detection and identification. This problem has been addressed to varying degrees of sophistication in the experimental arrangements used. It should be recalled in this connection that for elastic Coulomb scattering it is sufficient to specify both scattering angles in order to determine $R(t)$ uniquely, and that two-body final-state quasielastic events can also be defined in this manner. We can, for example, visualize that a time delay can be introduced by a quasielastic scattering that proceeds through an isolated resonance or through many overlapping resonances (Re 81, Rh 83). For such reactions and for other two-body final-state reactions that are focused in emission angle there is an obvious advantage in a two-counter array with good angular resolution. However, it should be noted that binary events may not, *a priori*, be the most interesting ones.

There are also some demanding capabilities required of the positron detector. On the one hand, a broad-band pass is desirable both in energy and solid angle to accommodate the breadth of the dynamic spectrum up to ~1.5 MeV and the need for overall high efficiency mentioned previously. On the other hand, there also may be a need for good energy resolution. The latter feature, obviously, is not essential for studying sub-Coulomb barrier collisions except insofar as it facilitates the unfolding of the measured spectrum and defines the energy well enough so that the dynamic aspects associated with the minimum energy transfer $(E_e + 2mc^2)$ can be tested. Good resolution only becomes important with the possibility that peak structure can occur when a favorable nuclear reaction produces the necessary time delay. Ultimately, the goal is to measure the intrinsic line shapes of such peaks if they appear. In the limiting case of a very long-lived superheavy nuclear complex ($>10^{-19}$ s), the intrinsic linewidth of the spontaneous positron line can be <10 keV. Unfortunately, achieving the necessary energy resolution to investigate such very narrow lines is difficult in the face of meeting the dominant requirement for broad-band features. The problem is further augmented by the need to filter out the effects from the large flux of electrons and gamma rays and to provide a clear signal to identify the positrons. These conflicting requirements are being addressed by arrangements which separate the functions of broad-band acceptance and energy measurement.

There are several particular points to consider in connection with the search for peak structure: (1) The peaks are expected to appear in the energy region below ~400 keV where detection is most difficult. Beside the problems associated with detecting low-energy positrons with high resolution, there are additional difficulties introduced by the fact that electron and gamma spectra tend to increase exponentially with decreasing energy and, therefore, so does the positron background related to these radiations. Therefore, not only does the spectrometer have to possess high efficiency for positron collection at low energies, but concurrently it must exhibit excellent background rejection in this most relevant energy region with a proficiency which increases with decreasing positron energy. Obviously, under these circumstances the signature for the positron is especially important. (2) The intrinsic line shape of the spectrometer is also an important quantity. It has to be well defined and understood if information is going to be extracted regarding the intrinsic width and line shape of the peak. (3) The large solid angle for positron collection referred to above has significant consequences for peak detection. Since the positrons are emitted from fast-moving sources, the intrinsic line can be both shifted in energy and broadened by the Doppler effect. If, as desired, the acceptance of positron emission angles is large relative to the direction of the moving source, the Doppler broadening can be appreciable unless the positron and source directions are both specified. In fact, it will be shown that for some detector arrangements used, Doppler broadening dominates the contributions to the measured instrumental linewidth. Therefore, unless it can be unfolded from the measured line shape, it determines the effective energy resolution of the spectrometer. Of course, unfolding the Doppler broadening component from the measured peak structure involves knowing both the velocity and direction of the emitter.

On the other hand, the Doppler energy shift and broadening can be used to advantage to address a crucial part of the problem of nuclear background. In Section 5.2.2 we emphasized the need to simultaneously monitor the gamma-ray and electron spectra from which nuclear positron backgrounds can be inferred; gamma ray and electron detectors should be an integral part of the experimental arrangement. But, in addition, Doppler-shifted and broadened line spectra may be utilized, in principle, to identify the nuclear or atomic origin of the positrons by providing a measurement of the emitter velocity. This follows from the fact that the atomic positrons are associated with the quasimolecular system moving along the beam direction with the center-of-mass velocity, while the nuclear positrons are emitted by the individual scattered or recoiling nuclei, and their velocity depends on the angle of scattering and the reaction Q. The Doppler effects in the two cases can be quite different if advantageous geometries are exploited. In principle, therefore, this type of measurement can provide a very effective test, and probably the most convincing one, of the origin of the emitting source. This point is explored in more detail below.

Therefore, it is apparent from the discussion in the last sections that the studies of positron production in superheavy collision systems involve experiments that are multidimensional. The interpretation of the measurements, and, in particular, the identification of the origin of the positrons, relies heavily on detecting essentially all emissions from the heavy-ion collision. These include, electrons, gamma rays, and the products of the nuclear scattering, in addition to the primary positrons. Moreover, the need for differential information, emphasized by the theory, leads to multiparameter measurements for each radiation. The parameters associated with positron, electron, and gamma-ray emission include the energy and the angle of emission relative to the reaction plane or the beam direction. For the nuclear scattering products measuring the scattering angles, energies, mass, and charge exchanged and the Q transferred can provide the information required to isolate the selective reactions related to the interesting features of the positron spectrum. These measurements have to be carried out under varied input conditions to explore the positron production characteristics as a function of Z_u, R_0, bombarding energy, and nuclear species. For example, as illustrated in Figure 5.8, the characteristic dependences of the major positron mechanisms on Z_u can be used to advantage. There is an obvious complexity in including all the parameters mentioned in a single experiment. Nevertheless, many aspects of this multiparameter space have been and continue to be explored.

The detection systems presently in use feature some version of the following capabilities:

(1) A magnetic transport system removes the positrons and electrons from the target vicinity to a region of low gamma-ray intensity. The transport system acts either solely as a large solid angle device with broad energy bandpass, or it performs a partial momentum selection.

(2) A measurement is made of the positron energy by an auxiliary counter. It either provides the principal energy measurement or is used to subdivide a broader energy selection made by the magnetic transport system.

(3) A filter, combining geometry and a magnetic field, is used to discriminate against electrons in the positron channel.

(4) An auxiliary identification of positrons is provided by annihilation

radiation. One of the spectrometers ("Orange") does not possess this feature but relies exclusively on magnetic field selection.

(5) The scattered ions are detected in heavy-ion avalanche counters. The detector arrays have so far been confined to two counters detecting binary events. The reasons are the initial interest in elastic scattering events and the prejudice of simplicity.

(6) Gamma-ray counters are used to monitor nuclear reactions.

(7) An electron channel is incorporated in some of the spectrometers allowing simultaneous measurements with positrons (EPOS and TORI). Separate measurements have been carried out in the "Orange" spectrometer by reversing the magnetic field.

(8) Monitoring of elastic scattering energy spectra in a high-resolution detector such as Si is used to follow the deterioration of the target. Target quality and deterioration may be an important consideration for nuclear reactions which exhibit a sharp resonancelike behavior on bombarding energy. If such reactions are responsible for notable features of the positron spectrum, then the degree to which these features stand out can depend on the average or even local target thickness. Measuring the elastic scattering peak shape can provide information on the target composition, although the sensitivity of the method is not clear.

Three spectrometers are currently in use. Since they represent improved versions of earlier devices, we will confine our discussion to these instruments.

5.3.1. The ORANGE Spectrometer

The basic features of the "Orange" spectrometer have been discussed in Section 4.3 in connection with the delta-electron experiments and are illustrated in Figure 4.8. The adaptation to positron emission studies involves mainly a modification of the positron (electron) detector at the focus of the spectrometer and a further development of the heavy-ion detector from the assembly described in Section 4.3. The configuration of the spectrometer and its auxiliary apparatus used recently in positron detection experiments at GSI, Darmstadt is shown schematically in Figure 5.14 (Ko 79, Ki 81, Cl 83, Cl 84).

The detector accepts a momentum interval of 14% over its length. If completely unobstructed so that access to the entire coil is available, the solid angle efficiency of the spectrometer is 15% of 4π in this momentum interval. Placing the heavy-ion detector in the position shown partially obstructs this solid angle acceptance so that typically the spectrometer has operated with approximately half this efficiency. The response function of the

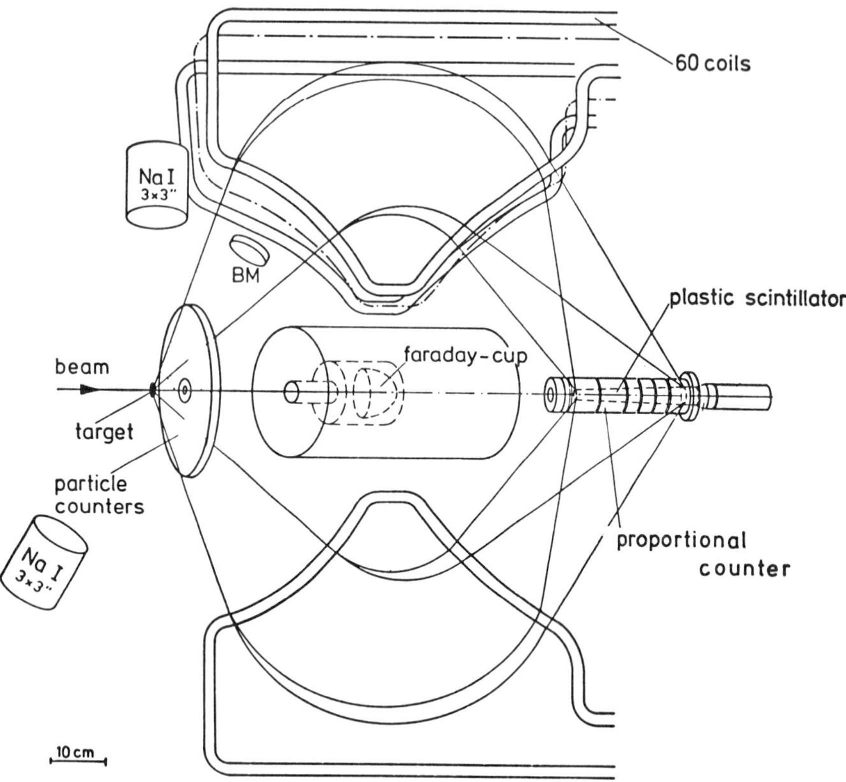

Figure 5.14. Schematic drawing of the experimental arrangement used with the iron-free "Orange" β-spectrometer [after (Ko 79, Ki 81, Cl 83)].

spectrometer is independent of the energy. The positron spectrum is accumulated by stepping the magnetic field, with normalization between steps provided by the elastic scattering intensity.

The positron detector is composed of a plastic scintillator, 12 in. long and shaped conically for good light collection into a photomultiplier. It is placed along the spectrometer symmetry axis; equal momentum bites translate into unequal lengths along the counter. The counter is thick enough to stop the positrons of interest producing a signal linearly proportional to the energy with a resolution of 17%. This resolution provides the discriminating signal against the continuous background. The background intensity increases with decreasing positron energy and sets a limit of >150–200 keV, below which measurements become very difficult. The main source of this background appears to be Compton scattering around the extensive shielding between the target, Faraday-cup areas, and the scintillation counter. Background suppression is aided by requiring a coincidence between the plastic scintillator and a transmission proportional counter which surrounds the plastic scintillator. The gas-proportional counter is subdivided into four (or six) rings to provide further momentum resolution. Each ring corresponds to a $\Delta P_{e^+}/P_{e^+} \simeq 0.03$ which for $E_{e^+} = 300$ keV translates to an energy resolution of $\simeq$15 keV.

As in the case of the delta-electron studies, the projectiles or recoil nuclei scattered from the target are detected by a parallel plate avalanche detector possessing an annular geometry centered at the beam axis. In one experiment it accepted scattering angles from 12° to 51°, so that, in that case, the positrons could be detected between 52.4° and 70° relative to the beam direction (Ki 81). The angle of scattering is measured by subdividing the anode into 16 concentric rings which can be read out independently to provide a high count-rate capability. Angular resolution of the order of 2° to 3° is achieved in this configuration, and kinematic coincidences can be detected over a limited range of ±5° around 45° scattering. With this angular resolution, it is not possible to use the kinematic coincidences in this small angular range to distinguish projectile and target recoil, unless the collision partners are very asymmetric.

NaI detectors, located at 90° and 150° to the beam axis, monitor the gamma-ray yield in coincidence with the scattered ions. Electron emission can be studied with the same spectrometer by reversing the magnetic field, but, of course, data on electrons cannot be accumulated simultaneously with the positron studies. This later point is mentioned because such simultaneous data collection can be important if reproducibility of the target condition is in question. We have noted previously a possible reason of concern for the target condition. In this apparatus target deterioration is monitored with a Si detector mounted at 45° to the beam axis.

5.3.2. Solenoidal Transport System

This electron-positron spectrometer ("EPOS") (Bo 81, Gr 82, Sc 83b) separates the function of broad-band collection of positrons (or electrons) both in solid angle and energy from the intrinsic energy measurement. The basic scheme was first utilized in studying the rare process of internal pair creation in beta decay (Gr 56), and in in-beam electron spectroscopy to study Coulomb excitation of nuclear states (Bu 66). The broad-band transport property is provided by a solenoidal magnetic field which focuses the positrons emitted from the target onto a high-resolution Si(Li) detector located in a region of low background. The Si(Li) detector then is the energy-measuring device and reflects the intrinsic energy resolution and line shape of the system. An interesting property of this transport system is that, with little sacrifice to efficiency,. the distance between target and detector, and therefore the reduction of neutral particle background related to target excitation, can be as large as money can buy by merely increasing the size of the system.

More specifically, the spectrometer exploits the two basic features of the solenoidal transport system, namely, the inverted spiraling directions of the oppositely charged electrons and positrons and the property that charged particles emitted from a target placed in a cylindrical symmetric magnetic field will always return to an axis contained by the point of emission and the field lines. This axis, therefore, is the natural choice for the geometric configuration of the charged-particle detector used to detect the emissions from the target (Gr 56, Bu 66). A detector placed along the field lines containing the target has the clear advantage of optimizing the efficiency while simultaneously providing high discrimination against particles which originate from points off the axis formed by the target and detector.

The arrangement used in EPOS is illustrated in Figure 5.15. The

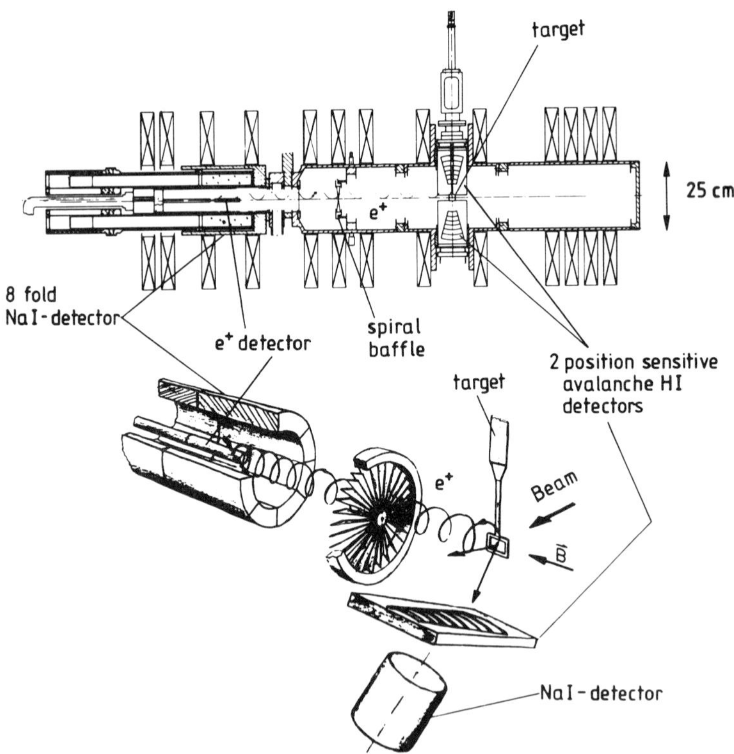

Figure 5.15. The experimental arrangement used with the EPOS solenoidal transport system. The lower panel is a perspective view of the main components. In the latter only one of the two heavy-ion detectors is shown for clarity (Bo 81, Gr 82, Sc 83b).

solenoidal magnetic field is oriented perpendicular to the beam axis and is produced by 13 "pancake" coils. The magnetic field is approximately uniform along the solenoid axis except in two areas where it is designed to maximize the transmission of positrons to the detector while suppressing the electron flux. One of these areas is used to increase the total efficiency by creating a magnetic mirror field on the side of the target opposite to the detector to reflect positrons (and electrons) towards the detector which are initially emitted in the other direction. With the magnetic mirror all positrons emitted with an emission angle relative to the solenoid axis less than $\theta_{max} = \pi - \sin^{-1}(B_T/B_0)^{1/2}$ are focused to the detector. Typical values for the field B_T at the position of the target and for the maximum mirror field B_0 result in $\theta_{max} < 137°$, corresponding to a solid angle of approximately 0.8 of 4π. Departure from a uniform field is also produced at a point about halfway between the target and the detector where a helical baffle is inserted with a spiraling direction set for positrons. The baffle performs the function of filtering out electrons coming from the target. We have seen that their intensity can exceed the positrons by several orders of magnitude in the energy region of interest. The rejection of electrons by the baffle and the transmission of positrons through the baffle are both enhanced by creating a depression in the magnetic field strength in the vicinity of the baffle. As shown in the left-hand panel of Figure 5.16, the electron rejection efficiency varies with energy, increasing sharply for low-energy electrons. This is just the functional dependence on the energy desired since the delta-electron production increases exponentially at low energies. With this arrangement the piling up of electrons onto positron events due to electron leak-through to the detector has been reduced to negligible proportions. Significant

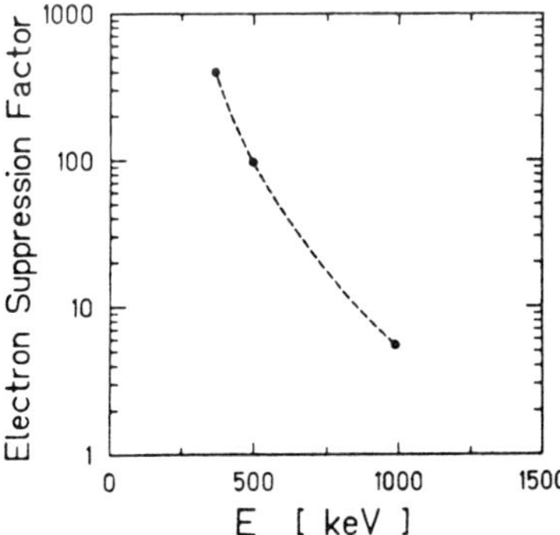

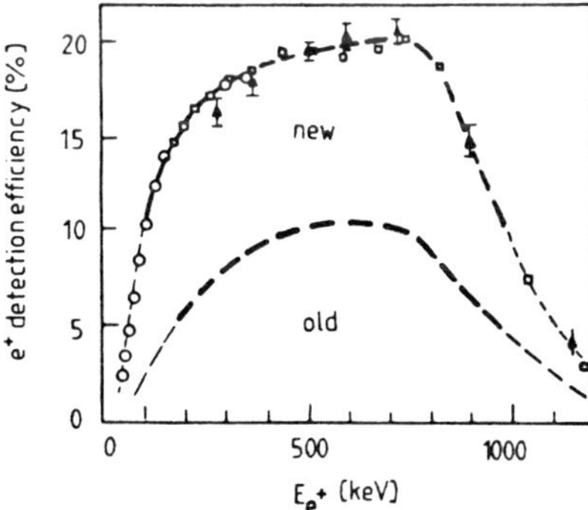

Figure 5.16. The left and right panels show the electron suppression factor and the positron detection efficiency, respectively, as a function of the positron energy for the EPOS system. The detection efficiency is shown for two designs of the baffle and Si(Li) detector geometry used in experiments (Bo 81, Gr 82, Co 84).

production of pairs by high-energy electrons from the target striking the detector can also be discounted.

The Si(Li) counter is 1 cm in diameter and 10 cm long. It is cooled to liquid-nitrogen temperatures and possesses an intrinsic energy resolution of $\simeq$10 keV at $E_{e^+} \simeq 300$ keV with a peak-to-total ratio of $\simeq$0.7. As noted above, the pencillike detector and the axial geometry used greatly reduce the background due to positrons or electrons that are scattered on the way to the detector from the target or that originate in places other than the target. Thus, the coaxial arrangement of magnetic field, detector, and target greatly enhances the function of the baffle by rejecting positrons and electrons scattered or created in the baffle off-axis. Similarly, positrons and electrons created by gamma rays in the vacuum chamber walls have little chance for detection, while the small diameter counter presents little direct profile to gamma rays and neutrons from the target.

To enhance the function of the baffle, positive identification of positrons is achieved by requiring a coincidence between the Si(Li) detector and at least one of the two 511-keV positron annihilation quanta. The annihilation radiation is detected in an eight-segment cylindrical NaI detector array surrounding the positron counter coaxially. By summing the signals of the eight separate NaI crystals and requiring a total energy sum $\gtrsim$ 440 keV, the NaI efficiency attains a value of $\simeq$58% per positron detected in the Si(Li) counter.

Figure 5.16 shows total efficiencies for detecting target produced positrons at a selected magnetic field for two designs of the baffle and Si(Li) counter that have been used in experiments. The achievement of a high efficiency over a broad band of energies enables this experimental arrangement to study, at one magnetic field setting, essentially the whole range of positron energies suggested by theory to be of interest. The loss of efficiency at low energy is due to the interception of positrons with a small gyration radius by obstacles along the target–detector axis such as the inner core of the baffle. The high-energy cutoff in efficiency is determined by the maximum gyration radius that is accepted by the inner diameter of the NaI array. The dimensions and magnetic field can obviously be scaled to widen the bandpass and decrease the direct solid angle subtended by the Si(Li) counter at the target without loss in peak efficiency. As noted before, the limitation is the cost.

Because of the importance of measuring the kinematics of the nuclear scattering, this experimental arrangement has concentrated on defining the binary events very well. Towards this purpose both the scattered projectile and the recoiling target nucleus are detected in two parallel-plate avalanche detectors with continuous delay line readout to determine the scattering angle θ (Fu 78). The detectors are positioned symmetrically with respect to the beam axis, with the plane bisecting the two counters perpendicular to the solenoid axis. The angular acceptance of each detector is $15° \leqslant \theta \leqslant 75°$ with a constant azimuthal angle interval $\Delta\phi = 60°$. At $\theta = 45°$ scattering, the angular resolution is $<1°$. In addition to the particle scattering-angle information, the time-of-flight difference between both particles as well as rough Z information is obtained via the energy loss of the particles in the detector.

It had been noted previously that precise scattering angle information on the two-body final state which can distinguish the projectile from the recoiling target in an asymmetric collision determines uniquely the impact parameter (or equivalently the distance of closest approach R_0) in a Coulomb scattering event. Figure 5.17 illustrates the excellent separation between forward and backward scattering events in the U + Pb collision obtained with these detectors. The detection system is capable of making this separation for a difference of 15 mass units. Information can also be obtained from the two-particle angle correlations on energy loss and/or mass transfer due to nuclear reactions. For example, in a symmetric collision system elastic scattering is characterized by $(\theta_1 + \theta_2) = 90°$, while quasielastic scattering leads to $(\theta_1 + \theta_2) < 90°$. For asymmetric collisions,

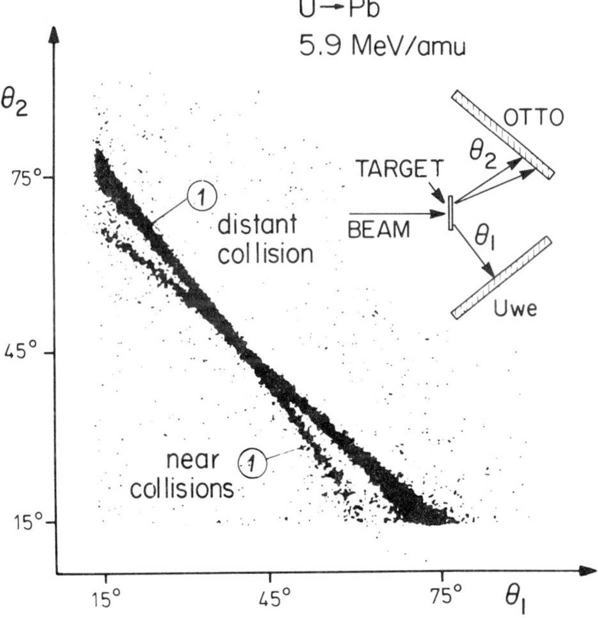

Figure 5.17. Response of the heavy-ion detectors used in the EPOS system to elastic U + Pb scattering. The intensity is shown as a function of the two scattering angles θ_1 and θ_2 observed in the lower and upper detector, respectively. The two elastic branches, representing distant and close collisions, can be clearly separated in this system. An event in the branch marked 1 corresponds to U being observed in the lower counter (Bo 81, Gr 82, Co 84).

$(\theta_1 + \theta_2)$ at constant $(\theta_1 - \theta_2)$ separates forward from backward elastic scattering, while the symmetry of the counters about the beam line translates to a symmetry of both elastic and inelastic events about $(\theta_1 - \theta_2) = 0$. Mass transfer also leads to a distortion of the kinematic (θ_1, θ_2) correlations from those observed in elastic scattering. The angular resolution achieved with good target and beam conditions is capable of an energy loss resolution of $\simeq$20 MeV at $\theta_1 = \theta_2 \approx 45°$. Using such information, the kinematic coincidences can be exploited together with the time of flight and rough Z identification to select desired reaction channels and to discriminate against backgrounds from the carbon target backing and/or covering foils.

The experiments with the EPOS positron detection system make use of NaI detectors to monitor the gamma rays emitted by the target, and, in the latest measurements, provision is also made to measure electron emission spectra concurrently with the positrons and gamma-rays. We have already emphasized the importance of making all the measurements under common target and beam conditions to ensure that the selection of the heavy-ion scattering via the two-body reaction kinematics is simultaneously correlated to all three radiation sources. The NaI detectors are positioned both forward and backward of the beam direction for purposes of checking on the Doppler effect and neutron emission, which is preferentially forward. It is found that the comparison between forward and backward counters very effectively identifies the neutron contribution to the gamma-ray spectra.

The simultaneous counting of electrons and positrons is achieved by exploiting the electrons emitted from the target that are not reflected by the magnetic mirror field toward the Si(Li) counter that counts the positrons. These electrons are also detected in a planar Si(Li) counter with intrinsic resolution similar to the positron counter. This resolution is somewhat degraded to ~20 to 30 keV by an absorber foil used to filter out the high flux of low-energy delta electrons. The transmission to the electron Si(Li) is on the order of 0.5% for the energies of interest above ~1200 keV, so that the ratio of positron to electron transmissions for electron energies $E_{e^-} \simeq E_{e^+} + 2mc^2$ is approximately 50. Since electron internal conversion of a nuclear transition exceeds internal pair conversion of the same transition by an even larger factor, typically two or three orders of magnitude, the data collection rate for electrons is still very adequate in spite of the small solid angle for electron acceptance. The selection of a small angular cone of $<15°$ around the perpendicular to the beam does have the beneficial effect of limiting the Doppler broadening so that for an electron with energy $E_{e^-} \simeq E_{e^+} + 2mc^2$ it is very similar to the corresponding broadening for the positron with energy E_{e^+}.

The EPOS system, therefore, samples many of the experimental parameters that theory suggests may be relevant. In addition, the target condition is monitored with the energy analysis of elastically scattered ions in a Si detector, using the energy spectrum also to keep track of the relative abundance of target contaminants such as carbon and oxygen and the primary target material. Postirradiation examinations with a proton microprobe are also carried out to study the changes in target composition produced by the heavy-ion bombardment.

5.3.3. S-Shape Transport System

An electron and positron transport system, TORI (Ka 80b), employing an S-shaped guiding magnetic field, has recently been developed to provide a facility to simultaneously study both electron and positron emission. A view of the experimental arrangement is given in Figure 5.18. As in the EPOS system the positrons and electrons are transported to a region of low background. However, in this case the S-shaped guiding field is used to spacially separate the electrons and positrons by producing oppositely directed drifts for the negative and positive particles perpendicular to the plane shown in Figure 5.18. This separation, after the first leg of the S-shape, is used to stop or count the electrons while the positrons continue on in the second leg to be detected in a planar Si(Li) counter which is surrounded by NaI detectors to register the annihilation radiation. The maximum positron efficiency reached to date is $\sim 13\%$ of 4π for positrons emitted by the target,

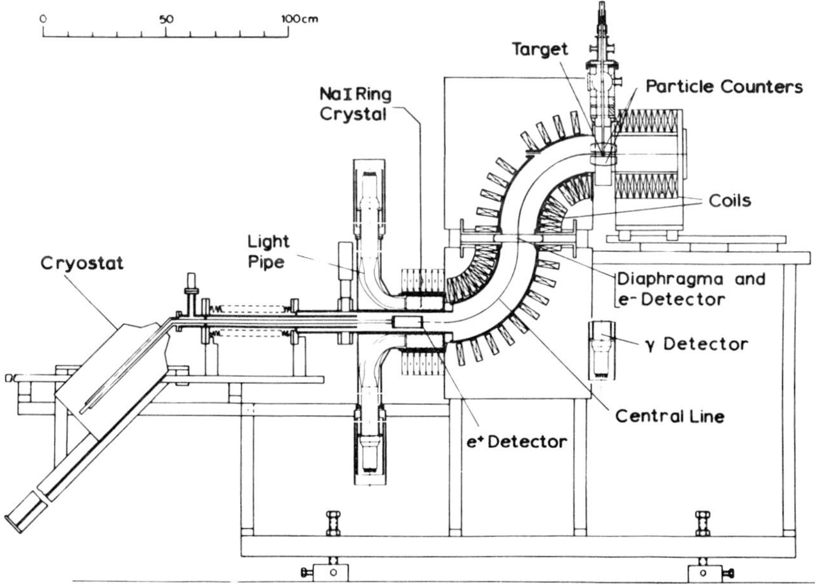

Figure 5.18. Side view of the TORI experimental arrangement (Ka 80b). The beam enters perpendicular to the plane drawn.

with the possibility of increasing this efficiency substantially. The separation of electrons and positrons is energy dependent, being poorest at low energies. Finite geometries of beam, target, etc. limit the efficiency for good electron–positron separation at low energies. In a configuration where the efficiency remains high above 1 MeV, it falls off sharply below 300 keV. Doppler broadening and other characteristics are similar to those for the EPOS system. The positrons (electrons) are measured in coincidence with the scattered ions detected in two θ-sensitive parallel-plate avalanche detectors that cover a range $20° \leqslant \theta \leqslant 70°$ with respect to the beam. The θ resolution is about 2°. As in the other systems the gamma rays are monitored by NaI detectors.

The three experimental arrangements described are in some sense complimentary and in other ways they duplicate capabilities. The differences are in the emphasis placed (1) on the geometry of positron collection relative to the beam direction; (2) on the ultimate positron energy resolution that can be achieved, which is partly coupled to the Doppler shift and broadening due to geometry; (3) on the need for broad-band characteristics, which is coupled to beam and target stability and to the data accumulation rate; and (4) on the clarity of the signature for the positron which determines the low-energy band width that can be explored.

The geometry utilizing the detection of electrons in a cone cylindrically symmetric about the beam axis ("Orange" spectrometer presently) has the advantage of a well-defined Doppler shift and small broadening if the angles of emission in θ and ϕ are traced for both the positron and scattered heavy ions and if the emitter velocity is known. This geometry then leads, in principle, to the best measured resolution. Conversely, the ϕ dependence of the positron emission can provide a measurement of the velocity of the emitter for a peak in the spectrum since for center-of-mass emission the Doppler effect is independent of ϕ, and for emission from the individual nuclei the Doppler effect depends on the angle between the positron and the plane of scattering. Such a measurement, of course, has to rely on sufficient data accumulation in these low count-rate experiments, especially since different fractions of the spectrum may not have the same origin and therefore different velocities. The lack of a clear identification signal for the positron, exclusive of the magnetic field selection, tends to reduce the low-energy effectiveness of the "Orange" spectrometer arrangement in a region where the spontaneous emission signal may appear for some of the collision systems such as U + U.

The EPOS and TORI systems feature high efficiency and broad energy band characteristics. The EPOS system has high efficiency with clear positron identification down to energies of ~100 keV, which covers most of the range of interest in the hunt for spontaneous positron emission. The advantages of broad-band capability are obvious. The penalty however, is a maximum Doppler broadening of spectrum lines if positrons are collected within the 2π angle of emission relative to the direction of motion of the emitting system. As noted earlier, the effective energy resolution is, therefore, limited in both transport systems to the Doppler broadening. But viewed in more detail, it is found that since the broadening depends sensitively on the velocity of the emitter, the latter quantity can again be measured by a shape analysis of the Doppler-broadened line. Such an analysis, thus, again provides the potential for resolving the question of whether the positrons in the peak are associated with the center of mass or the individual nuclei. Of course, a coaxial geometry between the beam and positron emission can also be employed in the EPOS and TORI systems, although this has not been attempted as yet.

In the following section we gather together the results of measurements obtained with the Orange, EPOS, and TORI systems. We will also include in the discussion some of the results from the first generation of experiments (Ba 78c, Ko 79); but for details of the apparatus used in the latter the reader is referred to the original papers.

5.4. Experimental Results

The data on positron production in heavy-ion collisions has been accumulating since 1976, when the first measurements were carried out at GSI, Darmstadt, following the availability of the first Pb and U beams that could reach energies sufficient to reach the Coulomb barrier for the heaviest targets such as Pb, U, Cm, and Cf. The experiments, thus far, have been exclusive to that laboratory. A natural division of the results is provided by separating the discussion into a consideration of first the gross and then the detailed features of positron production that have emerged from the measurements. As would be expected, this sequence also represents the chronological development of the data. It is of some interest to follow the chronological development of the experiments since the results have not always been anticipated and have focused emphasis on considerations in theory.

5.4.1. Gross Features of Positron Production

In qualitative terms, the positron production processes exhibit a particular sensitivity to a few variables. These include the total charge of the colliding system, the distance of minimum approach between nuclear centers, R_0, the value of the critical internuclear separation, R_{cr}, the relative velocity of the collision, and the effect of nuclear time delay. In an interrelated way, they determine the strength of the central perturbing field and the potential binding of the quasimolecular states, the spacial extent of this state, the

Fourier frequencies inherent in the collision, and the time spent in the collision. Of course, a contribution from spontaneous positron emission has to be sought at small R_0, well within the critical radius, as a deviation from dynamic processes acting alone. The first experiments attempted to explore these intuitive features, in essence looking for qualitative effects which could signal a departure from expected dynamic behavior. In these first explorations, the initial measurements on the $^{208}Pb + ^{208}Pb$ collision system played an important role in establishing the first demonstration of positron production from the QED processes not associated with nuclear decay (Ba 78c).

As a first measurement this presents an especially convincing case since the determination of the nuclear background for the $^{208}Pb + ^{208}Pb$ system is particularly straightforward due to the exceptional stability of ^{208}Pb to nuclear excitation. The separation of nuclear and atomic processes is readily achieved in this case. For collisions below the Coulomb barrier, the nuclear source of positrons is associated with a well-known excitation mechanism and known energy levels so that it can be simulated exactly; the dominant source of nuclear positrons is from the Coulomb excited 3^-(2.614)-MeV state which decays by internal pair conversion. It happens that this state is also populated in the decay of ^{212}Pb, and its decay is again the principal source of positrons from the radioactive source. Therefore, measuring the ratio of positrons to 2.614 gamma rays emitted by ^{212}Pb placed in the position of the target and multiplying this ratio by the intensity of the 2.614-MeV gamma ray observed in-beam directly yields the positron intensity from the Coulomb excitation of the 3^- state, independent of any efficiency factors which cancel out in the comparison. Only a small additional correction for the weak excitation of the 2^+ state and a well-known correction for the angular distribution of the gamma rays from Coulomb excitation have to be applied to the data.

Figure 5.19 displays the measured differential cross sections for scattering at $45° \pm 10°$ and $25.5° \pm 4.5°$. While the bombarding energy was varied in the experiment for the 45° scattering, the cross sections are plotted as a function of the distance of minimum approach, $2a$, in a head-on collision and also as a function of the minimum internuclear separation, R_0, for 45° scattering. We note that even for a symmetric collision system the latter quantity is defined for 45° scattering. But since projectile and target recoils cannot be differentiated, the calculated cross sections shown in Figure 5.19 are symmetrized with respect to the two scattered particles to allow comparison with experimental data at the two angles measured. The measurement integrates over the positron spectrum up to a few MeV, which covers the range that theory predicts to be interesting. The information obtained in this experiment is limited, but it is sufficient to demonstrate a number of points:

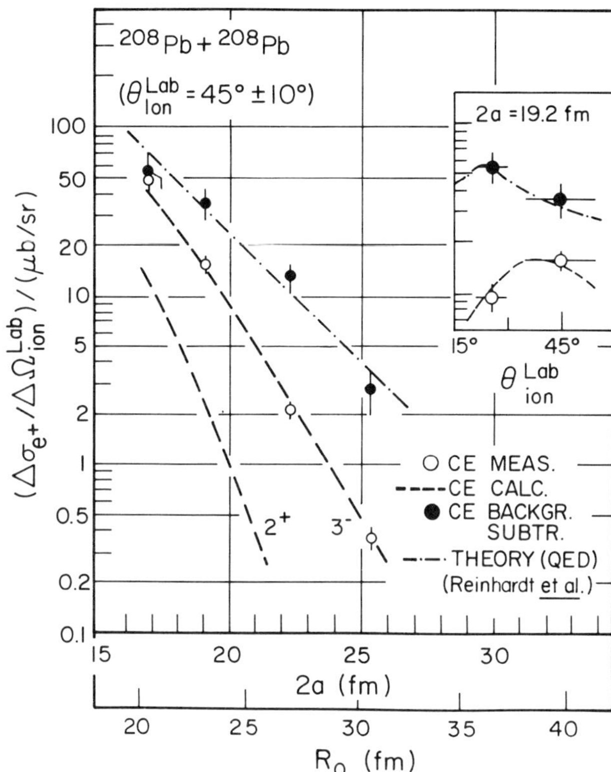

Figure 5.19. Differential positron cross sections as a function of R_0 for $^{208}Pb + ^{208}Pb$ (Ba 78c). Also shown is the directly measured nuclear related background from the excitation of the 3^- state in ^{208}Pb and Coulomb excitation (CE) calculations for both the 3^- and 2^+ states. The inset shows measured and calculated angular distributions.

(1) Even for a low $Z_u = 164$ it is possible to observe positron production in excess of the nuclear background. It is clear that the nuclear component is only a fraction of the total positron yield, particularly at the large R_0 values. A comparison between calculated nuclear positron yields from Coulomb excitation and internal pair conversion (Sc 78) with the directly measured intensities shows excellent agreement. The inset in Figure 5.19 illustrates that the differential cross section with respect to $\theta_{\rm ion}^{\rm lab}$ for the excess positron intensity over the Coulomb-excitation contribution possesses a more forward-peaked angular distribution than that for positrons from the deexcitation of the 3^- state. It is also evident that the increasing deviation between the positrons not associated with Coulomb excitation and the nuclear positron intensities as R_0 increases demonstrates that there are no other significant undetected backgrounds from nuclear reactions. (2) The

positron production in excess of nuclear background is at a level which is consistent with processes involving positron emission by the very strong time-varying electric fields present in the quasimolecular collision system. The good agreement of theory (Re 78b) with the experimental intensity and, in particular, the observed rapid falloff of the cross section with R_0 stresses this point and establishes that the Coulomb ionization description of pair production is substantially correct. Being a subcritical system, only the dynamic processes are expected to contribute with the direct excitation amplitude supplying the largest fraction.

The Pb + Pb system turns out to be an isolated case where dealing with the nuclear background is straightforward. The background situation changes with the attempt to extend the information on the dynamic processes to systems with higher Z_u such as U + Pb, or to study systems that can exhibit supercritical binding. The U nucleus then becomes involved either as a projectile or target as well as other heavier target nuclei whose nuclear structure closely resembles U. All these nuclei are deformed with low-lying collective states soft to excitation. The background conditions encountered in these experiments are, therefore, very different and much more complex than those found in the Pb + Pb measurements. Furthermore, the construction and composition of these targets also introduces other elements which compound the difficulties. As outlined in Section 5.2.2, the approach to this problem has been to deduce this background from in-beam gamma-ray spectra measured under the kinematic conditions used to obtain the positron spectra (Me 77b). The particular empirical procedure adopted in the experiments we will be discussing grew out of studies correlating the positron and gamma-ray yields in a number of nuclear collision systems using U as a common projectile. Various scattering systems with a variety of kinematic conditions were investigated in several experiments (Ko 79, Ki 81, Bo 81, Bo 82, Ba 81, Gr 80b), including a number of systems with low Z_u where emission of positrons is expected to be confined to nuclear sources.

Figure 5.20 (Ba 81, Gr 82) displays the essential feature that has been brought out by these studies. The figure plots an effective normalized ratio of positrons to gamma-ray yields, where the positrons are integrated over all detected positrons and the gamma-ray yield is integrated over the energies appropriate for producing these positrons. These ratios clearly exhibit a dual behavior as a function of Z_u. For the lower Z_u region they appear to be independent of the nuclear species, while the higher Z_u collision systems exhibit a prominent increase in this ratio with Z_u which correlates to the scattered particle number N_p rather than to the number of gamma rays in coincidence with scattered particles, $N_{\gamma p}$. Therefore, a dual source of positrons is indicated which can be expressed in a correlation to both N_p and $N_{\gamma p}$:

$$N_{e^+p}(\theta, Z_u) = P_{e^+}(\theta, Z_u)\, N_p + CN_{\gamma p} \tag{5.11}$$

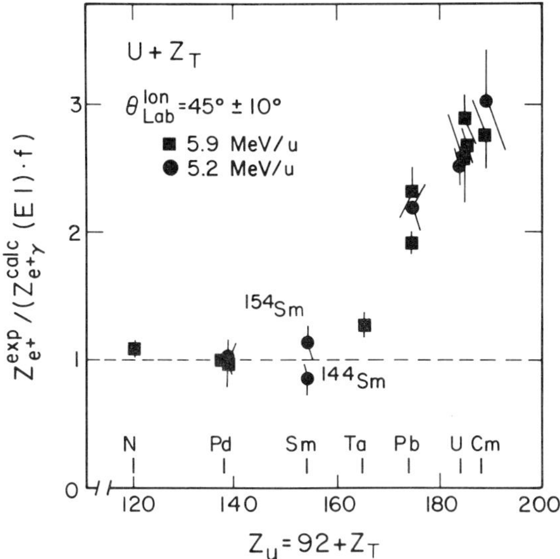

Figure 5.20. The ratio of measured positron yields to those calculated from the gamma-ray spectra. The positrons are integrated over all detected positrons and the gamma-ray yield is integrated over the energies appropriate to producing these positrons. $E1$ multipolarity is assumed for the transitions, and the factor f, close to unity, normalizes the ratio to unity for the lowest Z_u systems (Ba 81, Gr 82).

where C represents an empirical pair conversion coefficient which is apparently independent of Z_u, and can be obtained by fitting all the data to this equation. A straightforward interpretation of these observations follows by associating the component correlated to the gamme rays with a nuclear origin while assigning to the heaviest systems a new source of positrons which contributes in excess to the nuclear source present in the light systems. In fact, calculations indicate that atomic positron production is expected to become comparable to the measured nuclear background at a Z_u value in the neighborhood of 160.

This simple correspondence between the total gamma-ray and positron yields for $Z_u < 160$ reflects, upon more detailed examination, the similarity among the gamma-ray and also the positron energy distributions. Examples of some gamma-ray spectra obtained in an early measurement (Ko 79, Ki 81, Gr 80b) are shown in Figure 5.21. All the spectra exhibit an exponential falloff with gamma-ray energy with very similar slopes. The detection angle for the scattered ion also has little effect on the slope, as these and other more detailed data (Bo 82) show. We note that the small differences in shape that are observed lead to inconsequential changes in the positron energy distribution when Equation (5.10) is used to generate positron spectra for any fixed multipolarity or multipolarity mix. The

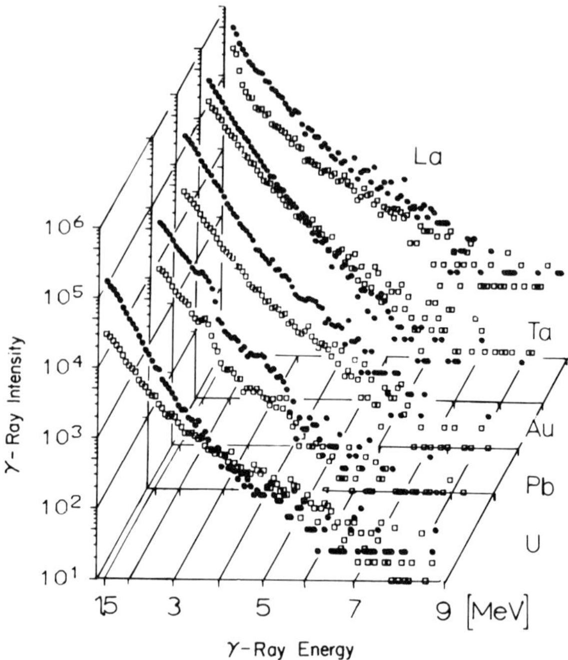

Figure 5.21. Gamma-ray spectra in coincidence with particles scattered between 16.1° and 20.2°, and between 26.9° and 32.3° from 5.9 MeV/u U colliding with U, Pb, Ta, and La targets (Ko 79, Ki 81, Gr 80b).

similarity in shape of all the positron spectra indicates that the multipolarity mix and the angular distribution cannot be very different in all the lower Z_u systems. An example of such a measured positron spectrum (folded with the efficiency and energy response of the spectrometer) from the $^{238}U + ^{165}Ho$ collision system is shown in Figure 5.22 (Bo 81, Bo 82). Also shown are spectra derived with Equation (5.10) from a simultaneously measured gamma-ray spectrum assuming $E1$ and $E2$ transitions. The $E1$ multipolarity appears to reproduce the measured spectrum very well, except at the lowest energies where the deviation increases with decreasing positron energy. In view of the low-lying collective excitations in these nuclei, this tendency for $E2$ transitions to contribute more prominently to the lower-energy part of the spectrum is not surprising.

Because of the uncertainty introduced by the unknown multipolarity, the procedure to determine the nuclear background for systems with $Z_u > 160$ has not resorted to calculating the positron spectrum absolutely from the measured gamma-ray distributions, but has followed

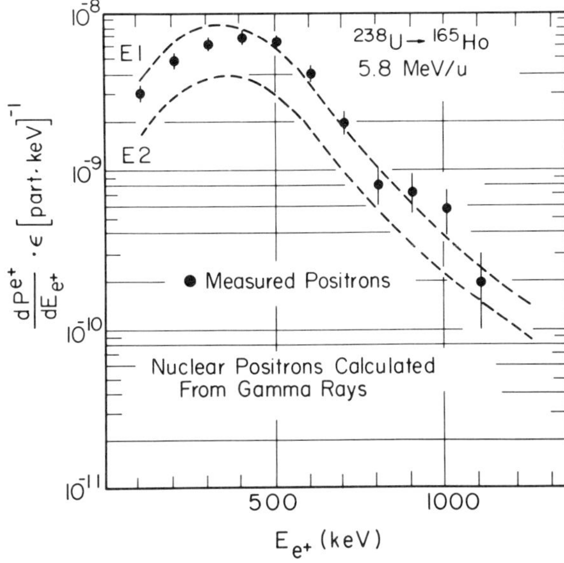

Figure 5.22. Measured positron spectrum from 5.8 MeV/u $^{238}U + {}^{165}Ho$ collisions compared with spectra calculated from simultaneously measured gamma-ray spectra assuming $E1$ or $E2$ multipolarity for the transitions. The calculated spectra are corrected for the efficiency and energy response of the spectrometer ε. (Bo 81, Bo 82).

Equation (5.11) and used the data for $Z_u < 160$ to extract an effective average internal pair conversion coefficient, $C(E_\gamma)$, taking advantage of a common multipolarity mix for all the collision systems implied by the close similarity among the gamma-ray spectra and among the corresponding positron spectra. The empirically deduced coefficient, thus, averages over the unknown mixture of multipolarities, including the effect of E0 transitions that are not observed in the gamma-ray spectra. Utilizing these coefficients for systems with $Z_u > 160$, of course, tacitly implies that this multipolarity mix can be extrapolated to the heavier systems, and that the gamma-ray angular distributions also do not change with Z_u. Although the validity of these assumptions cannot be established independently, they are reasonable to make since all the nuclei participating in the collisions have, on the average, similar nuclear structures. Indeed, this contention appears to be supported by the observation that the gamma-ray spectra for U + U and U + Pb, shown in Figure 5.21, have slopes closely paralleling those of the lighter systems where we found similar slopes are associated with similar positron spectra. In fact, the close resemblance of all gamma-ray spectra for all systems reduces the procedure for calculating the nuclear background to scaling the measured positron spectra for $Z_u < 160$ by the intensity ratio of gamma rays observed. What is more, the background is then obtained independently of the experimental factors such as the efficiences for positron and gamma-ray detection. It bears noting that the latent uncertainty that can be introduced by incomplete knowledge of the multipole mix probably does not appreciably prejudice the shape of the nuclear positron spectrum deduced by this method, and it is estimated that the intensity is uncertain by approximately 20%. In this context it should also be noted that the nuclear background constitutes ~30% of the positron production in U + U and ~20% in U + Cm at 5.9 MeV/u collisions.

In addition to providing this study of the nuclear background, these data also demonstrate one of the more striking qualitative features of positron production predicted by theory (Section 5.2.1b), the remarkable increase of the yield with Z_u. This is evidenced in the yield correlated with N_p of Equation (5.11) which describes the steep rise in the positron yield for $Z_u > 160$ over that expected from nuclear internal conversion as it is extrapolated from the positron to gamma-ray ratio measured for $Z_u < 160$ (Figure 5.20). More specifically, as illustrated in Figure 5.23 (Gr 82), for constant R_0 and relative velocity v, positron production in superheavy collision systems is found to increase as Z_u^{21}. It should be noted that, in contrast to this result, under similar conditions the $1s\sigma$ vacancy production decreases by a factor of 2 when Z_u increases from 164 to 184. This early result from the positron experiments provided an important confirmation of

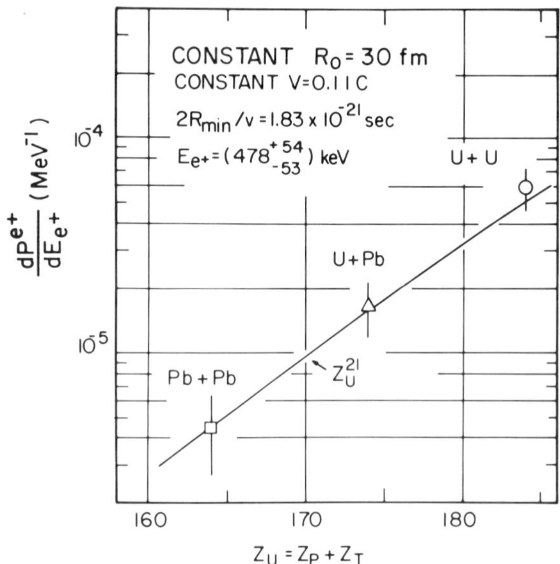

Figure 5.23. The dependence of the positron emission probability on the total charge Z_u of the collision system (Ko 79, Gr 82).

one of the principal characteristics of dynamic positron creation singled out by theory.

The same experiments also produced other results which define some of the general features of the dynamic processes. As examples of some of the early data, Figure 5.24 and 5.25 display how positron creation probabilities behave as a function of the parameters R_0 and v. In Figure 5.24 the collision velocity is kept constant and a differential slice of ~100 keV centered around 500 keV kinetic energy is detected, while in Figure 5.25 the scattering angle is kept constant at 45° ± 10° and positron energies below ~1 MeV are accepted. The data in Figure 5.24 are compared with calculations by Reinhardt *et al.* (Re 78b), using both perturbation theory and a coupled-channel approach. It is evident that the strong dependence of the excitation probability on R_0 found in the Pb–Pb data is reproduced in the U–Pb and U–U systems also, with essentially the same slope. The dependence on the relative velocity is common to all three systems. We also see that the collision time $\sim R_0/v$ very nicely unifies the data, as suggested by Equation (5.4). In fact the first experiments did not reveal any surprises

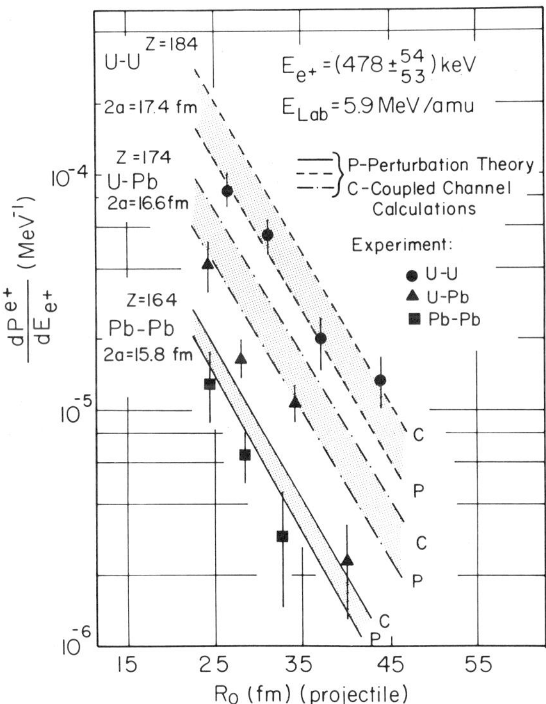

Figure 5.24. Positron production probabilities at 5.9 MeV/u as a function of R_0 (Ko 79, Gr 80b, Gr 82). The positron energy is kept constant at $(478 \pm ^{54}_{53})$ keV.

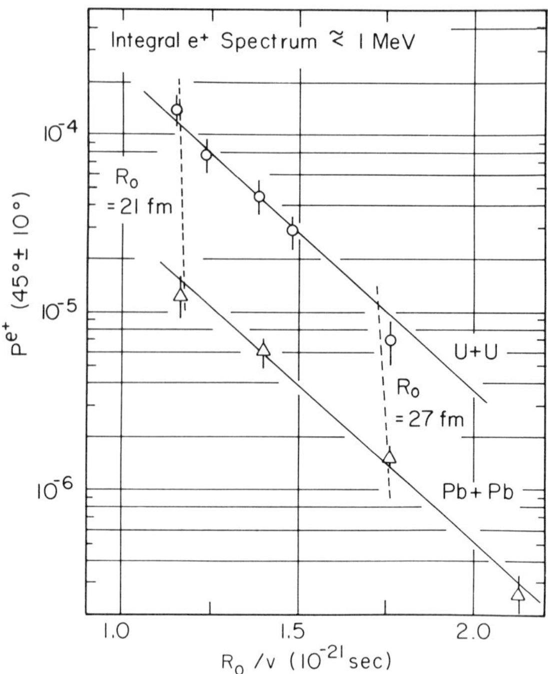

Figure 5.25. Positron emission probability as a function of R_0/v for U + U and U + Pb collisions (Ba 81, Gr 82). The scattering angle is fixed at 45° ± 10° and the relative velocity, v, is varied to determine R_0.

which could be considered as anomalously deviating from the theoretical positron production rates dominated by the dynamic mechanisms, although the U + U system could be expected to produce supercritical binding for some of the bombarding conditions used. Within the limited range of projectile energies, positron energies, scattering angles, and fairly loose impositions on the reaction kinematics, the early experiments were able to establish the following general points:

(i) Atomic QED processes account for the major fraction of the total positron creation cross section for collision systems with $Z_u > 164$, whereas for $Z_u < 160$, the dominant contribution to the positron yield is associated with the pair decay of nuclear excited states.

(ii) The positron creation probability from the quasiatomic processes exhibits a rapid exponential falloff with increasing R_0/v. This behavior parallels that found for the ionization probability of inner-shell electrons and reflects the same mechanism of Coulomb ionization responsible for the electron excitation processes in the superheavy collision systems. In this case, the minimum energy transfer is given by the energy gap $2mc^2$ for pair creation which sets the scale for the spatial confinement of this process, as it did for inner-shell vacancy formation and delta-electron emission, through $\exp(-\alpha R_0 \Delta E/\hbar v)$.

(iii) The almost exponential increase of the positron creation probability as a function of Z_u cannot be accounted for without invoking a nonperturbative treatment of positron production in the quasiatomic picture of $(Z_T + Z_p)$ acting in unison to generate very strong electric fields and, therefore, the relativistic effects that follow.

These principal features have been explored and developed in more detail in some more comprehensive measurements with the EPOS system (Bo 82, Gr 83). They include a study of the full positron energy distribution

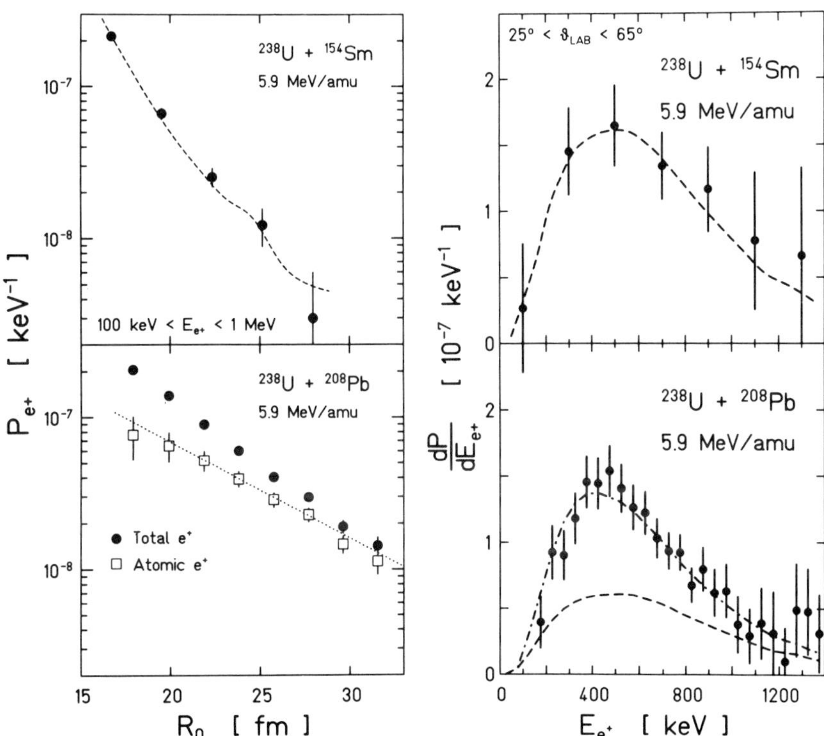

Figure 5.26. Positron production probabilities for the subcritical $^{238}U + {}^{154}Sm$ and $^{238}U + {}^{208}Pb$ collision systems showing the dependence on R_0 for the energy integrated interval between 100 keV and 1.0 MeV and the energy spectra for the laboratory scattering angle integrated interval between 25° and 65° (Bo 82, Gr 83). The dashed lines are calculated from the gamma-ray spectra and represent the nuclear component. The dotted line is obtained from atomic theory based on Rutherford scattering (Mu 83b) and the dot-dashed line is obtained from same theory plus nuclear background added (EPOS data).

produced by binary scattering events over a broad range in scattering angles. As was noted in the description of the EPOS system, the binary (θ_1, θ_2) kinematic correlations can be used to provide a clear selection of elastic scattering. Figures 5.26 and 5.27 illustrate some of the results obtained from measurement with ^{238}U projectiles at bombarding energies close to the Coulomb barrier. The data have been preselected for elastic scattering events or other events that overlap with elastic scattering kinematics. The latter have been treated in the angular distribution variables as if they corresponded to elastic scattering. The production probabilities in Figures 5.26 and 5.27 are determined by normalizing the rate of positron production in coincidence with the correlated particle scatterings to the rate of these binary scatterings alone for common kinematic conditions. The

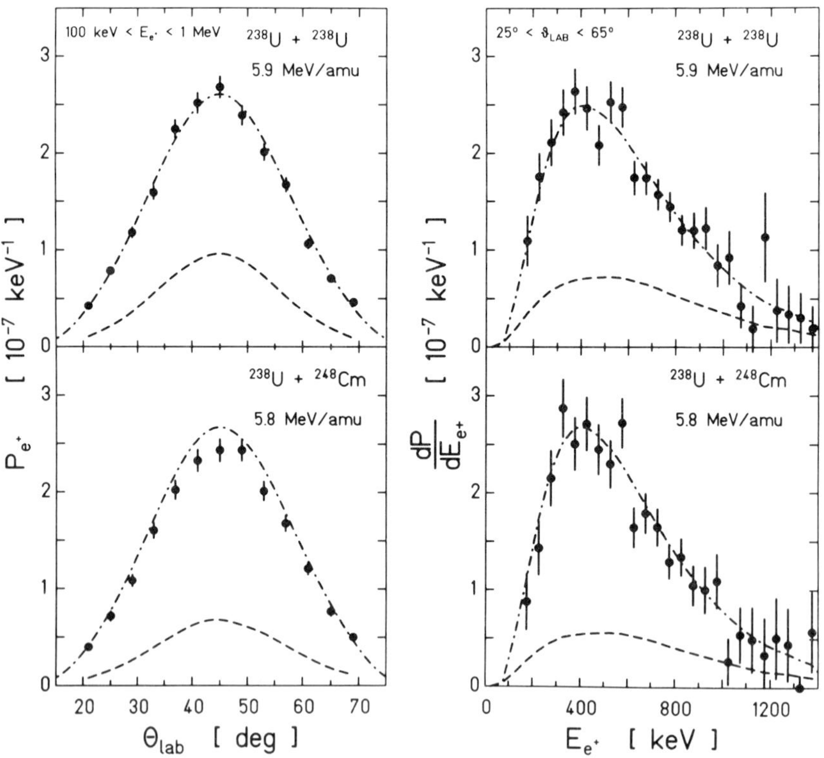

Figure 5.27. Positron production probabilities for the supercritical $^{238}U + ^{238}U$ and $^{238}U + ^{248}Cm$ collision systems showing the energy-averaged $(100\ keV < E_{e+} < 1.0\ MeV)$ angular distributions and the angle averaged $(25° < \theta_{lab} < 65°)$ kinetic energy distributions. The dashed lines are calculated from the gamma-ray spectra and represent the nuclear component. The dot-dashed lines are a sum of the nuclear component and theory based on Rutherford scattering (Mu 83b) (EPOS data).

energy response function of the detector and the transport efficiency is folded out of the data. It should be noted that for the two lightest systems involving ^{208}Pb and ^{154}Sm targets, the kinematic resolution is more than sufficient to distinguish forward and backward scattering in the center-of-mass system. In these cases, since R_0 is clearly defined, the production probability is plotted against this variable which is most directly connected to the physics. For the other symmetric and nearly symmetric systems, $^{238}\text{U} + {}^{238}\text{U}$ and $^{238}\text{U} + {}^{248}\text{Cm}$, where $\theta_{\text{c.m.}}$ and $(\pi - \theta_{\text{c.m.}})$ scattering cannot either be distinguished in principle (U + U) or clearly separated (U + Cm), the data is plotted against θ_{lab} for one of the counters. To compare with the data for the latter cases, therefore, the theoretical probabilities have to be symmetrized with respect to $\theta_{\text{lab}} = \theta_{\text{proj}}$ and $\theta_{\text{lab}} = \theta_{\text{targ}}$, i.e.,

$$\frac{dP_{e^+}}{dE_{e^+}} = \frac{\left[\dfrac{d\sigma^R}{d\Omega}\dfrac{dP_{e^+}}{dE_{e^+}}\right]_{\theta_{\text{proj}}} + \left[\dfrac{d\sigma^R}{d\Omega}\dfrac{dP_{e^+}}{dE_{e^+}}\right]_{\theta_{\text{targ}}}}{\left[\dfrac{d\sigma^R}{d\Omega}\right]_{\theta_{\text{proj}}} + \left[\dfrac{d\sigma^R}{d\Omega}\right]_{\theta_{\text{targ}}}} \tag{5.12}$$

This averaging tends to emphasize the distant collisions, the degree depending on $dP_{e^+}(R_0)/dE_{e^+}$.

To bring out the general features, the data have been grouped into a positron energy distribution averaged over scattering angles between $25° < \theta_{\text{lab}} < 65°$ and an angular distribution integrated over positron energies between 100 keV and 1 MeV. As illustrated in Figure 5.20, positron production in the $^{238}\text{U} + {}^{154}\text{Sm}$ system is dominated by nuclear processes. Figure 5.26, therefore, provides a measure of the ability to determine the nuclear background using the procedure outlined previously and the $^{238}\text{U} + {}^{165}\text{Ho}$ gamma-ray and positron emission spectra as references. We see that the spectrum calculated in this manner and this R_0 dependence reproduce the measurements very well. The nuclear backgrounds for the other three systems have been calculated with the same procedure. It is apparent that after including the nuclear background (indicated as dashed lines in the figures), the theory (Mu 83b), based on pure Rutherford trajectories, constitutes a good fit to the measurements for the subcritical systems $^{238}\text{U} + {}^{208}\text{Pb}$ $(Z_u = 174)$ as well as for the supercritical systems $^{238}\text{U} + {}^{238}\text{U}$ $(Z_u = 184)$ and $^{238}\text{U} + {}^{248}\text{Cm}$ $(Z_u = 188)$. The agreement is well within the estimated normalization uncertainty of ~30%. The calculations not only account for the average features of the shape and the absolute emission probability, but they provide an equally good fit to the angular distributions. For the case of the $^{238}\text{U} + {}^{208}\text{Pb}$ system, where forward and backward scattering is resolved, the theory is tested, with excellent results, down to very small internuclear separations, $R_0 = 18$ fm. Actually, this measurement represents the only such test available to date, and we note that it is for a system where supercritical binding cannot be achieved for any of the measured scattering angles.

From these measurements it can be concluded, therefore, that, at least for a selection of collision events which are dominated by Rutherford scattering, the general aspects of the energy spectra and angular distributions of positrons emitted from heavy-ion–atom collisions appear to be well understood. The theory meets with similar success in describing other measurements, such as, for example, those presented in Figure 5.28 (Ba 81) where a series of spectra are obtained for more restricted scattering angles at $45° \pm 10°$, again at a bombarding energy close to the Coulomb barrier. It bears emphasis that, in essence, all these experiments featuring Coulomb trajectories probe almost exclusively the dynamically induced positron production processes since the effect of spontaneous positron emission on the energy integrated angular distributions or on the angle integrated energy distributions is expected to be small. In this respect it is apparent that the

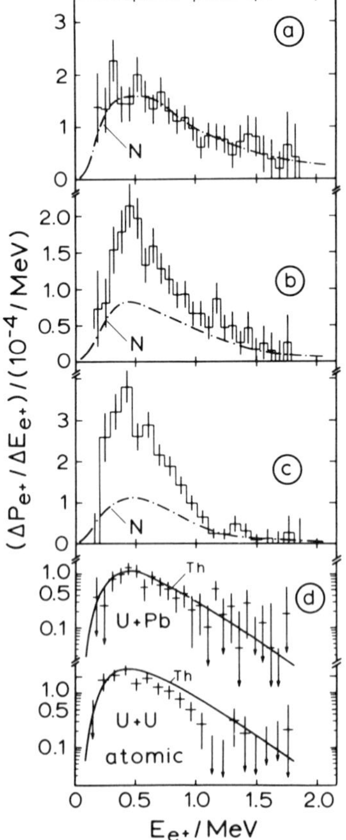

Figure 5.28. Positron spectra for (a) the nuclear U + Pd, (b) the subcritical U + Pb, and (c) the supercritical U + U systems at a U beam energy of 5.9 MeV/u and $\theta_{\text{lab}} = 45° \pm 10°$ (Ba 81). The curves "N" are derived from gamma-ray spectra assuming $E1$ multipolarity. In (d) the semilogarithmic presentations of the spectra after "N" has been subtracted are compared with theory based on Rutherford scattering (Re 81b).

theory correctly predicts that, for collisions dominated by Coulomb scattering, there is no qualitative effect to mark the transition from subcritical to supercritical status.

However, the theory becomes less specific in its predictions as we move away from the Coulomb barrier into a region where nuclear reactions can become significant. In Section 5.2.1c it was pointed out that under these circumstances the role played by spontaneous positron emission depends on the details of the time development of the collision trajectory $R(t)$. Because of this connection of time delay with the detection of the spontaneous emission process, there is, obviously, some interest in documenting how nuclear reactions affect the general features of the positron spectrum in a situation where nuclear reactions are known to dominate the scattering events. Several measurements have pursued this question (Ba 83), most recently with the TORI instrumentation (Kr 83).

The experiments have involved studying $^{238}U + ^{238}U$ collisions with projectile energies of 8.4 and 7.5 MeV/u. Such encounters are considerably above the Coulomb barrier and have been associated with deep-inelastic collisions where kinetic energy and angular momentum are dissipated into internal degrees of freedom. The interesting aspect of this reaction mechanism is that it is also associated with finite reaction times. Deep-inelastic collisions have been described in terms of a number of models. Characteristically they predict strong deviations from a Coulomb trajectory and reaction times that increase with energy dissipation. For example, the microscopic friction model by Schmidt *et al.* (Sc 78d) is able to explain large observed kinetic energy losses up to 30% and predicts delay times up to $\sim 1.1 \times 10^{-21}$ s for U + U collisions at 7.5 MeV/u. From the discussion in Section 5.2.1c, it is evident that this time delay would not be long enough to allow a spontaneous line spectrum to emerge above the continuum distribution, but, nevertheless, it is still of some interest since it may lead to observable effects connected with the dynamic positron emission. How these effects will manifest themselves in detail will depend on the specific $R(t)$ which, presently, can only be modeled.

To single out the inelastic collisions, the experiments measure positrons in coincidence with the fission fragments at a laboratory angle $35° < \theta_{\text{lab}} < 55°$ (Ba 83). The ring plastic scintillator counter used to detect the reaction products relies on the difference between the broad energy distribution of events connected with fission and the full-energy peak from the quasielastic events to distinguish the two event types. It was found that at projectile energies of 5.9, 7.5, and 8.4 MeV/u, the quasielastic contribution was 100%, 30% and 0%, respectively, at these detector angles. The nuclear background is handled by evaluating the multipolarity mixture in the measured gamma-ray spectrum from a measurement of reference spectra in U + Pd collisions at the appropriate bombardment energies. Figure 5.29

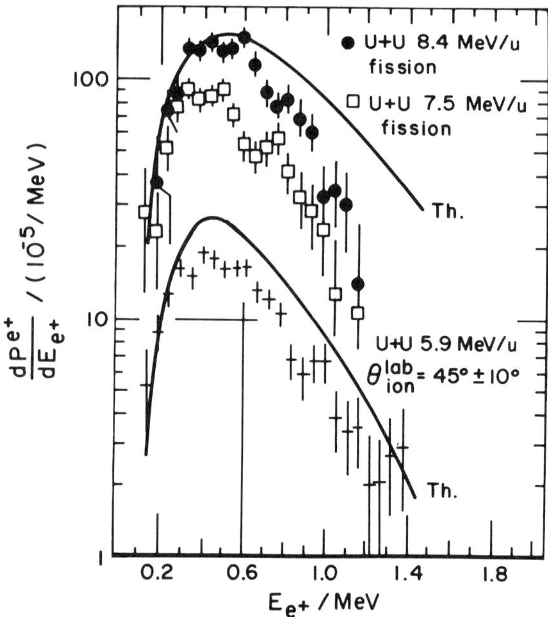

Figure 5.29. Atomic positron spectra for U + U scattering systems (Ba 83, Ba 82). Spectra at 8.4 and 7.5 MeV/u have been accumulated in coincidence with fission fragments. Rutherford scattering trajectories have been assumed in theoretical calculations (Re 81b) with $\theta_{\text{c.m.}} = 85.2°$ at 8.4 MeV/u and with $\theta_{\text{c.m.}} = 90 \pm 20°$ at 5.9 MeV/u.

displays the spectra observed at 5.9, 7.5, and 8.4 MeV/u. They are corrected for background and detector efficiency. With the exception of a possible fluctuation at 750 keV in the spectrum accumulated at 7.5 MeV/u, the spectra are manifestly smooth. The spectrum for 5.9 MeV/u, which is associated with Coulomb scattering, is, as in the other experiments, reproduced in shape by calculations based on Rutherford scattering (Re 81b), although the predicted intensity is somewhat too large with respect to the data. The latter discrepancy may be partly due to the omission of screening for the calculation shown.

The interesting feature in these data is that the shapes of the spectra taken in coincidence with fission fragments are not consistent with the predictions of Rutherford scattering. They drop off more steeply with increasing positron energy than predicted by theory for Rutherford trajectories. It is not a large effect so that, obviously, the background subtraction becomes an important consideration. Nevertheless, the effect is well established and it can be explained by a modification of the nuclear motion due to a time delay of approximately 10^{-21} s. More quantitatively, the microscopic friction model (Sc 78d) referred to previously has been employed (Mu 83b) to make a good fit to the data. This particular modeling of the reaction path $R(t)$ indicates that an effective average delay time of $\sim 1.5 \times 10^{-21}$ s explains the data. Therefore, it appears that at least a qualitative understanding of the general features of the spectra also extends to the situation where a time delay has an effect on the spectra, although, as expected from previous experience with deep-inelastic reactions, these time delays are too short to produce a clear signature for spontaneous positron emission. It bears noting, however, that these experiments suggest that positron spectra may, indeed, be useful in measuring nuclear reaction times and also the form and the distribution of the time delay. It is a point that should be pursued in more reaction studies.

Summarizing the results on the gross features of positron emission from the superheavy collision systems, it is apparent that with the success achieved in understanding these data it becomes concurrently clear that a signature for spontaneous positron emission is not readily forthcoming, at least not in gross effects that can be measured. As anticipated by theory, nothing unexpected occurs in the gross features as Z_{cr} and R_{cr} are passed. In the main, the dynamic processes predominate both for collisions near the Coulomb barrier and for nuclear reactions above the barrier which involve short reaction times. It is evident that the search for a clear signature for spontaneous positron emission has to be sought in finer details among this dominating background which is always present. If such a signature is available in the data already presented, then it appears that it has to be associated with a small part of the cross section.

Noting this limitation and the experimental confirmation of the basic general features of theory, it is apparent from the discussion in Section 5.2.1c that a clear signal for spontaneous positron emission is inseparable from the formation of a superheavy complex which exists on a time scale sufficient for the development of a narrow peak. It was also pointed out in Section 5.2.1c that lifetimes of the order of 10^{-20} s could lead to a very prominent signal even under conditions where the desirable nuclear reactions constitute a small fraction of all the collisions. Under these circumstances any effect reflected in the total spectrum could be small, as observed, but could still appear greatly enhanced in selected correlations between the spectral distribution and kinematics of scattering. Such arguments lead naturally into a search for effects in the spectrum that are focused in bombarding energy and in restricted reaction kinematics. The most recent experiments have pursued such measurements.

5.4.2. Search for Peak Structure

The obvious question to be addressed concerns the conditions required to form metastable nuclear complexes in these giant systems. Except for the rather short reaction times that are associated with deep inelastic collisions above the Coulomb barrier, there is no evidence, independent of the experiments we are discussing, that a superheavy complex with a sufficiently long lifetime $>10^{-20}$ s can be formed. In this connection we note that, so far, the search for superheavy elements has not been fruitful. Also, because of the large Coulomb energies involved, conventional wisdom argues against such a possibility. Therefore, there appears to be little guidance, experimental or theoretical, in the choice of bombarding conditions except, possibly, the notion that at Coulomb barrier bombarding energies, where the nuclei gently touch and begin to interact through nuclear forces, the small amount of energy and angular momentum available to be released into intrinsic excitation can help the formation of rather cold complexes which may stay together for a long period. [The cold rearrangement of nucleons to fuse nuclei has recently been exploited successfully for fusion near the Coulomb barrier in lighter systems (Ar 82).] In fact, this intuitive approach to the question of where to look in bombarding energy only supplies post-factum justification to the same suggestion provided by the experimental data itself.

For example, if the spectrum from U + Cm collisions presented in Figure 5.27 is examined more closely before the averaging that occurs in unfolding the efficiency and resolution functions, then it becomes evident that there are some features in the U + Cm spectrum which are not present in the spectrum based on Rutherford scattering alone. This raw spectrum is shown in Figure 5.30 (Gr 82, Gr 83) together with the theoretical spectrum

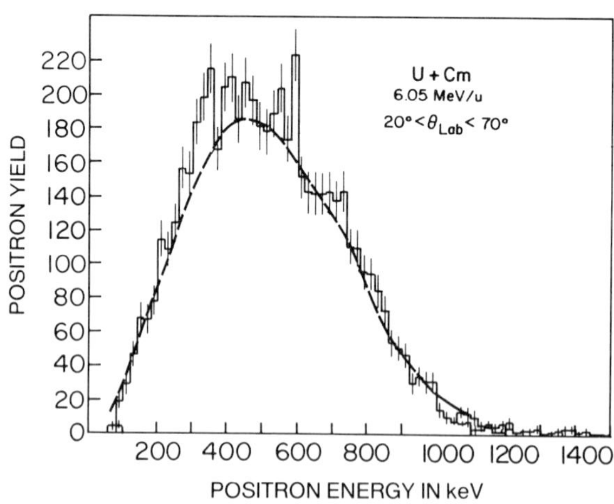

Figure 5.30. Angle integrated ($20 < \theta_{\text{lab}} < 70°$) spectrum obtained from 6.05 MeV/u U + Cm collisions before correcting for transport efficiency and Si(Li) response (Gr 82, Gr 83). A shape comparison is made with theory (dashed line) based on Rutherford scattering added to the nuclear background calculated from the gamma-ray spectrum (EPOS data).

folded with the efficiency and resolution functions. Although the overall theoretical shape is followed by the data, there appear to exist statistically significant deviations from this envelope at the lower energy region of the spectrum.

As will be discussed in more detail below, this particular structure found in the U + Cm system by the EPOS system is, in fact, very closely tied to bombarding energies in the vicinity of the Coulomb barrier, almost following a resonancelike behavior (Sc 83b). Structure in the positron spectra has also been observed in the U + U system at bombarding energies near the Coulomb barrier both with the "Orange" and "EPOS" detector systems (Ki 81, Bo 81, Gr 82, Be 82c, Cl 83, Cl 84). A feature that appears to characterize all these observations is the correlation of the structure with kinematic constraints. This is strikingly illustrated in the studies carried out with the U + Cm collision system which represents the largest combined nuclear charge, $Z_u = 188$, investigated to date (Sc 83b). As emphasized in Figure 5.8, from the point of view of providing the strongest signal for spontaneous positron emission, the highest Z_u is most desirable for reasons of both maximizing the probability of emission and the kinetic energy of the positrons.

Figure 5.31 displays two positron spectra from U + Cm collisions obtained in a measurement with the projectile energy at 6.05 MeV/u (Sc 83b). Under the assumption that the binary scattering collision partners follow elastic scattering trajectories, the kinematic region associated with part (a) overlaps with backward scattered projectiles between $100° < \theta_{\text{c.m.}} < 130°$, whereas, in part (b) the kinematic cut covers the forward elastic scattering of the projectiles into angles $50° < \theta_{\text{c.m.}} < 80°$. This selection is made by choosing two regions of $(\theta_1 + \theta_2)$ for a common slice of $-35° \leqslant (\theta_1 - \theta_2) \leqslant -10°$ (see Section 5.3.2). It should be noted, however, with regard to the definition of scattering angles, that in this nearly symmetric collision a clear separation is not achieved between the close and distant collisions, as was found in the case of U + Pb shown in Figure 5.17. Although it is difficult to establish quantitatively the mixing of forward and backward scattering at each laboratory scattering angle, a study of gamma-ray emission as a function of $(\theta_1 + \theta_2)$ for fixed $(\theta_1 - \theta_2)$ indicates that there is a good degree of separation of near and far collisions in the region of scattering angles corresponding to the kinematic cuts taken in Figure 5.31. The particularly large forward–backward emission asymmetry for gamma rays provides a sensitive measure of this separation.

The contrast in the spectra found in parts (a) and (b) of Figure 5.31 is striking. A smooth spectrum is observed in part (b) with no apparent structure. The unknown mixture of forward and backward angles precludes any absolute comparison with theoretical yields, but, nevertheless, the shape of this spectrum can be very well reproduced by the dynamic theory based

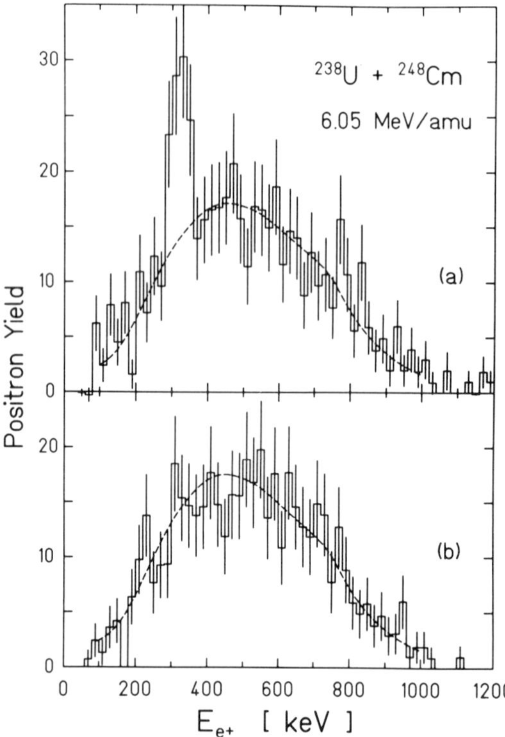

Figure 5.31. Positron energy spectra observed for $^{238}U + {}^{248}Cm$ collisions at a projectile energy of 6.05 MeV/u (Sc 83b). Panels (a) and (b) refer to two kinematic conditions specified in text. The dashed lines represent the theoretical distributions for dynamic positron production based on Rutherford scattering together with nuclear background, folded with the positron efficiency and normalized to the region outside the peak (EPOS data).

on Rutherford scattering superimposed on a nuclear background deduced from the simultaneously recorded gamma-ray spectrum. (The theoretical yields have been folded with the positron detection efficiency and resolution responses). On the other hand, a well-defined peak emerges with the choice of scattering angles represented by Figure 5.31(a) at a mean energy of $E_{e^+} = 316 \pm 10$ keV with a narrow width of ~80 keV. Dynamic theory together with nuclear background yield an excellent fit to the smooth part of the spectrum excluding the peak region with a normalized $\chi^2 = 1.01$. Within the 200-keV interval occupied by the peak, the excess peak intensity deviates from this fit to the continuous part of the spectrum by ~5 standard deviations. The probability of fitting the entire spectrum with dynamic theory plus background is at a confidence level of <0.1%. The peak, therefore, appears to be established statistically. Furthermore, the strong correlation of the peak intensity with the choice of scattering angles excludes the possibility that the appearance of the peak can be traced to instrumental effects associated with the positron detection method used.

A number of other observations regarding the peak are also of special interest:

(1) As will become evident from the analysis of the peak presented in the following section, a significant broadening of the peak produced by the Doppler effect ensures that the intrinsic width of the peak is surely less than the ~80 keV width observed, implying that the source of the peak exists for times longer than $\sim 10^{-20}$ s.

(2) An excitation function measurement, covering projectile energies slightly below and above the nominal Coulomb barrier, shows that the peak structure is connected with a narrow interval in bombarding energy between 6.0 and 6.2 MeV/u. A large part of this interval can be accounted for by the mean energy loss in the target, implying that the excitation of the peak structure is associated with a narrow resonancelike behavior. It is important to note that for all the conditions of projectile energy and scattering angle where the peak is detected, its mean energy and width remain substantially constant.

(3) Further insight into the association of the peak with the kinematic selection made in Figure 5.31 is obtained by studying in more detail the angular distributions connected with the positron energy intervals that both include (280–360 keV) and exclude (460–800 keV) the peak. This leads to the following interesting observation (Sc 83b): whereas the events associated with positrons outside the peak region follow (θ_1, θ_2) correlations exhibited by Rutherford scattering properly weighted by theoretical dynamic positron emission probabilities, for the positrons within the energy interval occupied by the peak the (θ_1, θ_2) distribution behaves quite differently. Most notably, it breaks the intrinsic symmetry about $(\theta_1 - \theta_2) = 0$ called for by the symmetric arrangement of heavy-ion detectors used in the "EPOS" geometry. The pattern of the events observed can be explained by taking into account the effect of the solenoid's magnetic field on the trajectories of the scattered ions with the provision that the ionic charge states associated with the positrons in the peak differ from the mean velocity-dependent equilibrium charge states which are found to be characteristic of the Rutherford scattering events producing the underlying dynamic spectra. This anomalous behavior seems to single out the peak from the other part of the spectrum and again suggests that the peak does not originate from Rutherford scattering. Moreover, within the context of this interpretation, it becomes clear that the contrasting peak to dynamic background ratios exhibited by the spectra in Figures 5.31a and 5.31b are not necessarily associated with backward and forward elastic scattering; in large part they reflect the magnetic field redistribution of the kinematic correlations for the quasielastic events and, therefore, do not directly represent the angular distribution of the peak events. The peak in Figure 5.31a becomes particularly prominent above background because its (θ_1, θ_2) distribution happens to overlap a kinematic region of backward elastic scattering where the intensity of dynamic positron

production is small. In fact, kinematically the (θ_1, θ_2) correlations are consistent with a reaction which leads to a near binary exit channel with two approximately equal masses and with very little Q loss, <20 MeV, from the incoming channel.

Surveying experiments further, it is found, as mentioned previously, that significant structure is also observed in positron spectra recorded for U + U collisions. For example, it is interesting to find that, in the "EPOS" experiments, a peak pattern again emerges in the same kinematic region where the peak was found in the U + Cm data, while, outside this region, kinematic cuts appear to lead to smooth spectra (Gr 82). It bears noting in this connection that, since U + U is a symmetric collision system, backward and forward scattering cannot, in principle, be distinguished in this case. Again, without invoking some unusual mechanism, such as that proposed to explain the angle–angle correlations in the U + Cm data, it is difficult to explain the (θ_1, θ_2) pattern observed for the structure-related events in U + U. Therefore, again these peak structures seem to be associated with unusual events. Spectra for the two kinematic regions are shown in Figure 5.32. Unlike the U + Cm data the peak pattern exhibited in Figure 5.32a appears to be composed of several structures of almost equal intensity. There is also some suggestion in data not displayed herein that this kind of interesting multiple peak pattern may also be present in the U + Cm collisions, although in the latter case the spectrum is dominated by the 316-keV peak so that it is difficult to draw conclusions with the limited statistical accuracy available.

Extensive measurements have been carried out with the "Orange" spectrometer on U + U collisions at several bombarding energies near the Coulomb barrier (Ki 81, Cl 83, Cl 84, Be 82c). These experiments again explore the correlations between the positron spectrum and the heavy-ion scattering angles. As noted in Section 5.3.1, the scattering-angle acceptance of the spectrometer ranges from 12° to 51°, and except for the region 45° ± 5° where kinematic coincidences are recorded, the scattering event is specified only with the detection of one scattered particle. Figure 5.33 (Be 82c, Cl 84) displays spectra obtained at several projectile energies and scattering-angle regions. In particular, this figure compares two spectra measured at one projectile energy, 5.73 MeV/u in coincidence with different scattering angles. The more forward angle data, $26° \leqslant \theta \leqslant 38°$, is substantially uneventful with possibly only minor structure content. On the average, this spectrum exhibits the typical bell-shape distribution characteristic of a combination of positron production from Rutherford trajectories and nuclear background with which we have become very familiar. However, the scattering angles $38° \leqslant \theta \leqslant 52°$ project out a structured spectrum. Most prominent is a peak centered at a mean energy of 308 ± 15 keV with a width ≃60 keV. For comparison with other experiments we note that this mean

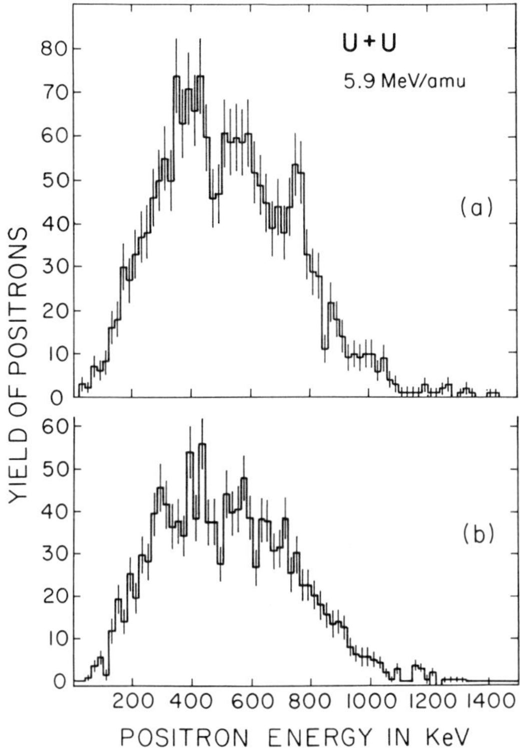

Figure 5.32. Positron energy spectra for 5.9 MeV/u $^{238}U + ^{238}U$ collisions subject to the same kinematic conditions specified for Figure 5.31 (Gr 82) (EPOS data).

energy incorporates a Doppler shift correction which has been computed assuming that the positron is emitted from the center-of-mass system (quasimolecule). For the geometry of the "Orange" spectrometer experimental arrangement, the Doppler shift is $\simeq 20$ keV for 300-keV positrons emitted from the center of mass. Another notable point is that the Doppler broadening for 300-keV positrons is of the order of 10 keV for center-of-mass emission, but at least 60 keV if the positrons originate from the individual final-state nuclei. Therefore, the observed width of 60 keV either represents the intrinsic width of the line or the Doppler broadening. A comparison of the line profile of positrons emitted in and out of the plane of scattering, in principle, could answer this question. Such measurements are now in progress.

Figure 5.33 also shows other spectra measured at higher projectile energies of 5.9 and 6.2 MeV/u. The scattering angle regions shown have been chosen to select cuts in the data that best exhibit peak structure. As found at

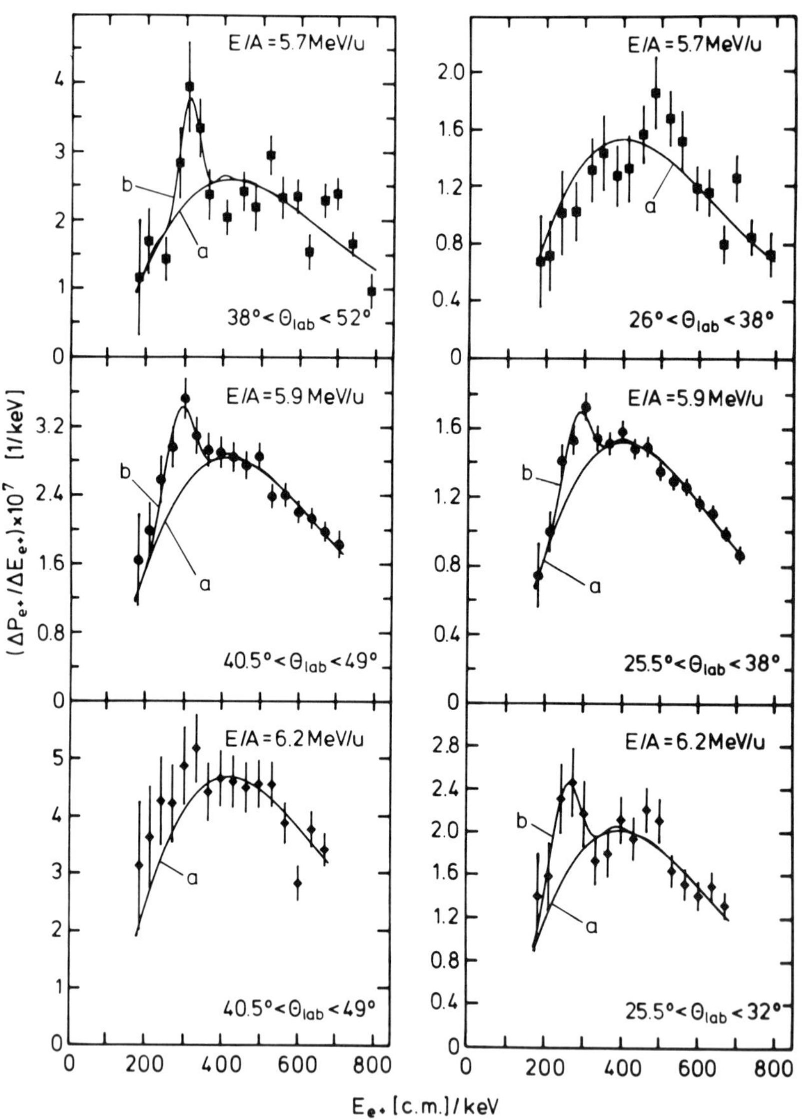

Figure 5.33. Positron energy spectra from $^{238}U + ^{238}U$ collisions at 5.7, 5.9, and 6.2 MeV/u and at scattering angles as indicated (Cl 84, Be 82c). Fitted to the data are calculated distributions for dynamic positron emission (Re 81b, Mu 83b) normalized to the data, curves (a), together with an analytic expression (Re 81, Mu 83b) for emission of spontaneous positron lines, curves (b) ("Orange" data).

5.73 MeV/u, in each case one peak or enhancement dominates over a less structured spectrum that on the average can be fit by theory (Re 81) based on Rutherford scattering and nuclear background [curve (a)]. Increasing the projectile energy and changing the detection angle seems to affect the position and width of the peak. Generally, the structure tends to widen with increasing projectile energy and shift to lower energies at the smaller scattering angles. The mean energy appears to vary within a range of ~250 to ~310 keV. The minimum width observed is ≃60 keV in the 5.73-MeV/u data, but can be more than twice as large (Cl 83).

Another interesting feature of these data is the association of narrow regions in scattering angles with the emergence of the peak structure above the underlying background. A selectivity connected with narrow kinematic constraints seems to be equally characteristic of the "EPOS" data on U + U collisions. For example, the data on 5.9 MeV/u collisions used to derive the spectra shown in Figure 5.32 also yield the spectra shown in Figures 5.34a

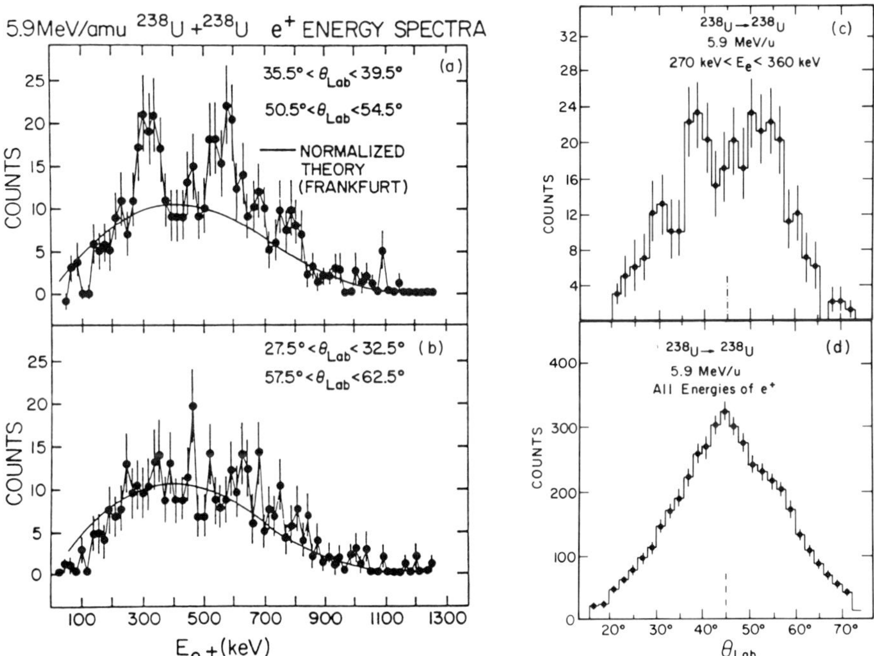

Figure 5.34. Positron energy spectra observed from $^{238}U + ^{238}U$ at 5.9 MeV/u after taking kinematic cuts that overlap with inelastic collisions with a small energy loss of ~20 MeV (Bo 81): (a) sum of two angular regions $35.5° \leqslant \theta \leqslant 39.5°$ and $50.5° \leqslant \theta \leqslant 54.5°$; (b) sum of $27.5° \leqslant \theta \leqslant 32.5°$ and $57.5° \leqslant \theta \leqslant 62.5°$. The spectra are not corrected for efficiency and the response of the Si(Li) detector. Panels (c) and (d) show, respectively, the measured angular distributions for positrons energies including the peak at 320 keV and for positron energies $100\ \text{keV} \leqslant E_{e^+} \leqslant 1000\ \text{keV}$ (EPOS data).

and 5.34b when the (θ_1, θ_2) scattering correlations are searched to optimize peak structure (Bo 81). Again two kinematic cuts are made: (a) a sum of $35.5° \leqslant \theta \leqslant 39.5°$ and $50.4° \leqslant \theta \leqslant 54.5°$ which contains the peak structure, and, for contrast, (b) a sum of $27.5° \leqslant \theta \leqslant 32.5°$ and $57.5° \leqslant \theta \leqslant 62.5°$. (Here conjugate angular regions symmetric about 45° have been added.) A further selection is also imposed on $(\theta_1 + \theta_2)$ to pick out slightly inelastic collisions which enhance the peak structure. It is apparent that positrons from Rutherford scattering and nuclear decay can account for the data in part (b), but part (a) seems to exhibit two peaks. The lower-energy peak, centered at ~320 keV, appears to be common to both experiments. The peak centered at ~550 keV is more prominent in the "EPOS" data, but could also be present in the "Orange" spectrometer spectra. (Differences in the two spectra can be due to the detail with which the kinematics are defined in the two experiments.) The events associated with the peak near ~300 keV are focused at similar laboratory angles in both experiments. To be more specific, the angular dependence of the 320-keV peak intensity obtained in the "EPOS" measurement is more fully displayed in Figures 5.34c and 5.34d. These plots compare the positron production probabilities as a function of scattering angle for positron energies $100 \text{ keV} \leqslant E_{e^+} \leqslant 1000 \text{ keV}$ (d) and $270 \text{ keV} \leqslant E_{e^+} \leqslant 360 \text{ keV}$ (c). The latter energy cut, which includes the peak at 320 keV, obviously, enhances the angular regions adjacent to 45°.

Considering all these data, there seems to be little doubt that there is peak structure in the positron spectra emitted in U + Cm and U + U collisions near the Coulomb barrier. In the U + Cm system, the structure seems to have one prominent peak at 316 ± 10 keV which is excited within a narrow range of projectile energies. It maintains a constant mean energy and width over the range of projectile energies and scattering angles at which it has been detected. The fact that common peak structures are found in the U + U system with two very dissimilar experimental arrangements precludes that they are generated spuriously by the apparatus. However, in apparent contrast to U + Cm, the situation in U + U appears more complex, involving the presence of multiple peaks in some spectra and the variation of peak position and width with bombarding energy and scattering angle.

With the structure found in the supercritical collision systems, of course, it is of great interest also to examine collision systems such as U + Pb ($Z_u = 174$) and Pb + Th ($Z_u = 172$) where supercritical binding is not expected to be present. Figures 5.35 and 5.36 show, respectively, some data obtained with the "EPOS" system on 5.9-MeV/u U + Pb collisions (Gr 82) and with the "Orange" spectrometer on 5.9-MeV/u Pb + Th collisions (Cl 84). As demonstrated in Figure 5.17, excellent separation of close and distant collisions is achieved in the "EPOS" measurement of U + Pb. To make a direct comparison with the U + Cm data, the two

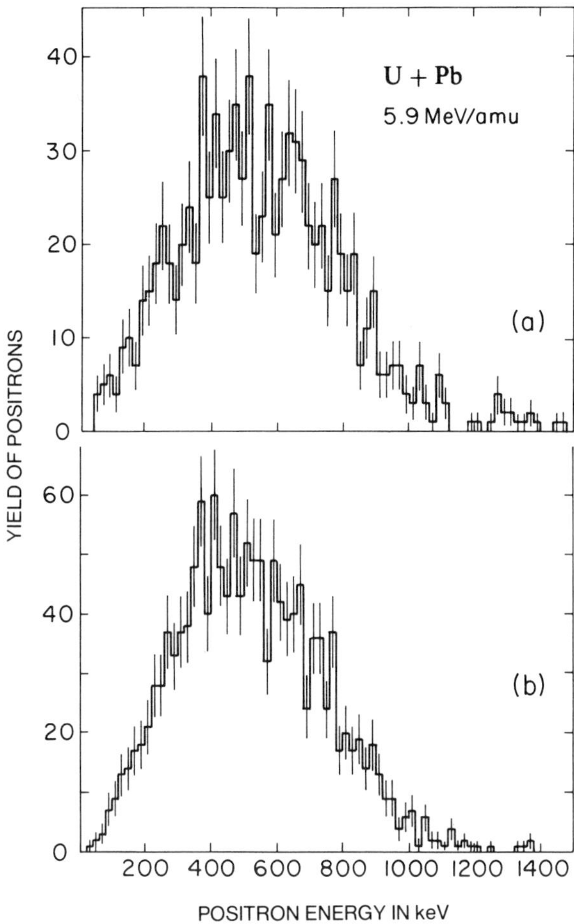

Figure 5.35. Positron energy spectra from $^{238}U + ^{208}Pb$ collisions at 5.9 MeV/u subject to the same kinematic conditions specified in Figure 5.31 (Gr 82) (EPOS data).

spectra displayed in Figure 5.35 were taken in coincidence with projectiles scattered into the same elastic scattering center-of-mass scattering angles that are associated with the spectra shown in Figures 5.31. If structure is present in these spectra, certainly it is much less prominent here than in the U + Cm and U + U collision systems. A similar result occurs for the Pb + Th spectrum obtained in coincidence with particles scattered into $34° \leqslant \theta \leqslant 53°$. It should be noted that one of the difficulties encountered in studying the U + Pb and the Pb + Th systems is that $\gtrsim 50\%$ of the total positron yield is contributed by the smooth background from pair conversion of nuclear

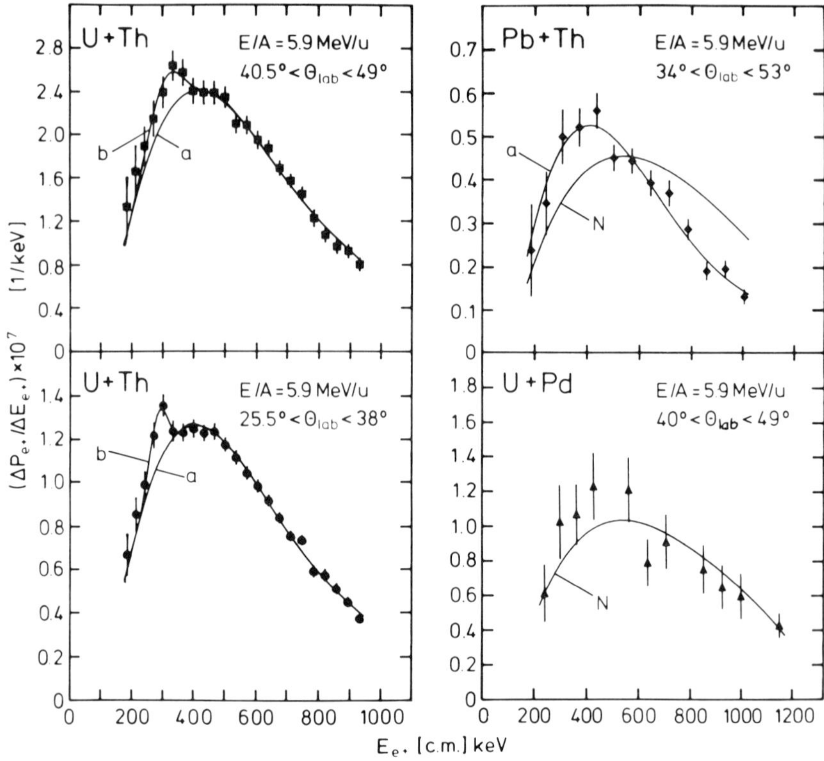

Figure 5.36. Positron energy spectra for U + Pd, Pb + Th and U + Th collisions at 5.9 MeV/u bombarding energy and scattering angles as indicated (Cl 84). Curves (a) and (b) represents similar fits to those in Figure 5.33, and curve (N) shows the subtracted nuclear background for the Pb + Th system. The spectrum for U + Pd is scaled with the ratio of the high-energy gamma-ray yield for U + U to that for U + Pd ("Orange" data).

gamma-ray transitions. It, therefore, reduces the sensitivity to observing structure of atomic origins in these cases. However, presently, there does not seem to be any evidence that structure exists in these two subcritical systems.

5.4.3. Origin of Positron Peak Structures

Whatever the sources of the peaks observed in U + U and U + Cm collisions, it is apparent that the explanation must be sought outside the scope of the theory for positron creation based on Rutherford scattering alone since this mechanism does not allow for narrow peak structure in the positron spectrum. Details of the effects observed differ among experiments due to differences in the detection schemes and in the measurement conditions utilized, but the broad conclusions are similar in the following sense: all the experiments carried out to date indicate that there are positron production mechanisms in superheavy collision systems which are coupled to time scales that are long compared to those associated with the Coulomb trajectories. We noted previously that, even discounting Doppler broadening, the narrow peak widths observed dictate lifetimes for the positron emitter which are more than an order of magnitude longer than the Coulomb scattering time of $\simeq 10^{-21}$ s.

Spontaneous positron emission from a long-lived giant nuclear composite system is one of the candidate mechanisms for producing a narrow positron peak. Unfortunately, as was pointed out in Section 5.2.2, internal pair conversion processes in the final-state nuclei proffer a less interesting nuclear alternative with a spectrum of lifetimes which readily exceeds the lower limit required. The experimental effort is presently being directed at exploring these atomic and nuclear alternatives.

Particularly the question of peak origin is being addressed with two experimental approaches: (i) One of these exploits the measured width and shape of the positron peak together with the influence of the Doppler effect on the peak shape to provide information on the source of the peak. The line shape discriminates between some emission processes, and the Doppler shift and broadening, under favorable circumstances, supply a measure of the emitter velocity. These two properties can be combined for an effective diagnostic technique. (ii) In complimentary measurements a search is being conducted for other radiations which should accompany nuclear transitions producing positrons. It was noted in Section 5.2.2 that these include gamma rays and electrons and that it is possible to predict the yield of the latter from the positron yield through readily calculable branching ratios. In these measurements the issue is the threshold for detection.

The following section discusses some of these considerations together with additional observations pertinent to the question of peak origin. For this purpose we turn mainly to the U + Cm system and use these data to demonstrate the type of measurements and analyses being carried out (Sc 83b). We also consider the data on U + U as well as new data on the U + Th and Th + Cm systems to present the interesting current experimental situation.

5.4.3a. Nuclear Background. In Section 5.2.2 we listed two nuclear processes that could be responsible, in a significant way, for the production of a peak structure in the positron spectrum: one of these, normal internal pair conversion (IPC), leads to a triangular-shaped energy spectrum (Figure 5.12) with a sharp cutoff at the maximum energy ($E_x - 2mc^2$). The other, which involves the capture of the pair electron into an empty K shell, produces monoenergetic positron emission (MPC) with an energy ($E_x - 2mc^2 + |E_B^K|$). The line-shape studies argue strongly against the possibility that internal pair conversion of both kinds are responsible for the positron peak in the U + Cm system. Further, compelling evidence for excluding IPC and MPC as the source of the peak appears to be supplied by the investigations of the gamma-ray and electron spectra. Let us first consider the information derived from the line-shape studies.

In addition to the intrinsic line profile which reflects the emission process, contributions to the line shape are also made by the Doppler effect and the detector response function. In the "EPOS" system the Doppler effect can play a particularly important role since the symmetric detection of positrons emitted parallel and antiparallel to the emitter motion results in a maximum Doppler broadening. In fact the Doppler line shape is very simple since for a particular momentum p^*_{e+} (keV/c) in the emitting system, the probability for positrons emitted at the laboratory kinetic energy E_{e+} is given by

$$\frac{dP_{e+}}{dE_{e+}} = \left[\frac{dP_{e+}}{d(\cos\theta^*)}\right]\left[\frac{1}{\gamma\beta p^*_{e+}}\right] \tag{5.13}$$

where $dP_{e+}/d(\cos\theta^*)$ is the angular distribution of the positrons in the emitting system, and the quantities $\beta = v/c$ and $\gamma = (1-\beta^2)^{-1/2}$ refer to the motion of the emitter relative to the laboratory. It is evident, therefore, that isotropic emission in the source system produces a uniform spectrum with a width $\Delta E_{e+} = 2\beta\gamma p^*_{e+}$ and a centroid offset from E^*_{e+} by the energy $(\gamma-1)(E^*_{e+} + mc^2)$. (We may recall that isotropic emission is expected within the monopole approximation.) Typically for 6.05-MeV/u U + Cm collisions with $E^*_{e+} \simeq 315$ keV and $\beta \simeq 0.056$, $\Delta E_{e+} \simeq 72$ keV and $E_{e+} - E^*_{e+} \simeq 1.5$ keV. Some distortion of this simple shape is produced due to the acceptance of the transport system. The solenoid axis is oriented perpendicular to the mean scattering plane defined by the two counters so that the positrons with maximum Doppler shift have the largest solid angle for detection. This geometry, together with the transport efficiency as a function of positron energy and emission angle, produces a line shape with the maximum intensity concentrated at the maximum Doppler energy shifts $\pm\gamma\beta p^*_{e+}$. Thus, in the case of an intrinsic line width much narrower than $2\beta\gamma p^*_{e+}$, the Doppler broadening, modified by the detector resolution function (FWHM $\simeq$ 10 keV), provides a clear measurement of the emitter velocity. For some favorable scattering angles this velocity can be utilized to identify the emitting source. For example, Figure 5.37 correlates the Doppler broadening with the emitter velocity and the scattering angle of the emitter (Co 84). It shows that an intrinsically narrow line emitted by the U + Cm system from the center of mass can, in principle, be differentiated from emission by the final-state particles scattered into the angular region specified in Figure 5.31(a) or into 45° where a peak has also been observed. Obviously the sensitivity of the measurement depends on how much of the total width is contributed by the intrinsic line and on the uncertainty in the

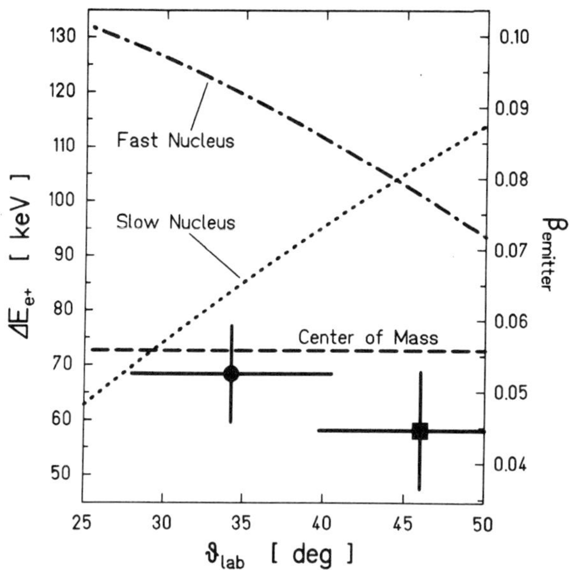

Figure 5.37. The Doppler broadening, ΔE_{e^+}, expected in the EPOS system for a ~300-keV monoenergetic positron line spectrum emitted by the slow and fast nuclei or the center of mass in the binary scattering event U + Cm at 6.05 MeV/u as a function of laboratory detection angle. The notation "fast" or "slow" only refers to detection angles less than 45°. The Doppler broadening for center-of-mass emission is independent of the detection angle. The figure also shows linewidths measured at two laboratory scattering angles. In the latter the width is extracted from the best fit to the Doppler broadened line shape (Co 84).

width measurement. A salient identifying feature is particularly provided by the fact that the Doppler broadening associated with center-of-mass emission is independent of the angle of scattering.

The measured peak width in U + Cm is, within measurement error, consistent with the Doppler broadening of ≃72 keV expected for emission by the center of mass. Then even a crude measurement of the width substantially eliminates normal internal pair conversion as the source of the peak since it is much too narrow to fit the much broader peak expected. A more specific analysis of the line shape anchors this conclusion more quantitatively (Sc 83b). For IPC the linewidth is dominated by the intrinsic shape which has a half-width of ~150 keV for all multipolarities (Sc 83). Correcting this spectral profile by Doppler broadening and detector resolution broadens it further to a width in excess of ~200 keV. Figure 5.38(a) (Sc 83b) illustrates an attempt to fit the calculated line shape to the data by varing the transition energy, the intensity, and the continuous distribution. As in Figure 5.31a, the smooth dashed line in the plot represents a fit to the smooth part of the spectrum with the dynamic shape and nuclear

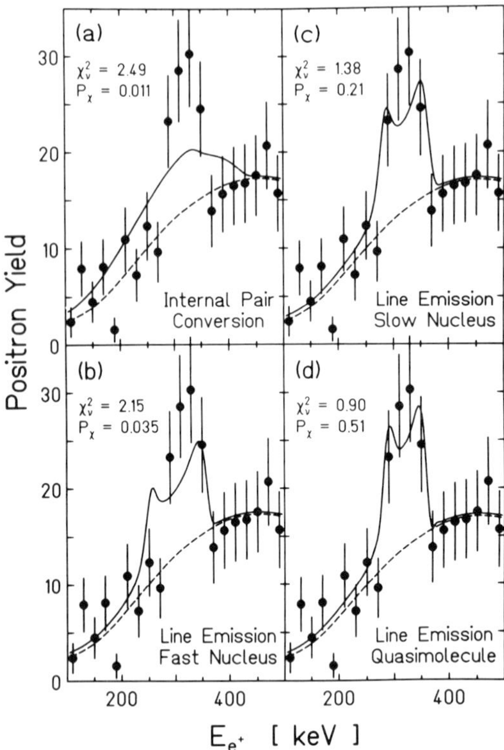

Figure 5.38. Maximum-likelihood fits to the energy spectrum of Figure 5.31a assuming origins for the peak structure as indicated. The fit involves varying the peak intensity, its position, and the continuous distribution of Figure 5.31a. The fit is made to the entire spectrum of Figure 5.31a, and the χ^2 is quoted for the 200-keV region centered at the peak (Sc 83b) (EPOS data).

background, corrected for the proper response functions. A chi-squared test for the 200-keV region surrounding the peak shows that IPC of any multipolarity is a very unlikely match to the data at a confidence level of better than 98%.

Unlike the situation for IPC, for the monoenergetic positrons expected from the MPC process the measured peak would be dominated by the Doppler broadening. Fits have also been made to the spectrum in Figure 5.31a assuming that monoenergetic positrons are emitted by the two ejectiles (Sc 83b). Figures 5.38b and 5.38c show the best fits to the data obtained for the two possibilities where it is assumed that positrons are emitted from either the fast or the slow collision products corresponding to the kinematics imposed by Figure 5.31a. The spectrum emitted by the fast fragment again produces a poor fit to the data, rejecting this possible source at a confidence level of better than 95%. Emission by the slow fragment leads to a fit which cannot be rejected on statistical grounds only. The statistical accuracy achieved is not sufficient to distinguish emission by the slow fragment from emission by the center of mass. However, additional data on U + Cm, obtained recently with particle detectors of improved angular resolution so that the angular region near 45° scattering with slightly negative Q can be studied, yield a peak at the same mean energy and peak width found in Figure 5.31a (Co 84). These data are displayed in Figure 5.39. As shown in Figure 5.37, the linewidths deduced for the two scattering angle regions appear to favor a Doppler broadening independent of scattering angle and, therefore, emission by the combined system.

As noted above, the gamma-ray and electron emission measurements also support the view that the peak positrons are not nuclear in origin. In these measurements a search is made for gamma-ray and electron lines at positions and with intensities predicted by the peak positron intensity. The limit on the sensitivity of such measurements is the accuracy with which the positron intensity can be determined. We present the information obtained from the gamma-ray spectra first.

If it is assumed that IPC is responsible for the positron peak, Doppler-shifted gamma-ray lines are expected at 1406 and 1272 keV, depending on which final-state nucleus emits the gamma ray. The expected gamma-ray intensity is readily calculated from the measured probability of positron emission, $P_{e^+} = (2.5 \pm 0.5) \times 10^{-5}$, using Equation (5.8). Figure 5.40a compares the intensity of gamma rays necessary to explain the positron peak as an IPC process to the gamma-ray spectrum measured with the same

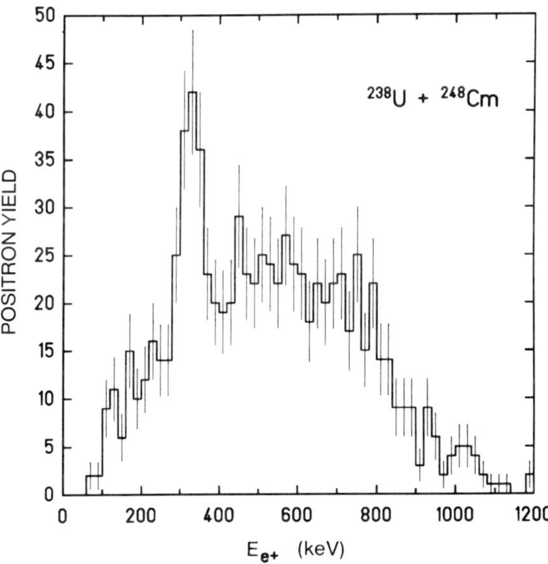

Figure 5.39. Spectral distribution of positrons from $^{238}U + ^{248}Cm$ collisions in coincidence with ions scattered near 45° in the laboratory. The mean bombarding energy is 6.08 MeV/u. As shown in Figure 5.37, the Doppler width is compatible with emission of a narrow line by the center of mass (Co 84) (EPOS data).

kinematic conditions as the peak (Co 84). The worst case situation is assumed where the positron intensity is divided equally between the two emitters. The calculated gamma-ray peaks are obtained using the internal pair conversion coefficient, β, given by Schluter *et al.* (Sc 83), and include the effects of Doppler broadening and detector resolution. They are plotted

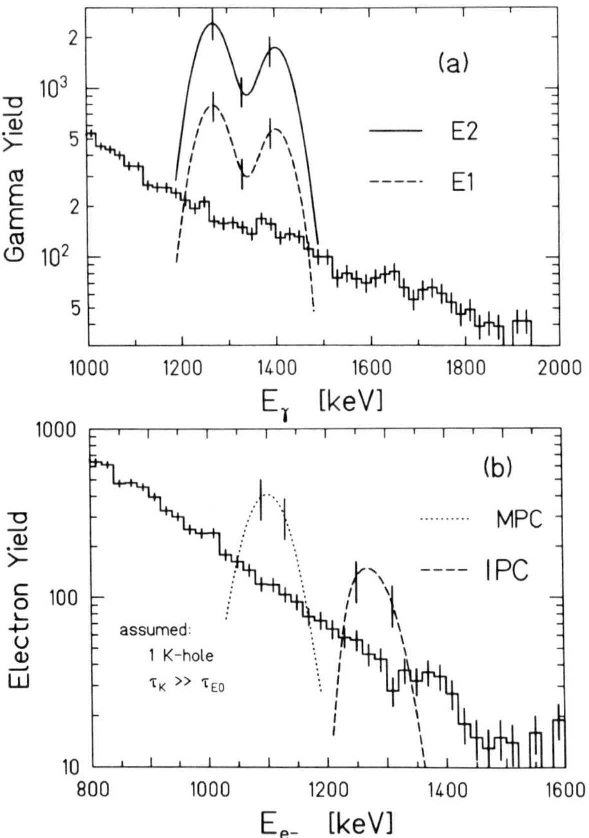

Figure 5.40. (a) Gamma-ray from $^{238}U + ^{248}Cm$ collisions measured simultaneously with the positrons subject to the same kinematic conditions that produce the positron peak (Co 84). The intensity of the two peaks is calculated from the intensity of the positron peak using Equation (5.8). The two peaks are associated with the two scattered ions as sources. The comparison is made for $E1$ and $E2$ multipolarities; other multipolarities result in more intense peaks. (b) Electron spectrum from $^{238}U + ^{248}Cm$ collisions measured simultaneously with the positrons and subject to the same kinematic conditions that produce the positron peak. The electron peaks are calculated from the positron peak intensity assuming $E0$ multipolarity both for the normal (IPC) and monoenergetic (MPC) internal pair conversion process (Co 84) (EPOS data).

here without adding them to the measured spectra. It is clear that the gamma-ray intensity required to explain the yield for the positron peak far surpasses the measured intensity even for the most strongly converted $E1$ transition. For this case the positron intensity would reflect with the minimum sensitivity in the gamma-ray spectrum. The fact that the predicted intensity exceeds the measured unstructured gamma-ray spectrum by more than four standard deviations appears to preclude an $E1$ or any higher multipole IPC as the mechanism for producing the positron peak.

Considering the MPC process, again using the appropriate branching ratios given by Schluter *et al.* (Sc 83), leads to an almost identical conclusion quantitatively. The only internal pair conversion process which is not deemed improbable by the gamma-ray measurements as the source for the peak positrons is internal pair conversion of an intense E0 transition. In fact, it was pointed out in Section 5.2.2, Figure 5.12, that the most favorable conditions for observing monochromatic positrons from MPC occurs for E0 transitions if the required vacancy is available at the time of transition. The electron spectra provide the opportunity to search for these E0 transitions and, indeed, they can be used effectively to argue against this mechanism for producing positron peaks.

For the IPC process an electron internal conversion line from the K shell in the U atom will appear at a mean energy of 1276 keV, while for the MPC process the equivalent line will be at 1106 keV. Both lines will be Doppler broadened in "EPOS" to $\simeq$90 keV widths.

Some results obtained from the electron measurements are displayed in Figure 5.40b. It shows an electron spectrum measured simultaneously with the positrons, again, subject to the same kinematic cuts that produce the positron peak. The spectrum exhibits the familiar smooth profile seen in the delta-electron studies discussed in Section 4 without any significant structure present at the interesting energy regions. The peaks shown, superimposed (not added) onto this spectrum, correspond to the normal and monoenergetic line spectra deduced from the positron intensity. All necessary instrumental line shapes have been folded into the calculations which use electron conversion coefficients supplied by Schlüter *et al.* (Sc 83). Of course, the intensity of the electron line varies with the occupancy of the atomic orbital at the time of conversion. For orientation, one K vacancy is assumed present indefinitely after the collision. It is clear that under such circumstances it would be very improbable to miss the electron line.

However, the question of vacancy availability during the conversion process is, obviously, an important consideration. Both internal conversion processes decrease with increasing x in Equation (5.9). What is more, electron internal conversion in the MPC process not only decreases for increasing x, but the monoenergetic positron line also grows simultaneously. A quantitative analysis shows (Co 84) that in order to explain both the positron peak intensity and the structureless electron spectrum with a confidence level of 95% requires that there are ~1.8 vacancies available on the average for both internal conversion mechanisms. This is a rather unrealistic situation given that an empty K shell is needed for the nuclear lifetime $\tau_N(\text{E0}) \simeq 10^{-12}$ s, and that the lifetime of a K vacancy in a normal atom is $\sim 10^{-17}$ s. The sources of K vacancies are ionization in the collision, typically <10%, and internal converison of nuclear transitions which may have a high multiplicity. But irrespective of the vacancy production mechanism, in order for them to be sustained, the atomic state would have to be substantially devoid of electrons during a period $\sim \tau_N$. Mean atomic collision times of $\sim 10^{-17}$ s, which characterized the capture time of electrons into outer shells, quench the high degree of ionization. Although atomic cascade times would slow down the filling of the K vacancies, even prolonging the neutral atom lifetime of 10^{-17} s by several orders of magnitude falls short of meeting the time matching discrepancy between the nuclear and atomic lifetimes. It is very difficult, therefore to visualize a situation where $[P_K \, \tau_K/(\tau_N + \tau_K)] > 1.8$ since for $P_K = 2$ (vacancies), $\tau_K/\tau_N \simeq 10$. It should also be noted that the MPC line is always accompanied by the broader energy distribution associated with both pair members emitted into continuum states. There is no evidence for this component. Nuclear monoenergetic positron emission has to be considered an improbable process.

Therefore, within the context of the existing data on U + Cm the nuclear processes do not seem to provide a likely explanation for the positron peak. We have seen that the gamma-ray data rule out internal pair conversion for all gamma-emitting multipoles ($L \geqslant 1$). E0 conversion also offers only a very improbable alternative. Moreover, the line-shape studies are consistent with these findings. A search for gamma-ray and internal electron conversion lines corresponding to the $\simeq$310-keV positron peak found in the "Orange" spectrometer U + U data has also been reported (Be 82b, Cl 84). Some results are displayed in Figure 5.41a and 5.41b. Although the exact circumstances are a little different in this case, the conclusions are similar to those found in U + Cm: it is difficult to explain the 310-keV peak in U + U by nuclear conversion processes.

Although not exhaustive, the negative results obtained in all existing tests for nuclear sources of the positron peak open up to serious consideration the possibility that spontaneous positron production is being observed. Some of the observations presented appear to encourage this viewpoint. However, data have also been presented which may not be readily incorporated into any straightforward explanation of the positron peaks within the context of spontaneous positron emission and the theoretical models involved. This interesting situation is considered next.

5.4.3b. Spontaneous Positron Emission. The following points, developed

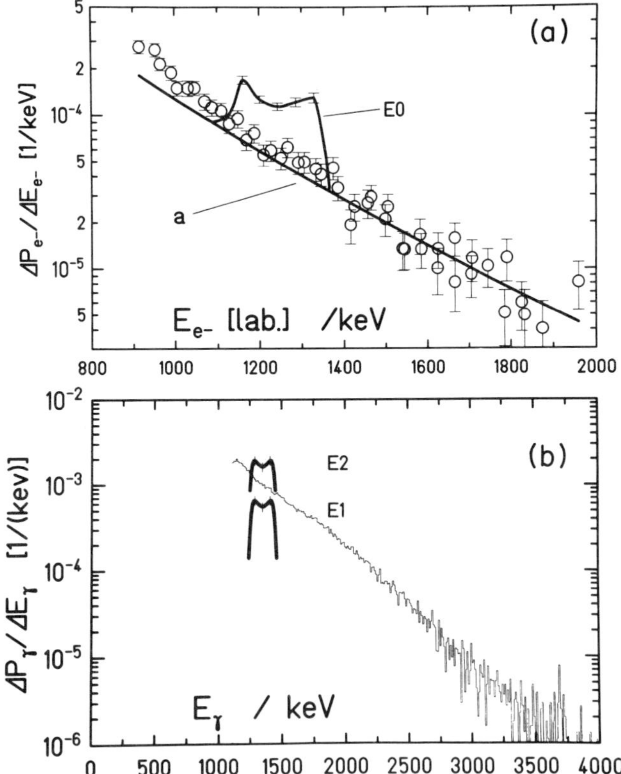

Figure 5.41. Electron and gamma emission probabilities from U + U collisions at 5.9 MeV/u and scattering angles between 40.5° and 49°. The double humped Doppler broadened lines would be expected to be superimposed on the observed spectra under the assumption that the positron lines would originate from nuclear decays in the separated U-fragments for multipolarities as indicated (Cl 84) ("Orange" data).

in the U + Cm data, appear to be consistent with spontaneous positron emission:

(i) The mean peak energy of 316 ± 10 keV coincides with the kinetic energy calculated (Re 81) for positrons spontaneously emitted from the composite U + Cm nuclear system at the minimum internuclear separation reached in head-on collisions.

(ii) The mean energy and width of the peak found in U + Cm seems to be independent of the projectile scattering angle. The Doppler broadening averaged over equal contributions from the fast and slow ejectiles detected always exceeds the measured width. Within error the measured widths are consistent with the emission of a Doppler broadened line from a source moving with the center-of-mass velocity.

(iii) The events associated with the positron peak exhibit angle–angle correlations that distinguish them from the other scatterings producing the continuous part of the positron spectrum. The mechanism for separation seems to involve ionization states for the final-state nuclei which differ from the usual velocity-dependent equilibrium mean charge states connected with Rutherford scattering.

(iv) The peak in U + Cm is excited only within a very narrow band of projectile energies compatible with the target thickness.

The observations on the line shape are particularly suggestive that spontaneous positron emission can be responsible for the peak observed in U + Cm. In fact, as demonstrated in Figure 5.38d, the best fit to the line shape is obtained with the assumption that a narrow positron line spectrum is emitted by the combined U + Cm system moving with the center-of-mass velocity. The intrinsic width emerging from this analysis is <40 keV. Then if, indeed, the peak represents spontaneous positron emission, the parent nuclear supercritical charge must exist for a time compatible with this intrinsic width. As discussed in Section 5.2.1c, the translation of the width into a lifetime depends on the time development of the nuclear reaction, but, clearly, times greater than 10^{-20} s are involved. Assuming the formation of such a metastable giant nuclear complex, Müller *et al.* (Mu 83b) have provided a more quantitative analysis of the spectrum in 5.31a.

In their analysis they employ the schematic sticking model outlined in Section 5.2.1c which simulates the time evolution of the reaction with a sharp time interval T. The spectrum combines a fraction q of reactions with positron production from Rutherford scattering events. The line profile has the functional dependence on T given in Equation (5.5). To fit the width and position of the peak requires a time delay $T \gtrsim 6.5 \times 10^{-20}$ s and an inter-nuclear separation during the reaction of $\sim$16 fm. It should be noted that Doppler broadening as well as the resolution and efficiency responses of the "EPOS" system have not been taken into account in this fit. This has no major consequences on the conclusions except the extracted reaction time now becomes a lower limit for this model. It also bears repeating that a fixed reaction time is not a very realistic representation of the time development of the nuclear reactions, and other time distributions such as an exponentially decaying state may be more suitable. The latter would reduce the mean reaction time deduced from the peak width, as noted in Section 5.2.1c, but the essential feature of the long reaction time to fit the data, $>10^{-20}$ s, remains.

The other interesting point to emerge from this analysis is obtained by comparing the observed probability for producing the peak with the theoretical calculations. This leads to the estimate that only a fraction $q \simeq 2 \times 10^{-3}$ of delayed collisions per elastically scattered ion is sufficient to produce the peak observed. Thus, these are very rare events which could explain why there is no previous indication for the formation of such composite systems: they could have readily escaped detection in conventional scattering experiments involving measurements of the kinematics and properties of the reaction products. Indeed, the observations (iii) and (iv) listed above single out the peak positron events as being unusual. In particular, the pronounced resonancelike behavior may reflect the formation of the metastable nuclear complex which provides the parent supercritical charge to spark the vacuum. The gentle formation for long lifetimes and then the breakup of the system into primarily binary components without any major energy loss and mass transfer makes these events special. In this sense, spontaneous positron emission may provide a very sensitive probe to select out these unusual reactions and thus provide access to new nuclear physics.

The first observation listed above also appears to provide a link with the spontaneous emission process. There happens to be close agreement between the mean energy of the peak in U + Cm and the theoretical calculations for $1s\sigma$ binding in the quasimolecule when the two nuclei barely touch at approximately 17 fm and the ionization state of the system is at a reasonable value of +50 (see Figure 5.5). Of course, in making such comparisons cognizance has to be taken of the sensitivity of the $1s\sigma$ binding energy to the internuclear separation, the shape of the composite system, and the screening effect of the electrons. We previously brought attention to the influence the nuclear dimensions have on the collapsing wave functions and binding energies. Nevertheless, in the U + Cm system the positron peak is located at a plausible energy compatible with calculations for spontaneous positron emission.

Unfortunately, this correspondence of the peak energy with binding energy of the $1s\sigma$ state does not seem to extend to the other systems measured. In U + U the most prominent peak again occurs at $\simeq$300 keV, whereas Figure 5.5 indicates that for touching spheres and a reasonable $\sim 50^+$ ionization state the mean energy of the spontaneous positron peak is predicted to be located at $\simeq$180 keV. Therefore, associating both this peak and the pronounced peak observed in U + Cm with spontaneous positron emission from a ground-state configuration of the giant intermediate nucleus clearly leads to the difficulty that we have to exclude any simple scaling of the $1s\sigma$ binding energy with the combined charge of the collision system. In order for the $1s\sigma$ binding energy in U + U to accommodate the peak at $\sim$300 keV, a drastically different nuclear configuration is required where the internuclear separation diminishes to $\simeq$12 fm or, alternatively, a situation where the ionization state increases so that the giant nuclear complex is completely stripped of electrons. Obviously, both situations represent extremes for which there are presently no detailed explanations.

The questions raised by the attempt to associate the peak in U + U as well as the peak in U + Cm with spontaneous positron emission are further compounded by the results of recently reported measurements on other systems with supercritical binding. Figure 5.36 presents data on the U + Th system measured with the "Orange" spectrometer (Cl 84). Peak structures again appear near ~300 keV whereas a diving energy of ~130 keV is predicted by calculations (Re 81). The widths and prominence of these structures are very similar to those found in U + U by this apparatus. The Th + Cm system has been measured with the "EPOS" system specifically to trace the scaling with Z_u (Co 84). In this case, simply adjusting the distance of minimum approach so that the nuclear overlap parallels the conditions encountered in the U + Cm collisions predicts that the positron peak should shift down to a convenient energy of ~250 keV where the detection efficiency of the "EPOS" apparatus is high. The Th + Cm system offers the particular advantage of not including the U nucleus which is common to all the other collision systems studied and, therefore, it can be suspect as the source of a common peak at ~300 keV. It also has the added advantage of sufficient mass asymmetry to allow a good separation of forward and backward elastic scattering events. Figure 5.42 shows that a very striking peak also appears in the Th + Cm spectrum at projectile energies of 6.00 and 6.04 MeV/u. Its emergence above the continuous background also displays the sensitivity to the bombarding energy observed with the U + Cm system (Sc 83b). The bombarding energy at which the peak appears roughly corresponds to the distance of minimum approach where the nuclear surfaces marginally overlap. The width and other characteristics of this peak are very similar to those found in the U + Cm collisions.

It is apparent that as a principal feature all the supercritical collision systems studied have in common a peak structure centered at approximately 300 keV. This pattern of near degeneracy in peak energies stands out as the most salient result of these investigations. We have already pointed out that in order to incorporate such a pattern of peak energies into the context of spontaneous positron emission involving the formation of a giant metastable nuclear complex formed from the target and projectile charges, we must necessarily invoke radically different charge configurations or ionization states for the compound systems. Of course, the evaporation of charged particles in the formation of the giant system together with the excessive sensitivity of the $1s\sigma$ binding energy to the charge configuration can allow considerable latitude in the location of the supercritical state. Yet the near degeneracy in energy is probably more than a coincidence. Perhaps in the context of forming a giant system there may be the possibility of forming an exceptionally stable common nuclear complex in all the collision systems.

In reaching for an explanation of the data, it bears emphasizing that, exclusive of spontaneous positron emission, it is difficult to invent a mechanism that can produce the narrow positron peaks observed, particularly since the emission appears to be by the combined system in

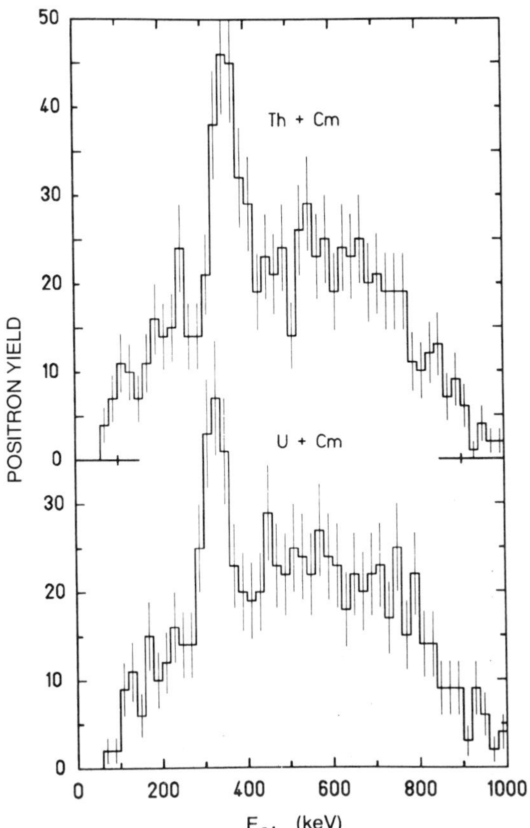

Figure 5.42. $^{232}Th + {}^{248}Cm$ and $^{238}U + {}^{248}Cm$ positron spectra obtained at average bombarding energies of 6.02 and 6.08 MeV/u, respectively. The kinematic conditions for scattering are similar in the two cases (Co 84) (EPOS data).

U + Cm and Th + Cm collisions (Co 84).Trivial origins for the peaks, such as nuclear decay from the individual nuclei, are rejected by the measurements with both experimental arrangements. We can consider the possibility of the internal conversion of an exceptionally fast nuclear transition ($\tau \lesssim 10^{-19}$ s) in the composite nuclear complex. In this case the peak may be narrow and a very large internal conversion coefficient could preclude detecting the transition in the gamma-ray and electron spectra. But if this process exists, so must spontaneous positron emission. In the most recent experiments with the "EPOS" system particular care has been taken to maintain a high efficiency for detecting positrons with energies as low as 100 keV so that the range of energies predicted for spontaneous positron emission could be probed. To date, as already described, no simple scaling with Z_u has been observed.

There is clearly latitude to explain the data in the context of spontaneous positron emission. But, presently, this must be considered an open experimental task and, in general, the origin of the narrow peak structures must be viewed as an open question. One direction to pursue this question involves an understanding of the nuclear physics associated with the possible formation of giant nuclear systems. Towards this goal, the excitation of the peaks must be studied in great detail over a broad range of bombarding energies, and, particularly, the species comprising the final scattering states have to be identified and their energy measured. Such experiments are presently beginning. As was mentioned previously, positron spectroscopy may provide the probe for studying giant nuclear complexes if such objects are being formed in these experiments.

6. Summary

The Introduction outlined the broad range of phenomena which presently delineate the field of high-energy atomic and molecular physics. In this article we have only been able to make contact with a small fraction of the ongoing research. The concentration has been on the investigations involving superheavy collision systems. They have been singled out because they particularly represent new directions which can provide access to a previously unexplored domain of atomic physics where relativistic effects play a dominant role. Specifically, the quantum mechanics of strong binding can be examined as a leading effect and the fundamental problem of how strong external fields can rearrange the structure of the vacuum state can be tested in possibly the only such experimental opportunity available to us. In connection with the latter, the search for spontaneous positron emission has become the central experimental issue. This quest has spawned many of the investigations discussed in this article.

Since the source of the strong electric fields is to be found in the formation of the superheavy quasimolecule (or quasiatom), the quasimolecule and its properties have been at the focus of many of the studies. In this regard we have included in our discussion the work on low-Z systems since they have been particularly useful in testing the theoretical framework of the quasimolecular description for electron excitation and radiation in the collision. Indeed, our basis for considering the quasimolecule as a meaningful concept has been developed with the experiments on low-Z systems. With examples of characteristic x-ray studies we have considered the couplings leading to excitation and ionization of inner shells. We have particularly documented the important role played by electron promotion.

The studies of MO x-ray emission cited have provided some of the more direct evidence for quasimolecule formation. Much of this model's framework developed for the lower-Z systems has been successfully carried over to the higher-Z systems.

However, we have also seen that in many respects the superheavy collision system is both theoretically and experimentally disassociated from the rest of the ion–atom collision studies. The singular behavior which distinguishes the superheavy system is the transfer of very large energy and momentum in the excitation processes due to the strong binding and the high degree of localization of the inner-shell electrons. We have seen these properties manifest themselves in the large cross sections for K-shell ionization and in the concentration of the ionization at the small internuclear separations. Both these characteristics are essential to the observation of spontaneous positron emission. The high-momentum components of the initial-state wave functions are most graphically displayed in the high-energy component of the delta electron spectrum, as our comparison of low-Z and high-Z collision spectra convincingly demonstrates. All these effects can only be produced by the combined Z_u of the target and projectile acting together so that with these studies we have found that the signatures for the formation of superheavy quasiatomic systems are clearly observed.

The investigations into positron emission from the superheavy collision systems have confirmed the dynamic mechanism for positron production considered theoretically. Particularly, the success achieved by theory in reproducing the continuous spectra demonstrates that the details of positron production associated with the time-varying fields produced by the nuclear motion are well understood. It is especially striking to verify the $\sim Z_u^{20}$ dependence of the cross section which reflects the strong fields and the effects of the wave function collapsing about the nuclear centers produced by the relativistic effects. This aspect of positron production provides an especially clear test of the theory of electronic excitations in this highly relativistic situation for the wave functions.

However, with the studies seeking spontaneous positron creation it has become apparent also that a new dimension of questions has to be addressed. Not anticipated was the observation of the narrow peaks in the positron spectra from superheavy systems colliding at bombarding energies close to the Coulomb barrier. The peaks appear in four collision systems. They cannot be attributed to any trivial effect reflecting nuclear transitions from the individual final-state nuclei, and they seem to be associated with emission by the combined system. The narrowness of the peaks implies a parent state which lives longer than 10^{-20} s. The observations that the scattering kinematics associated with the peaks in the U + Cm and Th + Cm differ from Rutherford scattering and that these peaks are produced in narrow intervals of bombarding energy emphasize the special nature of these events.

However, the absence of a simple scaling with Z_u presently precludes any straightforward association of the peaks with spontaneous positron emission from a giant metastable nuclear composite system. To reconcile all the data into this context we would have to introduce the complexity of drastically changing nuclear configurations or the formation of a similar superstable giant nucleus in each collision. But the experiments do open the scenarios for forming giant nuclei to serious consideration and with them the possibility of engaging in a spectroscopy of such objects.

It is apparent, therefore, that the four questions posed in the Introduction have only begun to be answered. With respect to the possibility of utilizing superheavy collision systems to extract information on superheavy atomic species, we have found that the dynamic excitation mechanisms provide information on bound-state and continuum wave functions and, in an approximate way through the average energy transfer, on the binding energies of inner-shell states. These types of measurements do not provide an energy level diagram but they do establish some of the basic relativistic effects being sought. The other questions regarding the existence of supercritical binding, the possibility of employing superheavy collision systems to observe spontaneous positron emission, and the possibility of exploiting the positron spectrum as a probe for the formation and properties of superheavy nuclear species all center on understanding the nature and source of the narrow positron lines that have been observed. The absence of sources of known low-energy line positron spectra in nature which can explain the observed peaks makes further pursuit of these and other related questions particularly exciting.

Acknowledgments

We are indebted to many people for contributions to the work reviewed in this chapter and for discussions and materials supplementing the published results. We are particularly grateful for stimulation and collaboration to H. Backe, H. Bokemeyer, D. A. Bromley, K. Bethge, T. Cowan, C. K. Davis, H. Folger, A. Gruppe, E. Liarokapis, J. J. O'Brien, J. Schweppe, D. Schwalm, H. P. Trautvetter, and T. Zouros. We thank P. Kienle and C. Kozhuharov for many discussion of their work and for supplying their data. Discussions on theoretical aspects with W. Greiner, B. Müller, U. Müller, J. Reinhardt, Theo de Reus, and G. Soff are especially acknowledged, and we wish to thank them for supplying unpublished calculations. The assistance of Rita Bonito and Annette Maita in preparing the manuscript, and Sandra Sicignano and Mike Jakubowicz in preparing the figures is particularly appreciated.

References

(Ab 76) T. Åberg, K. A. Jamison, and P. Richard, *Phys. Rev. Lett.* **37**:63 (1976).

(Af 65) V. V. Afrosimov, Yu. S. Gordeev, M. H. Panov, and N. V. Fedorenko, *Sov. Phys. Tech. Phys.* **9**:1248, 1256, 1265 (1965).

(Al 58) S. K. Allison, *Rev. Mod. Phys.* **30**:1137 (1958).

(Am 75) P. A. Amundsen and L. Kocbach, *J. Phys. B* **8**:L122 (1975).

(Am 76) P. A. Amundsen, *J. Phys. B* **9**:971 (1976).

(Am 76b) P. A. Amundsen, L. Kocbach, and J. M. Hansteen, *J. Phys. B* **9**:L203 (1976).

(Am 78) P. Amundsen, *J. Phys. B* **11**:L737 (1978).

(An 76) J. U. Anderson, E. Laegsgaard, M. Lund, and C. D. Moak, *Nucl. Instrum. Methods* **132**:507 (1976).

(An 76b) R. Anholt and T. K. Saylor, *Phys. Lett.* **A56**:455 (1976).

(An 77) R. Anholt and W. E. Meyerhof, *Phys. Rev. A* **16**:190 (1977).

(An 77b) R. Anholt, W. E. Meyerhof, and A. Salin, *Phys. Rev. A* **16**:951 (1977).

(An 78) R. Anholt, *Z. Phys.* **A289**:41 (1978).

(An 78b) R. Anholt, *Z. Phys.* **A288**:257 (1978).

(An 78c) R. Anholt, *Phys. Rev. A* **17**:976 (1978).

(An 78d) R. Anholt, *Phys. Rev. A* **17**:983 (1978).

(An 79) R. Anholt, *Z. Phys.* **A291**:289 (1979).

(An 79b) R. Anholt, private communication (1979).

(An 79c) R. Anholt, H. H. Behncke, S. Hagmann, P. Armbruster, F. Folkmann, and P. H. Mokler, *Z. Phys.* **A289**:349 (1979).

(An 79d) R. Anholt, W. E. Meyerhof, and Ch. Stoller, *Z. Phys.* **A291**:287 (1979).

(An 79e) C. H. Annett, B. Curnutte, and C. L. Cocke, *Phys. Rev. A* **19**:1038 (1979).

(An 79f) R. Anholt, W. E. Meyerhof, and Ch. Stoller, *Z. Phys.* **A291**:287 (1979).

(An 80) R. Anholt, Ch. Stoller, and W. E. Meyerhof, *J. Phys. B* **13**:3807 (1980).

(An 81) R. Anholt, *Phys. Lett.* **84A**:62 (1981).

(An 82) R. Anholt, *Nucl. Instrum. Methods* **198**:567 (1982).

(An 83) R. Anholt, "Molecular Orbital X-ray Production by Low-Velocity Highly Stripped Ions," Progress Report 1981–82, Stanford University (1983); and to be published.

(Ar 62) P. Armbruster, *Z. Phys.* **166**:341 (1962).

(Ar 64) P. Armbruster, E. Röckl, H. J. Specht, and A. Vollmer, *Z. Naturforsch.* **19**:1301 (1964).

(Ar 78) P. Armbruster, H. H. Behncke, S. Hagmann, D. Liesen, F. Folkmann, and P. Mokler, *Z. Phys.* **A288**:277 (1978).

(Ar 82) P. Armbruster, GSI Report, 82-5, preprint (1982).

(Ba 11) C. G. Barkla, *Phil Mag.* **22**:396 (1911).

(Ba 58) D. R. Bates and R. McCarrol, *Proc. R. Soc. London Ser. A* **245**:175 (1958).

(Ba 59) J. Bang and J. M. Hansteen, *K. Dan. Vidensk. Selsk. Mat.-Fys. Medd.* **31**:No. 13 (1959).

(Ba 71) G. Basbas, W. Brandt, and R. Laubert, *Phys. Lett.* **A34**:277 (1971).

(Ba 71b) G. Basbas, W. Brandt, R. Laubert, A. Ratkowski, and A. Schwarzchild, *Phys. Rev. Lett.* **27**:171 (1971).

(Ba 72) W. Bambynek, B. Crasemann, R. W. Fink, H. U. Freund, H. Mark, C. D. Swift, R. E. Price, and V. P. Rao, *Rev. Mod. Phys.* **44**:716 (1972).

(Ba 72b) M. Barat and W. Lichten, *Phys. Rev. A* **6**:211 (1972).

(Ba 73) G. Basbas, W. Brandt, and R. Laubert, *Phys. Rev. A* **7**:983 (1973).

(Ba 73b) G. Basbas, W. Brandt, and R. H. Ritchie, *Phys. Rev. A* **7**:1971 (1973).

(Ba 78) H. Backe, L. Richter, R. Willwater, E. Kankeleit, E. Kuphal, Y. Nakayama, and B. Martin, *Z. Phys.* **A285**:159 (1978).

(Ba 78b) H. Backe, private communication (1978).

(Ba 78c) H. Backe, L. Handschug, F. Hessberger, E. Kankeleit, L. Richter, F. Weik, R. Willwater, H. Bokemeyer, P. Vincent, Y. Nakayama, and J. S. Greenberg, *Phys. Rev. Lett.* **40**:1443 (1978).

(Ba 78d) G. Basbas, W. Brandt, and R. Laubert, *Phys. Rev. A* **17**:1655 (1978).

(Ba 79) J. Bang and J. M. Hansteen, *Phys. Lett.* **72A**:218 (1979).

(Ba 81) H. Backe, W. Bonin, E. Kankeleit, M. Kramer, R. Krieg, V. Metag, P. Senger, N. Trautmann, F. Weik, and J. B. Wilhelmy, in (Gr 81), p. 107 (1981).

(Ba 81b) J. Bang and J. M. Hansteen, *Phys. Scr.* **22**:609 (1981).

(Ba 82) H. Backe, Les Houches Lectures, 1982; calculations by G. Soff, private communication.

(Ba 83) H. Backe, P. Senger, W. Bonin, E. Kankeleit, M. Krämer, R. Kreig, V. Metag, N. Trautmann, and J. B. Wilhelmy, *Phys. Rev. Lett.* **50**:1838 (1983).

(Be 63) F. Beck, H. Steinwedel, and G. Sussmann, *Z. Phys.* **171**:189 (1963).

(Be 73) H. D. Betz, F. Bell, H. Panke, G. Kalkoffen, M. Solz, and D. Evers, *Phys. Rev. Lett.* **30**:358 (1973).

(Be 75) F. Bell, H. D. Betz, H. Panke, E. Spindler, W. Stehling, and M. Kleber, *Phys. Rev. Lett.* **35**:841 (1975).

(Be 75b) H. D. Betz, F. Bell, H. Panke, W. Stehling, E. Spindler, and M. Kleber, *Phys. Rev. Lett.* **34**:1256 (1975).

(Be 76) W. Betz, G. Soff, B. Müller, and W. Greiner, *Phys. Rev. Lett.* **37**:1046 (1976).

(Be 77) H. A. Bethe and E. E. Salpeter, *Quantum Mechanics of One- and Two-Electron Atoms*, Plenum Press, New York (1977).

(Be 79) E. Berdemann, F. Bosch, H. Bokemeyer, M. Clemente, F. Güttner, P. Kienle, W. Koenig, C. Kozhuharov, B. Martin, B. Povh, H. Tsertos, and Th. Walcher, GSI-Annual Report, p. 100 (1979).

(Be 79b) H. H. Behncke, P. Armbruster, F. Folkmann, S. Hagmann, J. R. MacDonald, and P. H. Mokler, *Z. Phys.* **A289**:333 (1979).

(Be 81) O. Benka, R. L. Watson, and R. A. Kenefick, *Phys. Rev. Lett.* **47**:1202 (1981).

(Be 82) F. Bell, G. Trollman, H. D. Betz, and E. Spindler, *Phys. Rev. A* **26**:72 (1982).

(Be 82b) E. Berderman, F. Bosch, M. Clemente, P. Kienle, W. Koenig, C. Kozhuharov, H. Tsertos, and W. Wagner, Jahresbericht (1982), Technischen Universität, Munchen, p. 58.

(Be 82c) E. Berdermann, M. Clemente, P. Kienle, H. Tsertos, W. Wagner, F. Bosch, and C. Kozhuharov, G.S.I., Scientific Report (1982), p. 147.

(Bh 73) C. P. Bhalla, *Phys. Rev. A* **8**:2877 (1973).

(Bh 75) C. P. Bhalla, *Phys. Rev. A* **12**:122 (1975).

(Bo 48) N. Bohr, *Mat.-Fys. Medd. Dan. Vidensk. Selsk.* **18**, No. 8 (1948).

(Bo 78) F. Bosch, H. Krimm, B. Martin, B. Povh, Th. Walcher, and K. Traxel, *Phys. Lett.* **78B**:568 (1978).

(Bo 80) F. Bosch, D. Liesen, P. Armbruster, D. Maor, P. H. Mokler, H. Schmidt-Böcking, and R. Schuch, *Z. Phys.* **A296**:11 (1980).

(Bo 81) H. Bokemeyer, K. Bethge, H. Folger, J. S. Greenberg, H. Grein, A. Gruppe, S. Ito, R. Schulé, D. Schwalm, J. Schweppe, N. Trautmann, P. Vincent, and M. Waldschmidt, in (Gr 81), p. 273 (1981), and to be published.

(Bo 82) H. Bokemeyer, H. Folger, H. Grein, Y. Kido, T. Cowan, J. S. Greenberg, J. Schweppe, K. Bethge, A. Gruppe, R. Merten, Th. Odenweller, K. E. Stiebing, D. Schwalm, P. Vincent, and N. Trautmann, G.S.I. Scientific Report (1982), 83-1, p. 146, to be published.

(Br 66) W. Brandt, R. Laubert, and I. Sellin, *Phys. Rev.* **151**:56 (1966).

(Br 66b) W. Brandt, R. Laubert, and I. Sellin, *Phys. Lett.* **21**:518 (1966).

(Br 70) W. Brandt and R. Laubert, *Phys. Rev. Lett.* **24**:1037 (1970).

(Br 72) J. S. Briggs and J. Macek, *J. Phys. B* **5**:579 (1972).

(Br 73) W. Brandt, Proc. Int. Conf. Inner-Shell Ionization Phenomena and Future Applications, Atlanta, Georgia (1972); (U.S. AEC, Oak Ridge, Tennessee, 1973), p. 948; see also references contained herein.

(Br 73b) J. S. Briggs and M. R. Hayns, *J. Phys. B.: At. Mol. Phys.* **6**:514 (1973).

(Br 73c) J. S. Briggs and J. Macek, *J. Phys. B* **6**:982, 2484 (1973).

(Br 74) J. S. Briggs, Atomic Energy Research Establishment Tech. Report No. #594 (1974).

(Br 75) J. S. Briggs and K. Taulbjerg, *J. Phys. B.: At. Mol. Phys.* **8**:1909 (1975).

(Br 75b) J. S. Briggs, *J. Phys. B.: At. Mol. Phys.* **8**:L485 (1975).

(Br 76) W. Brandt and K. W. Jones, *Phys. Lett.* **57A**:35 (1976).

(Br 76b) J. S. Briggs, *Rep. Prog. Phys.* **39**:217 (1976).

(Br 77) J. S. Briggs and K. Dettmann, *J. Phys. B* **10**:1113 (1977).

(Br 78) S. J. Brodsky and P. J. Mohr, in *Structure and Collisions of Ions and Atoms* (I. A. Sellin, ed.), Springer, Berlin (1978), Topics in Current Physics, Vol. 5, p. 3.

(Br 79) J. S. Briggs, J. H. Macek, and K. Taulbjerg, *J. Phys. B* **12**:1457 (1979).

(Bu 66) G. A. Burginyon and J. S. Greenberg, *Nucl. Instrum. Methods* **41**:109 (1966).

(Bu 72) E. H. S. Burhop and W. N. Asaad, in "The Auger Effect," *Adv. At. Mol. Phys.* **8**:163 (1972).

(Bu 81) P. Bürgy, G. Bonani, E. Morenzoni, M. Stöckli, Ch. Stoller, M. Suter, and W. Wölfli, *Z. Phys.* **A300**:105 (1981).

(Ch 13) J. Chadwick, *Phil. Mag.* **25**:193 (1913).

(Ch 14) J. Chadwick and A. S. Russel, *Phil. Mag.* **27**:112 (1914).

(Ch 69) B.-H. Choi and E. Merzbacher, *Phys. Rev.* **177**:233 (1969).

(Ch 70) B.-H. Choi and E. Merzbacher, *Phys. Rev. A* **1**:299 (1970).

(Ch 75) M. H. Chen, B. Crasemann, and D. L. Mathews, *Phys. Rev. Lett.* **34**:1309 (1975).

(Cl 75) D. L. Clark, T. K. Li, J. M. Moss, G. W. Greenless, M. E. Cage, and J. H. Broadhurst, *J. Phys. B* **8**:L378 (1975).

(Cl 83) M. Clemente, Ph. D. thesis, Technische Universität, Munchen (1983).

(Cl 84) M. Clemente, E. Berdermann, P. Kienle, H. Tsertos, W. Wagner, C. Kozhuharov, F. Bosch, and W. Koenig, Proc. Int. Conf. on Nuclear Physics, Florence, Italy, p. 693 (1983); *Phys. Lett.* **137B**:41 (1984).

(Co 34) W. M. Coates, *Phys. Rev.* **46**:542 (1934).

(Co 67) C. A. Coulson and A. Joseph, *J. Quantum Chem.* **1**:337 (1967).

(Co 73) C. L. Cocke and R. Randall, *Phys. Rev. Lett.* **30**:1016 (1973).

(Co 76) C. L. Cocke, R. R. Randall, S. L. Varghese, and B. Curnette, *Phys. Rev. A* **14**:2026 (1976).

(Co 80) C. L. Cocke, "Coincidence Experiments for Studying Impact-Parameter-Dependent and Charge-Changing Processes," (Ri 80), p. 303.

(Co 84) T. Cowan, H. Backe, M. Begemann, K. Bethge, H. Bokemeyer, H. Folger, J. S. Greenberg, H. Grein, A. Gruppe, M. Klüver, D. Schwalm, J. Schweppe, K. E. Stiebing, N. Trautmann, and P. Vincent, private communication, and to be published (1984).

(Cr 71) D. S. F. Crothers, *Adv. Phys.* **20**:405 (1971).

(Cr 73) D. S. F. Crothers, *J. Phys. B* **6**:1418 (1973).

(Cr 75) See review papers in *Atomic Inner-Shell Processes* (Bernd Crasemann, ed.), Academic Press, New York (1975).

(Cr 82) *X-Ray and Atomic Inner-Shell Physics* (Bernd Crasemann, ed.), American Institute of Physics, New York (1982).

(Da 74) C. K. Davis and J. S. Greenberg, *Phys. Rev. Lett.* **32**:1215 (1974).

(Da 82) S. Datz, J. L. Duggan, L. C. Feldman, E. Laegsgard, and J. U. Andersen, *Phys. Rev. A* **9**:192 (1974); *Physics of Electronic and Atomic Collisions* (S. Datz, ed.), North-Holland, Amsterdam (1982).

(De 64) Y. N. Demkov, *Sov. Phys. JETP* **18**:138 (1964).

(De 81) T. de Reus, U. Müller, J. Reinhardt, P. Schlüter, K. H. Wietschorke, B. Müller, W. Greiner, and G. Soff, in (Gr 81) (1981).

(De 83) T. H. J. de Reus, J. Reinhardt, B. Müller, W. Greiner, G. Soff, and U. Müller, GSI preprint, pp. 18–83, July (1983).

(Do 71) H. G. Dosch, J. H. D. Jensen, and V. F. Müller, *Phys. Norv.* **5**:2 (1971).

(Ei 76) J. Eichler, U. Wille, B. Fastrup, and K. Taulberg, *Phys. Rev. A* **14**:707 (1976).

(Ev 65) E. Everhart and Q. C. Kessel, *Phys. Rev. Lett.* **14**:247 (1965).

(Fa 65) U. Fano and W. Lichten, *Phys. Rev. Lett.* **14**:627 (1965).

(Fa 74) B. Fastrup, E. Bøving, G. A. Larsen, and P. Dahl, *J. Phys. B* **7**:L206 (1974).

(Fo 74) F. Folkmann, C. Gaarde, T. Huus, and K. Kemp, *Nucl. Instrum. Methods* **116**:487 (1974).

(Fo 75) F. Folkmann, K. O. Groeneveld, R. Mann, G. Nolte, S. Schumann, and R. Spohr, *Z. Phys.* **A275**:229 (1975).

(Fo 76) F. Folkmann, P. Armbruster, S. Hagmann, G. Kraft, P. H. Mokler, and H. J. Stein, *Z. Phys.* **A276**:15 (1976).

(Fo 76b) C. Foster, T. P. Hoogkamer, P. Woerlee, and F. W. Saris, *J. Phys. B: At. Mol. Phys.* **9**:1943 (1976).

(Fo 77) R. J. Fortner, P. H. Woerlee, S. Doorn, Th. P. Hoogkamer, and F. W. Saris, *Phys. Rev. Lett.* **39**:1322 (1977).

(Fr 75) W. Frank, P. Gippner, K.-H. Kaun, H. Sodan, and Yu. P. Trekyakov, *Phys. Lett.* **59B**:41 (1975).

(Fr 76) W. Frank, K. H. Kaun, P. Manfrass, N. V. Pronin, and Yu. P. Tretyakov, *Z. Phys.* **A279**:213 (1976).

(Fr 78) W. Frank, K. H. Kaun, and P. Manfrass, *Z. Phys.* **A287**:149 (1978).

(Fr 77) B. Fricke and G. Soff, *At. Data Nucl. Data Tables* **19**:84 (1977).

(Fr 79) W. Fritsch and U. Wille, *Jpn J. Appl. Phys.* **17**: Suppl. 17-2, 387 (1979); W. Fritsch and U. Wille, *J. Phys. B* **12**:L335 (1979).

(Fu 78) P. Fuchs, H. Emling, E. Grosse, D. Schwalm, H. J. Wollersheim, and R. Schulze, GSI, Darmstadt, Jahresbericht (1977), GSI-J-1-78 (1978).

(Fu 73) L. P. Fulcher and A. Klein, *Phys. Rev. D* **8**:2455 (1973).

(Fu 74) L. P. Fulcher and A. Klein, *Ann. Phys.* (*N.Y.*) **84**:335 (1974).

(Ga 73) J. D. Garcia, R. J. Fortner, and T. M. Kavanagh, *Rev. Mod. Phys.* **45**:111 (1973).

(Ga 76) M. Gavrila and J. E. Hansen, *Phys. Lett.* **58A**:158 (1976).

(Ga 77) R. K. Gardner, T. J. Gray, P. Richard, C. Schmiedekamp, K. A. Jamison, and J. M. Hall, *Phys. Rev. A* **15**:2202 (1977).

(Ga 77b) G. Gaukler, H. Schmidt-Böcking, R. Schuch, R. Schulé, H. J. Specht, and I. Tserruya, *Nucl. Instrum. Methods* **141**:115 (1977).

(Ge 61) S. S. Gershtein and V. D. Krivchenkov, *JETP* (*USSR*) **40**:1491 (1961).

(Ge 69) S. S. Gershtein and Ya. B. Zel'dovich, *Lett. Nuovo Cimento* **1**:835 (1969).

(Ge 70) S. S. Gershtein and Ya. B. Zel'dovich, *Sov. Phys. JETP* **30**:358 (1970).

(Ge 73) S. S. Gershtein and Ya. B. Zel'dovich, *Lett. Nuovo Cimento* **6**:593 (1973).

(Gr 56) J. S. Greenberg and M. Deutsch, *Phys. Rev.* **102**:415 (1956).

(Gr 74) J. S. Greenberg, C. K. Davis, and P. Vincent, *Phys. Rev. Lett.* **33**:473 (1974).

(Gr 76) T. J. Gray, P. Richard, K. A. Jamison, J. M. Hall, and R. K. Gardner, *Phys. Rev. A* **14**:1333 (1976).

(Gr 76b) K. O. Groeneveld, B. Kolb, J. Schader, and K. D. Sevier, *Z. Phys.* **A277**:13 (1976).

(Gr 76c) M. Gros, B. Müller, and W. Greiner, *J. Phys. B* **9**:1849 (1976).

(Gr 77) J. S. Greenberg, H. Bokemeyer, H. Emling, E. Grosse, D. Schwalm, and F. Bosch, *Phys. Rev. Lett.* **39**:1404 (1977).

(Gr 77b) J. S. Greenberg, P. Vincent, and W. Lichten, *Phys. Rev. A* **16**:964 (1977).

(Gr 80) T. J. Gray, "Target Ionization and X-ray Production for Ions Incident Upon Solid Targets," [Ri 80], pp. 193–278.

(Gr 80b) J. S. Greenberg, *Electronic and Atomic Collisions* (N. Oda and K. Takayanagi, ed.), North-Holland, Amsterdam (1980), p. 351.

(Gr 81) *Proceedings of NASI Conference on the Quantum Electrodynamics of Strong Fields*, Lahnstein/Rhein (W. Greiner, ed.), Plenum Press, New York (1981).

(Gr 82) J. S. Greenberg, *X-Ray and Atomic Inner-Shell Physics–1981* (B. Crasemann, ed.), American Institute of Physics, New York (1982), p. 173.

(Gr 83) J. S. Greenberg, Proceedings of Conf. on *Nuclear Physics with Heavy-Ions* (Peter Braun-Munzinger, ed.), Harwood Academic Publishers, New York (1983), p. 419.

(Gr 83b) W. Greiner, Int. Conf. on Nuclear Physics, Florence, Italy (1983).

(Gu 78) G. Guillaume, P. Fintz, F. C. Jundt, K. W. Jones, and B. M. Johnson, *Phys. Lett.* **68A**:39 (1978).

(Gu 82) F. Güttner, W. Koenig, B. Martin, B. Povh, H. Skapa, J. Soltani, T. Walcher, F. Bosch, and C. Kozhuharov, *Z. Phys.* **A304**:207 (1982).

(Gy 74) M. Gyulassy, *Phys. Rev. Lett.* **33**:921 (1974).

(Ha 73) J. S. Hansen, J. C. McGeorge, D. Nix, W. D. Schmidt-Ott, I. Unus, and R. W. Fink, *Nucl. Instrum. Methods* **106**:365 (1973).

(Ha 73b) J. S. Hansen, *Phys. Rev. A* **8**:822 (1973).

(Ha 75) J. M. Hansteen, O. D. Johnsen, and L. Kocbach, *At. Data Nucl. Data Tables* **15**:305 (1975).

(Ha 75b) J. M. Hansteen, in *Advances in Atomic and Molecular Physics* (D. R. Bates, ed.), Academic Press, New York (1975), Vol. 11, p. 299.

(Ha 78) H. Hartung and B. Fricke, *Z. Phys.* **A288**:345 (1978).

(Ha 79) S. Hagmann, C. L. Cocke, E. Justiniano, J. R. MacDonald, and H. Schmidt-Böcking, *J. Phys. B* **12**:L395 (1979).

(Ha 81) J. Hall, P. Richard, T. J. Gray, C. D. Lin, K. Jones, B. Johnson, and D. Gregory, *Phys. Rev. A* **24**:2416 (1981).

(Ha 82) See also S. Hagmann, C. L. Cocke, J. R. MacDonald, P. Richard, H. Schmidt-Böcking, and R. Schuch, *Phys. Rev. A* **25**:1918 (1982).

(He 36) W. Heisenberg and H. Euler, *Z. Phys.* **98**:714 (1936).

(He 65) G. Herzberg, *Spectra of Diatomic Molecules*, Van Nostrand, New York (1965), p. 315.

(He 72) K. Helfrich and H. Hartmann, *Theor. Chim. Acta* **24**:271 (1972).

(He 76) K. H. Heinig, H. U. Jaeger, H. Richter, and H. Woittennik, *Phys. Lett.* **60B**:240 (1976).

(He 78) G. Heiligenthal, W. Betz, G. Soff, B. Müller, and W. Greiner, *Z. Phys.* **A285**:105 (1978).

(He 81) U. Heinz, W. Greiner, and B. Müller, *Phys. Rev. A* **23**:562 (1981).

(He 83) K. H. Heinig, H. U. Jaeger, K. H. Kaun, H. Richter, and H. Woittennek, "$2p\sigma$ Molecular Orbital Radiation of Colliding Atoms: Dynamic-Model Predictions Versus Experimental Data," (Gr 81), p. 335.

(Ho 75) F. Hopkins, *Phys. Rev. Lett.* **35**:270 (1975).
(Ho 80) F. Hopkins, "Ion-Induced X-ray Spectroscopy," (Ri 80), p. 355 (1980).
(Hu 40) F. Hund, *Z. Phys.* **117**:1 (1940).
(Hu 53) T. Huus and J. H. Bjerregard, *Phys. Rev.* **92**:1579 (1953).
(Is 76) K. Ishii, S. Morita, and H. Tawara, *Phys. Rev. A* **13**:131 (1976).
(It 82) S. Ito, P. Armbruster, H. Bokemeyer, F. Bosch, H. Emling, H. Folger, E. Grosse, R. Kulessa, D. Liesen, D. Maor, D. Schwalm, J. S. Greenberg, R. Schulé, and N. Trautmann, GSI Scientific Report 1981, p. 141 (1982).
(Ja 57) D. Jamnik, and C. Zupanic, *Kl. Dan. Vidensk. Selsk. Mat.-Fys. Medd.* **31**, No. 2 (1957).
(Ja 66) M. Jammer, *The Conceptual Development of Quantum Mechanics*, McGraw-Hill, New York (1966).
(Ja 75) D. H. Jakubassa and M. Kleber, *Z. Phys.* **A273**:29 (1975).
(Ja 76) D. H. Jakubassa and M. Kleber, *Z. Phys.* **A227**:41 (1976).
(Ja 76b) D. H. Jakubassa, *Phys. Lett.* **58A**:163 (1976).
(Ja 79) D. H. Jakubassa, *Z. Phys.* **A293**:281 (1979).
(Ja 80) D. H. Jakubassa and K. Taulbjerg, *J. Phys. B* **13**:757 (1980).
(Jo 76) B. M. Johnson, K. W. Jones, and D. J. Pisano, *Phys. Lett.* **59A**:21 (1976).
(Jo 79) B. M. Johnson, K. W. Jones, W. Brandt, F. C. Jundt, G. Guillaume, and T. H. Kruse, *Phys. Rev. A* **19**:81 (1979).
(Ju 74) F. C. Jundt, H. Kubo, and H. E. Gove, *Phys. Rev. A* **10**:1053 (1974).
(Ka 77) K. H. Kaun, P. Manfrass, and W. Frank, *Part. Nuclei (Moskow)* **8**:1246 (1977).
(Ka 80) E. Kankeleit, *Nukleonika* **25**:253 (1980).
(Ka 80b) E. Kankeleit, R. Köher, M. Kollatz, M. Krämer, R. Krieg, P. Senger, and H. Backe, GSI-Annual Report, p. 195 (1980), and to be published.
(Ke 73) Q. C. Kessel and B. Fastrup, "The Production of Inner-Shell Vavancies in Heavy Ion-Atom Collisions," *Case Studies in Atomic Physics Vol. III*, North-Holland, Amsterdam (1973), pp. 137–213.
(Ke 76) H. P. Kelly, *Phys. Rev. Lett* **37**:386 (1976).
(Ki 73) P. Kienle, M. Kleber, B. Povh, R. M. Diamond, F. S. Stephens, E. Grosse, M. R. Maier, and D. Proetel, *Phys. Rev. Lett.* **31**:1099 (1973).
(Ki 78) J. Kirsch, W. Betz, J. Reinhardt, G. Soff, B. Müller, and W. Greiner, *Phys. Lett.* **72B**:298 (1978).
(Ki 79) J. Kirsch, W. Betz, J. Reinhardt, B. Müller, W. Greiner, and G. Soff, *Z. Phys.* **A292**:227 (1979).
(Ki 80) J. Kirsch, Ph. D. thesis, U. Frankfurt (1980).
(Ki 81) P. Kienle, in (Gr 81), p. 293.
(Ki 81b) M. Kimura and W. R. Thorson, *Phys. Rev. A* **24**:3019 (1981).
(Kl 29) O. Klein, *Z. Phys.* **53**:157 (1929).
(Kl 75) M. Kleber and D. H. Jakubassa, *Nucl. Phys.* **A252**:152 (1975).
(Ko 77) C. Kozhuharov, P. Kienle, D. H. Jakubassa, and M. Kleber, *Phys. Rev. Lett.* **39**:540 (1977).
(Ko 79) C. Kozhuharov, P. Kienle, E. Berdermann, H. Bokemeyer, J. S. Greenberg, Y. Nakayama, P. Vincent, H. Backe, L. Handschug, and E. Kankeleit, *Phys. Rev. Lett.* **48**:376 (1979).
(Ko 81) C. Kozhuharov and W. Koenig, private communication.
(Ko 81b) C. Kozhuharov, *Quantum Electrodynamics of Strong Fields* (W. Greiner, ed.), Plenum Press, New York (1981), pp. 317–333.
(Ko 82) C. Kozhuharov, in *Physics of Electronic and Atomic Collisions* (S. Datz, ed.), North-Holland, Amsterdam (1982), pp. 179–193.

(Kr 74) G. Kraft, P. H. Mokler, and H. J. Stein, *Phys. Rev. Lett.* **33**:476 (1974).

(Kr 78) H. Krimm, Universität Heidelberg, 1978 Diplomarbeit; F. Guttner, Universität Heidelberg, 1979 Diplomarbeit.

(Kr 83) R. Krieg, E. Bozek, E. Kankeleit, M. Krämer, U. Meyer, H. Oeschler, and P. Senger, private communication (1983).

(La 32) L. Landau, *Phys. Z. Sowjjetunion* **2**:46 (1932).

(La 71) F. P. Larkins, *J. Phys. B* **4**:L29 (1971).

(La 71b) F. P. Larkins, *J. Phys. B* **4**:14 (1971).

(La 76) R. Laubert, R. S. Peterson, J. P. Forester, K. H. Liao, P. M. Griffin, H. Hayden, S. B. Elston, D. J. Pegg, R. S. Thoe, and I. A. Sellin, *Phys. Rev. Lett.* **36**:1574 (1976).

(Le 69) H. Levy and W. R. Thorson, *Phys. Rev.* **181**:244 (1969).

(Le 69b) H. Levy and W. R. Thorson, *Phys. Rev.* **181**:252 (1969).

(Le 71) C. F. Lebeda, W. R. Thorson, and H. Levy, *Phys. Rev. A* **4**:901 (1971).

(Le 77) W. N. Lennard, I. V. Mitchell, J. S. Forster, and D. Phillips, *J. Phys. B* **190**:2199 (1977).

(Le 78) W. N. Lennard, I. V. Mitchell, and D. Phillips, *J. Phys. B* **11**:1283 (1978).

(Le 78b) C. M. Lederer and V. S. Shirley, *Table of Isotopes*, 7th ed., Appendix III, Wiley, New York (1978).

(Le 81) W. H. Lennard, I. V. Mitchell, G. C. Ball, and P. H. Mokler, *Phys. Rev. A* **23**:2260 (1981).

(Le 81b) W. N. Lennard, I. V. Mitchell, and D. Phillips, *J. Phys. B* **14**:2215 (1981).

(Li 63) W. Lichten, *Phys. Rev.* **131**:229 (1963).

(Li 67) W. Lichten, *Phys. Rev.* **164**:131 (1967).

(Li 71) D. A. Liberman, D. T. Cromer, and J. T. Weber, *Comp. Phys. Commun.* **2**:107 (1971).

(Li 76) W. Lichten, Proc. 2nd Int. Conf. on Inner Shell Ioniz., Phen., Invited Papers, Freiburg, p. 249 (1976), and references herein.

(Li 77) V. I. Lisin, M. S. Marinov, and V. S. Popov, *Phys. Lett.* **69B**:141 (1977).

(Li 78) C. D. Lin, S. C. Soong, and L. N. Tunnell, *Phys. Rev. A* **17**:1646 (1978).

(Li 80) C. D. Lin and L. N. Tunnell, *Phys. Rev. A* **22**:76 (1980).

(Li 80b) V. I. Lisin, M. S. Marinov, and V. S. Popov, *Phys. Lett.* **91B**:20 (1980).

(Li 80c) D. Liesen, P. Armbruster, F. Bosch, S. Hagmann, P. H. Mokler, H. Schmidt-Böcking, R. Schuch, J. B. Wilhemy, H. J. Wollersheim, *Phys. Rev. Lett.* **44**:983 (1980).

(Lo 62) R. Lombard and F. Rys, *Nucl. Phys.* **31**:163 (1962).

(Lo 68) R. Lombard, C. Perdrisat, and J. H. Brunner, *Nucl. Phys.* **A110**:41 (1968).

(Lu 74) M. Lund, thesis, University of Aarhus, Denmark (1974).

(Lu 77) W. L. Luken, J. S. Greenberg, and P. Vincent, *Phys. Rev. A* **15**:2305 (1977).

(Lu 78) H. O. Lutz, W. R. McMurray, R. Pretorius, R. J. van Reenan, and I. J. van Heerden, *Phys. Rev. Lett.* **40**:1133 (1978).

(Lu 79) N. Luz, S. Sackmann, and H. O. Lutz, *J. Phys. B* **12**:1973 (1979).

(Ma 73) J. R. MacDonald, M. D. Brown, and T. Chiao, *Phys. Rev. Lett.* **30**:471 (1973).

(Ma 74) J. H. Macek and J. S. Briggs, *J. Phys. B* **7**:1312 (1974).

(Ma 75) D. H. Madison and E. Merzbacher, in *Atomic Inner-Shell Processes* (B. Crasemann, ed.), Academic Press, New York (1975), Vol. I, p. 1.

(Ma 75b) M. S. Marinov and V. S. Popov, *Sov. J. Nucl. Phys.* **20**:641 (1975).

(Ma 80) J. R. MacDonald, "Charge Dependence of Atomic Inner-Shell Cross Sections," (Ri 80) (1980).

(Mc 69) E. J. McGuire, *Phys. Rev.* **185**:1 (1969).

(Mc 70) M. R. C. McDowell and J. P. Coleman, *Introduction to the Theory of Ion-Atom Collisions*, North-Holland, Amsterdam (1970).

(Me 58) E. Merzbacher and H. W. Lewis, *Handbuch der Physik*, Springer-Verlag, Berlin (1958), Vol. 34, p. 166.

(Me 73) W. E. Meyerhof, *Phys. Rev. Lett.* **31**:1341 (1973).

(Me 73b) W. E. Meyerhof, T. K. Saylor, S. M. Lazarus, W. A. Little, B. B. Triplett, and L. F. Chase, Jr., *Phys. Rev. Lett.* **30**:1279 (1973).

(Me 74) W. E. Meyerhof, *Phys. Rev. A* **10**:1005 (1974).

(Me 75) W. E. Meyerhof, T. K. Saylor, and R. Anholt, *Phys. Rev. A* **12**:2641 (1975).

(Me 76) E. Merzbacher, *Proceedings of the Second Int. Conf. on Inner-Shell Ionization Phenomena*, Freiburg (W. Melhorn and R. Brenn, eds.), (1976), p. 1.

(Me 76b) W. E. Meyerhof, R. Anholt, T. K. Saylor, S. M. Lazarus, A. Little, and L. F. Chase, *Phys. Rev. A* **14**:1653 (1976).

(Me 76c) See papers in *Proceedings of the Second Int. Conf. on Inner-Shell Ionization Phenomena*, Freiburg (W. Melhorn and R. Brenn, eds). (1976).

(Me 77) W. E. Meyerhof and Knud Taulbjerg, *Ann. Rev. Nucl. Sci.*, ed., Segrè, **27**:279 (1977).

(Me 77b) W. E. Meyerhof, R. Anholt, Y. El Masri, I. Y. Lee, D. Cline, F. S. Stephens, and R. M. Diamond, *Phys. Lett.* **69B**:41 (1977).

(Me 77c) W. E. Meyerhof, R. Anholt, and T. K. Saylor, *Phys. Rev. A* **16**:169 (1977).

(Me 78) W. E. Meyerhof, R. Anholt, J. Eichler, and A. Salop, *Phys. Rev. A* **17**:108 (1978), and references herein.

(Me 78b) W. E. Meyerhof, *Phys. Rev. A* **18**:414 (1978).

(Me 79) W. E. Meyerhof, D. L. Clark, Ch. Stoller, E. Morenzoni, W. Wölfli, F. Folkmann, P. Vincent, P. H. Mokler, P. Armbruster, *Phys. Lett.* **70A**:303 (1979).

(Mi 67) M. H. Mittleman and L. Wilets, *Phys. Rev.* **154**:12 (1967).

(Mo 14) H. G. J. Moseley, *Phil. Mag.* **26**:1024 (1913); **27**:703 (1914).

(Mo 65) E. Moll and E. Kankeleit, *Nukleonika* **7**:180 (1965).

(Mo 72) P. H. Mokler, H. J. Stein, and P. Armbruster, *Phys. Rev. Lett.* **29**:827 (1972).

(Mo 78) P. H. Mokler and F. Folkmann, "X-Ray Production in Heavy Ion–Atom Collisions," *Structure and Collisions of Ions and Atoms* (Sellin, ed.), Springer-Verlag, Berlin (1978), pg. 201.

(Mo 80) P. H. Mokler, W. N. Lennard, and I. V. Mitchell, *J. Phys. B* **13**:4607 (1980).

(Mo 81) P. J. Mohr, in *Proceedings of NASI Conference on the Quantum Electrodynamics of Strong Fields*, Lahnstein/Rhein (W. Greiner, ed.), Plenum Press, New York (1981).

(Mo 81b) E. A. A. Morenzoni, Ph. D. thesis, ETH Zurich (1981).

(Mo 83) P. H. Mokler and D. Liesen, "X-Rays from Superheavy Collision Systems," *Progress in Atomic Spectroscopy*, Part C (Beyer and Kleinpoppen, eds.), Plenum Press, New York (1983), pg. 321.

(Mu 72) B. Müller, H. Peitz, J. Rafelski, and W. Greiner, *Phys. Rev. Lett.* **28**:1235 (1972).

(Mu 72b) B. Müller, J. Rafelski, and W. Greiner, *Z. Phys.* **257**:62, 183 (1972).

(Mu 73) B. Müller, J. Rafelski, and W. Greiner, *Phys. Lett.* **47B**:5 (1973).

(Mu 73b) B. Müller, J. Rafelski, and W. Greiner, *Nuovo Cimento* **18A**:551 (1973).

(Mu 75) B. Müller, Proc. 9th Int. Conf. on Physics of Electronic and Atomic Collisions, Seattle, Univ. of Washington Press, Abstracts, p. 481 (1975).

(Mu 76) B. Müller, *Ann. Rev. Nucl. Sci.* **26**:351 (1976).

(Mu 76b) B. Müller and W. Greiner, *Z. Naturforsch.* **31a**:1 (1976).

(Mu 78) B. Müller, G. Soff, W. Greiner, and V. Ceausescu, *Z. Phys.* **A285**:27 (1978).

(Mu 83) B. Müller, J. Reinhardt, W. Greiner, and G. Soff, *Z. Phys.* **A311**:151 (1983).

(Mu 83b) U. Müller, G. Soff, T. de Reus, I. Reinhardt, B. Müller, and W. Greiner, *Z. Phys.* **A313**:263 (1983), and private communication.

(Ni 67) V. S. Nikolaev, *Sov. Phys. JETP* **24**:847 (1967).

(No 80) G. Nolte, J. Volpp, R. Schuch, H. J. Specht, W. Lichtenberg, and H. Schmidt-Böcking, *J. Phys. B* **13**:4599 (1980).

(Ob 80) J. J. O'Brien, E. Liarokapis, and J. S. Greenberg, *Phys. Rev. Lett.* **44**:386 (1980).

(Od 80) *Electronic and Atomic Collisions* (N. Oda and K. Takayanagi, eds.), North-Holland, Amsterdam (1980).

(Og 72) G. N. Ogurtsov, *Rev. Mod. Phys.* **44**:1 (1972).

(Ol 72) R. E. Olson, *Phys. Rev. A* **6**:1822 (1972).

(Pe 62) C. F. Perdrisat, H. J. Leisi, and J. H. Brunner, *Nucl. Phys.* **31**:157 (1962).

(Pe 73) H. Peitz, B. Müller, J. Rafelski, and W. Greiner, *Lett. Nuovo Cimento* **8**:37 (1973).

(Pe 78) H. Peitz and W. Greiner, *Phys. Unserer Zeit* **9**, No. 6 (1978).

(Pf 80) E. Pflanz, R. Schuch, J. Neumann, G. Nolte, and H. Schmidt-Böcking, Contributed paper, International Conference on X rays and Inner Shell Ionization, Sterling (1980).

(Pi 69) W. Pieper and W. Greiner, *Z. Phys.* **218**:327 (1969).

(Po 45) I. Pomeranchuk and I. Smordinsky, *J. Phys. USSR* **9**:97 (1945).

(Po 71) V. S. Popov, *Sov. Phys. JETP* **12**:235 (1971).

(Po 71b) V. S. Popov, *Sov. Phys. JETP* **32**:526 (1971).

(Po 73) V. S. Popov, *Sov. J. Nucl. Phys.* **17**:332 (1973).

(Po 74) V. S. Popov, *Sov. Phys. JETP* **38**:18 (1974).

(Po 74b) V. S. Popov, *Sov. J. Nucl. Phys.* **19**:81 (1974).

(Pr 80) G. Presser, E. Scherer, and J. Stahler, *Z. Phys.* **A295**:27 (1980).

(Ra 71) J. Rafelski, L. P. Fulcher, and W. Greiner, *Phys. Rev. Lett.* **27**:958 (1971); W. Greiner spoke of superheavy quasimolecules at GSI (Darmstadt) seminars in 1969/70.

(Ra 74) J. Rafelski, B. Müller, and W. Greiner, *Nucl. Phys.* **B38**:585 (1974).

(Ra 76) J. Rafelski and B. Müller, *Phys. Lett.* **65B**:205 (1976).

(Ra 76b) R. R. Randall, J. A. Bednar, B. Curnutte, and C. L. Cocke, *Phys. Rev. A* **13**:204 (1976).

(Ra 78) J. Rafelski, L. P. Fulcher, and A. Klein, *Phys. Rep.* **38C**:227 (1978).

(Ra 78b) J. Rafelski, B. Müller, and W. Greiner, *Z. Phys.* **A285**:49 (1978).

(Re 69) D. Rein, *Z. Phys.* **221**:423 (1969).

(Re 76) J. Reinhardt, G. Soff, and W. Greiner, *Z. Phys.* **A276**:285 (1976).

(Re 77) J. Reinhardt and W. Greiner, *Rep. Prog. Phys.* **40**:219 (1977).

(Re 78) J. Reinhardt, V. Oberacker, B. Müller, W. Greiner, and G. Soff, *Phys. Lett.* **78B**:183 (1978).

(Re 78b) J. Reinhardt, B. Müller, G. Soff, and W. Greiner, (specific calculations for this case supplied) (1978).

(Re 79) J. Reinhardt, B. Müller, W. Greiner, and G. Soff, *Phys. Rev. Lett.* **43**:1307 (1979).

(Re 80) J. Reinhardt, G. Soff, B. Müller, and W. Greiner, *Progr. Part. Nucl. Phys.* **4**:547 (1980).

(Re 81) J. Reinhardt, U. Müller, B. Müller, and W. Greiner, *Z. Phys.* **A303**:73 (1981).

(Re 81b) J. Reinhardt, B. Müller, and W. Greiner, *Phys. Rev. A* **24**:103 (1981).

(Re 83) J. Reinhardt, B. Müller, W. Greiner, and U. Müller, *Phys. Rev. A* **28**:2558 (1983).

(Re 85) J. Reinhardt and W. Greiner, "Heavy-ion Atomic Physics—Theoretical,"

Treatise on Heavy-Ion Science (D. A. Bromley, ed.), Plenum Press, New York (1985), Chap. 1, this volume.

(Rh 83) M. J. Rhoades-Brown, V. E. Oberacker, M. Seiwart, and W. Greiner, *Z. Phys.* **A310**:287 (1983).

(Ri 78) See review papers in *Methods of Experimental Physics—Accelerators in Atomic Physics* (P. Richard, ed.), Academic Press, New York (1978).

(Ri 80) P. Richard, ed., *Methods of Experimental Physics*, Vol. 17, Academic Press, New York (1980).

(Ri 80b) See (Ri 80), p. 216.

(Ru 13) E. Rutherford and H. Richardson, *Phil. Mag.* **25**:722 (1913).

(Ru 72) M. E. Rudd and J. H. Macek, in "Mechanisms of Electron Production in Ion–Atom Collisions," *Case Stud. At. Phys.* **3**: No. 2 (1972).

(Sa 32) F. Sauter, *Z. Phys.* **69**:742 (1931); **73**:547 (1932).

(Sa 72) F. W. Saris, W. F. Van der Weg, H. Tawara, and R. Laubert, *Phys. Rev. Lett.* **28**:717 (1972).

(Sa 74) S. Sackmann, H. O. Lutz, and J. Briggs, *Phys. Rev. Lett.* **32**:805 (1974).

(Sa 77) F. W. Saris and T. P. Hoogkamer, *Atomic Physics*, Vol. 5 (R. Marrus, Mlt. Prior, and H. A. Shugart, eds.), Plenum Press, (1977), p. 507.

(Sc 40) L. I. Schiff, H. Snyder, and J. Weinberg, *Phys. Rev.* **57**:315 (1940).

(Sc 51) J. Schwinger, *Phys. Rev.* **82**:664 (1951).

(Sc 72) H. W. Schnopper, H. D. Betz, J. P. Delvaille, K. Kalata, A. R. Sohval, K. W. Jones, and H. E. Wegner, *Phys. Rev. Lett.* **29**:898 (1972).

(Sc 72b) D. Schwalm, A. Bamberger, P. G. Bizzeti, B. Povh, G. A. P. Engelbertnik, J. W. Olness, and E. K. Warburton, *Nucl. Phys.* **A192**:449 (1972).

(Sc 74) H. W. Schnopper *et al.*, *Phys. Lett.* **A47**:61 (1974).

(Sc 77) R. Schuch, H. Schmidt-Böcking, R. Schulé, and I. Tserruya, *Phys. Rev. Lett.* **39**:79 (1977).

(Sc 78) P. Schlüter, G. Soff, and W. Greiner, *Z. Phys.* **A286**:149 (1978).

(Sc 78b) H. Schmidt-Böcking, W. Lichtenberg, R. Schuch, J. Volpp, and I. Tserruya, *Phys. Rev. Lett.* **41**:859 (1978).

(Sc 78c) H. Schmidt-Böcking, R. Schuch, I. Tserruya, R. Schulé, H. J. Specht, and K. Bethge, *Z. Phys.* **A284**:39 (1978).

(Sc 78d) R. Schmidt, V. D. Tonew, and G. Wolschin, *Nucl. Phys.* **A311**:247 (1978).

(Sc 79) P. Schlüter and G. Soff, *Atom. Data and Nucl. Data Tab.* **24**:509 (1979).

(Sc 79b) R. Schuch, G. Nolte, H. Schmidt-Böcking, and W. Lichtenberg, *Phys. Rev. Lett.* **43**:1104 (1979).

(Sc 79c) R. Schuch, G. Nolte, B. M. Johnson, and K. W. Jones, *Z. Phys.* **A293**:91 (1979).

(Sc 80) R. Schuch, G. Nolte, and H. Schmidt-Böcking, *Phys. Rev. A* **22**:1447 (1980).

(Sc 80b) R. Schuch, G. Gaukler, G. Nolte, K. W. Jones, and B. M. Johnson, *Phys. Rev. A* **22**:2513 (1980).

(Sc 81) R. Schuch, Invited Talk XII Inter. Conf. on Phys. Electron and Atomic Collisions, Gatlinburg (1981).

(Sc 81b) R. Schuch, H. Schmidt-Böcking, and I. Tserruya, *Phys. Lett.* **82A**:453 (1981).

(Sc 82) H. Schmidt-Böcking, R. Anholt, R. Schuch, P. Vincent, K. Stiebing, and H. U. Jäger, *J. Phys. B: Atom. Mol. Phys.* **15**:3057 (1982).

(Sc 82b) H. Schmidt-Böcking, H. Baumann, K. Bethge, A. Gruppe, W. Schadt, R. Schuch, and K. E. Stiebing, *Z. Phys.* **A304**:177 (1982).

(Sc 83) P. Schlüter, T. de Reus, J. Reinhardt, B. Müller, and G. Soff, *Z. Phys.* **A314**:297 (1983).

(Sc 83b) J. Schweppe, A. Gruppe, K. Bethge, H. Bokemeyer, T. Cowan, H. Folger, J. S. Greenberg, H. Grein, S. Ito, R. Schulé, D. Schwalm, K. E. Stiebing, N. Trautmann, P. Vincent, and M. Waldschmidt, *Phys. Rev. Lett.* **51**:2261 (1983).

(Se 72) K. D. Sevier, *Low Energy Electron Spectroscopy*, Wiley-Interscience, New York (1972).

(Se 73) V. SethuRaman, W. R. Thorson, and C. F. Lebeda, *Phys. Rev. A* **8**:1316 (1973).

(Se 81) W. D. Sepp, B. Fricke, and T. Morovic, *Phys. Lett.* **81A**:258 (1981).

(Sh 64) S. Shimizu, Y. Nakayama, H. Hirata, and H. Mazaki, *Nucl. Phys.* **54**:265 (1964).

(Si 20) M. Siegbahn, *Ark. Mat. Astron. Fys.* **14**: No. 9 (1920).

(Sl 49) L. A. Sliv, *Dokl. Akad. SSSR* **64**:521 (1949); *Zh. Eksp. Teor. Fiz.* **25**:7 (1953).

(Sm 74) K. Smith, H. Peitz, B. Müller, and W. Greiner, *Phys. Rev. Lett.* **32**:554 (1974).

(So 74) G. Soff, B. Müller, and J. Rafelski, *Z. Naturforsch.* **29A**:1267 (1974).

(So 77) G. Soff, J. Reinhardt, B. Müller, and W. Greiner, *Phys. Rev. Lett.* **38**:592 (1977).

(So 78) G. Soff, W. Betz, B. Müller, W. Greiner, and E. Merzbacher, *Phys. Lett.* **65A**:19 (1978).

(So 78b) G. Soff, J. Reinhardt, W. Betz, and J. Rafelski, *Phys. Scr.* **17**:417 (1978).

(So 79) G. Soff, W. Greiner, W. Betz, and B. Müller, *Phys. Rev. A* **20**:169 (1979).

(So 79b) G. Soff, J. Reinhardt, B. Müller, and W. Greiner, *Z. Phys.* **A292**:211 (1979); *Phys. Rev. Lett.* **43**:1981 (1979).

(So 80) G. Soff, J. Reinhardt, B. Müller, and W. Greiner, *Z. Phys.* **A294**:137 (1980).

(So 81) G. Soff, J. Reinhardt, and W. Greiner, *Phys. Rev. A* **23**:701 (1981).

(So 81b) G. Soff, B. Müller, and W. Greiner, *Z. Phys.* **A299**:189 (1981).

(So 81c) G. Soff, P. Schlüter, and W. Greiner, *Z. Phys.* **A303**:189 (1981).

(So 82) G. Soff, P. Schlüter, B. Müller, and W. Greiner, *Phys. Rev. Lett.* **48**:1465 (1982).

(So 82b) G. Soff, T. de Reus, B. Müller, and W. Greiner, *Phys. Lett.* **88A**:398 (1982).

(So 82c) G. Soff, U. Müller, T. de Reus, P. Schlüter, A. Schäfer, J. Reinhardt, B. Müller, and W. Greiner, Preprint GSI-82-17 and *Atomic Physics of Highly Ionized Atoms*, Cargese, France, June 1982.

(So 83) G. Soff, T. de Reus, U. Müller, J. Reinhardt, P. Schlüter, K. H. Wietschorke, B. Müller, and W. Greiner, "Spectroscopy of Superheavy Quasimolecules," (Gr 81), p. 233 (1983).

(Sp 65) H. J. Specht, *Z. Phys.* **185**:301 (1965).

(Sp 79) E. Spindler, H. D. Betz, and F. Bell, *Phys. Rev. Lett.* **42**:832 (1979).

(St 58) W. E. Stephens and H. Staub, *Phys. Rev.* **109**:1196 (1958).

(St 70) H. Storm and H. I. Israel, *Nucl. Data Tables* **A7**:565 (1970).

(St 75) N. Stolterfoht, D. Schneider, D. Burch, B. Aagaard, E. Bøving, and B. Fastrup, *Phys. Rev. A* **12**:1313 (1975).

(St 77) C. Stoller, W. Wölfli, G. Bonani, M. Stöckli, and M. Suter, *J. Phys. B* **10**:L347 (1977).

(St 77b) C. Stoller, W. Wölfli, G. Bonani, M. Stöckli, and M. Suter, *Phys. Rev. A* **15**:990 (1977).

(St 78) N. Stolterfoht, in "Structure and Collision of Ions and Atoms," *Topics in Modern Physics V*, Chap. 5, ed., (I.A. Sellin, ed.), Springer-Verlag, Berlin (1978).

(St 78b) C. Stoller, W. Wölfli, G. Bonani, E. Morenzoni, and M. Stöckli, *Z. Phys.* **A287**:33 (1978).

(St 80) C. Stoller, E. Morenzoni, W. Wölfli, W. E. Meyerhof, F. Folkmann, P. Vincent, P. H. Mokler, and P. Armbruster, *Z. Phys.* **A297**:93 (1980).

(St 81) C. Stoller, R. Anholt, and W. E. Meyerhof, *Z. Phys.* **A302**:95 (1981).

(St 81b) J. Stähler and G. Presser, *J. Phys. B* **14**:4093 (1981).

(St 82) K. Stiebing, H. Schmidt-Böcking, W. Schadt, K. Bethge, P. H. Mokler, F. Bosch, D. Liesen, P. Vincent, R. Schuch, and S. Hagmann, GSI Report-82-14 (1982).

(Ta 75) K. Taulbjerg and J. S. Briggs, *J. Phys. B: At. Mol. Phys.* **8**:1895 (1975), and references herein.

(Ta 76) K. Taulbjerg, J. S. Briggs, and J. Vaaben, *J. Phys. B* **9**:1351 (1976).

(Ta 78) J. A. Tanis and S. M. Shafroth, *Phys. Rev. Lett.* **40**:1174 (1978).

(Ta 81) J. A. Tanis, S. M. Shafroth, J. E. Willis, and J. R. Mowat, *Phys. Rev. A* **23**:366 (1981).

(Ta 82) H. Tawara, P. Richard, and K. Kawatsura, *Phys. Rev. A* **26**:154 (1982).

(Th 69) W. R. Thorson and H. Levy, *Phys. Rev.* **181**:230 (1969).

(Th 77) W. R. Thorson and J. H. Choi, *Phys. Rev. A* **15**:550 (1977).

(Th 79) J. Theis, J. Reinhardt, and B. Müller, *J. Phys. B* **12**:L479 (1979).

(To 82) T. Tomodo and H. A. Weidenmüller, *Phys. Rev. A* **26**:162 (1982).

(To 82b) T. Tomoda, *Phys. Rev. A* **26**:174 (1982).

(To 83) T. Tomoda and H. A. Weidenmüller, *Phys. Rev. C* **28**:739 (1983).

(To 83b) T. Tomoda, preprint MPI H-1983-V8.

(Tr 76) H. P. Trautvetter, J. S. Greenberg, and P. Vincent, *Phys. Rev. Lett.* **37**:202 (1976).

(Ts 76) See also, I. Tserruya, H. Schmidt-Böcking, R. Schulé, K. Bethge, R. Schuch, and H. J. Specht, *Phys. Rev. Lett.* **36**:1451 (1976).

(Ts 78) I. Tserruya, H. Schmidt-Böcking, and R. Schuch, *Phys. Rev. A* **18**:2482 (1978).

(Ts 83) I. Tserruya, R. Schuch, H. Schmidt-Böcking, J. Barrette, Wang Da-Hai, B. M. Johnson, M. Meron, and K. W. Jones, *Phys. Rev. Lett.* **50**:30 (1983).

(Va 81) J. Vaaben and K. Taulbjerg, *J. Phys. B: At. Mol. Phys.* **14**:1815 (1981).

(Vi 32) J. P. Vinti, *Phys. Rev.* **42**:632 (1932).

(Vi 77) P. Vincent, Ph. D. thesis, Yale University (1977).

(Vi 78) P. Vincent, C. K. Davis, and J. S. Greenberg, *Phys. Rev. A* **18**:1878 (1978).

(Vi 79) P. Vincent and J. S. Greenberg, *J. Phys. B* **12**:L641 (1979).

(Vi 81) P. Vincent, "Continuum X-Ray Emission in Adiabatic Heavy Ion-Atom Collisions," (Gr 81), p. 359; P. Vincent, "Molecular Orbital X-Ray Emission in Heavy Ion Collisions," *Inner Shell and X-ray Physics of Atoms and Solids* (D. J. Fabian, H. Kleinpoppen, L. M. Watson, eds.), Plenum Press, New York (1981), p. 117.

(Vo 29) J. von Neumann and E. P. Wigner, *Z. Phys.* **30**:467 (1929).

(Vo 61) V. V. Voronkev and N. N. Koleznikov, *Sov. Phys. JETP* **12**:136 (1961).

(Wa 74) R. L. Watson, F. E. Jenson, and T. Chiao, *Phys. Rev. A* **10**:1230 (1974).

(Wa 77) R. L. Watson, A. K. Leeper, B. I. Sonobe, T. Chiao, and F. E. Jenson, *Phys. Rev. A* **15**:914 (1977).

(Wa 78) A. Warczak, D. Liesen, J. R. MacDonald, and P. H. Mokler, *Z. Phys.* **A285**:235 (1978).

(We 33) V. Weisskopf, *Phys. Z.* **34**:1 (1933).

(We 58) F. G. Werner and J. A. Wheeler, *Phys. Rev.* **109**:126 (1958).

(Wi 63) R. Wiener, C. Chasman, P. Harihar, and C. S. Wu, *Phys. Rev.* **130**:1069 (1963).

(Wi 79) K-H. Wietschorke, B. Müller, W. Greiner, and G. Soff, *J. Phys. B: At. Mol. Phys.* **12**:L31 (1979).

(Wo 75) W. Wölfli, C. Stoller, G. Bonani, M. Suter, and M. Stöckli, *Phys. Rev. Lett.* **35**:656 (1975).

(Wo 78) W. Wölfli, E. Morenzoni, C. Stoller, G. Bononi, and M. Stöckli, *Phys. Lett.* **68A**:217 (1978).

(Wo 78b) W. Wölfli, C. Stoller, G. Bonani, M. Stöckli, M. Suter, and W. Däppen, *Z. Phys.* **A286**:249 (1978).

(Ze 32) C. Zener, *Proc. R. Soc. London Ser. A* **137**:696 (1932).

(Ze 72) Ya. B. Zel'dovitch and V. S. Papov, *Sov. Phys. Usp.* **14**:673 (1972).

(Zu 54) C. Zupancic and T. Huus, *Phys. Rev.* **94**:206 (1954).

3

INDREK MARTINSON is Professor of Atomic Physics at the University of Lund, Lund, Sweden. He did both his undergraduate and graduate work at the University of Stockholm, receiving the Fil. Dr. degree in 1971. From 1968 through 1970 he was a Research Associate at the University of Arizona in Tucson, and from 1970 through 1975 a Research Scientist at the Research Institute for Physics in Stockholm. He took up his present position in 1975.

Beam-Foil Spectroscopy

INDREK MARTINSON

1. Introduction

The energy levels and spectra of free atoms and ions have been studied for many years. The physical quantities of interest include transition wavelengths, excitation and ionization energies, as well as fine- and hyperfine structure separations. More recently it has also become possible to make accurate determinations of the lifetimes of excited states in atoms and ions and their related quantities, atomic transition probabilities. All this experimental information is of great interest in atomic theory, for instance for understanding the electron correlation and relativistic effects on atomic structure. There are also important applications of the atomic data in astrophysics as well as in plasma physics and research towards controlled thermonuclear fusion.

A variety of experimental techniques have been developed for research in modern atomic physics. One of the most versatile and powerful methods uses beams of heavy ions from accelerators. This method, usually called beam-foil spectroscopy (BFS) or sometimes fast-beam spectroscopy (FBS), was discovered in the early 1960's by Kay (Ka 63) and Bashkin (Ba 64). These authors independently realized the significance of directing ions from, e.g., Van de Graaff accelerators through a thin carbon foil and making spectral studies of the light emitted from the subsequent decay processes. The initial work thus showed that the light predominantly originated from the accelerated ions (and not from ejected foil particles or rest-gas excitation). By accelerating various elements different line spectra were obtained. The foil-excited ion beam forms a light source for atomic spectroscopy. Unlike most other such sources BFS also possesses excellent time resolution which

INDREK MARTINSON • Department of Physics, University of Lund, S-223 62 Lund, Sweden.

allows studies of atomic lifetimes. This possibility was already realized in the early work (Ba 64, Ka 63).

The method of BFS has subsequently developed into a standard technique in contemporary atomic physics and it has yielded valuable experimental data such as wavelengths, excitation and ionization energies, fine- and hyperfine structure, Landé g factors, atomic lifetimes and transition probabilities, autoionization rates, and Lamb shifts. International conferences have been held in 1967, 1970, 1972, 1975, and 1978 and their proceedings are available (Ba 68, Ma 70, Ba 73, Se 76, De 79). There is also a book (Ba 76) and a number of detailed review articles (e.g., Ba 74, An 74, Ma 74, Ma 74a, Co 76, Se 76a, Be 77, An 79, Pe 80) which summarize the work that has been carried out and the progress made. The material has already been well systematized in these reviews (of which the most recent ones, An 79 and Pe 80, are particularly thorough), and this fact eases the demands on the present review. We will be somewhat less comprehensive than the majority of previous review articles, to which we can refer for more detailed discussions.

2. The Beam-Foil Light Source

The principle of BFS is illustrated in Figure 1. A heavy-ion accelerator produces a beam of fast ions which are magnetically analyzed (to ensure a chemically and isotopically pure beam) and then directed through a thin (typically $10\,\mu g/cm^2$ thickness) exciter foil, usually made of carbon. Because of collisions with the foil atoms the fast ions may undergo further ionization (electron capture is also possible at lower ion velocities), and they often emerge from the foil in excited electronic states. These decay on the downstream side of the foil by the emission of light in the visible, ultraviolet, or x-ray regions. As schematically shown in Figure 1 also electron emission is possible. The foil excited ion beam can be considered as a light source for atomic spectroscopy in a wide sense (i.e., including electron spectroscopy). There are several interesting and quite attractive properties which characterize this spectroscopic source. It has a wide applicability. Practically all elements can be accelerated in modern ion accelerators, and indeed more than 70 different chemical elements have already been studied by beam-foil investigators. By varying the accelerator energy it is possible to reach many ionization degrees. The presently available BFS results include neutrals as well as more than 40 times ionized atoms. The excitation is largely unselective, which means that very many levels in any ion of interest can be studied. (As discussed below, this broad-band excitation has also certain drawbacks, particularly in connection with lifetime measurements.) The acceleration to a well-defined energy and subsequent magnetic analysis usually guarantees a chemically and isotopically pure beam, and the BFS

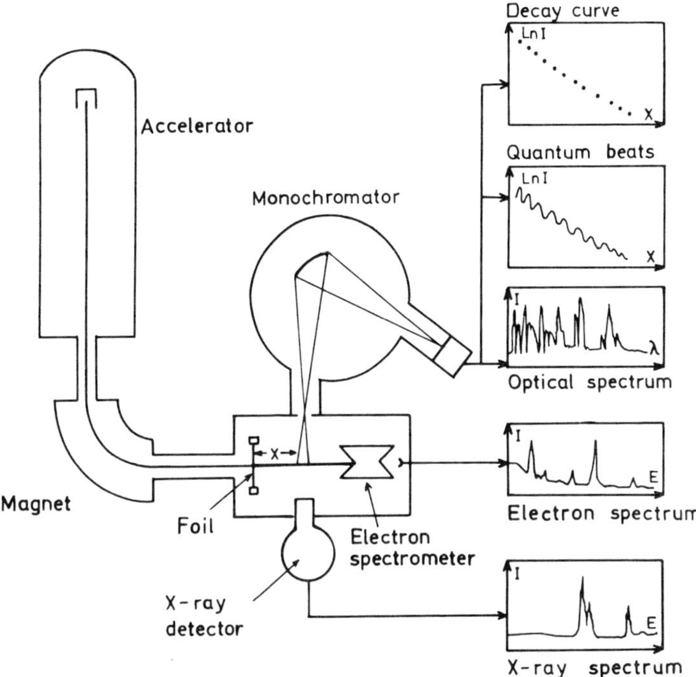

Figure 1. Experimental arrangement for beam-foil studies. Ions from the accelerator are mass-analyzed and sent through a thin foil. The radiation emitted on the downstream side is studied by optical, x-ray, or electron spectroscopy. Lifetime and quantum beat experiments are also indicated.

light source is therefore free from impurities which are present in other spectroscopic light sources. For example, if we accelerate a beam of B^+ ions and select the mass-11 beam we can be convinced that all transitions observed in our spectra belong to transitions in various ions of B. The only exception is here that certain transitions in atomic C, from sputtered, excited foil atoms, are observed (Be 70). The rest-gas pressure in the target chamber is typically 10^{-6} torr (or lower) and the decay of the excited states of the fast ions is therefore not influenced by secondary collisions and excitations. Perhaps the most important property of the foil-excited beam is its spatial (and temporal) resolution. The place of excitation is extremely well defined and—since the velocity of the ions is usually known with an uncertainty of 1 %–2 % or less—the time scale of the excited ions is accurately known. Thus, by measuring the intensity decays of spectral lines it is possible to determine level lifetimes using the very simple relation

$$I(x) = I(0) \exp(-x/v\tau) \tag{1}$$

$I(0)$ and $I(x)$ are the counting rates close to the foil and at a variable distance x along the beam, respectively, v is the velocity of the ions, and τ is the lifetime of the level under study. The lifetimes of excited states in atoms and ions are typically in the range 10^{-8}–10^{-11} s, and such short times can be determined by BFS in a straightforward way. For example, ^{16}O ions of 10 MeV energy have a velocity of 11 mm/ns, and even a lifetime of 10^{-10} s corresponds to a distance that can be measured quite accurately. As will be shown later, the dynamic range of BFS is approximately from 1 ps to 1 μs, and even shorter lifetimes can sometimes be determined.

The excitation process in BFS is quite abrupt (typically 10^{-15} s or lower) and as a result of this the atoms or ions are not prepared into eigenstates (solutions to the time-independent Schrödinger equation) but into linear, coherent superpositions of eigenstates. Interference effects between these states may cause periodic modulations (quantum beats) on the exponential decay curves (Figure 1). The ion–foil interaction frequently results in a nonstatistical population of magnetic sublevels. Effects such as orientation (vector polarization) or alignment (tensor polarization) can be observed by studying the polarization of the emitted light.

While most of the work that has been performed so far has used relatively straightforward BFS techniques, several important modifications have also been developed in recent years. The fast ions can be excited in a gas cell, and this method is also applicable to studies of atomic spectra and lifetimes. Furthermore, the ions from the accelerator may be excited by a laser (e.g., a tunable dye laser) which leads to selective population of the excited states.

The most obvious properties of the BFS light source have been briefly mentioned in this section. Additional (and more detailed) information can be found in the following section. We also refer to the various reviews mentioned above, and to an early review by Bickel (Bi 67) for complementary and thorough discussions.

3. Experimental Methods

A comprehensive survey of experimental methods can be found in the monograph edited by Bashkin (Ba 76). A wealth of important information, particularly valuable to scientists who plan to set up beam-foil experiments, is available in recent review articles (Be 77, An 79, Pe 80).

3.1. Accelerators

Bashkin (Ba 76) points out that "any accelerator of positive ions can be used for BFS." The present developments have confirmed this statement. Indeed, such work has been carried out with low-energy isotope separators, Cockcroft-Walton accelerators, Van de Graaff and other electrostatic accelerators (single-ended as well as tandem machines), cyclotrons and heavy-ion linear accelerators. Beam particle energies ranging from 10 keV to more than 700 MeV have already been used in BFS.

In the low-energy region small and relatively inexpensive ion accelerators can be used. A simple low-energy accelerator, equipped with a radiofrequency (rf) ion source, has been described by Chin-Bing *et al.* (Ch 70). Jensen and Veje (Je 74) report the construction of a low-energy accelerator, equipped with a universal ion source and a Wien filter for mass analysis. Isotope separators have been frequently used in the early BFS work (e.g., Bi 69) and a similar machine is being used for beam-laser experiments (Ar 75). Multiply ionized species can frequently be accelerated in such machines and Curtis *et al.* (Cu 73) thus discuss lifetime measurements using triply ionized Mn and Tm from a low-energy accelerator.

There have been recently important developments in obtaining multiply ionized atoms from ions sources. Bannenberg *et al.* (Ba 78) describe a 320-kV ion accelerator equipped with a pulsed ion source which gives good yields of up to 9–10 times ionized atoms.

Commercial heavy-ion accelerators with maximum voltages in the range 350–450 kV have been used by several groups, e.g. (Lu 73, Pi 73, Ga 75). These machines are usually equipped with universal ion sources and the ion energy can be further extended by accelerating multiply ionized species. A similar machine which goes up to 600 kV, was constructed at the University of Aarhus many years ago. It is being frequently used for BFS experiments (An 69).

Much work in BFS has been carried out with commercial single-ended Van de Graaff accelerators, e.g., the various high voltage machines with voltages up to 400 kV, 2 MV, and 5.5 MV (Ba 76). These are usually provided with rf-ion sources and this fact limits the number of elements that can be accelerated. However, these accelerators have also been equipped with thermionic ion sources whereby the number of elements is increased. Oona and Bickel (Oo 76) obtained beams of metal ions from an rf-ion source by fabricating a section of the exit channel from the desired metal. This simple but efficient method has been successfully used by several groups (Ke 79, Ba 80).

Most tandem accelerators were earlier limited to elements which easily form negative ions, e.g., C, N, O, F, S, Cl, Br, and I. In recent years ion sources of the sputtering type (Mi 74) have been developed for these machines. BFS work on many highly ionized atoms has thereby become possible.

Very high energies for BFS experiments have been reached in heavy-ion

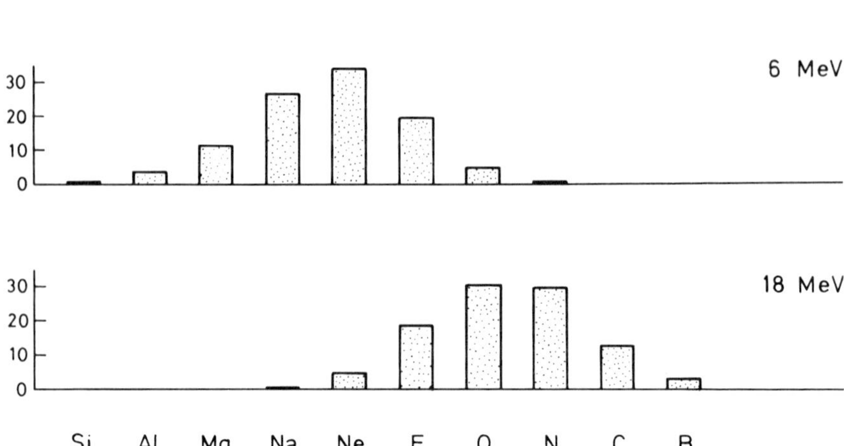

Figure 2. Two examples of equilibrium charge distributions, for Cl ions. At 6 MeV the most probable charge state is Cl^{7+} (Ne-like Cl, or Cl VIII) whereas O-like Cl X and N-like Cl XI dominate at 18 MeV. The data are from (Be 72).

linear accelerators, e.g., those in Orsay (Bu 80), Berkeley (Di 80), and Darmstadt (Do 78), and much additional work along such lines is expected.

As already noted a very large number of ionization states is available in BFS experiments. By varying the accelerator energy the yield of a desired ionization state can be optimized.

An example of charge distributions is shown in Figure 2. The data are based on the work of Betz (Be 72). Using the modified Bohr criterion, Stelson (St 76) has calculated the ionization degrees that can be reached with high-energy machines. The energies required to ionize very heavy ions so as to create vacancies in L shells and K shells are found to be as high as 30 MeV/amu and 200–300 MeV/amu, respectively.

3.2. Spectrometers and Detectors

The particle density in the BFS light source is typically $10^5/\text{cm}^3$, which is several orders of magnitude lower than in most other spectroscopic light sources. Relatively fast optical equipment (with a corresponding sacrifice of high spectral resolution) is usually needed for recording the spectra.

Depending on the wavelength (energy) of the detected photons, various optical instruments are being used by BFS investigators. In the so-called air region ($\lambda > 2000$ Å, where $1\text{ Å} = 10^{-10}$ m), relatively simple scanning monochromators, e.g., of the Czerny–Turner type, are frequently used. The focal lengths of the instruments range from about 0.25 to 2.0 m. Below 2000 Å vacuum instruments must be used. These are usually of the normal-incidence or of Seya-Namioka design (Sa 67) and are appropriate for the region 300–4000 Å (40–3 eV). Grazing-incidence spectrometers are needed for investigations of even shorter-wavelength regions. Here, also, both commercial and home-made instruments (with 1–3 m focal widths) have been used (see, e.g., He 67, Be 71), and the wavelengths covered range from about 20 to 600 Å (600–20 eV).

All these spectrometers are usually equipped with photomultipliers or open detectors (e.g., Channeltrons) at the exit slit and pulse-counting techniques are generally being employed. In the early experiments the beam-foil spectra were frequently registered photographically (Ba 66), but nowadays the photoelectric detection methods are usually preferred, because of their linearity and higher sensitivity. However, most beam-foil experiments use single-channel methods, which are rather archaic in accelerator-based research. Multichannel detection techniques must therefore be developed to reduce the time needed for data taking. This is of course particularly important when the work is performed at large and heavily scheduled accelerators. Such detectors, for the vacuum ultraviolet region, have recently been successfully tested in beam-foil experiments (Li 80, Li 81).

Multichannel systems, e.g., using Si (Li) solid-state detectors, are readily available in the x-ray region, and these have been used in a number of BFS investigations—see, e.g., (Ma 72). When higher spectral resolution is needed crystal spectrometers are necessary. Much work to refine the latter for BFS has been carried out by Silver and collaborators (Tr 79, Ar 80) at Oxford. High spectral resolution has also been achieved with a Doppler-tuned x-ray spectrometer (Sc 73) which takes advantage of the (usually troublesome) Doppler effects in the beam-foil light source.

The kinetic energies of the electrons emitted by the foil-excited ions may range from tens of eV to several keV. A variety of electron spectrometers have been used to analyze these electrons. Several groups have employed electron spectrometers of the cylindrical mirror type (e.g., Br 75, Se 76b). Rødbro *et al.* (Rø 77) used an electrostatic parallel-plate spectrometer whereas a commercial double-focusing ESCA spectrometer has been utilized by Schneider *et al.* (Sc 77).

The electronics and data-taking equipment in BFS experiments are usually relatively straightforward and simple. As an illustration of this we show the system which is presently being used at the University of Lund (Figure 3). Here the foil-excited beam (from a 3-MV Pelletron tandem accelerator) is viewed by two monochromators, one of which is used for the vacuum uv and the other for the visible region. The pulses from the photon detectors at the exit slits of these monochromators are amplified, discriminated, and fed into a minicomputer. The latter also directs the

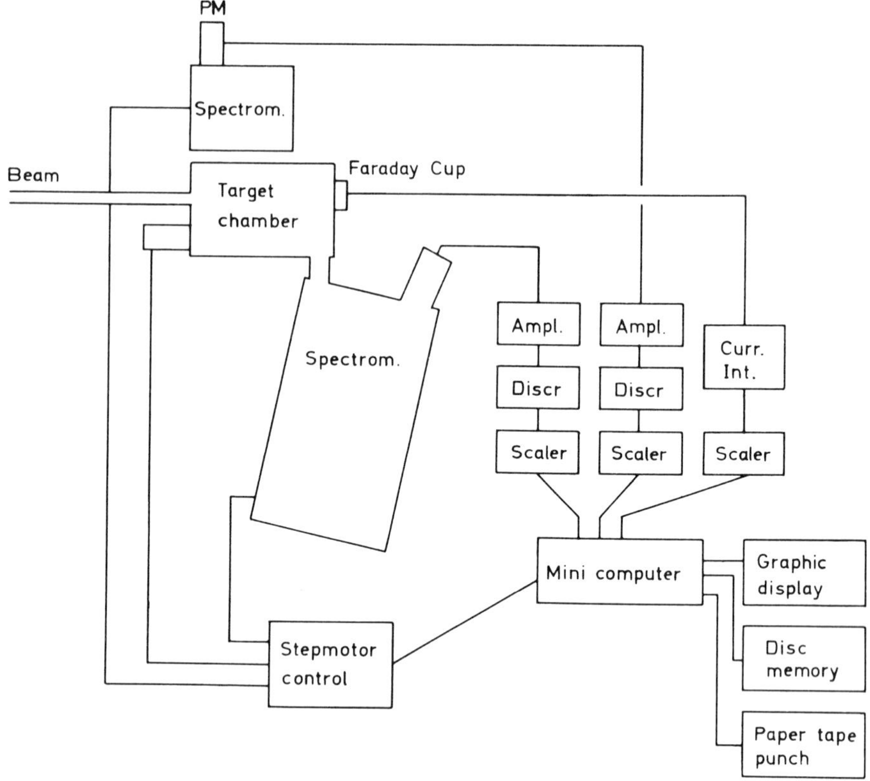

Figure 3. Typical experimental arrangement on the target side. The foil-excited beam is viewed by two optical spectrometers in order to cover different wavelength regions.

experiments by scanning the monochromators and moving the foil along the beam.

3.3. Targets

The target foils are usually made of carbon. It is not very difficult to make such foils by evaporation (De 60). Several companies also offer carbon foils. Foils of other materials have also been tried, but most investigators prefer those of carbon. The foil breakage is a distressing problem in accelerator physics (e.g., concerning stripper foils in tandem accelerators) and it also plays an important role in BFS, particularly when beams of low-energy heavy ions are used. Systematic studies of such effects have been reported by several authors (Yn 74, Li 78, Sa 79). It is often important to know the foil thickness rather accurately so as to compute the energy loss experienced by the ions in the foil. Several methods have been developed for this purpose (Ba 76). However, there also occurs a thickening of the foil during ion bombardment which has been quantitatively studied (Bi 74, Du 76) and which may cause problems in precision work.

In this chapter we have limited ourselves to discussions of a few general characteristics of BFS. More specific information is provided in the forthcoming sections.

4. Atomic Energy Level Studies

4.1. Experimental Problems

Early beam-foil studies of atomic spectra frequently yielded optical transitions that had not been reported in the previous literature on atomic spectroscopy. Several authors (Ba 66, Le 67, Fi 68, Le 69, and others) thus discovered new transitions in multiply ionized N and 0, only a fraction of which could be classified (i.e., assigned to definite energy levels in ionized N and 0), whereas the majority of the lines were left unidentified. Similar studies, for Na (Br 68) and S (Be 70a), were also reported in the literature.

Many of these new transitions were explained by the fact that the spectra of multiply ionized atoms may often be incompletely known (or sometimes totally unknown). Beam-foil studies could then provide some new spectroscopic material. However, it was soon also realized that the beam-foil excitation may populate states which are difficult to excite in other light sources. Some of the new transitions were explained in this way.

The linewidths could be 10–20 Å in many of the early beam-foil spectra in the visible region, which meant that wavelengths could only be determined with ± 2 or ± 3 Å uncertainties—although data of much better quality were also sometimes reported (De 68). The linewidths were largely caused by sensitivity problems which demanded the use of small and fast monochromators and wide slits. Furthermore, the light is emitted by moving ions, and Doppler effects must therefore be considered. When the foil-excited beam is viewed at an angle of 90° the first-order Doppler shift vanishes, whereas the second-order shift, proportional to v^2/c^2, must be corrected for. There is also a Doppler broadening of the spectral lines, due to the finite acceptance angle α of the spectrometer. This broadening is approximately given by

$$\Delta\lambda \sim \lambda \alpha v/c \tag{2}$$

With α equal to 0.05 rad, λ 5000 Å, and v/c 2 % we obtain a line broadening of 5 Å (orders of magnitude larger than in many other light sources), and this clearly rules out precision spectroscopy.

Efforts have been made to eliminate this drawback. By refocusing the spectrometer to a moving source it is possible to reduce much of the line broadening (St 71, Le 73). In another system, developed by Kay and coworkers (Ka 73), the beam is viewed by two cylindrical lenses. This method, which also improves the resolution, has not yet been widely used.

As shown in Figure 1 only a small section of the beam is observed at a time and much intensity is therefore lost. To overcome this serious drawback Stoner and Leavitt (St 73) developed an optical system (consisting of a cylindrical lens) which allows a much larger amount of light to be collected. In a typical case the intensity was increased by a factor of 40. However, the design sacrificed the spatial resolution, which means that the method is not useful in lifetime studies. Subsequently Bergkvist (Be 76) discovered an ingenious method which both leads to a very large intensity gain (more than 70 times in the first tests) and preserves the spatial resolution. In this geometry the entrance slit of the spectrometer is replaced by a specially designed coupling lens. This design has later been modified by Jelley *et al.* (Je 77).

The ultimate wavelength resolution in BFS is limited by scattering processes in the foil which lead to a divergent beam on the downstrem side. To reduce this effect very thin foils must be used.

At present it is fair to state that high-quality beam-foil spectra can be routinely obtained. Thus Livingston *et al.* (Li 80) have recently obtained 0.5–0.8 Å linewidths in the region 600–900 Å and they determined wavelengths with relative errors as low as 10^{-4}. Similar results have been reported by other groups, e.g., Armour *et al.* (Ar 79). Figure 4 gives an example of modern beam-foil spectrum (Ga 77).

After a wavelength measurement the observed transitions should be assigned to the proper ionization states. This is usually achieved by studying the intensity of spectral lines as a function of beam energy, i.e., recording spectra at several energies. This method, introduced by Kay (Ka 65), assumes that the excitation functions for levels within a given ion are similar. Exceptions to this have been found, however (Du 70). Another method consists of applying a transverse electrostatic field to the excited beam (Ma 65). The optic axis of the spectrometer is parallel to the field direction. The lines are then red or blue shifted depending on the polarity of the electrostatic field. This shift is proportional to the charge of the ion. A similar concept by Oona and Bickel (Oo 76a) is based on separating the beam by an electrostatic field and focusing the spectrometer on each charge state individually. The electrostatic methods are more direct than the intensity-energy studies and they should perhaps be tested more often. However, they are difficult to apply to short-lived states.

The resolution problems in electron spectroscopy of foil-excited ions are rather similar to those discussed above. With foil targets the linewidths are

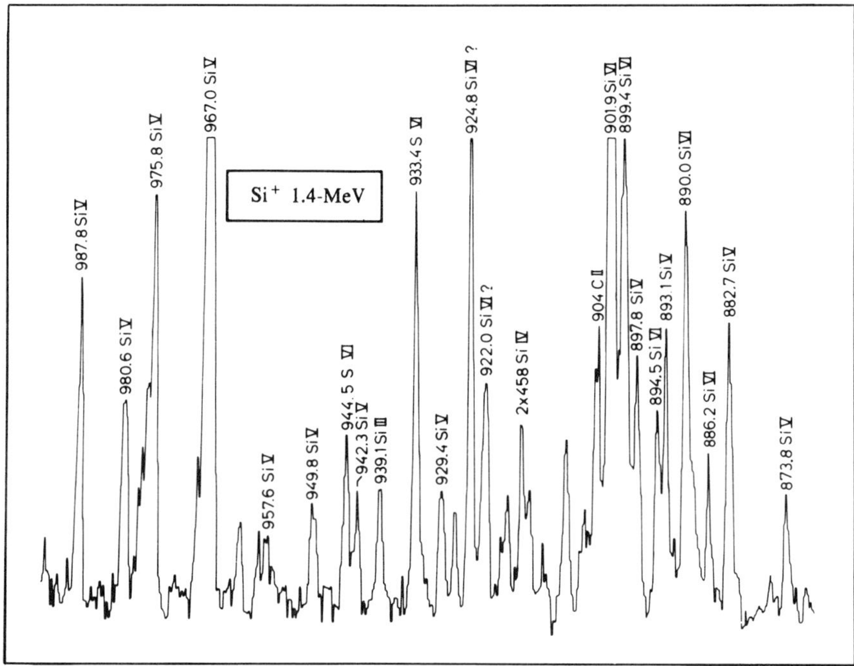

Figure 4. Beam-foil spectrum of foil-excited Si, obtained with 1.4-MeV Si^+ ions. The wavelengths and ionization degrees of the strongest lines are indicated. (Ga 77)

usually much larger than the resolution of the electron spectrometer used. This is mostly due to multiple scattering of the ions in the target foil. Gaseous targets are therefore frequently used to otain narrower lines.

4.2. Results of Spectral Studies

The early beam-foil results, mentioned above, have stimulated more quantitative studies of atomic energy levels by BFS, and these investigations have resulted in a considerable amount of valuable material. Although atomic spectroscopy is one of the oldest areas of modern physics, it has not been exhausted. As noted, there are many systems for which the experimental information is far from complete. Even some light and relatively fundamental systems, e.g., O II, O III, and F IV–F VIII (we use conventional spectroscopic notation, according to which O II means singly and O III doubly ionized oxygen, etc) are rather fragmentarily known. It is thus not surprising that BFS studies have led to new spectroscopic information. There are also two categories of atomic energy levels which are very strongly excited at the beam-foil interaction, namely, hydrogenlike states with high n and l quantum numbers (Rydberg states) and multiply excited (inner-shell excited) configurations. For systematic purposes we may divide the beam-foil results into these three groups, i.e., (a) previously incompletely studied systems, (b) hydrogenlike levels, and (c) multiply excited levels. More detailed discussions of such beam-foil results can be found in previous literature (Ma 76, Pe 80).

4.2.1. *Previously Incompletely Studied Systems*

Beam-foil investigations have yielded new spectroscopic data for many light atoms and ions. Even in a spectrum as fundamental as that of Li II new levels have been located by BFS experiments (Be 72a), and similar data are also available for other He-like ions, such as B IV (To 74) and C V (Ja 80). In the latter case, about 30 new C V lines were classified. Since atomic theory is very accurate for He-like systems it is quite important to obtain high-quality experimental data for comparison and confirmation.

Similar examples of new beam-foil data can be found for the Be I isoelectronic sequence (i.e., Be I, B II, etc). For Be I Hontzeas *et al.* (Ho 72) located several levels of the type $2pnl$ ($n = 3, 4, 5...$) by observing transitions of the type $2snl$–$2pnl$ in beam-foil spectra. Similar transitions, in B II, were found by Martinson *et al.* (Ma 70a) and Berry and Subtil (Be 74). It has been noted that BFS is very efficient at populating configurations of the type $2pnl$ in light atomic systems.

Interesting examples of beam-foil results can also be found in the case of Ne-like ions. Here the systems from Ne I to P VI have been well studied by classical atomic spectroscopy (e.g., by using spark light sources). For S VII new spectroscopic results were obtained in the BFS studies by Garnir *et al.* (Ga 78, Ga 79), while Jupén (Ju 82) has studied the system Cl VIII which was practically unknown when the beam-foil investigation began. The results are shown in Figure 5, which shows that the analysis is now rather complete. It is thus clear that BFS is a good method for atomic energy-level investigations. Identifications of transitions in Ne-like Fe (Fe XVII) have recently been proposed by Buchet *et al.* (Bu 80), who used high-energy beams from the Orsay heavy-ion accelerator ALICE.

Beam-foil data are also available for ions belonging to the Na I and Mg I isoelectronic sequences. In the Na-like ions (which have only one valence electron) the spectra of S VI (Dy 78), Cl VII (Ba 71, Bh 74), and Ar VIII (Li 81) have been investigated by BFS.

The Mg-like ions have the electron configuration $2p^6 3snl$ or $2p^6 3pnl$ $^{1,3}L$. Interesting examples of configuration interaction between the two valence electrons can be found here. Until recently the spectra of S V and higher Mg-like ions have been rather poorly known. Beam-foil studies have therefore resulted in new energy-level data for S V (Dy 78), Cl VI (Ba 71, Bh 74), and Ar VII (Li 81).

Much interest has recently been focused on the Cu I and Zn I isoelectronic sequences, largely because optical transitions in Mo XIV and Mo XIII—which are members of these sequences—have been observed in

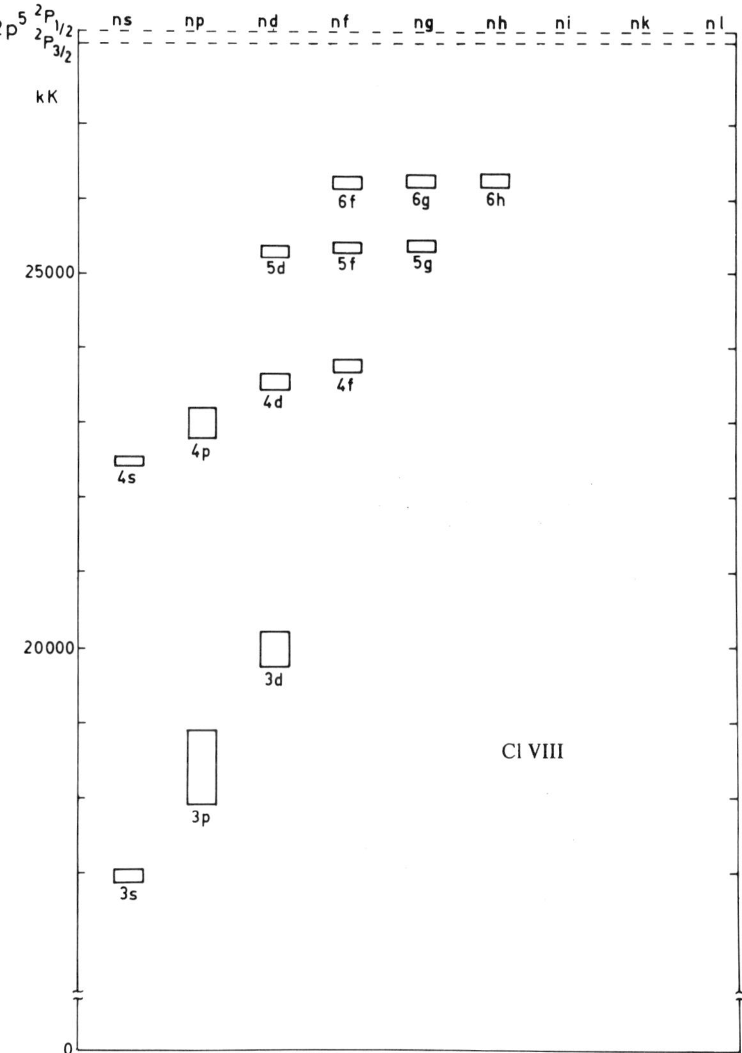

Figure 5. Energy level diagram for Cl VIII (Ne I isoelectronic sequence), according to the beam-foil study of Jupén (Ju 82). The ground state is $2p^6\ ^1S_0$, while the excited levels are of the type $2p^5\ nl$.

Tokamak discharges (Hi 76). Beam-foil studies were therefore undertaken to provide energy-level and wavelength data for various Cu- and Zn-like ions so as to permit extrapolations to Mo XIV and Mo XIII. Thus Denne *et al.* (De 79a) have successfully combined beam-foil and classical spectroscopy to study the interaction between the $4s4d\ ^1D$ and $4p^2\ ^1D$ terms in Zn-like Ga II. Valuable beam-foil results for Cu-like Br VII (Li 81a) and Kr VIII (Li 80a) have also appeared recently. Such work has also been extended to heavier ions, and energy-level data for multiply ionizied I (On 79) and Xe (Ke 80, Ba 81) are already available.

The vuv region has been investigated in most of the beam-foil studies

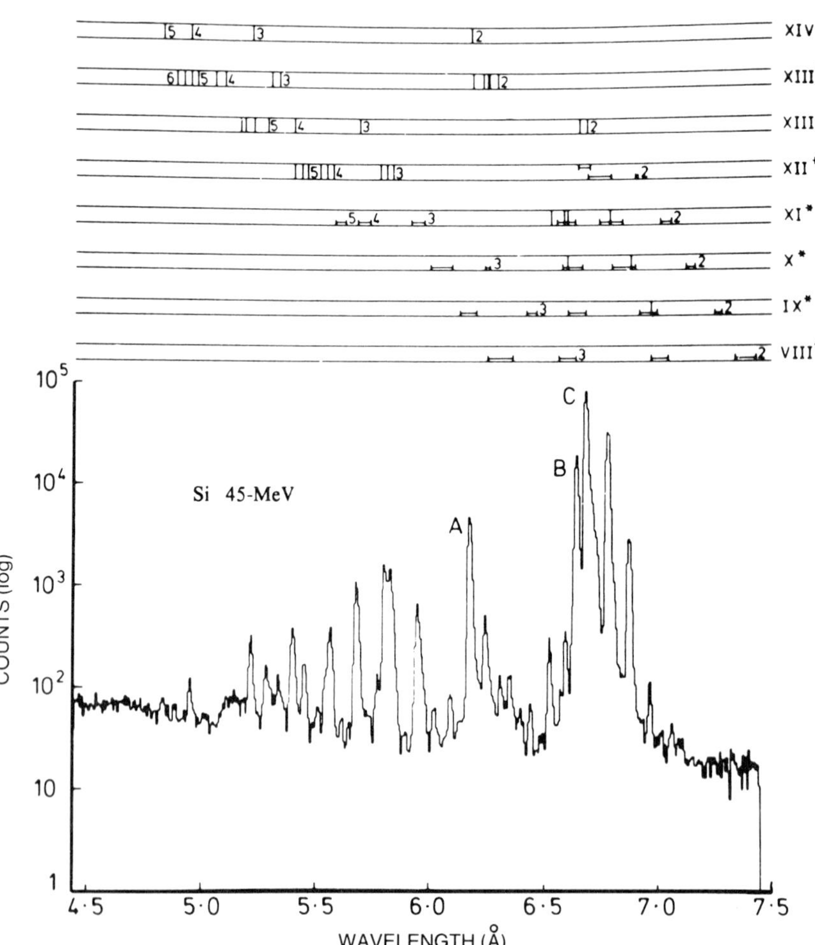

Figure 6. X-ray spectrum of Si, obtained with 45-MeV Si ions (Tr 79, Tr 79a). The line assignments are indicated.

mentioned above. However, a wealth of important spectral data can also be obtained by using ion beams of tens of MeV energy and studying the x-ray region. Thus Träbert *et al.* (Tr 79, Tr 79a) have sent 35–45-MeV Si ions through carbon foils and measured the x-ray spectra (3–9 Å) using a curved-crystal x-ray spectrometer. One of their spectra is shown in Figure 6. Many new transitions in He-like and Li-like Si were identified as well as lines belonging to lower ionization stages of Si. Similar results, for Mg and Al, have more recently been obtained by Armour *et al.* (Ar 80). This method (which yields high-resolution x-ray spectra of foil-excited ions) seems to be quite powerful and well-suited for detailed studies of the structure of highly ionized species.

4.2.2. Hydrogenlike Systems

The term value T_{nl} (the energy difference between the ionization energy E_∞ and the excitation energy E_{nl}) of an atomic energy level can generally be written as

$$T_{nl} = R\zeta^2/n^{*2} \tag{3}$$

where R is the Rydberg constant, ζ the spectrum number (frequently also called the "net charge of the core," $\zeta = 1$ for neutral atoms, $\zeta = 2$ for singly ionized species, etc.), and n^* the effective quantum number, i.e., $n^* = n - \delta$, where δ is the quantum defect (Ed 64). For sufficiently high values of the azimuthal quantum number l the orbit of the excited electron does not penetrate the core, formed by the passive electrons. The term values can then be written as

$$T_{nl} = T^0_{nl} + \Delta p \tag{4}$$

where T^0_{nl} is the relativisitically corrected hydrogenic term value

$$T^0_{nl} = \frac{R\zeta^2}{n^2}\left[1 + \frac{\alpha^2\zeta^2}{n^2}\left(\frac{n}{l+1/2} - \frac{3}{4}\right)\right] \tag{5}$$

and Δp the polarization energy, which is usually expressed as

$$\Delta p = \alpha_d R\langle r^{-4}\rangle + \alpha_q R\langle r^{-6}\rangle \tag{6}$$

Here α_d and α_q are the dipole and the quadrupole polarizabilities of the core and $\langle r^{-4}\rangle$ and $\langle r^{-6}\rangle$ the quantum-mechanical expectation values of r^{-4} and r^{-6}. These have been tabulated in the literature (Ed 64, Bo 74).

Much interest has recently been focused on the determination of α_d and α_q (Sh 79, Cu 80), and wavelength measurements are here of great value. Using transitions between nonpenetrating orbits and Equations (4)–(6) the

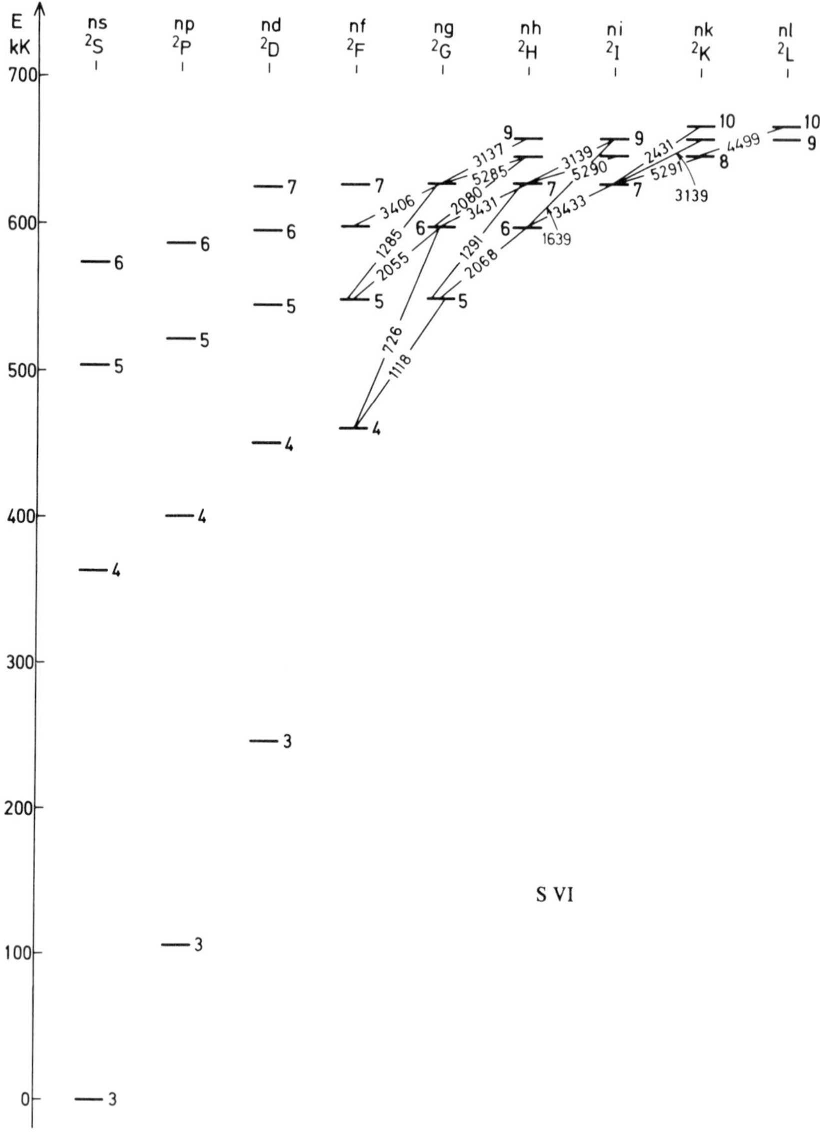

Figure 7. The energy level structure of S VI (Na I isoelectronic sequence). The transitions in the hydrogenlike part of the spectrum ($l \geqslant 3$) observed in a beam-foil experiment (Dy 78) are indicated by wavelength (in Å units).

values for α_d and α_q can be determined. The polarization formula is most easily applied to systems with only one valence electron (e.g., Li-like and Na-like ions), but it can also be extended to more complex cases.

When the fast ions leave the foil, electron capture into high-lying states may have a high probability. These states subsequently decay by photon emission. Beam-foil spectra may therefore be dominated by strong radiative transitions between high-lying states (including the so-called yrast states).

In several of the early beam-foil experiments transitions between hydrogenlike levels appeared although they were usually not classified. Somewhat later identifications were given (Dr 70, Ba 71, Le 72, Ha 73, Bu 74). Some results, for S VI, are shown in Figure 7. The hydrogenlike transitions in this ion, observed by BFS, are indicated by wavelength (Dy 78). Similar studies have recently been performed for a large number of ionic systems.

4.2.3. *Multiply Excited States*

Investigations of atomic spectra deal mostly with states formed by the excitation of one outer electron. In several light sources also multiply excited states may be populated, however. Consider as an example the system He I where the ground state is $1s^2\ {}^1S$ and the singly excited states are of the type $1snl\ {}^1L,\ {}^3L$. The latter converge to the ionization limit (He II ground state) at 24.6 eV above the $1s^2$ ground state. The doubly excited states, $2snl$ or $2pnl$ have high excitation energies with the lowest state $2s^2\ {}^1S$ lying 57.9 eV above the He I ground state and thus more than 33 eV above the He II ground state. Such doubly excited configurations (often denoted by A^{**}) usually autoionize by the electrostatic interaction

$$A^{**} \rightarrow A^{+} + e^{-} \tag{7}$$

with typical transition rates of 10^{13}–$10^{15}\ \mathrm{s}^{-1}$. The selection rules for this process are $\Delta J = 0$ (and $\Delta S = \Delta L = 0$ within the LS coupling) and no parity change. When there are no final states that fulfill these requirements autoionization via spin–orbit, spin–other-orbit or spin–spin interactions may take place. However, the probabilities of these processes are orders of magnitude lower than for the electrostatic (Coulomb) autoionization. Radiative decays (with 10^8–$10^{10}\ \mathrm{s}^{-1}$ rates) then form an alternate decay possibility and the multiply excited states can be investigated by photon spectroscopy.

It was found several years ago that BFS is very efficient in populating multiply excited states (Bi 69a, Bu 69), the decay of which could be observed by photon spectroscopy in emission. Those first studies dealt with Li I where configurations such as $1s2snl$ and $1s2pnl\ {}^4L$ (these may also be called inner-shell excited states since one 1*s* electron is excited, but we will include them in the discussion of multiply excited states) were studied. These levels lie more than 50 eV above the Li II ground state, but because of the $\Delta S = 0$ rule they cannot autoionize into the doublet continuum. Somewhat later doubly excited levels in He I were also studied by BFS (Be 71, Kn 73) and the work has subsequently been extended to many members of the He I and Li I isoelectronic sequences. A thorough review on multiply excited states in BFS has been written by Berry (Be 75).

4.2.3a Photon Spectroscopy. The multiply excited states can be studied both by photon and by electron spectroscopy. In the latter case it is also possible to observe such states which autoionize rapidly.

In the He I isoelectronic sequence configurations such as $2pnp\ ^1P$, ^3P, $2pnd\ ^1D$, ^3D, and others with $l = L$ do not autoionize by the Coulomb interaction, but they make radiative transitions to lower levels. In the beam-foil spectra of He there are many lines close to 300 Å which are transitions from the doubly excited nonautoionizing $2pnl$ configurations to the $1snl$ system of He I (Be 71, Kn 73). Transitions between doubly excited levels ($2pnl$ and $2pn'l'$) which appear above 2000 Å have also been observed in BFS (Be 72b, Be 73, Is 78). In the most recent study, Brooks and Pinnington (Br 80) determined accurate wavelengths of more than ten He I** transitions above 2000 Å.

The beam-foil studies of He I** have thus yielded detailed information about the nonautoionizing levels and the work largely confirms theoretical studies of excitation energies (Li 77). Moreover, the results complement experimental data for autoionizing states obtained by exciting He atoms with synchroton radiation (Ma 63). Transitions in the doubly excited Li II system have also been observed by BFS (Bu 73). More recently Andersen *et al.* (An 80) studied the doubly excited states of Be III, using 200–600-keV Be^+ beams. About 10 levels ($2pnp$ or $2pnd$) were found in Be III** and some minor disagreements with theory (Li 77) could be noted.

Following the early work on doubly excited levels in Li I (Bi 69a, Bu 69) many beam-foil experiments have been undertaken to study the quartet levels in the Li I isoelectronic sequence. Results have thus been obtained for Li I–Ar XV (Be 75, Se 76b, To 79). The possibility of carrying out such systematic studies for many ions is one of the main merits of BFS. The most recent results, for Li I**, are shown in Figure 8, (from Mannervik (Ma 81)). There are interesting examples of configuration interaction in such doubly excited systems. For instance, the odd ^4P levels may be mixtures of $2snp$, $2pns$, and $2pnd$ configurations. The theoretical problems in describing these levels may be formidable; theory has on the other hand also been vital for understanding the experimental results. The corresponding system in Be II** which was investigated by Hontzeas *et al.* (Ho 72) has recently been revised by Mannervik *et al.* (Ma 81a). One of their spectra is reproduced in

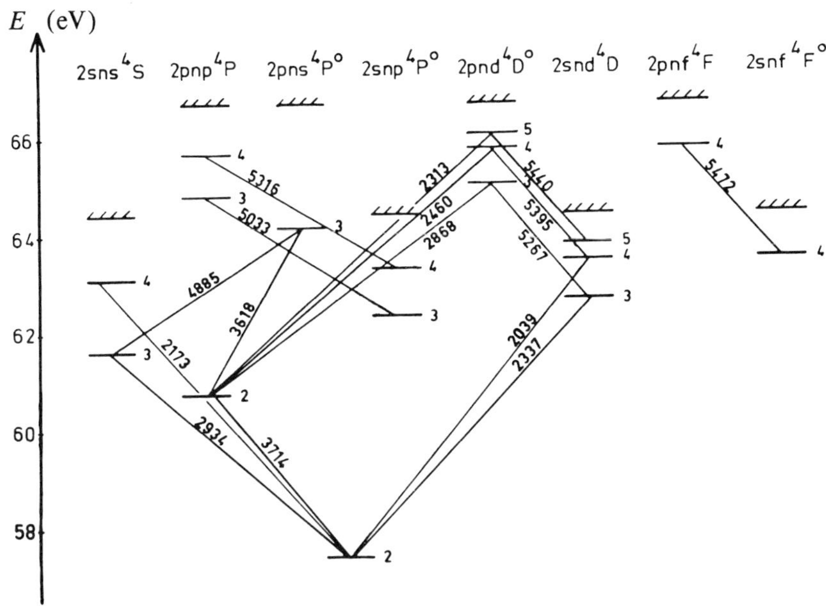

Figure 8. Transitions observed in the $1s2snl$ and $1s2pnl\ ^4L$ system of Li I (Ma 81). The wavelengths are given in Å. These levels lie more than 50 eV above the Li II ground state.

Figure 9, which gives an indication of the intensities of lines from doubly excited levels in beam-foil spectra. These lines have not appeared in other light sources. Extensive studies of Li-like quartets have also been performed for B III–F VII (Be 75, To 79). In all such photon spectra the $1s2s2p\ ^4P$–$1s2p^2\ ^4P$ transitions are particularly strong. (We use here traditional spectroscopic notation, i.e., $1s2p^2\ ^4P$ is the upper state). Livingston and Berry (Li 78a) have been able to resolve this multiplet in beam-foil excited C IV, N V, and O VI. They have thus determined the fine structure separations of both ^4P terms. In order to reproduce the experimental results very advanced theoretical calculations (which include both electron correlation and relativistic effects) have been undertaken (Ch 78).

In the early beam-foil studies of Li I** a line at 3489 Å was observed. For many years its origin was a mystery until Bunge (Bu 80a) made very elaborate configuration-interaction calculations and showed that it arises from the $1s2s2p^2\ ^5P$–$1s2p^3\ ^5S$ transition in the *negative* Li ion (Li^-). This is the first known example of a transition in emission between two bound states in a negative ion. The predictions were confirmed in subsequent, independent BFS studies (Ma 80, Br 80a). Using electrostatic fields the authors definitely

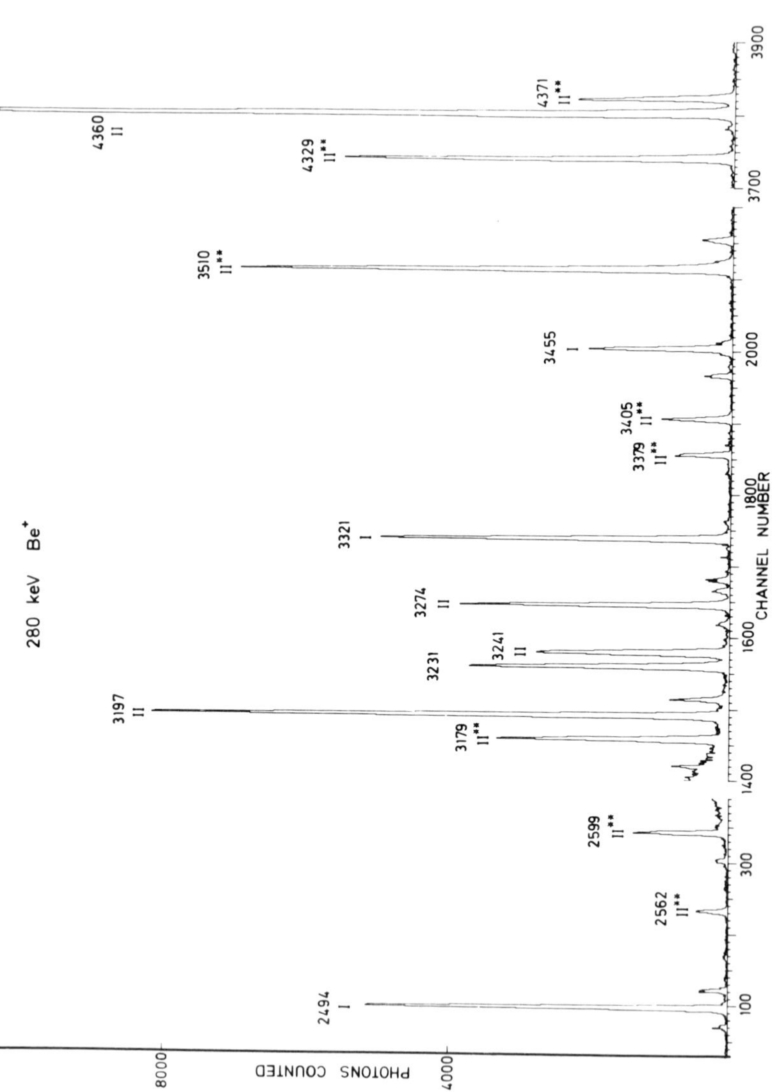

Figure 9. Beam-foil spectrum of Be, obtained with 280-keV Be^+ ions (Ma 81a). The wavelengths (in Å) and ionization states of the strongest lines are given. The transitions denoted with II** belong to the quartet system of Be II.

established the Li^- origin for the 3489 Å line. Lifetime measurements were also made which agreed very well with the theoretical $1s2p^3\ ^5S$ lifetime (Bu 80a). Isoelectronic analysis further showed that the ^5P–^5S transition in Be I–F VI also appears in beam-foil spectra (Br 80a). For O V it has also been possible to measure the fine structure of the $1s2s2p^2\ ^5P$ term (Li 80b) and compare it to theory.

4.2.3b Electron Spectroscopy. Studies of the spectra of the electrons emitted by the multiply excited ions provide information which complements photon spectroscopy. By making observations close to the foil one can also investigate the decays of strongly autoionizing levels, whereas electron spectra taken at distances from the foil (that correspond to some ns) mostly reveal the decay of states with longer lifetimes, i.e., those which autoionize via the magnetic spin–orbit or spin–spin interactions.

The first such beam-foil measurements were undertaken by Sellin, Pegg, and collaborators (Se 71, Se 76b). Their spectra showed the autoionization electrons from the decay of $1s2s2p\ ^4P$, $1s2p^2\ ^4P$, and other levels in Li-like O and F. Detailed studies of O, F, Cl, and Ar ions (Pe 73) subsequently resulted in energies of many multiply excited levels.

Efforts have been made to reach as high resolution as possible. Gaseous targets have therefore frequently been used. The absence of multiple collisions results in higher resolution. Comprehensive studies of autoionizing

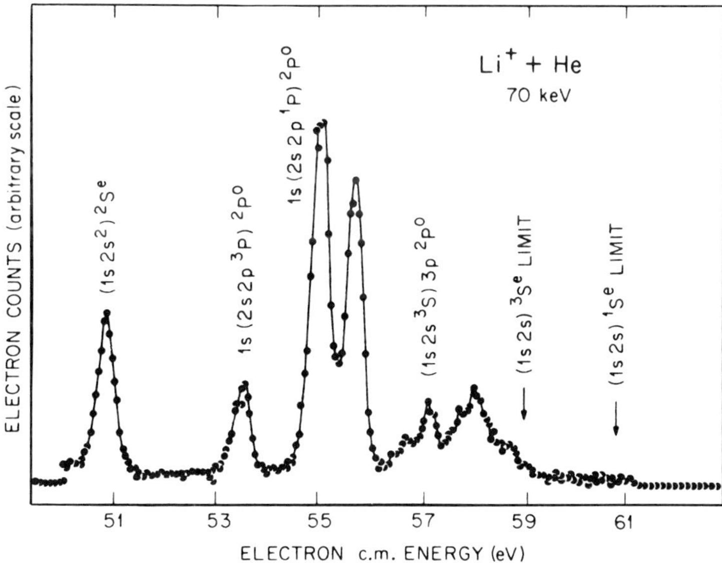

Figure 10. Electron spectrum of doubly excited Li, obtained by bombarding He gas with 70 keV Li^+ ions (Pe 75). The transitions observed arise when inner-shell excited states in Li decay by autoionization.

states in Li–0 have been made in recent years using beam-foil and beam-gas spectroscopy. For example, Bruch *et al.* (Br 75) have been able to observe *triply excited* levels in Li I ($2s^2 2p\ ^2P$, $2s2p^2\ ^2S$ and ^2D). The energies of these levels were determined quite accurately. Figure 10 (from Pe 75) shows an electron spectrum, obtained by sending Li ions into He gas. Bruch *et al.* (Br 75, Rø 77, Rø 79) have made extensive studies of multiply excited levels in light atoms such as Be, B, and C. Among the results are energies for states with high spin, and the decay of $1s2s2p^2\ ^5P$ in four-electron systems (mentioned above) has been studied by electron spectroscopy.

5. Lifetime Measurements

5.1. Experimental Problems

The BFS method is a very simple and direct one for measurements of the lifetimes of excited states in atoms and ions [Figure 1 and Equation (1)]. This is one of the main advantages of the BFS technique. In contrast to most other lifetime techniques— which are only applicable to neutral, singly, or (sometimes) doubly ionized atoms—BFS can be successfully used to determine lifetimes in highly ionized atoms. The range of lifetimes that can be studied is illustrated in Figure 11 (according to Br 73 and An 79). Two of

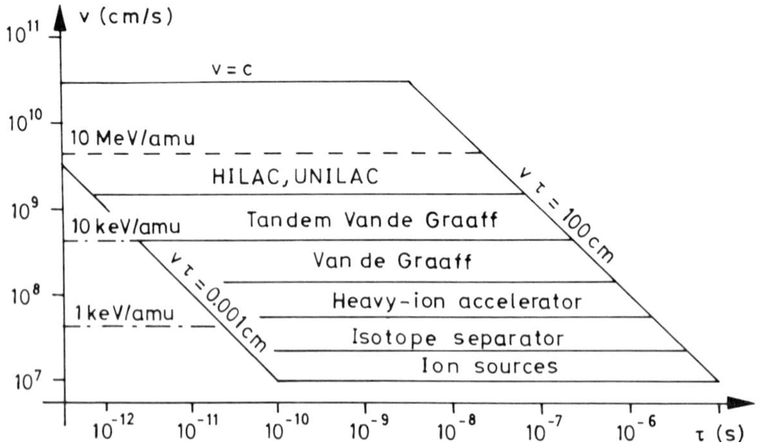

Figure 11. The range of atomic lifetimes that can be conveniently measured with various accelerators (An 79, Br 73). The lowest velocity is that needed to penetrate a foil. The spatial resolution is assumed to be 10^{-3} cm while the maximum measurable decay length is taken as 100 cm.

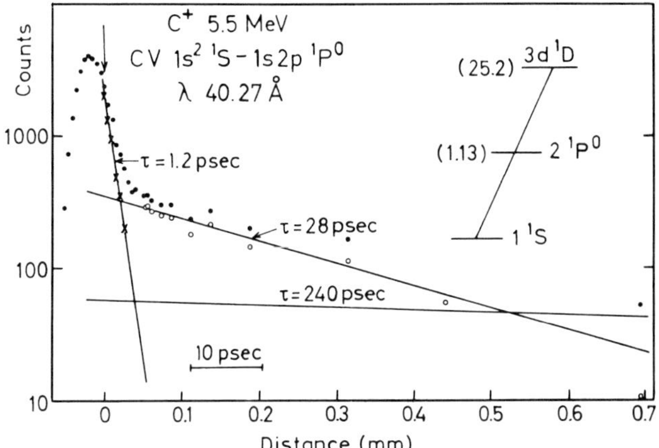

Figure 12. Example of the measurement of a very short lifetime (Kn 76). Note that the 28-ps component roughly corresponds to the theoretical lifetime for the most probable cascading level.

the limits are based on the assumptions that the spatial resolution is 10^{-3} cm while the length of the chamber is taken as 100 cm. The horizontal limits are fairly obvious. A fine example of a measurement of a very short lifetime is given in Figure 12 (Kn 76).

In most lifetime measurements the decay curves are obtained by stepwise translation of the foil along the beam with a precision screw. Sometimes the spectrometer is moved instead. By optical methods it is also possible to record a decay curve without moving the foil or spectrometer (Pi 70). Various normalization procedures have been developed to compensate for fluctuations of the ion beam, the simplest method being charge normalization on the beam with a Faraday cup. At low velocities (where a large fraction of the beam may be neutralized in the foil), and also when the foil breakage is a substantial problem, optical normalization methods are usually prefered (As 76, An 79).

While it is often easy to obtain decay curves of quite high quality their analysis may present substantial difficulties. A number of experimental problems must be mastered before reliable lifetimes can be obtained by BFS.

Cascading from higher-lying states into the level studied is perhaps the most serious obstacle. Since the BFS excitation is largely nonselective, many levels can be populated in an atom or ion. Their decay into lower levels may thus complicate lifetime measurements. The experimental decay curves must frequently be decomposed into several exponentials, of the form

$$I(x) = \sum_i I_i(0) \exp(-x/v\tau_i) \qquad (8)$$

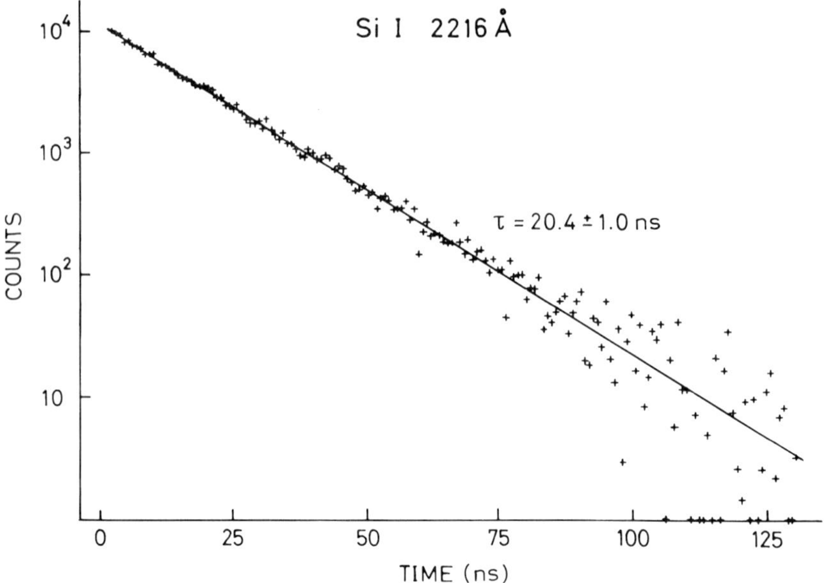

Figure 13. A decay curve (for the $3p^3\ ^3D$ level in Si I) which shows practically no cascading effects (Ba 80a).

where τ_i are the lifetimes of the various levels involved in—or influencing—the observed decay.

A decay curve which apparently shows no cascading is displayed in Figure 13. (Ba 80a). However, it should be remembered that even a curve that appears as a single exponential may be a sum of several exponentials, as already pointed out by Mitchell and Zemansky (Mi 34).

Most decay curves encountered in BFS show rather pronounced multiexponential structure. An example is displayed in Figure 14 (En 81). Here we observe both a "growing-in," which indicates cascading from levels with lifetimes shorter than the level under direct study , and further out on the decay curve also cascading from longer-lived states is evident.

Even with powerful multiexponential fitting programs such as HOMER (Ir 74) and DISCRETE (Pr 76) it will often be very difficult to obtain the physically correct decomposition. Already a decomposition into three exponentials implies the determination of six parameters (three lifetimes and three intercepts at time $t = 0$). There are several examples in the literature that illustrate the dangers and systematic errors which may arise in such analyses (Wi 70, Cu 70).

It would clearly be best to use intensity-calibrated detection systems and to measure all important cascade transitions explicitly and then subtract

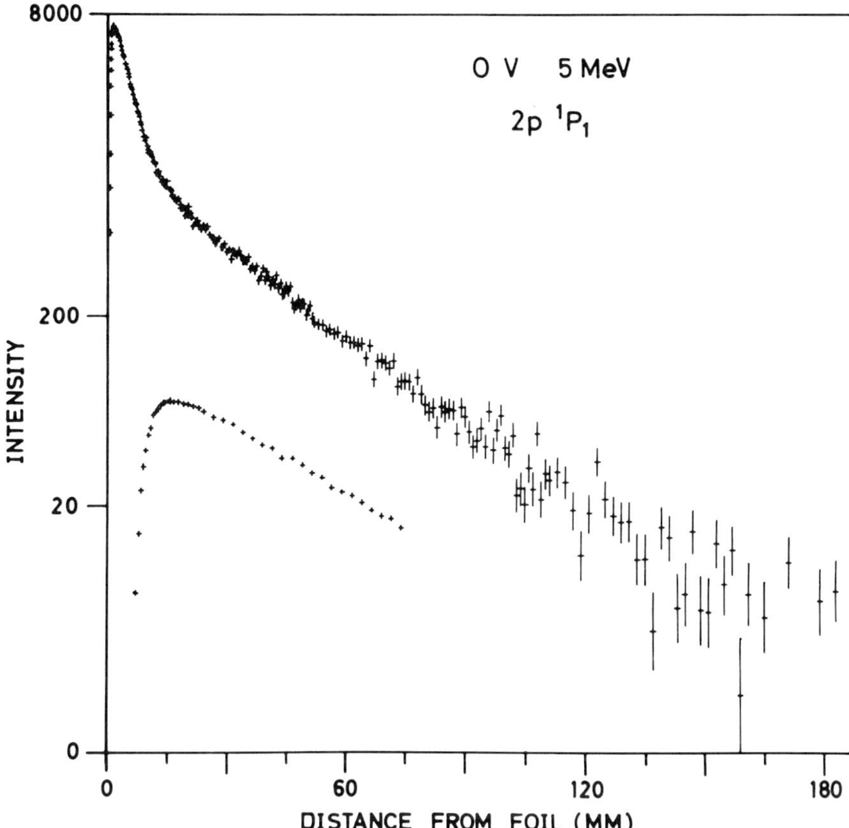

Figure 14. A fairly complex decay curve. Such curves are frequently encountered in beam-foil experiments. The $2s2p\ ^1P$ level in O V is fed from a short-lived higher state ($2p^2\ ^1S$), and this growing-in is best seen in the lower curve, which has an expanded abscissa. Furthermore, there are pronounced effects of cascading from a long-lived state ($2p^2\ ^1D$). From Engström *et al.* (En 81).

these from the observed primary decay curve. In most cases this is very difficult to accomplish, however. The various transitions often lie in quite different wavelength regions and this complicates intensity calibrations. Curtis (Cu 71, Cu 76), has therefore developed an effective method usually called ANDC (arbitrarily normalized decay curves) which only requires the principal cascades to be measured while intensity calibration is not necessary. This method has been tested in quite a number of cases and the results are very encouraging. Thus, for the example shown in Figure 14, multiexponential curve-fitting programs yield a lifetime value of 0.41 ± 0.03 ns, whereas an ANDC analysis results in 0.33 ± 0.02 ns (En 81).

The latter value is in excellent agreement with the theoretically predicted lifetime, 0.341 ns (Hi 80). Since theory is considered very accurate for Be-like ions, the ANDC value is clearly superior. The example also illustrates the risks of simple curve fits.

The cascading has been found to be particularly troublesome in atomic spectra with one or two valence electrons. Pinnington *et al.* (Pi 79) and Livingston *et al.* (Li 80a) have here compared multiexponential fitting and ANDC analysis, one of the conclusions being that ANDC techniques yield good lifetime data even in the presence of quite heavy cascading.

An alternative method for analysis consists of constructing "synthetic" decay curves using theoretical level lifetimes and various models for the foil excitation process (Cr 76, Yo 78, Bu 80, De 81). Comparisons between the experimental and theoretical curves then yield information about the extent of cascading and permit the extraction of reliable primary lifetimes.

Cascading was overlooked or underestimated in the majority of early BFS work, and such data therefore frequently give too optimistic error estimates. This has been criticized in the literature and remedies have been suggested (Wi 70, Cu 76, Yo 78, Cu 79).

When the beam energies are in the MeV range, the energy loss in the foil is usually small on a relative scale and it can be corrected for quite easily (No 70). At lower ion energies, the energy loss may amount to tens of per cent of the incoming energy which makes precise determinations of the velocity of the foil-excited particles more difficult. Electrostatic analyzers are frequently used together with low-energy machines for this purpose (Bi 69, An 79). It is also possible to apply optical Doppler-shift methods for velocity determinations (Pi 70, Sø 76, Hu 79).

Because of the resolution problems in BFS, spectral line blends may distort lifetime measurements. This problem is particularly serious in line-rich systems, such as the ions of iron-group and rare-earth elements. The spectral structure must further be sufficiently well known before it is worth undertaking decay studies. In a lifetime measurement for a Sc II level ($z\ ^1P^0$) Buchta *et al.* (Bu 71) obtained a value of 5.5 ± 0.5 ns, which did not agree with theory (Vi 76). Only later was it found that the Sc II line was blended with a Sc III transition (Pa 76), the wavelength of which was not known at the lifetime measurement.

5.2. Experimental Modifications

Because of these problems in lifetime studies, by BFS, important modifications have been developed. These include (a) beam-laser techniques, (b) alignment transfer studies, and (c) coincidence methods.

The principle of the beam-laser method, introduced by Andrä *et al.* (An 73), is shown in Figure 15. The ions from the accelerator are selectively

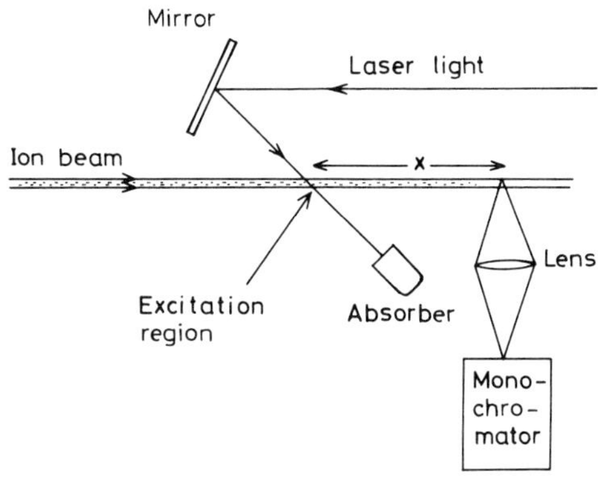

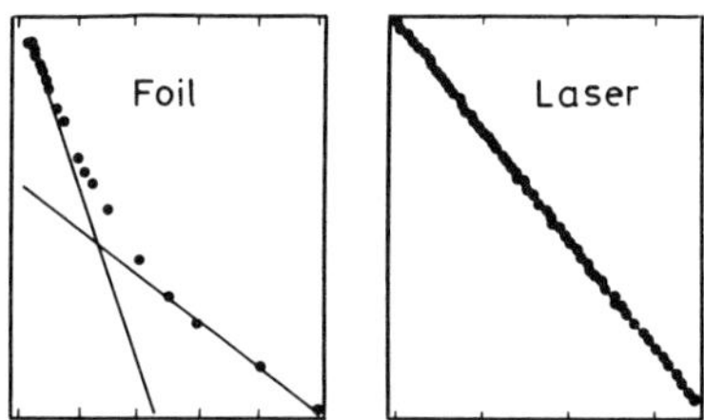

Figure 15. The upper part shows a typical geometry for ion-laser experiments. The decay curves (both for the same level, $3p\ ^2P$ in Na I) illustrate the difference between using a foil (left) and the laser (right), from Harde (Ha 76).

excited with monochromatic laser light. In the first experiments an Ar^+ laser was used to excite Ba^+ ions from the accelerator. The ground state in Ba II is $6s\ ^2S_{1/2}$ while the $6s\ ^2S_{1/2}$–$6s\ ^2P_{3/2}$ transition has a wavelength of 4554 Å. The Ar^+ laser emits a strong line at 4545 Å. By Doppler tuning, a resonance was found with the Ba^+ line and the $6p\ ^2P_{3/2}$ level was thus selectively excited. The first measurments yielded a lifetime of $\tau(^2P_{3/2}) = 6.25 \pm 0.06$ ns (An 73), while later refinements result in $\tau(^2P_{3/2}) = 6.312 \pm 0.016$ ns (An 76), one of the most accurately known atomic lifetimes.

Tunable dye lasers have also been used for such measurements (Ha 76), and this widens the applicability of the method. An interesting example is the La II study of Arnesen *et al.* (Ar 77). Here La^+ ions from an isotope separator were excited from their ground state by laser light while the spontaneous decay into another configuration was measured. Since the excitation and detection occurred at different wavelengths no stray-light problems (which may be troublesome in ion-laser experiments) were encountered.

Such beam-laser measurements can also be performed on neutral atoms, by first neutralizing the ions from the accelerator in a gasous target. Accurate data for, e.g., Na I (Ha 76) and Sr I (An 75) have been obtained in this way.

Another interesting experimental modification consists of using parallel ion and laser beams (collinear beam geometry). The resonance can be tuned in by changing the beam velocity with a variable longitudinal electric field (Wi 77). This method has recently yielded some very accurate lifetimes in Rb II (Ce 80).

The beam-laser methods have additional advantages besides being cascade-free. Line blends are absent and the time scale after excitation is much better known than in BFS experiments; there are no energy-loss problems in the foils. The main drawback is the limitation of the present lasers. For example, work in the far uv is not yet possible. The method has only been applied to neutral and singly ionized atoms. A useful bibliography can be found in the review of Andrä (An 79).

A reduction (although not elimination) of cascading can also be achieved by combining fast-beam and Hanle-effect techniques. In the classical Hanle-effect experiments (Za 67) atoms in a resonance cell are excited with continous polarized resonance radiation from a lamp.

An external variable magnetic field is applied and the intensity pattern of the fluorescence radiation is measured. The width of such a Hanle curve equals the product of the Landé g factor and the lifetime. This method was modified to fast beams by Carré *et al.* (Ca 71), who made lifetime measurements for He I. Church *et al.* (Ch 71) report lifetime measurements in Ar by this method, whereas Andersen, Poulsen, and collaborators (An 76a, Er 80) measured lifetimes in heavier systems. One of the basic assumptions in such Hanle-effect studies using fast ions is that cascading is largely reduced because transfer of alignment from higher levels is not very probable (Du 73).

Coincidence methods are well known in nuclear spectroscopy and they should also be applicable to atomic problems. The effects of cascading could then be effectively eliminated. Unfortunately the very low transmission in optical spectrometers is a highly limiting factor. Despite these problems Masterson and Stoner (Ma 73) have performed coincidence studies on the foil-excited beam and determined a lifetime in O II. Because of low counting rates the method has not been widely applied.

An interesting way of reducing cascading by making measurements at as low an ion energy as possible (for exciting a given level) has been developed by Andersen *et al.* (An 79a).

5.3. Special Methods

Lifetimes below 10^{-12} s are very difficult to measure in the majority of cases. However, the method of nonproportional yield of x rays permits determination of lifetimes in the 10^{-14}-s region. In this technique, introduced by Betz *et al.* (Be 74a), the yield of x rays is measured as a function of foil thickness. It can be shown that the observed intensity of x rays depends

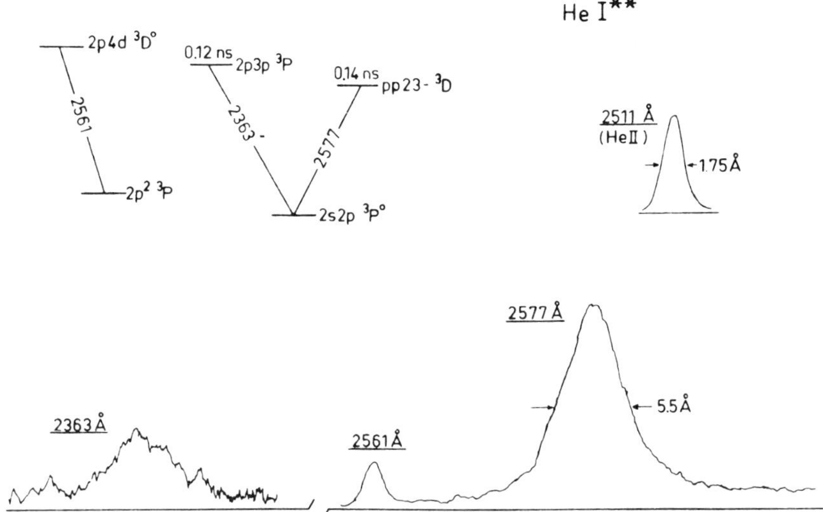

Figure 16. Determination of a very short lifetime in He I** from linewidth measurements (St 76a). The lines at 2363 and 2577 Å have a common lower state, $2s2p\ ^3P^0$, which autoionizes by Coulomb repulsion whereas the upper states decay by photon emission. A line-profile measurement yields the $2s2p\ ^3P^0$ lifetime. Transitions between nonautoionizing levels are also shown, for comparison of linewidths.

linearly on lifetime. Using this method Varghese *et al.* (Va 76) determined the lifetime of the $1s2p\ ^1$P level in He-like Si and Cl, obtaining values in good agreement with theory.

It may sometimes also be possible to measure lifetimes from linewidths. Several investigators have (Be 72b, St 76a, Is 78) thus determined the lifetime of the $2s2p\ ^3P$ level in He I** by determining the width of a spectral line at 2577 Å for which $2s2p\ ^3P$ is the final state. An example of the data is shown in Figure 16. Results close to 6×10^{-14} s have thereby been obtained. Similar data are also available for doubly excited Li I (Ma 81).

5.4. Results of Lifetime Measurements

The lifetime τ_i of an excited level in an atom or ion is related to atomic transition probabilities A_{if} as follows:

$$\tau_i^{-1} = \sum_f A_{if} \tag{9}$$

where the summation includes all final states f that are connected with i. A lifetime measurement may therefore only yield the sum of transition probabilities. If individual values A_{if} are needed, the relative intensities (branching ratios) of the various decay channels must also be determined.

For "allowed" transitions ($E1$ transitions) in atoms the probability is expressed as

$$A_{if} = (4\alpha\omega^3/3c^2)\,|\langle\phi_i|\,Q\,|\phi_f\rangle|^2 \tag{10}$$

where $\hbar\omega$ is the photon energy, α the fine-structure constant, and c the speed of light, ϕ_i and ϕ_f the wave functions on the levels involved in the decay, and $Q = \Sigma r_j$ the dipole operator.

It is often convenient to use the oscillator strength (or f value), numerically related to A_{if} as follows:

$$f = 1.499 \times 10^{-16}\lambda^2(g_i/g_f)\,A_{if} \tag{11}$$

Here λ is the wavelength of the transition (in Å) and g_i and g_f the statistical weights of the levels.

Experimental measurements of lifetimes (and transition probabilities) provide useful tests of the wave functions and approximations used in quantum-mechanical calculations of atomic structure.

A very important problem in atomic theory concerns the correlation of electrons in atoms and ions (Hi 75). The f values are particularly sensitive indicators of this effect. Calculations that neglect correlation may thus give f values which are wrong by order of magnitude. Even in a small and fundamental system such as B I the correlation effects can change f values by a factor of 5 or more from the single-electron values. For detailed discussions of f value calculations we refer to several reviews (Hi 75, We 73, Hi 77).

To systematize the lifetime (and f value) data it is useful to consider certain regularities. For example, the following relation holds for f values:

$$f = f_0 + f_1(1/Z) + f_2(1/Z)^2 + \cdots \tag{12}$$

where Z is the nuclear charge, f_0 the hydrogenic value and f_1 and f_2 constants. By displaying f versus $1/Z$ for a given transition smooth curves are obtained and interesting regularities can be displayed (Wi 68).

A considerable amount of activity in BFS has been devoted to such studies. Among the first experimental investigations of this kind were those by Berkner *et al.* (Be 65), who measured the lifetime of the $2p\ ^2P$ level in Li-like C, N, O, and F. Some more recent data for the $2s\ ^2S$–$2p\ ^2P$ transition in Li-like ions is shown in Figure 17. Here the BFS results are compared with theoretically predicted values—the latter also include relativistic effects. The experimental data for Fe XXIV and Kr XXXIV—obtained in high-energy experiments at the Berkeley SUPER-HILAC (Di 78, Di 80)—are particularly valuable since they confirm the relativistic calculations (Ki 76).

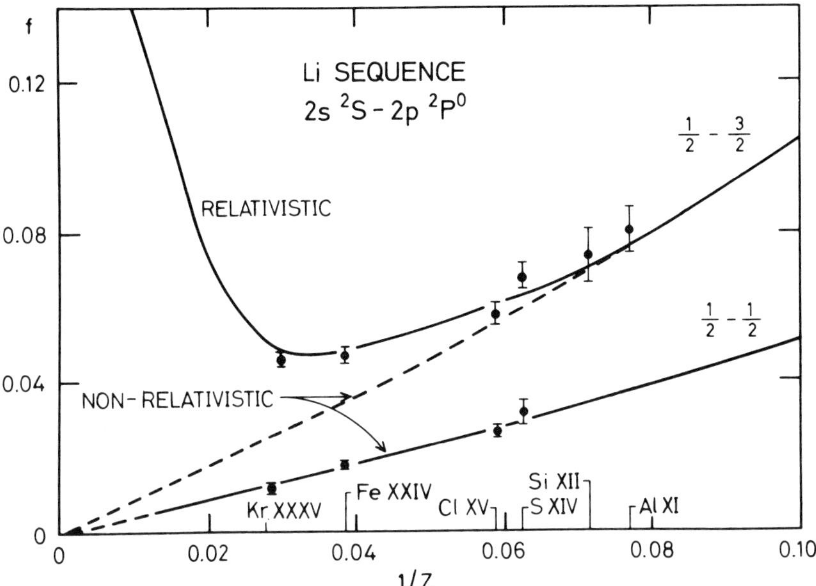

Figure 17. Oscillator strengths for the $2s\ ^2S_{1/2}$–$2p\ ^2P_{1/2,3/2}$ transitions in Li-like ions. Note that the relativistic effects are nicely confirmed by the Fe XXIV (Di 78) and Kr XXXIV (Di 80) experiments.

The power of BFS is also nicely illustrated by the fact that the $2p\ ^2P$ lifetime has been determined for many ions ranging from Li I to Kr XXXIV.

The correlation between the two valence electrons in Be-like ions complicates theoretical calculations of f values. Some experimental data for the $\Delta n = 0$, $n = 2$ transitions in Be-like ions are shown in Figure 18. These data originate from the work of Pegg *et al.* (Pe 79) and Träbert *et al.* (Tr 80) and are based on BFS experiments using tandem accelerators. The agreement is quite satisfactory with theoretical data (Ar 76, Ki 76).

While these examples display rather regular f-value curves, cases have also been found where the f value for a given transition varies strongly from one ion to another. For example, theory (We 69) predicted an extremely low f value for the $2p\ ^2P$–$3s\ ^2S$ transition in C II whereas the f values for the same transition in B I and N III were found to be much higher. The reason for this was explained as a destructive interference, caused by the interaction of two levels of the same parity and angular momentum ($3s\ ^2S$ and $2p^2\ ^2S$) in C II. This prediction has been experimentally verified by BFS (Ma 70b, Bu 73a).

The f values are very sensitive indicators of such interactions in atomic systems. In the spectrum of Al II the series $3snf\ ^3F$ is perturbed by a term

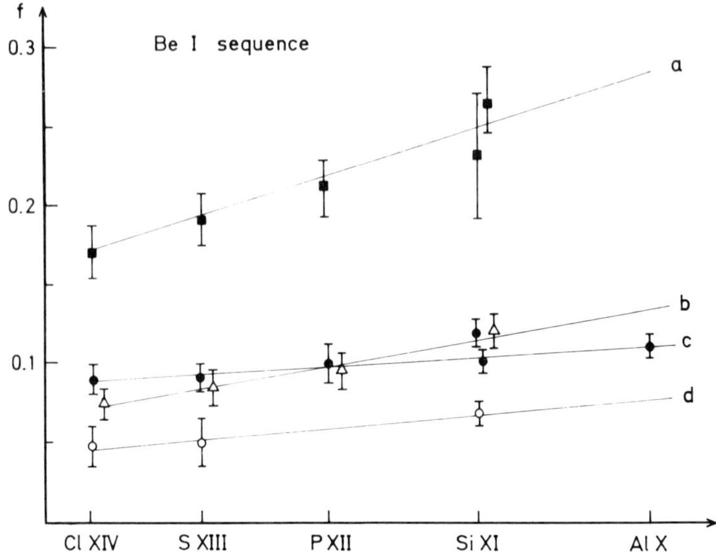

Figure 18. Experimental socillator strengths for the $2s^2\ ^1S$–$2s2p\ ^1P$ (a), $2s2p\ ^3P$–$2p^2\ ^3P$ (b), $2s2p\ ^1P$–$2p^2\ ^1D$ (c), and $2s2p\ ^1P$–$2p^2\ ^1S$ (d) transitions in Be-like ions (Pe 79, Tr 80).

$3p3d\ ^3F$ (Ed 64) and this interaction strongly alters the f values, as theoretically found by Weiss (We 70). Experimental evidence was first provided by Andersen *et al.* (An 71) who measured the lifetimes of several ^3F levels in Al II. There are several other interesting cases which have been explored in BFS investigations, e.g., the interaction between the $3snd\ ^1D$ series ($n = 3, 4...$) and the $3p^2\ ^1D$ term in Mg-like ions and the interaction between the $3s^2nd\ ^2D$ series and the $3s3p^2\ ^2D$ term in Al-like ions. These have also been studied theoretically (Fr 68, Si 73, Fr 81). The main importance of BFS lies in the possibility of investigating such interactions for several values of Z. The f-value regularities have been thoroughly discussed in a recent review (Wi 79).

Level lifetimes in quite heavy atoms and ions have also been determined by BFS. The majority of such studies have been performed with the 600-kV accelerator of the University of Aarhus. For example, Andersen *et al.* (An 72) report systematic studies of transition probabilities for Ag I, Cd I, Au I, and Hg I isoelectronic sequences. The experimental methods and problems in such work have been discussed by Sørensen (Sø 76).

We will present additional lifetime data in connection with discussions of the application of BFS to astrophysics and plasma physics.

5.4.1. Forbidden Transitions

The well-known selection rules for electric dipole ($E1$) transitions in the LS-coupling approximation ($\Delta L = 0, \pm 1, \Delta S = 0$, parity change) can often be violated. The "forbidden" transitions which then take place have usually low probabilities in neutral and mildly ionized light atoms but these rates increase drastically with Z. Since BFS can be extended to highly ionized atoms it has been relatively easy to obtain good experimental material for forbidden transitions in multiply ionized atoms. The processes that have been investigated include magnetic dipole ($M1$), electric quadrupole ($E2$), and magnetic quadrupole ($M2$) decays as well as two-photon ($2E1$) transitions and $E1$ decays induced by the spin–orbit interaction ("intercombination lines") and the hyperfine interaction.

Examples of forbidden transitions in hydrogenlike (a) and heliumlike (b) systems are shown in Figure 19. For more detailed discussions of these systems, including both theory and experiment, we refer to the reviews by Marrus and Mohr (Ma 76a, Ma 78).

The $2s\ ^2S_{1/2}$ level in H I and hydrogenlike ions decays by two-photon emission. The photons have a continuous energy distribution with the sum equaling the $1s$–$2s$ energy difference. This process was theoretically studied by Breit and Teller (Br 40). A more recent calculation (Kl 69) yields the theoretical $2s\ ^2S_{1/2}$ transition rate of $8.2294\ Z^6\ \mathrm{s}^{-1}$ for the $2E1$ decay of $2s\ ^2S_{1/2}$. For very high Z ($Z > 40$) the $M1$ transition to the $1s\ ^2S_{1/2}$ ground state also plays a role (Ma 78). The theoretical calculations of the two-

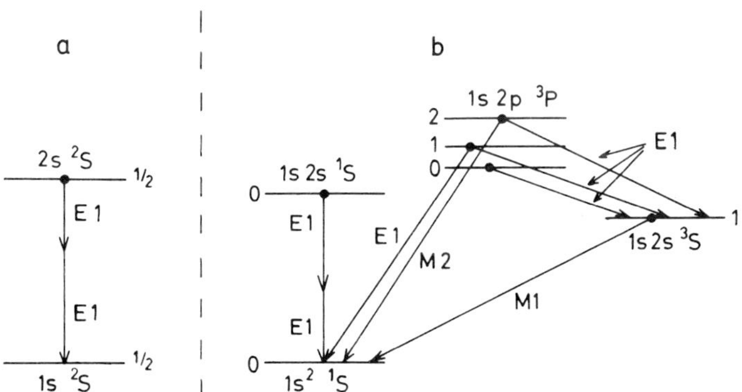

Figure 19. Allowed and forbidden lines in H-like (a) and He-like (b) systems. The $2s\ ^2S_{1/2}$ and $1s2s\ ^1S_0$ states decay by two-photon emission. The $1s2s\ ^3S_1$ level depopulates by a relativistic $M1$ process. For low Z all three $1s2p\ ^3P$ levels predominantly make allowed decays to $1s2s\ ^3S_1$. As Z increases the spin-forbidden $1s^2\ ^1S_0$–$1s2p\ ^3P_1$ transition becomes very probable. The $^3P_2$ level also decays by $M2$ transitions to the ground state. For systems with a nuclear spin the hyperfine-induced $E1$ transition $1s^2\ ^1S_0$–$1s2p\ ^3P_0$ also is possible.

photon decay were nicely confirmed by Marrus and Schmieder (Ma 72) who studied this process in hydrogenlike Ar XVIII, using a 400-MeV Ar beam from the Berkeley HILAC. It was found that the photons had a continuous energy distribution with the sum corresponding to the $1s$–$2s$ energy difference of 3.33 keV. The lifetime measurement yielded $\tau(2s) = 3.54 \pm 0.25$ ns, in perfect agreement with the theoretical value of 3.57 ns. The authors also studied this decay in S XVI. There are also beam-foil data for the $2s\ ^2S_{1/2}$ decay in H-like O and F (Co 74). In all cases theory and experiment are in excellent agreement. It is quite important to check theory in this particular case, because of the relevance to Lamb-shift measurements (see below).

Such $2E1$ decays have also been investigated in He-like systems, where the $1s2s\ ^1S_0$ level can only decay by two-photon emission (Figure 19b). Here, also, BFS experiments have provided interesting experimental data. For Ar XVII Marrus and Schmieder (Ma 72) obtained a $1s2s\ ^1S_0$ lifetime of 2.3 ± 0.3 ns, in very good accord with the theoretical value of 2.35 ns (Ma 78).

We will now consider the triplet states in He-like systems (Figure 19). The lowest of these, $1s2s\ ^3S_1$ was originally assumed to decay by two-photon emission to the $1s^2\ ^1S_0$ ground state, but astrophysical observations (Ga 70) indicated that a relativistic $M1$ decay (with the rate proportional to Z^{10}) is the main decay mechanism. This interpretation was confirmed by Marrus and Schmieder (Ma 72), who observed a single-photon transition corresponding to the expected $1s2s\ ^3S_1$ excitation energy in He-like Si XIII, S XV, and Ar XVII. The $2s\ ^3S_1$ lifetime has been measured for S XV, Cl XVI, Ar XVII, Ti XXI, V XXII, and Fe XXV. Initially the experiments and theory differed by 20 %–25 % for Cl XVI and Ar XVII, while agreement was good for higher-Z systems. These disagreements led to careful reconsideration of the experimental situation, and new measurements have bridged the gap. The experimental problems were most likely due to the low resolution of the solid-state x-ray detectors used, which caused blends from satellite transitions. The $1s2s\ ^3S_1$ lifetime is now one of the most extensively studied cases. References to all BFS work and theory are found in the review by Marrus and Mohr (Ma 78).

The $1s2p\ ^3P_{0,1,2}$ levels in He-like ions also possess some interesting decay channels. For low Z the dominant decay for all three states is the allowed transition to $1s2s\ ^3S_1$ which in He I has the probability of about $10^7\ \mathrm{s}^{-1}$. This grows approximately linearly with $Z - 1$. Other decay modes become important when Z increases. For instance, the $^3P_1$ level also decays to the $1s^2\ ^1S_0$ ground state by so-called spin-forbidden $E1$ transitions (intercombination lines). These violate the $\Delta S = 0$ rule and are caused by the mixing of $1s2p\ ^3P_1$ with $1s2p\ ^1P_1$ and—to a lesser extent—with higher $^1P_1$ levels. The transition probability of the intercombination line is approximately proportional to Z^{10}. Already for quite low Z this is the dominant decay channel of the $^3P_1$ level. The first measurements of the $1s^2\,{}^1S_0$–$1s2p\,{}^3P_1$ transition probability were made by Sellin *et al.* (Se 68), who used 6–42-MeV N and 0 ions from a tandem Van de Graaff accelerator. From the measured $1s2p\,{}^3P_1$ decay probability the rate of the allowed $^3S_1$–$^3P_1$ branch was subtracted. This amounts to 33 % in N VI and 13 % in O VII. The $^3P_1$ lifetime has subsequently been measured for He-like F, Si, and S. For the latter two elements the $^3P_1$ lifetime is in the ps range and careful experimental designs are necessary to obtain good results (Va 76a, Ar 81).

Besides decaying to $1s2s\,{}^3S_1$ the $1s2p\,{}^3P_2$ level also decays by magnetic quadrupole ($M2$) transitions to the $1s^2\,{}^1S_0$ ground state. The latter process scales approximately as Z^8. For Ar XVI the allowed $E1$ and the forbidden $M2$ decay are equally probable, as experimentally shown by Marrus and Schmieder (Ma 72). For higher Z ions the $M2$ process is the dominating decay form. Experimental values for the $M2$ transition rate have also been obtained for He-like S, Cl, Ti, V, and Fe. In all cases the theoretical and experimental data are in agreement. (Ma 78).

The $1s2p\,{}^3P_0$ level can "normally" only decay to $1s2s\,{}^3S_1$. However, in systems with a nuclear spin $I > 0$, transitions from $1s2p\,{}^3P$ $(J=0, F=I)$ to $1s^2\,{}^1S$ $(J=0, F=I)$ become possible. The rate of this hyperfine-induced $E1$ decay has been calculated by Mohr (Mo 76). Similar decay modes may also occur for the $1s2p\,{}^3P_2$ level.

The first experimental evidence of this hyperfine effect was obtained by Gould *et al.* (Go 74) who measured the $^3P_2$ lifetime in He-like V XXII $(I=7/2)$. Here the $^3P_2$ level is split into five hyperfine components ($F=3/2$, 5/2, 7/2, 9/2, and 11/2), three of which may take $E1$ transitions to the $F=7/2$ ground state. The decay measurements yielded a curve that clearly deviated from a single exponential and this was explained as being due to the hyperfine effect. However, because of the complex decay curves no quantitative information could be obtained. Mowat *et al.* (Mo 75) measured the $1s2p\,{}^3P_0$ and $1s2p\,{}^3P_2$ lifetimes in ^{19}F VIII $(I=1/2)$, but their experimental uncertainties (about ± 10 %) were as large as the predicted hyperfine quenching (Mo 76) and no definite conclusions could be drawn. In a subsequent beam-foil study Engström *et al.* (En 80) determined the $^3P_2$ and $^3P_0$ lifetimes in F VIII to 1 %–2 % accuracy, from which the $1s^2\,{}^1S_0$–$1s2p\,{}^3P_0$ probability was found to be $(1.42 \pm 0.20) \times 10^7\ \mathrm{s}^{-1}$, in excellent agreement with Mohr's calculated value $1.2 \times 10^7\ \mathrm{s}^{-1}$. In a similar experiment Denne *et al.* (De 80) measured the $^3P_2$ and $^3P_0$ lifetimes in He-like ^{27}Al XII $(I=5/2)$. For the $^3P_0$ level the lifetime 4.8 ± 0.2 ns was obtained, confirming the theoretical value 4.71 ns (Mo 76). Figure 20 shows the experimental and theoretical results for the $1s2p\,{}^3P_0$ lifetimes in He-like ions. Note that the shortening due to hyperfine quenching is particularly pronounced for Sc and

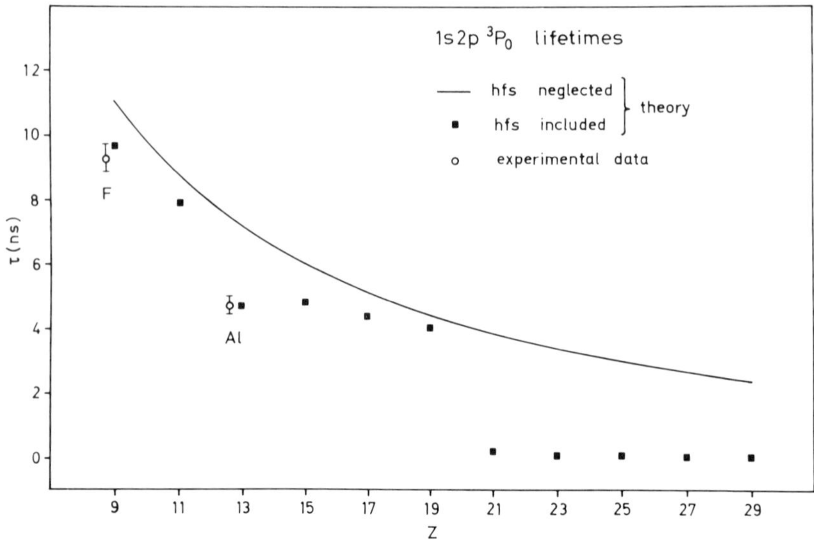

Figure 20. Theoretical lifetimes for the $1s2p\ ^3P_0$ level in He-like ions (Mo 76). The shortening of lifetimes because of hyperfine-induced decay modes is clearly seen. It has been confirmed for He-like F (En 80) and Al (De 80).

higher ions. It would clearly be desirable to perform additional measurements in such systems.

To summarize this discussion of two-electron spectra we show in Figure 21 the lifetimes obtained by BFS for the $1s2p\ ^3P_{0,1,2}$ levels, from He I to Kr XXXV. This is another illustration of the power of BFS for systematic studies.

Forbidden processes have also been quantitatively studied in three-electron systems. The decay of the lowest-level $1s2s2p^4P$ of the quartet system is particularly interesting. Of the three fine-structure components $^4P_{1/2}$ and $^4P_{3/2}$ autoionize via the spin–orbit and the spin–spin interaction, whereas only the latter decay mode is possible for the third component $^4P_{5/2}$. The lifetime of this level is therefore much longer than that of the other fine-structure levels. This lifetime has been systematically studied by BFS for many ions from Be II to Ar XVI (Se 76a). The $^4P_{5/2}$ level may also decay by $M2$ processes to the $1s^2\ ^2S_{1/2}$ ground state (Se 76a). There has been a systematic difference of about 15 % between the experimental and theoretical lifetimes for $^4P_{5/2}$. Cascades have been considered as creating experimental problems (Do 78), and cascade corrections give indeed better agreement. However, theoretical refinements may also be necessary (Se 76a, Se 76b).

The beam-foil method has also yielded transition probabilities for spin-

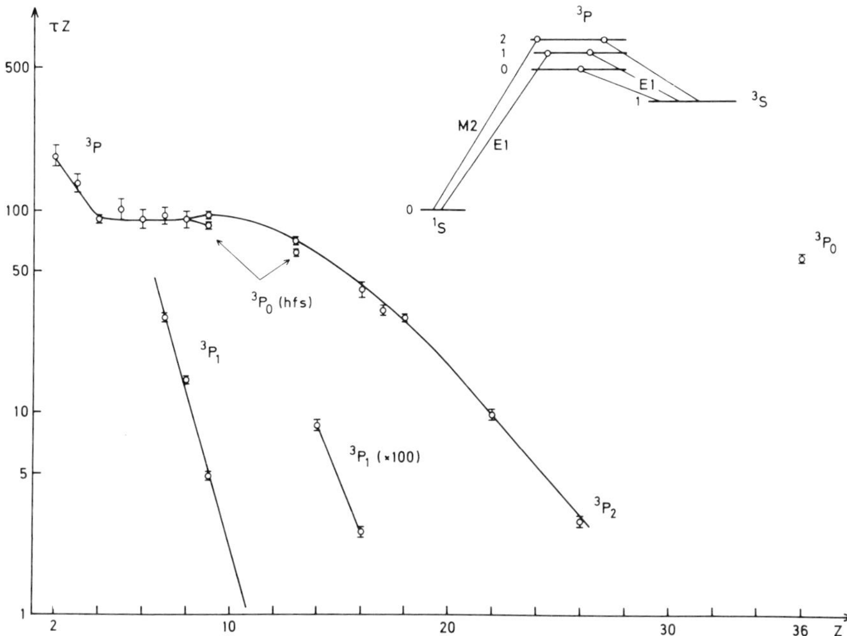

Figure 21. Lifetimes (multiplied by Z) of the $1s2p\ ^3P_{0,1,2}$ levels in the He I isoelectronic sequence. All data shown here originate from beam-foil work. Note the effects of the spin-forbidden decay mode on the $^3P_1$ lifetime and the shortening of the $^3P_2$ lifetime because of the $M2$ decay channel. For systems with nonzero nuclear spin, the $^3P_0$ lifetime is also shortened.

forbidden decays in Be-like ions. Thus Dietrich *et al.* (Di 78, Di 80) have determined the lifetime of the $2s2p\ ^3P_1$ level in Be-like Fe and Kr. This level mixes with $2s2p\ ^1P_1$ and decays to the $2s^2\ ^1S_0$ ground state. The experimental lifetimes were found to be in good agreement with theory. A similar effect, for the $2s3p\ ^3P_1$ level in Be-like N, O, and F has been studied by Engström *et al.* (En 79). The mixing is much stronger here than for the $2s2p\ ^3P_1$ level and the spin–orbit interaction effect is measurable for much lower values of Z.

6. Quantum-Beat Experiments

6.1. General Comments

Quantum beats are well-known phenomena in several areas of physics. They have been investigated for about 20 years in the case of atomic systems (Al 63, Co 64). The principle in such experiments consists of exciting the atoms with a short pulse and making a time-resoved study of the subsequent decay.

The theory for quantum beat phenomena is well established (see, e.g., An 79) and we will not rewrite the formulas but only give a short discussion. If the excitation time Δt fulfills the relation $\Delta t < h/\Delta E$ then two (or more) levels in an atom which lie within the energy difference ΔE are *coherently* excited. This means that the wave function of the atom is a linear superposition of the wave functions of the individual levels and the phase difference is fixed. In the case of two excited levels we may write the wave function $\psi(t)$ as

$$\psi(t) = a_1\phi_1 \exp(-\alpha_1/2 - iE_1/\hbar)\, t + a_2\phi_2 \exp(-\alpha_2/2 - iE_2/\hbar)\, t \tag{13}$$

where a_1 and a_2 are the excitation amplitudes, ϕ_1 and ϕ_2 the eigenfunctions of the time-independent Schrödinger equation for the two states, $E_1 = \hbar\omega_1$ and $E_2 = \hbar\omega_2$ their excitation energies, and α_1 and α_2 the decay probabilities. This expression can of course be generalized if additional levels are coherently excited. We assume now that the two levels decay to a common final state f. In the dipole approximation the decay probability is proportional to $|\langle f|\, Q\, |\psi(t)\rangle|^2$, where Q is the dipole operator. When we have an ensemble of atoms the intensity of the light emitted can be written as

$$I(t) \sim [A + B\cos(\omega_{12}t)] \exp(-\alpha t) \tag{14}$$

where we have assumed that $\alpha_1 = \alpha_2 = \alpha$ and $E_1 - E_2 = \hbar\omega_{12}$. A and B are time-independent constant that can be expressed in terms of the eigenfunctions and coefficients in Equation (13). The Equation (14) implies that the observed exponential decay (with a lifetime $\tau = 1/\alpha$) is modulated with a frequency corresponding to the separation of the levels.

In a time-resolved experiment we may thus expect to observe periodic modulations (quantum beats). However, while it is easy to understand the excitation of an atom into a superposition of eigenstates and the derivation of the oscillating term (which comes from adding amplitudes before taking the square), it is not directly evident that we can observe quantum beats in an actual experiment when the light comes from an ensemble of atoms. This problem has been discussed in detail by Macek (Ma 69, Ma 70c), who shows that additional conditions (besides coherence) must be fulfilled for the observation of quantum beats. The total intensity is unmodulated, but quantum beats can be observed in the light emitted in a particular direction if there is *excitation asymmetry* (orientation or alignment), i.e., a nonstatistical population of magnetic substates. For example, in the case of the $2p\ ^2P_{1/2}$ and $^2P_{3/2}$ levels in H I (which both decay to $1s\ ^2S_{1/2}$) the intensity perpendicular to the beam is proportional to

$$I(t) \sim [5\sigma_0 + 4\sigma_1 + 4(\sigma_0 - \sigma_1)\cos\omega t] \exp(-t/\tau) \tag{15}$$

where the angular frequency ω corresponds to the ${}^2P_{3/2}$ and ${}^2P_{1/2}$ energy difference while σ_0 and σ_1 are the cross sections for exciting the $M_L = 0$ and $|M_L| = 1$ magnetic substates of the P levels, respectively. (With the usual axial symmetry in a beam-foil experiment the cross sections for exciting the $M_L = 1$ and $M_L = -1$ levels are equal.) The relation above shows that σ_0 and σ_1 must be different if quantum beats are to be observed.

The theory, based on density operators, has been further developed by Berry *et al.* (Be 72c), Andrä (An 79), Ellis (El 73), and others. There are great similarities to the theory for perturbed angular correlations in nuclear spectroscopy.

6.2. Zero-Field Oscillations

After some initial observations of such quantum beats in the absence of external fields (Ba 66a), Andrä (An 70) made the first quantitative studies of such phenomena in beam-foil excitation. Observing the intensity decay of the $1s2s\ {}^3S_1$–$1s3p\ {}^3P_{0,1,2}$ multiplet (3889 Å) in beam-foil excited He I, Andrä found a modulation frequency 655 ± 6 MHz, which agreed well with the precisely determined (by radiofrequency spectroscopy) energy difference between the ${}^3P_1$ and ${}^3P_2$ levels, 658.55 MHz. The experiment has later been substantially refined, and such results are shown in Figure 22, according to Wittmann *et al.* (Wi 72). The separation between ${}^3P_1$ and ${}^3P_2$ is easily observed. At very high spatial resolution the much higher energy separation between ${}^3P_0$ and ${}^3P_2$ (8771 MHz) also appears very nicely. Beats between ${}^3P_0$ and ${}^3P_1$ cannot be observed in the usual BFS geometry (El 73).

Several investigations of "zero-field" quantum beats have been undertaken by BFS (see An 79 for a detailed review). By studying the 3889-Å line in ^{3}He I the separations of hyperfine states were observed (Ti 73). Extensive studies of the $1snp\ {}^3P$ ($n = 2$–5) levels in ^{6}Li II and ^{7}Li II were reported by Berry *et al.* (Be 73a). In these ions with both fine and hyperfine structure (of approximately equal magnitude) the decay curves were quite complex. For example, the $1s2p\ {}^3P$ term splits into seven (or eight) different JF combinations in ^{6}Li (or ^{7}Li). The decay curves were therefore Fourier-analyzed and separated into several frequencies which provided information about fine and hyperfine separations.

The decay of the $1s2p\ {}^3P$ term in He-like ^{9}Be III has been carefully investigated by Poulsen and Subtil (Po 74), who used 500-keV Be ions. Five frequencies were obtained; these could be measured with 2 % uncertainties, which made it possible to determine the Fermi coupling constant with 1 % error. Similar measurements have been extended to levels in B (Po 75), C (Su 76), and N (Su 76a). For example, Subtil *et al.* (Su 76) report hyperfine-structure measurements for transitions in ^{13}C II and ^{13}C III, whereas Subtil *et al.* (Su 76a) studied hyperfine beats in ^{14}N III. The data were in both

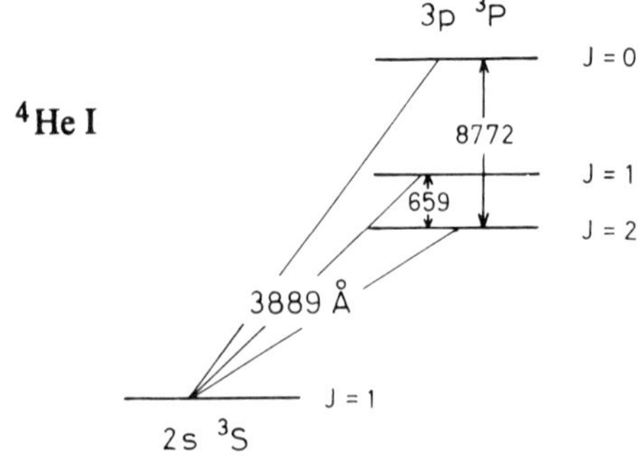

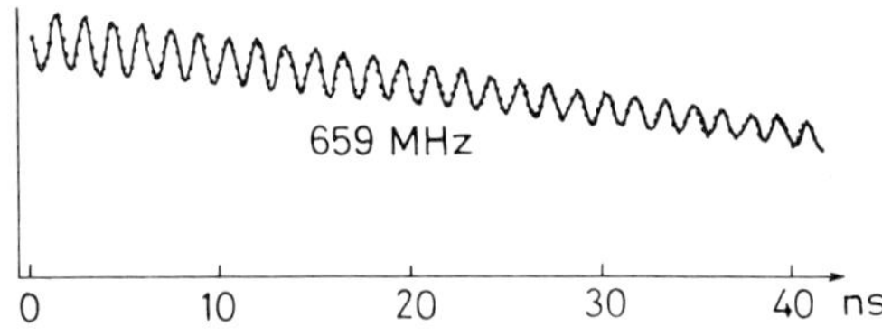

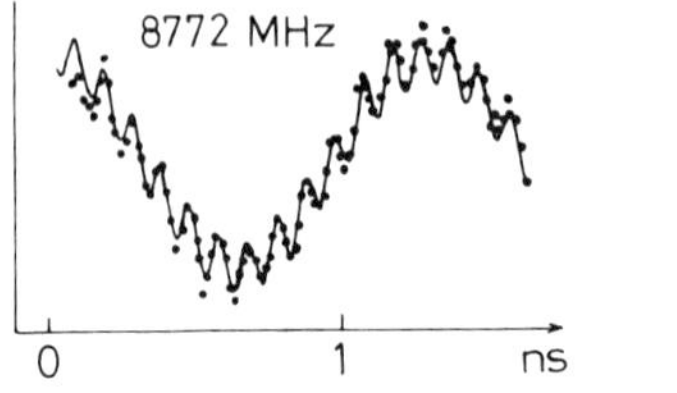

Figure 22. Examples of quantum beats. The three fine-structure levels, $J = 0, 1, 2$, of the $1s3p\ ^3P$ term in He I are coherently excited and they decay to the same final state, $1s2s\ ^3S_1$. The $J = 1$ and $J = 2$ separation (659 MHz) is easily observed from the quantum-beat pattern. In a high-precision experiment also the $J = 0$ and $J = 2$ quantum beats are observed. These data were obtained by Wittmann *et al.* (Wi 72).

cases compared with calculations based on the Hartree–Fock approximation. Differences were found between theoretical and experimental data, however. The most probable reason is the effect of core polarization, which was not accounted for in the calculations.

Very interesting data were obtained by Gaupp *et al.* (Ga 76), who investigated the decay of the doubly excited $1s2p^2\ ^4P$ level in Li I using high spatial resolution. The fine structure and hyperfine structure for ^{6}Li as well as ^{7}Li were determined.

Andrä (An 79) points out that the beautiful quantum-beat patterns are of little use for atomic theory unless the velocities of the foil-excited ions are very accurately known. The velocity uncertainty yields a corresponding uncertainty in energy separations. To check the mostly very accurate theoretical data high experimental precision is necessary. Gaupp *et al.* (Ga 76) measured the velocity of the Li ions using an electrostatic energy analyzer. Such an instrument was also used by Wittmann *et al.* (Wi 72), who calibrated it using the accurately known He I beat frequency of 658.55 MHz (mentioned above). By this technique the fine structure of higher $1snp\ ^3P$ levels in He I ($n = 4$–8) could be precisely determined.

The two-spectrometer method of Astner *et al.* (As 76a) represents another experimental development. Here the foil-excited He beam was viewed by two monochromators, one of which was set at the He I 3889-Å line while the other detected radiation from the $1snd\ ^3D$ levels of He I. The data were recorded simultaneously and the 658.55–MHz quantum beats on the 3889-Å line provided an accurately known time scale for the foil-excited He atoms. An example of such a simultaneous recording of two quantum-beat patterns is displayed in Figure 23. The measurement yielded so accurate fine-structure data for the $1snd\ ^3D$ levels that small but significant deviations from the existing theories could be spotted.

In passing we note that this two-spectrometer method for accurate determinations of the time scale after excitation also yields very precise lifetime data. In a decay measurement for the $1s3p\ ^1P_1$ level in He I the time scale was thus determined with the 658.55-MHz quantum beats. The resulting lifetime value for the $1s3p\ ^1P_1$ level, 1.7226 ± 0.0046 ns, is one of the most accurately determined atomic lifetimes (As 76).

Zero-field quantum beats have also been investigated by exciting the ion beam with laser light. Accurate measurements of the hyperfine structure of ^{23}Na I were thus reported by Krist *et al.* (Kr 77), whereas Wittmann (Wi 79a) has quite considerably improved the precision for the $1s2p\ ^3P$ fine and hyperfine structure when compared to the previous beam foil study (Be 73a).

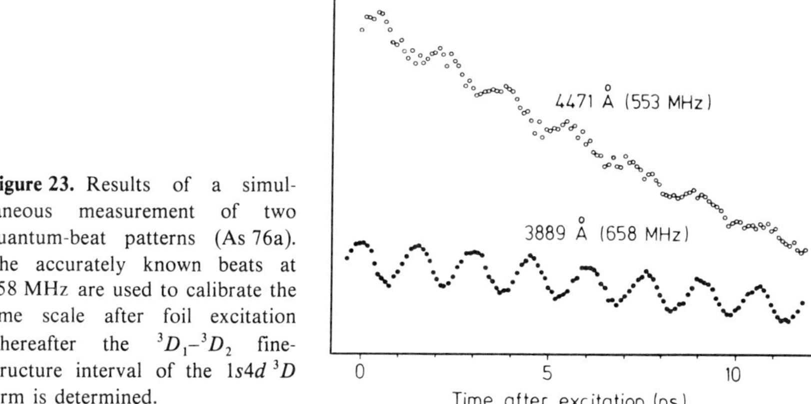

Figure 23. Results of a simultaneous measurement of two quantum-beat patterns (As 76a). The accurately known beats at 658 MHz are used to calibrate the time scale after foil excitation whereafter the $^3D_1$–$^3D_2$ fine-structure interval of the $1s4d\ ^3D$ term is determined.

Quantum beats in Ba II by laser excitation have been investigated by Andrä (An 75a) and Dufay *et al.* (Du 76a). In the latter study the ion and laser beams were collinear and the resonances were found by small postacceleration of the ions.

Many studies of zero-field quantum beats have been undertaken on small atoms for which the fine and hyperfine structure are accurately known. The principal value of this work has been the detailed insight into the beam-foil excitation mechanism that has been provided. Some of the results will be discussed in Section 9.2, where we also include the results of investigations using tilted foils and surfaces at grazing incidence.

6.3. Stark Beats

The very first observations of quantum beats with the beam-foil technique were made by Bashkin, Bickel, and collaborators (Ba 65, Bi 67a), who observed intensity oscillations for H I and He II lines when the foil-excited beam was sent through an electrostatic field. Subsequent experiments (Se 69, An 70a) clearly established that these oscillations occurred when states of different parity were coupled by the electrostatic field (Stark mixing). Andrä (An 70a) thus carefully measured the amplitude and frequency of the Stark beats in the Ly α radiation of H I as a function of the applied electric field. Pinnington *et al.* (Pi 73a) have investigated these Stark beats for He II and Li III. In all cases studied to date the observed frequencies are in good agreement with those expected from theory (Lü 51).

6.4. Zeeman Beats

Interesting results can be obtained by sending the foil-excited (or gas-excited) ion beam into a region of a magnetic field. This method allows measurements of Landé g factors, which are defined by $\boldsymbol{\mu} = -g\mu_B\mathbf{J}$ (where $\boldsymbol{\mu}$ is the magnetic moment, μ_B the Bohr magneton and $\mathbf{J}$ the total angular momentum). The well-known Landé g factor formula

$$g = 1 + [J(J+1) + S(S+1) - L(L+1)]/2J(J+1) \tag{16}$$

is valid in LS-coupling. Measurements of g are important in determining the intermediate-coupling wavefunctions, and they are often not available for ions. In the case of beam-foil experiments, the intensity of polarized light emitted in a given direction exhibits periodic modulation as a function of time, corresponding to the precession of $\boldsymbol{\mu}$ with respect to the applied magnetic field $\mathbf{B}$. These modulations can also be seen by varying the applied field $\mathbf{B}$ while keeping fixed the time t after excitation, the oscillating term being proportional to $\cos(2\mu_B gBt/\hbar)$ (Li 71). As shown by Church, Liu *et al.* (Li 72, Ch 73), g factors can be determined with only 1 %–2 % uncertainties. A detailed investigation of such Zeeman beats for ^{6}Li II and ^{7}Li II has been undertaken by Gaillard *et al.* (Ga 73), who also discuss the theory quite thoroughly.

It is also possible to perform time-integrated studies on the fast ions. Hereby the classical Hanle-effect investigations are combined with beam-foil or beam-gas spectroscopy (Ca 71, Po 75a).

7. Lamb-Shift Experiments

The studies of Lamb shifts for high-Z ions belong to the most interesting results obtained by BFS.

The excitation energies in the H I isoelectronic sequence depend—according to the relativistic Dirac theory—only on n and j quantum numbers. Thus states such as $2s\ ^2S_{1/2}$ and $2p\ ^2P_{1/2}$ should be exactly degenerate. However, it has been known for more than 30 years that this degeneracy is removed by the quantum electrodynamical effects. In H I the $2p\ ^2P_{1/2}$ level lies 1058 MHz (0.0353 cm^{-1}) below $2s\ ^2S_{1/2}$. This energy difference is usually called the $n = 2$ Lamb shift, often denoted S. The dominant contribution to the Lamb shift in H I is the so-called self-energy term (the electron emits and absorbs virtual photons), but there is also the effect of vacuum polarization correction to the Coulomb potential (Br 78). Theoretical calculations show that S can be expressed as a double sum in α (the fine-structure constant) and $Z\alpha$ with the first term being approximately proportional to $\alpha(Z\alpha)^4$. The summation should go to infinity. However, the first term accounts for 99.3 % of S in H I while its contribution is reduced to 90.9 % in O VIII (Le 73a) where thus higher-order terms play an important role. It should also be added that while the two most accurate calculations of S, by Erickson (Er 71) and Mohr (Mo 75a), only differ 4×10^{-3} % for $Z = 1$ their difference increases to 2 % for $Z = 18$ (hydrogenlike Ar). There are thus good justifications for Lamb shift studies in highly ionized atoms.

The H I and He II Lamb shifts have been measured by rf-spectroscopy, first by Lamb and collaborators (La 47, Tr 53). For an excellent discussion of such measurements we refer to the article by Beyer (Be 79). However, until very recently it has not been possible to apply these techniques to Li III and ions with higher Z. Here beam-foil techniques have been successfully used and such data are available for several ions from Li III to Ar XVIII.

These experiments are based on the field-quenching technique. Advantage is taken of the fact that the $2s\ ^2S_{1/2}$ and $2p\ ^2P_{1/2}$ levels have very different lifetimes, $0.122 \times Z^{-6}$ s and $1.60 \times 10^{-9}\ Z^{-4}$ s, respectively. In an electrostatic field E these levels are mixed and the $2s\,^2S_{1/2}$ lifetime will change according to

$$(\tau_{2s'})^{-1} = 3(\tau_{2p})^{-1}E^2a_0^2Z^{-2}\hbar^{-2}(S^2 + 1/4\tau_{2p}^2)^{-1} + (\tau_{2s})^{-1} \tag{17}$$

where τ_{2s} and τ_{2p} are the lifetimes in the absence of the field, a_0 the Bohr radius and S the $n = 2$ Lamb shift.

All the experiments are essentially based on the same idea, i.e., the hydrogenlike ions are first excited into the long-lived $2s\,^2S_{1/2}$ level and then they enter an electric field. The lifetime for the Stark-mixed compound state, $\tau_{2s'}$ is measured as a function of the external field strength E. The Lamb shift S is finally obtained using Equation (17).

In the first experiment of this kind Fan *et al.* (Fa 67) measured S for Li III using two different values of E. The result 63.031 ± 0.327 GHz was in good agreement with theory. This method was later applied to C VI (Mu 71), O VIII (Le 72a, La 72), and F IX (Mu 72). In these experiments the electrostatic field was replaced by a magnetic field **B**, equivalent to an electric field $(1 - v^2/c^2)^{1/2}\,\mathbf{v} \times \mathbf{B}$. The decay curves obtained in the O VIII experiment (Le 72a) are shown in Figure 24. More recently, Gould and Marrus (Go 78) used this method to determine the $n = 2$ Lamb shift in hydrogenlike Ar. The value 38.0 ± 0.6 THz (1 THz $= 10^{12}$ Hz) was obtained in this very difficult experiment. The experimental problems in these measurements are discussed by Leventhal (Le 73a) and Kugel and Murnick (Ku 77).

A variation of this method consists of inducing transitions between $2s\,^2S_{1/2}$ and $2p\,^2P_{3/2}$ by means of an infrared laser. Since the $2p\,^2P_{1/2,3/2}$ fine structure can be calculated very accurately without quantum electrodynamics, a determination of the $2s\,^2S_{1/2}$–$2p\,^2P_{3/2}$ separation gives indirect information about S. This method was used by Kugel *et al.* (Ku 75), who measured the $^2S_{1/2}$–$^2P_{3/2}$ separation in H-like F using a HBr laser. The resonance between the beam ions and the laser light was found by Doppler tuning (the angle between the beam direction and the laser light was changed in steps). The development of this method and its applications to Cl XVII has been described by Murnick *et al.* (Mu 79).

There are also other ways of observing resonances in the case of fast ions. In a preliminary experiment Hadeishi *et al.* (Ha 69) sent H atoms from an accelerator through a region with a spatially varying electrostatic potential. If the spatial period of this potential is d then the moving atoms experience an oscillating field with an frequency $\nu = v/d$, where v is the beam particle velocity. This method was later refined by Andrä (An 70b).

An interesting method which gives the Lamb shift from the anisotropy of the emitted Ly α radiation (transitions from $n = 2$ to $n = 1$) has been developed by Drake and collaborators (Dr 73, Dr 76). The anisotropy R is

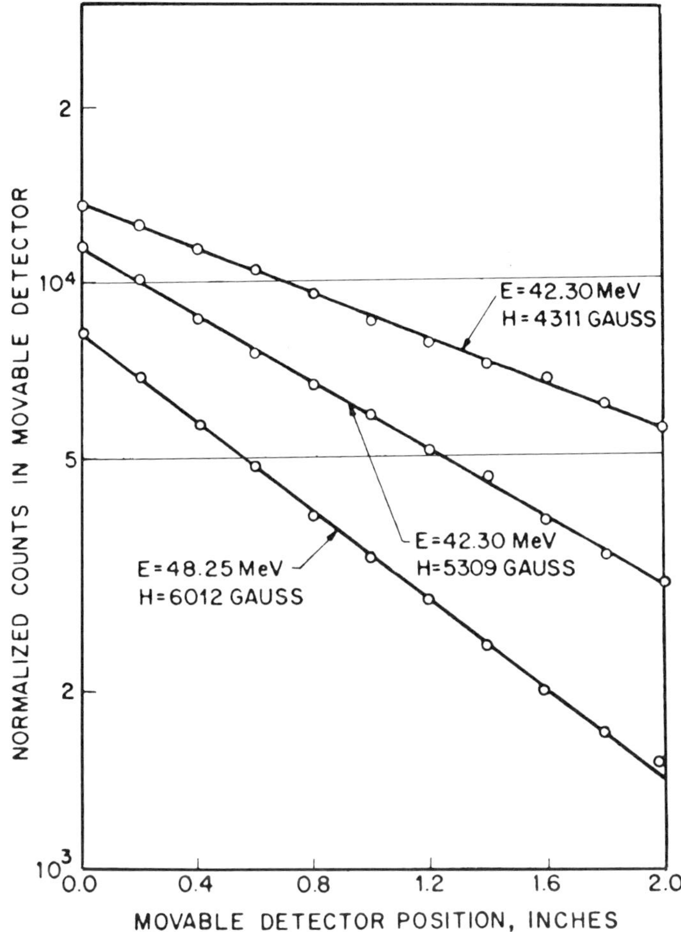

Figure 24. Results of Lamb-shift measurements in O VIII (Le 73a). The $2s_{1/2}$ decay time is determined in three different magnetic fields.

defined as $R = (I_{\parallel} - I_{\perp})/(I_{\parallel} + I_{\perp})$ where $I_{\parallel}$ and $I_{\perp}$ are the Ly α intensities in two perpendicular directions. Theoretical analyses show that S is a function of R. This method has recently yielded an accurate $n = 2$ Lamb shift for He II, the uncertainty being $2 \times 10^{-2}\,\%$ (Dr 79) and future applications to higher Z are of great interest. A selection of $n = 2$ Lamb shift data for H I–Ar XVIII has been displayed in Figure 25. The figure resembles that shown by Drake *et al.* (Dr 79). We also compare the experimental data to the two theoretical sets (Er 71, Mo 75a). New experiments for, e.g., H I (Lu 81) and Li III (Le 74) have been reported in recent years.

Interesting information about Lamb shifts can also be obtained from

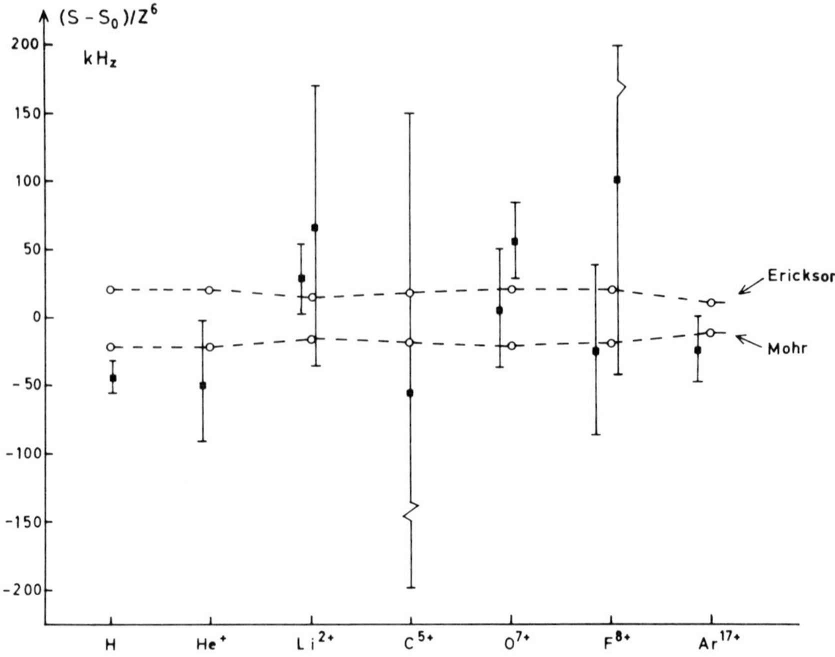

Figure 25. Comparison between theoretical and experimental Lamb shifts in hydrogenlike ions. The plot is similar to that shown by Drake *et al.* (Dr 79). The upper dotted line shows Erickson's theory (Er 71) and the lower that of Mohr (Mo 75a), while S_0 is the average of the two. Only a selection of all the experimental data is given. (See Be 79.)

two-electron spectra (He I isoelectronic sequence). Here the l states are no longer degenerate and the methods described above can therefore not be used. The quantum-electrodynamical (QED) effects are instead investigated by means of wavelength measurements which yield energy differences σ_{exp} between the levels studied. These energy differences have been quite accurately calculated, but in the majority of theoretical studies QED has not been considered. Information about such effects can then obtained by comparing theory and experiment, e.g., by assuming that the quantity $|\sigma_{\text{exp}} - \sigma_{\text{th}}|$ represents QED contributions.

The first measurements were here made by Berry *et al.* (Be 73b, Be 74b), who determined the $1s2s\ ^3S_1$–$1s2p\ ^3P_1$ wavelength in Li II and the $1s3s\ ^3S$–$1s3p\ ^3P$ wavelengths in N VI. In a more recent beam-foil experiment Berry *et al.* (Be 80) measured the wavelengths of the $1s2s\ ^3S_1$–$1s2p\ ^3P_{0,2}$ lines in Cl XVI with 3×10^{-5} to 2×10^{-4} accuracies, and this allowed quite precise determinations of QED contributions. Similar studies, for other He-like ions have also been reported (Ar 79, De 80, Li 80) and it is obvious that this method yields important information about Lamb shifts. One of the experimental problems concerns wavelength determination in BFS spectra for highly ionized atoms. There are frequently very few reference lines in the spectra that are useful for wavelength measurements.

8. Applications of Beam-Foil Results

8.1. Atomic Theory

In the preceding sections we have already emphasized the close connection between BFS and atomic theory. The experimental data obtained for energy levels (in particular multiply excited states), transition probabilities (allowed and forbidden radiative decays, autoionization etc), fine and hyperfine structure, Landé g values, radiative corrections, etc. have been of considerable relevance to modern atomic theory and the BFS studies have frequently stimulated new, advanced theoretical investigations. Inversely, many BFS investigations have been undertaken to check theoretical predictions and sometimes also to make detailed comparisons of various theoretical models. It is expected that this close interplay continues, it will be particularly useful when "exotic" systems, such as very highly ionized atoms are investigated.

8.2. Astrophysics

Work in modern astrophysics is highly dependent on reliable atomic data such as wavelengths, energy levels, and f values. One of the most interesting problems concerns the abundances of the chemical elements in the sun, stars, and interstellar medium. Such abundances yield information about, e.g., the creation of the solar and stellar systems, various pathways of nucleosynthesis, and processes such as convection, diffusion, mass loss, etc. (Bi 77). Many spectral lines are observed in the solar and stellar spectra. From the measured equivalent widths (intensities) of the lines the abundances of the various elements are derived if the f values are known. About 70 chemical elements have been identified in the sun, as can be seen from Figure 26 (from En 77). However, in many cases (e.g., for the rare-earth elements) the abundances are very poorly known because of the lack of good f values.

The elements Li, Be, and B have very low abundances; they are destroyed by spallation reactions. Beam-foil studies of lifetimes in Be I and Be II (Be 69) have provided useful data for the determination of the solar Be abundance.

The iron-group elements (Sc–Ni) have presented very intricate astrophysical problems. More than 10 years ago it was thus found that the

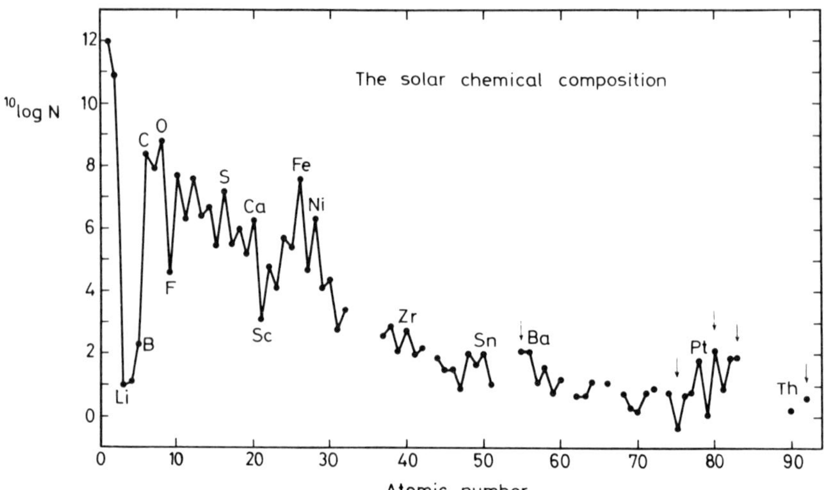

Figure 26. The chemical composition of the sun (En 77). About 70 elements have been identified in the solar spectra but the abundances are often quite uncertain. The abundances are on a logarithmic scale with the hydrogen abundance defined as $^{10}\log N_{\mathrm{H}} = 12.00$.

Fe abundance (relative to H) seemed to be about 10 times higher in the solar corona and chromosphere ($\log N_{\mathrm{Fe}} = 7.66$, on the scale with $\log N_{\mathrm{H}}$ defined as 12.00) than in the solar photosphere ($\log N_{\mathrm{Fe}} = 6.51$). No good explanation could be found, but the problem was solved when Whaling *et al.* (Wh 69) reported beam-foil measurements of Fe I lifetimes which differed by a factor of 10 or more from those earlier used to determine photospheric Fe abundance. These new data were in agreement with the results of other careful f-value experiments (Ga 69, Hu 68). Similar beam-foil studies have been performed for many iron-group elements and the solar abundance data have been revised in several cases. For example, the lifetime measurements for Mn II levels (Ma 77, Pi 74) support a change of the photospheric Mn abundance by a factor of 3. Valuable lifetime data for the important systems Cr I, Mn I, and Ni I have also been obtained by BFS (Co 71, Gr 79, Le 73b). Several authors have also studied the lifetimes for doubly ionized iron-group elements which are relevant for the determination of the interstellar abundances of these elements (An 77, Do 79). For additional discussions we refer to articles by Huber (Hu 77) and Biemont and Grevesse (Bi 77).

The solar abundances of the rare-earth elements present quite intricate problems. The theoretical calculations of f values are here uncertain (because of the complexity of the energy level structure) and experimental data are therefore of great value. Beam-foil studies have provided data for several elements, see, e.g., (Cu 73, La 76) which has called for changes of the previously assumed solar abundances (Bi 77).

The temperature in the solar corona is about 2×10^6 K and in the uv spectra strong transitions in, e.g., Na-like and Mg-like Fe XVI, Fe XV, Ni XVIII, and Ni XVII appear. The lifetimes of the relevant transitions have been determined by beam-foil investigators (Bu 80, Pe 78).

The forbidden transitions discussed in Section 5.4 are of importance for the interpretation of conditions in astrophysical plasmas. For example, the intensity ratio $I(1s^2\ {}^1S_0\text{–}1s2p\ {}^3P_1)/I(1s^2\ {}^1S_0\text{–}1s2s\ {}^3S_1)$ can be used to determine the electron density in the solar corona (Ga 70).

8.3. Plasma Physics and Fusion Research

One of the most serious problems in fusion research with magnetically confined high-temperature plasmas (e.g., Tokamaks) is caused by plasma impurities, e.g., heavy atoms, which enter the plasma when the light ions (hydrogen isotopes) interact with the surrounding wall or discharge limiter. The walls are usually made of stainless steel (Cr, Fe, Ni), whereas Mo has been a popular limiter material. In the hot plasma the heavy impurities become highly ionized. When the ions are excited by collisions with electrons they emit electromagnetic radiation. This is a serious contribution to the energy loss in plasmas (Hi 76, Dr 78). It is therefore vital to determine the impurity elements and their charge states and concentrations (and of course to reduce the impurity concentrations in various technical ways). For the diagnostics of impurities data such as wavelengths, f values, and collision cross sections are needed. Of particular interest are systems with a few valence electrons, e.g., the Li, Be, B, Na, Mg, Al, Cu, and Zn isoelectronic sequences. Transitions in highly ionized Ti, Cr, Fe, Ni, and Mo, belonging to these sequences have been observed in Tokamak discharges (Hi 76, Dr 78).

Much work has therefore been undertaken to explore the spectra mentioned above. The resonance lines in Na-like Mo XXXIII and Mg-like Mo XXXII appear in Tokamaks, and several theoretical studies have been made to determine the f values in these spectra. However, Froese Fischer (Fr 76) and others noted considerable differences between theory and BFS data in the case of Na-like ions. This was mostly assumed to be due to cascading processes in the experimental studies. More recent BFS work for Na-like Fe and Ni (Bu 80, Pe 78) tends to result in much better agreement with theory. This fact should imply that the theoretical f values are reliable for the Mo ions mentioned above— for which there are no experimental f values available.

However, beam-foil studies of Mo have already been performed. Using 22–200-MeV Mo ions Johnson *et al.* (Jo 80) registered spectra in the region 100–500 Å. These spectra showed similarities with those observed in Tokamaks. At the highest energy the average charge state was 28+ while it was 13+ at the lowest energy used. Similar results, at lower energies

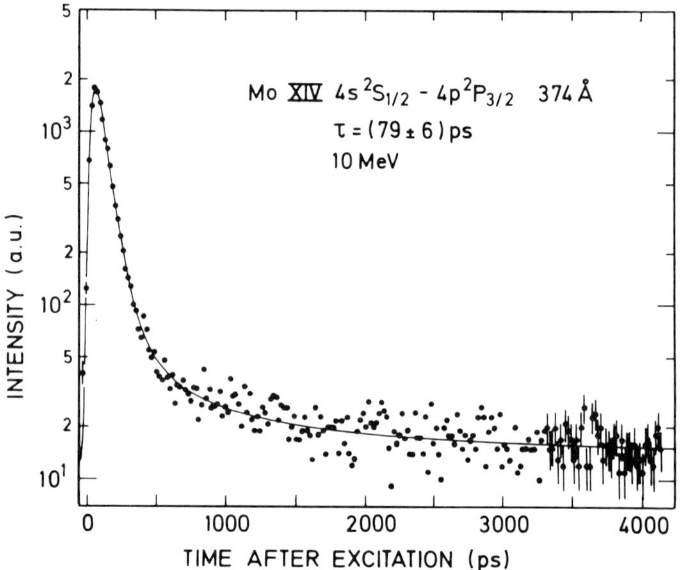

Figure 27. Result of a lifetime measurement for the $4p\ ^2P_{3/2}$ level in Cu I-like Mo XIV (De81).

(5–25 MeV), were obtained by Denne and Poulsen (De 81), who also made the first lifetime measurement for Cu-like Mo XIV. One decay curve is shown in Figure 27. The result underlines that theoretical f values are reliable for Cu-like ions.

Tokamak discharges (like astrophysical plasmas) may contain many elements, whereas—as already noted—the beam-foil source has very high chemical purity. Beam-foil spectra of Au and W have therefore been recorded and compared with Tokamak spectra to make clear which of these elements is more dominant in the latter case (Jo 79).

The importance of forbidden lines for plasma diagnostics (noted above) is also valid in the case of laboratory plasmas.

A thorough review of atomic data needs in fusion has recently appeared (Dr 81).

9. Excitation Mechanisms

In most of the work summarized in the previous sections the ion–foil, (or ion–gas) interaction has merely been used as a convenient way of exciting the atomic states of interest. Much effort has also been devoted to understanding more about the excitation process itself, and some of the results will be presented in this chapter.

9.1. Energy Level Populations

One of the most interesting questions concerns the initial population of the levels in beam-foil excitation. Besides the basic interest in learnmg more about excitation processes and light-source characteristics this problem is also of importance for understanding and correcting for cascading in lifetime measurements.

Many experiments have been performed on simple systems, e.g., H-like and He-like atoms and ions, where the atomic structure is well known.

From ion–atom collision theory the populations are expected to be proportional to $(2l+1)\,n^{-q}$ with $q \sim 3$. Support for this model has been found in several beam-foil experiments. For example, Davidson (Da 75), Dynefors *et al.* (Dy 76), and Heine *et al.* (He 79) found that the n^{-3} dependence agreed with experimental data. However, other values of q have also been proposed. Thus, Curtis (Cu 79) notes that certain decay curves for highly ionized levels show so prounced cascading tails that $q = 1$–2 must be assumed. However, such distributions diverge unless they are truncated for a certain n value. Many other studies of this problem have also appeared (Be 74a, Ho 76, El 80). These were stimulated by the appearance of such extensive cascading that certain decay curves seemed to follow a $t^{-3/2}$ law (Ri 73), which indicates extensive population of high-n levels in beam foil excitation.

Recent studies (An 81, Ba 82) show that the n^{-q} dependence is at best highly approximate. Certain n states seem to be preferentially populated as the result of a near-resonant charge transfer between the fast ions and foil atoms.

The l dependence (for a given n) has also been extensively investigated. Work in H I indicates that here s states are overpopulated in comparison to p states (Al 73), whereas in other systems the $2l+1$ dependence seems to be approximately valid. Indeed even more pronounced populations of high l (e.g., yrast) states have been noted (Le 73c). Heine *et al.* (He 79) also found that the l dependence varies with the energy. Other interesting data indicate irregularities in l populations (Ve 79). A model based on information theory has recently been applied to beam-foil excitations by Åberg and Goscinski (Åb 81). The n^{-3} dependence has thereby been shown for large n and a given l. The dependence of beam energy is also investigated. Such calculations are expected to provide much new insight into the excitation processes.

9.2. Coherence, Orientation, and Alignment

In our brief discussion of the zero-field quantum beats (Section 6.2) we mentioned the coherent excitation of fine- or hyperfine-structure levels of the same parity which decay to a common final state. Intensity oscillations (quantum beats) were then observed in time-resolved experiments. In the majority of these studies linearly polarized light (which shows the presence of alignment) was detected.

An interesting development consists of using a short gas cell to excite the ions into coherent states. Bromander *et al.* (Br 77) have thus demonstrated that zero-field quantum beats can also be observed in this case. In contrast to the foil-excitation case, single-collision conditions prevailed in this experiment, which makes it easier to compare the data with theoretical results for atomic collisions.

Also levels with different l quantum number but the same parity can be coherently excited in ion-beam experiments, as first predicted by Macek (Ma 70c). This has been experimentally verified by Burns and Hancock (Bu 71a), who found that the S and D levels were coherently excited in H I. This coherence is also present when a gas target is used (De 80a).

The excitation may also prepare an atom into a coherent superposition of states of opposite parity, as theoretically predicted by Eck (Ec 73). Experimental verifications have here been provided by Sellin *et al.* (Se 73) and Gaupp *et al.* (Ga 74), who found that the $2s\ ^2S_{1/2}$ and $2p\ ^2P_{1/2}$ levels in hydrogen are coherently excited in beam-foil experiments. Since these levels do not have a common final state (when $E1$ transitions are considered), the SP coherence can only be demonstrated with a external electric field. In order to distinguish this collision-induced coherence from the "ordinary" Stark mixing, the quantum beats in the Ly α radiation were measured with parallel as well as antiparallel electric fields and the results were subtracted. The difference still showed quantum beats which were due to the excitation coherence of $2s\ ^2S_{1/2}$ and $2p\ ^2P_{1/2}$. A detailed study of the excitation of $n = 2$ in hydrogen has recently been reported by Gabrielse (Ga 81), who determined the elements of the density matrix for a large energy range. The SP coherence has also been established in experiments using a gas target (Se 79), and here, also, the effect is quite pronounced.

In the "conventional" beam-foil geometry the beam direction is perpendicular to the foil surface. As suggested by Fano and Macek (Fa 73) and Ellis (El 73), interesting effects may be expected by tilting the foil relative to the ion beam. The first experiment of this kind was performed by Berry *et al.* (Be 74c), who directed 130-keV He^+ ions through a foil, the normal of which formed an angle with the beam direction. The light from a decay of a level in He I was found to show partial elliptical polarization, the degree of this polarization increasing when the tilt angle increased. The same effect was also established by quantum-beat techniques in a external magnetic field (Li 74). These experiments showed that both alignment and orientation were present (the latter effect manifests itself in the emission of circularly polarized light). The exit surface was found to play a very important role for

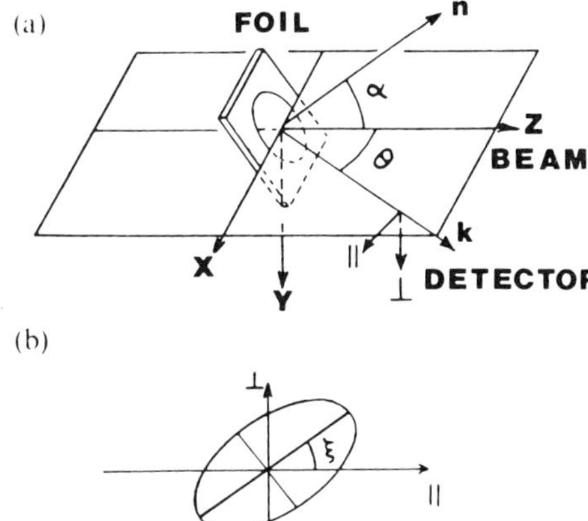

Figure 28. The tilted foil geometry and polarization ellipse (Be 74c).

the excitation of the transitions sutdied. One possible experimental geometry is shown in Figure 28. A considerable number of experiments have been performed in recent years to explore these phenomena. The degree of elliptical polarization has thus been investigated as a function of various quantum numbers (n, l, etc), degrees of ionization, beam velocity, but also of beam current, foil material, foil temperature, etc.; see, e.g., the article by Schectman *et al.* (Sc 79). Unfortunateley most of the experiments have been carried out in the absence of ultrahigh vacuum which is necessary for good surface studies; typical rest-gas pressures have usually been around 10^{-6} torr. Recently, some detailed experimental studies have been reported by Burns *et al.* (Bu 79), Schectman *et al.* (Sc 80), and Brooks and Pinnington (Br 80). In the latter experiment levels in doubly excited He I were studied. There are also several theoretical studies of the excitation process in tilted-foil situations (Ba 76a, Le 75, Lo 75).

Quantum beats can also be observed with the tilted-foil geometry. Figure 29 shows such beats between hyperfine levels in N IV (Be 75a). Note the presence of beats between the $F = 0$ and $F = 1$ states which are observable only when there is orientation (i.e., the tilted-foil geometry).

It was first shown by Andrä (An 75b, An 76b) that pronounced excitation asymmetry effects (orientation, circularly polarized light) can be observed when fast ions are scattered at surfaces. The orientation of the atomic levels is particularly pronounced when the ions impinge at nearly grazing incidence. In some cases the fraction $[I(\sigma^+) - I(\sigma^-)]/[I(\sigma^+) + I(\sigma^+)]$, where $I(\sigma^+)$ and $I(\sigma^-)$ are the intensities of left-hand and right-hand polarized light, was found to exceed 0.7. Following systematic studies of this effect (An 76b, Be 77a, To 78) as a possible tool for understanding surface properties and ion-surface interaction effects, this interesting technique has also been used to investigate fundamental atomic properties. Thus Andrä *et*

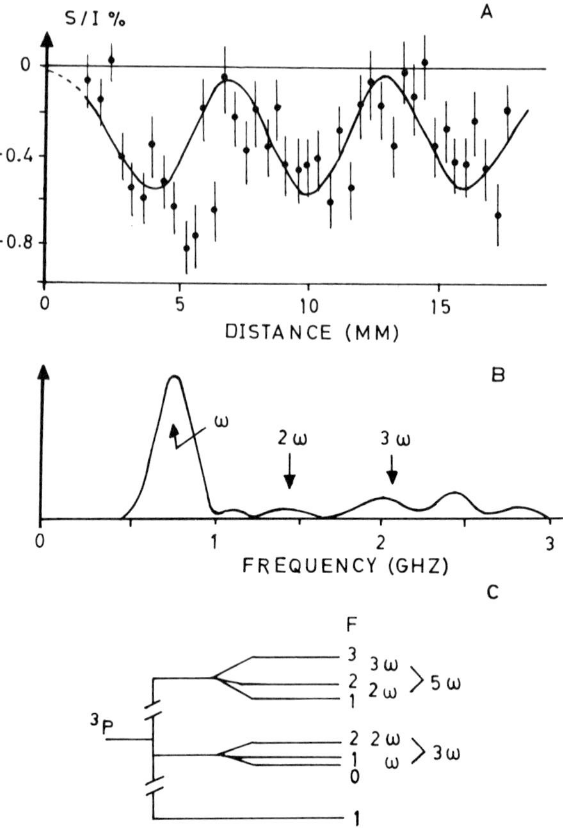

Figure 29. Zero-field quantum beats of the ^{14}N IV $3s\ ^3S$–$3p\ ^3P$ multiplet at 3480 Å, measured with a tilted foil (A). The Fourier transform is also shown (B) and the energy levels of the upper term (C). Note that the beats between $F = 0$ and $F = 1$ can only be observed in circular polarization (tilted-foil geometry). The data are from Berry *et al.* (Be 75a).

al. (An 76b) studied hyperfine beats for a level in ^{14}N II, and more recently Winter and Andrä (Wi 80) determined the hyperfine splittings for the $2p3p\ ^1P_1$ level in ^{14}N II. They obtained accurate values for the hyperfine constants from which even the quadrupole moment could be determined. The latter had an uncertainty of only 4 %.

This method also makes it possible to obtain polarized nuclei. The atomic orientation which is created at the ion–surface interaction can be transferred to the nucleus (via the hyperfine interaction). The nuclear orientation is indirectly observed by sending the scattered, excited ions through a foil at normal incidence. The presence of circularly polarized light after the passage of this analyzer foil shows the nuclear polarization. In a recent study of Li and N ions, nuclear polarizations as high as 20 % were observed. The ion-energy dependence of this effect has also been investigated (Wi 80a). The effect increased when the ions were scattered at two surfaces (Wi 79b). Similar experiments, using two foils (one of which is tilted relative to the beam), have also been carried out (Ra 79). These experiments may thus have important applications in nuclear physics.

The ion-foil excitation processes contain quite a number of intriguing problems, the solution of which may often lie very far ahead. While the modern theories for atomic structure are quite sophisticated and sometimes comparable with (or even superior to) experiments in accuracy, those for atomic interactions are far less quantitative, largely because of the greater complexities of the systems studied. Experimental measurements are of great value and they may frequently reveal most interesting effects.

Note Added in Proof

Several interesting reviews have appeared after the submission of the manuscript. The following references are particularly relevant:

1. E. J. Knystautas and R. Drouin, eds., "Fast Ion Beam Spectroscopy" (Proceedings of the Sixth International Conference on Fast Ion Beam Spectroscopy, Quebec, 1981), *Nucl. Instrum. Methods* **222**:1-402 (1982).
2. L. Liljeby, ed., "Production and Physics of Highly Charged Ions" (Proceedings of an International Symposium, Stockholm, 1982), *Phys. Scr.* **T3**:1-258 (1982).
3. R. Marrus, ed., "Atomic Physics of Highly Ionized Atoms", (Proceedings of a NATO Advanced Study Institute, Corscica, 1982), Plenum Press, New York, 1983.
4. H. G. Berry and M. Hass, *Ann. Rev. Nucl. Part. Sci.* **32**:1 (1982).

In 1984, another international conference on the physics of highly ionized atoms was held in Oxford. The proceedings will appear in *Nucl. Instrum. Methods* (N. J. Peacock and J. D. Silver, eds.).

Acknowledgments

I am grateful to S. Bashkin, L. J. Curtis, E. J. Knystautas, A. E. Livingston, D. J. Pegg, E. H. Pinnington, and E. Träbert for sending me their latest results. This work was supported by the Swedish Natural Science Research Council (NFR) and the National Swedish Board for Energy Source Development (NE).

References

(Åb 81) T. Åberg and O. Goscinski, *Phys. Rev. A* **24**:801 (1981).

(Al 63) E. B. Aleksandrov, *Opt. Spectrosc. (USSR)* **14**:233 (1963).

(Al 73) M. J. Alguard and C. W. Drake, *Phys. Rev. A* **8**:27 (1973).

(An 69) T. Andersen, K. A. Jessen, and G. Sørensen, *J. Opt. Soc. Am.* **59**:1197 (1969).

(An 70) H. J. Andrä, *Phys. Rev. Lett.* **25**:325 (1970).

(An 70a) H. J. Andrä, *Phys. Rev. A* **2**:2200 (1970).

(An 70b) H. J. Andrä, *Phys. Lett.* **32A**:345 (1970).

(An 71) T. Andersen, J. R. Roberts, and G. Sørensen, *Phys. Scr.* **4**:52 (1971).

(An 72) T. Andersen, A. Kirkegård Nielsen, and G. Sørensen, *Phys. Scr.* **6**:122 (1972).

(An 73) H. J. Andrä, A. Gaupp, and W. Wittmann, *Phys. Rev. Lett.* **31**:501 (1973).

(An 74) H. J. Andrä, *Phys. Scr.* **9**:257 (1974).

(An 75) H. J. Andrä, H.-J. Plöhn, W. Wittmann, A. Gaupp, J. O., Stoner, Jr., and M. Gaillard, *J. Opt. Soc. Am.* **65**:1410 (1975).

(An 75a) H. J. Andrä, in *Atomic Physics 4* (G. zu Putlitz, E. W. Weber, and A. Winnacker, eds.) Plenum Press, New York (1975), p. 635.

(An 75b) H. J. Andrä, *Phys. Lett.* **54A**:315 (1975).

(An 76) H. J. Andrä, in *Beam-Foil Spectroscopy* (I. A. Sellin and D. J. Pegg, eds.), Plenum Press, New York (1976), p. 835.

(An 76a) T. Andersen, O. Poulsen, and P. S. Ramanujam, *J. Quant. Spectrosc. Radiat. Transfer* **16**:521 (1976).

(An 76b) H. J. Andrä, R. Fröhling, H. J. Plöhn, and J. D. Silver, *Phys. Rev. Lett.* **37**:1212 (1976).

(An 77) T. Andersen, P. Petersen, and E. Biemont, *J. Quant. Spectrosc. Radiat. Transfer* **17**:389 (1977).

(An 77a) H. J. Andrä, H. J. Plöhn, A. Gaupp, and R. Fröhling, *Z. Phys.* **A281**:15 (1977).

(An 79) H. J. Andrä, in *Progress in Atomic Spectroscopy* (W. Hanle and H. Kleinpoppen, eds.), Plenum Press, New York (1979), p. 829.

(An 79a) T. Andersen, P. Eriksen, O. Poulsen, and P. S. Ramanujam, *Phys. Rev. A* **20**:2621 (1979).

(An 80) T. Andersen, S. M. Bentzen, and O. Poulsen, *Phys. Scr.* **22**:119 (1980).

(An 81) B. Andresen, B. Denne, J. O. Ekberg, L. Engström, S. Huldt, I. Martinson, and E. Veje, *Phys. Rev. A* **23**:479 (1981).

(Ar 75) A. Arnesen, A. Bengtsson, R. Hallin, S. Kandela, T. Noreland, and R. Lidholt, *Phys. Lett.* **53A**:459 (1975).

(Ar 76) L. Armstrong, Jr., W. R. Fielder, and D. L. Lin, *Phys. Rev. A* **14**:1114 (1976).

(Ar 77) A. Arnesen, A. Bengtsson, R. Hallin, J. Lindskog, C. Nordling, and T. Noreland, *Phys. Scr.* **16**:31 (1977).

(Ar 79) I. A. Armour, E. G. Myers, J. D. Silver, and E. Träbert, *Phys. Lett.* **75A**:45 (1979).

(Ar 80) I. A. Armour, B. C. Fawcett, J. D. Silver, and E. Träbert, *J. Phys. B* **13**:2701 (1980).

(Ar 81) I. A. Armour, J. D. Silver, and E. Träbert, *J. Phys. B* **14**:3563 (1981).

(As 76) G. Astner, L. J. Curtis, L. Liljeby, S. Mannervik, and I. Martinson, *Z. Phys.* **A279**:1 (1976).

(As 76a) G. Astner, L. J. Curtis, L. Liljeby, S. Mannervik, and I. Martinson, *J. Phys. B* **9**:L345 (1976).

(Ba 64) S. Bashkin, *Nucl. Instrum. Methods* **28**:88 (1964).

(Ba 65) S. Bashkin, W. S., Bickel, D. Fink, and R. K. Wangsness, *Phys. Rev. Lett.* **15**:284 (1965).

(Ba 66) S. Bashkin, D. Fink, P. R. Malmberg, A. B. Meinel, and S. G. Tilford, *J. Opt. Soc. Am.* **56**:1064 (1966).

(Ba 66a) S. Bashkin and G. Beauchemin, *Can. J. Phys.* **44**:1603 (1966).

(Ba 68) S. Bashkin, ed., *Beam-Foil Spectroscopy*, Gordon and Breach, New York (1968).

(Ba 71) S. Bashkin and I. Martinson, *J. Opt. Soc. Am.* **61**:1686 (1971).

(Ba 73) S. Bashkin, ed., *Beam-Foil Spectroscopy, Nucl. Instrum. Methods* **110**:1–522 (1973).

(Ba 74) S. Bashkin, in *Progress in Optics*, Vol. XII (E. Wolf, ed.), North-Holland, Amsterdam (1974), p. 289.

(Ba 76) S. Bashkin, ed., *Beam-Foil Spectroscopy* Springer, Berlin (1976).

(Ba 76a) Y. B. Band, *Phys. Rev. A* **13**:2061 (1976).

(Ba 78) J. G. Bannenberg, S. Doorn and F. W. Saris, in *Low-Energy Ion Beams*, Inst. Phys. Conf. Series No 38 (1978), p. 63.

(Ba 80) Y. Baudinet-Robinet, P. D. Dumont, H. P. Garnir, N. Grevesse, and E. Biemont, *J. Opt. Soc. Am.* **70**:464 (1980).

(Ba 80a) S. Bashkin, G. Astner, S. Mannervik, P. S., Ramanujam, M. Scofield, S. Huldt, and I. Martinson, *Phys. Scr.* **21**:820 (1980).

(Ba 81) S. Bashkin, R. Hallin, J. Leavitt, U. Litzén, and D. Walker, *Phys. Scr.* **23**:5 (1981).

(Ba 82) S. Bashkin, H. Oona, and E. Veje, *Phys. Rev. A* **25**:417 (1982).

(Be 65) K. Berkner, W. S. Cooper, S. N. Kaplan, and R. V. Pyle, *Phys. Lett.* **16**:35 (1965).

(Be 69) I. Bergström, J. Bromander, R. Buchta, L. Lundin, and I. Martinson, *Phys. Lett.* **28A**:721 (1969).

(Be 70) H. G. Berry, I. Martinson, and J. Bromander, *Phys. Lett.* **31A**:521 (1970).

(Be 70a) H. G. Berry, R. M. Schectman, I. Martinson, W. S. Bickel, and S. Bashkin, *J. Opt. Soc. Am.* **60**:335 (1970).

(Be 71) H. G. Berry, I. Martinson, L. J. Curtis, and L. Lundin, *Phys. Rev. A.* **3**:1934 (1971).

(Be 72) H. D. Betz, *Revs. Mod. Phys.* **44**:465 (1972).

(Be 72a) H. G. Berry, E. H. Pinnington, and J. L. Subtil, *J. Opt. Soc. Am.* **62**:767 (1972).

(Be 72b) H. G. Berry, J. Desesquelles, and M. Dufay, *Phys. Rev. A* **6**:600 (1972).

(Be 72c) H. G. Berry, J. L. Subtil, and M. Carré, *J. Phys. (Paris)* **33**:947 (1972).

(Be 73) H. G. Berry, J. Desesquelles, and M. Dufay, *Nucl. Instrum. Methods* **110**:43 (1973).

(Be 73a) H. G. Berry, J. L. Subtil, E. H. Pinnington, H. J. Andrä, W. Wittmann, and A. Gaupp, *Phys. Rev. A* **7**:1609 (1973).

(Be 73b) H. G. Berry and R. Bacis, *Phys. Rev. A* **8**:36 (1973).

(Be 74) H. G. Berry and J. L. Subtil, *Phys. Scr.* **9**:217 (1974).

(Be 74a) H. D. Betz, F. Bell, H. Panke, G. Kalkoffen, M. Welz, and D. Evers, *Phys. Rev. Lett.* **33**:807 (1974).

(Be 74b) H. G. Berry and R. M. Schectman, *Phys. Rev. A* **9**:2345 (1974).

(Be 74c) H. G. Berry, L. J. Curtis, D. G. Ellis, and R. M. Schectman, *Phys. Rev. Lett.* **32**:751 (1974).

(Be 75) H. G. Berry, *Phys. Scr.* **12**:5 (1975).

(Be 75a) H. G. Berry, L. J. Curtis, D. G. Ellis, and R. M. Schectman, *Phys. Rev. Lett.* **35**:274 (1975).

(Be 76) K.-E. Bergkvist, *J. Opt. Soc. Am.* **66**:837 (1976).

(Be 77) H. G. Berry, *Rep. Progr. Phys.* **40**:155 (1977).

(Be 77a) H. G. Berry, G. Gabrielse, and A. E. Livingston, *Phys. Rev. A* **16**:1915 (1977).

(Be 79) H. J. Beyer, in *Progress in Atomic Spectroscopy* (W. Hanle and H. Kleinpoppen, eds.), Plenum Press, New York (1979), p. 529.

(Be 80) H. G. Berry, R. DeSerio, and A. E. Livingston, *Phys. Rev. A* **22**:998 (1980).

(Bh 74) S. N. Bhardwaj, H. G. Berry, and T. Mossberg, *Phys. Scr.* **9**:331 (1974).

(Bi 67) W. S. Bickel, *Appl. Opt.* **6**:1309 (1967).

(Bi 67a) W. S. Bickel and S. Bashkin, *Phys. Rev.* **162**:12 (1967).

(Bi 69) W. S. Bickel, I. Martinson, L. Lundin, R. Buchta, J. Bromander, and I. Bergström, *J. Opt. Soc. Am.* **59**:830 (1969).

(Bi 69a) W. S. Bickel, I. Bergström, R. Buchta, L. Lundin, and I. Martinson, *Phys. Rev.* **178**:118 (1969).

(Bi 74) W. S. Bickel and R. Buchta, *Phys. Scr.* **9**:148 (1974).

(Bi 77) E. Biemont and N. Grevesse, *Phys. Scr.* **16**:39 (1977).

(Bo 74) K. Bockasten, *Phys. Rev. A* **9**:1087 (1974).

(Br 40) G. Breit and E. Teller, *Astrophys. J.* **91**:215 (1940).

(Br 68) L. Brown, K. Ford, V. Rubin, W. Trächslin, and W. Brandt, in *Beam-Foil Spectroscopy* (S. Bashkin, ed.) Gordon and Breach, New York (1968), p. 45.

(Br 73) J. Bromander, *Nucl. Instrum. Methods* **110**:11 (1973).

(Br 75) R. Bruch, G. Paul, and J. Andrä, *J. Phys. B* **8**:L253 (1975).

(Br 77) J. Bromander, L. Liljeby, and I. A. Sellin, *Z. Phys.* **A283**:299 (1977).

(Br 78) S. J. Brodsky and P. J. Mohr, in *Structure and Collisions of Ions and Atoms* (I. A. Sellin, ed.), Springer, Berlin (1978), p. 3.

(Br 80) R. L. Brooks and E. H. Pinnington, *Phys. Rev. A* **22**:529 (1980).

(Br 80a) R. L. Brooks, J. E. Hardis, H. G. Berry, L. J. Curtis, K. T. Cheng, and W. Ray, *Phys. Rev. Lett.* **45**:1318 (1980).

(Bu 69) J. B. Buchet, A. Denis, J. Desesquelles, and M. Dufay, *Phys. Lett.* **28A**:529 (1969).

(Bu 71) R. Buchta, L. J. Curtis, I. Martinson, and J. Brzozowski, *Phys. Scr.* **4**:55 (1971).

(Bu 71a) D. J. Burns and W. H. Hancock, *Phys. Rev. Lett.* **27**:370 (1971).

(Bu 73) J. P. Buchet, M. C. Buchet-Poulizac, H. G. Berry, and G. W. F. Drake, *Phys. Rev. A* **7**:922 (1973).

(Bu 73a) M. C. Buchet-Poulizac and J. P. Buchet, *Phys. Scr.* **8**:40 (1973).

(Bu 74) J. P. Buchet, M. C. Buchet-Poulizac, A. Denis, J. Desesquelles, and G. DoCao, *Phys. Scr.* **9**:221 (1974).

(Bu 79) D. J. Burns, R. D. Hight, and C. H. Greene, *Phys. Rev. A* **20**:404 (1979).

(Bu 80) J. P. Buchet, M. C. Buchet-Poulizac, A. Denis, J. Desesquelles, and M. Druetta, *Phys. Rev. A* **22**:2061 (1980).

(Bu 80a) C. F. Bunge, *Phys. Rev. Lett.* **44**:1450 (1980).

(Ca 71) M. Carré, J. Desesquelles, M. Dufay, and M. L. Gaillard, *Phys. Rev. Lett.* **27**:1407 (1971).

(Ce 80) P. Ceyzeriat, D. J. Pegg, M. Carré, M. Druetta, and M. L. Gaillard, *J. Opt. Soc. Am.* **70**:901 (1980).

(Ch 70) S. A. Chin-Bing, C. E. Head, and A. E. Green, *Am. J. Phys.* **38**:352 (1970).

(Ch 71) D. A. Church, M. Druetta, and C. H. Liu, *Phys. Rev. Lett.* **27**:1763 (1971).

(Ch 73) D. A. Church and C. H. Liu, *Nucl. Instrum. Methods* **110**:267 (1973).

(Ch 78) K. T. Cheng, J. P. Desclaux, and Y.-K. Kim, *J. Phys. B* **11**:L359 (1978).

(Co 64) A. Corney and G. W. Series, *Proc. Phys. Soc.* **83**:213 (1964).

(Co 71) C. L. Cocke, B. Curnutte, and J. H. Brand, *Astron. Astrophys.* **15**:299 (1971).

(Co 74) C. L. Cocke, B. Curnutte, J. R. Macdonald, J. A. Bednar, and R. Marrus, *Phys. Rev. A* **9**:2242 (1974).

(Co 76) C. L. Cocke, in *Methods of Experimental Physics*, Vol. 13 B, Academic Press, New York (1976), p. 213.

(Cr 76) R. J. S. Crossley, L. J. Curtis, and C. Froese Fischer, *Phys. Lett.* **57A**:220 (1976).

(Cu 70) L. J. Curtis, R. M. Schectman, J. L. Kohl, D. A. Chojnacki, and D. R. Shoffstall, *Nucl. Instrum. Methods* **90**:207 (1970).

(Cu 71) L. J. Curtis, H. G. Berry, and J. Bromander, *Phys. Lett.* **34A**:169 (1971).

(Cu 73) L. J. Curtis, I. Martinson, and R. Buchta, *Nucl. Instrum. Methods* **110**:391 (1973).

(Cu 76) L. J. Curtis, in *Beam-Foil Spectroscopy* (S. Bashkin, ed.) Springer, Berlin (1976), p. 63.

(Cu 79) L. J. Curtis, *J. Phys. (Paris)* **40**:C1–139 (1979).

(Cu 80) L. J. Curtis, *Phys. Scr.* **21**:162 (1980).

(Da 75) J. Davidson, *Phys. Rev. A* **12**:1350 (1975).

(De 60) G. Dearnalay, *Rev. Sci. Instrum.* **31**:197 (1960).

(De 68) A. Denis, J. Desesquelles, M. Dufay, and M. C. Poulizac, in *Beam-Foil Spectroscopy* (S. Bashkin, ed.), Gordon and Breach, New York (1968), p. 341.

(De 79) J. Desesquelles, ed. *Fast Ion Spectroscopy*, *J. Phys. (Paris)* **40**:C1–367 (1979).

(De 79a) B. Denne, U. Litzén, and L. J. Curtis, *Phys. Lett.* **71A**:35 (1979).

(De 80) B. Denne, S. Huldt, J. Pihl, and R. Hallin, *Phys. Scr.* **22**:45 (1980).

(De 80a) J. C. Dehaes and W. Singer, *Phys. Lett.* **75A**:276 (1980).

(De 81) B. Denne and O. Poulsen, *Phys. Rev. A* **23**:1229 (1981).

(Di 78) D. D. Dietrich, J. A. Leavitt, S. Bashkin, J. G. Conway, H. Gould, D. Mac Donald, R. Marrus, B. M. Johnson, and D. J. Pegg, *Phys. Rev. A* **18**:208 (1978).

(Di 80) D. D. Dietrich, J. A. Leavitt, H. Gould, and R. Marrus, *Phys. Rev. A* **22**:1109 (1980).

(Do 78) H.-D. Dohmann, D. Liesen, and H. Pfeng, *Z. Physik A* **285**:171 (1978).

(Do 79) J. S. Dolby, R. W. P. McWhirter, and C. J. Sofield, *J. Phys. B* **12**:187 (1979).

(Dr 70) M. Druetta, P. Ceyzeriat, and M. C. Poulizac, *C. R. Acad. Sci. Paris* **271B**:846 (1970).

(Dr 73) G. W. F. Drake and R. B. Grimley, *Phys. Rev. A* **8**:157 (1973).

(Dr 76) G. W. F. Drake and C. P. Lin, *Phys. Rev. A* **14**:1296 (1976).

(Dr 78) H. W. Drawin, *Phys. Rep.* **37**:125 (1978).

(Dr 79) G. W. F. Drake, S. P. Goldman, and A. van Wijngaarden, *Phys. Rev. A* **20**:1299 (1979).

(Dr 81) H. W. Drawin and K. Katsonis, eds, *Phys. Scr.* **23**:69–214 (1981).

(Du 70) M. Dufay, *Nucl. Instrum. Methods* **90**:15 (1970).

(Du 73) M. Dufay, *Nucl. Instrum. Methods* **110**:79 (1973).

(Du 76) P. D. Dumont, A. E. Livingston, Y. Baudinet-Robinet, G. Weber, and L. Quaglia, *Phys. Scr.* **13**:122 (1976).

(Du 76a) M. Dufay, M. Carré, M. L. Gaillard, G. Meunier, H. Winter, and Z. Zgainski, *Phys. Rev. Lett.* **37**:1678 (1976).

(Dy 76) B. Dynefors, I. Martinson, and E. Veje, *Phys. Scr.* **13**:308 (1976).

(Dy 78) B. I. Dynefors and I. Martinson, *Phys. Scr.* **17**:123 (1978).

(Ec 73) T. G. Eck, *Phys. Rev. Lett.* **31**:270 (1973).

(Ed 64) B. Edlén, in *Handbuch der Physik*, Vol. 27 (S. Flügge, ed.) Springer, Heidelberg (1964), p. 80.

(El 73) D. G. Ellis, *J. Opt. Soc. Am.* **63**:1232 (1973).

(El 80) D. G. Ellis, *Phys. Lett.* **80A**:375 (1980).

(En 77) O. Engvold, *Phys. Scr.* **16**:48 (1977).

(En 79) L. Engström, B. Denne, S. Huldt, J. O. Ekberg, L. J. Curtis, E. Veje and I. Martinson, *Phys. Scr.* **20**:88 (1979).

(En 80) L. Engström, C. Jupén, B. Denne, S. Huldt, Weng Tai Meng, P. Kaijser, U. Litzén, and I. Martinson, *J. Phys. B* **13**:L143 (1980).

(En 81) L. Engström, B. Denne, J. O. Ekberg, K. W. Jones, C. Jupén, U. Litzén, Weng Tai Meng, A. Trigueiros, and I. Martinson, *Phys. Scr.* **24**:551 (1981).

(Er 71) G. W. Erickson, *Phys. Rev. Lett.* **27**:780 (1971).

(Er 80) P. Eriksen and O. Poulsen, *J. Quant. Spectrosc. Radiat. Transfer* **23**:599 (1980).

(Fa 67) C. Y. Fan, M. Garcia-Munoz, and I. A. Sellin, *Phys. Rev.* **161**:6 (1967).

(Fa 73) U. Fano and J. Macek, *Revs. Mod. Phys.* **45**:553 (1973).

(Fi 68) U. Fink, G. N. McIntire, and S. Bashkin, *J. Opt. Soc. Am.* **58**:475 (1968).

(Fr 68) C. Froese Fischer, *J. Quant. Spectrosc. Radiat. Transfer* **8**:755 (1968).

(Fr 76) C. Froese Fischer, in *Beam-Foil Spectroscopy* (I. A. Sellin and D. J. Pegg, eds.) Plenum Press, New York (1976), p. 69.

(Fr 81) C. Froese Fischer, *Phys. Scr.* **23**:38 (1981).

(Ga 69) T. Garz and M. Kock, *Astron. Astrophys.* **2**:274 (1969).

(Ga 70) A. H. Gabriel and C. Jordan, *Phys. Lett.* **32A**:166 (1970).

(Ga 73) M. Gaillard, M. Carré, H. G. Berry, and M. Lombardi, *Nucl. Instrum. Methods* **110**:273 (1973).

(Ga 74) A. Gaupp, H. J. Andrä, and J. Macek, *Phys. Rev. Lett.* **32**:268 (1974).

(Ga 75) M. Gaillard, H. J. Andrä, A. Gaupp, W. Wittmann, H.-J. Plöhn, and J. O. Stoner, Jr., *Phys. Rev. A* **12**:987 (1975).

(Ga 76) A. Gaupp, M. Dufay, and J. L. Subtil, *J. Phys. B* **9**:2365 (1976).

(Ga 77) H. P. Garnir, A. E. Livingston, Y. Baudinet-Robinet, P. D. Dumont, E. Biémont, and N. Grevesse, *J. Opt. Soc. Am.* **67**:751 (1977).

(Ga 78) H. P. Garnir, Y. Baudinet-Robinet, P. D. Dumont, and M. Eidelsberg, *Phys. Scr.* **17**:463 (1978).

(Ga 79) H. P. Garnir, *J. Opt. Soc. Am.* **69**:916 (1978).

(Ga 81) G. Gabrielse, *Phys. Rev. A* **23**:775 (1981).

(Go 74) H. Gould, R. Marrus, and P. J. Mohr, *Phys. Rev. Lett.* **33**:676 (1974).

(Go 78) H. Gould and R. Marrus, *Phys. Rev. Lett.* **41**:1457 (1978).

(Gr 79) T. R. Greenlee and W. Whaling, *J. Quant. Spectrosc. Radiat. Transfer* **21**:55 (1979).

(Ha 69) T. Hadeishi, W. S. Bickel, J. D. Garcia, and H. G. Berry, *Phys. Rev. Lett.* **23**:65 (1969).

(Ha 73) R. Hallin, J. Lindskog, A. Marelius, J. Pihl, and R. Sjödin, *Phys. Scr.* **8**:209 (1973).

(Ha 76) H. Harde, in *Beam-Foil Spectroscopy* (I. A. Sellin and D. J. Pegg, eds.) Plenum Press, New York (1976), p. 859.

(He 67) L. Heroux, *Phys. Rev.* **153**:156 (1967).

(He 79) G. Heine, H. H. Bukow, and H.v. Buttlar, *J. Phys. (Paris)* **40**:C1–269 (1979).

(Hi 75) A. Hibbert, *Rep. Progr. Phys.* **38**:1217 (1975).

(Hi 76) E. Hinnov, *Phys. Rev. A* **14**:1533 (1976).

(Hi 77) A. Hibbert, *Phys. Scr.* **16**:7 (1977).

(Hi 80) A. Hibbert, *J. Phys. B* **13**:1721 (1980).

(Ho 72) S. Hontzeas, I. Martinson, P. Erman, and R. Buchta, *Phys. Scr.* **6**:55 (1972).

(Ho 76) F. Hopkins and P.v. Brentano, *J. Phys. B* **9**:775 (1976).

(Hu 68) M. Huber and F. L. Tobey, *Astrophys. J.* **151**:609 (1968).

(Hu 77) M. C. E. Huber, *Phys. Scr.* **16**:16 (1977).

(Hu 79) S. Huldt, L. J. Curtis, and R. M. Schectman, *J. Phys. (Paris)* **40**:C1–161 (1979).

(Ir 74) D. J. G. Irwin and A. E. Livingston, *Comp. Phys. Commun.* **7**:95 (1974).

(Is 78) K. Ishii and M. Tomita, *J. Phys. Soc. Jpn* **45**:230 (1978).

(Ja 80) C. Jacques, E. J. Knystautas, R. Drouin, and H. G. Berry, *Can. J. Phys.* **58**:1093 (1980).

(Je 74) K. Jensen and E. Veje, *Nucl. Instrum. Methods* **122**:511 (1974).

(Je 77) N. A. Jelley, J. D. Silver and I. A. Armour, *J. Phys. B* **10**:2339 (1977).

(Jo 79) K. W. Jones, *J. Phys. (Paris)* **40**:C1–197 (1979).

(Jo 80) B. M. Johnson, K. W. Jones, J. L. Cecchi, and T. H. Kruse, *Phys. Lett.* **78A**:61 (1980).

(Ju 82) C. Jupén, *Nucl. Instrum. Methods* **202**:25 (1982).

(Ka 63) L. Kay, *Phys. Lett.* **5**:36 (1963).

(Ka 65) L. Kay, *Proc. Phys. Soc.* **85**:163 (1965).

(Ka 73) L. Kay, B. Lightfoot, and G. Harding, *Nucl. Instrum. Methods* **110**:435 (1973).

(Ke 79) J. A. Kernahan and E. H. Pinnington, *Nucl. Instrum. Methods* **166**:313 (1979).

(Ke 80) J. A. Kernahan, E. H. Pinnington, J. A. O'Neill, J. L. Bahr, and K. E. Donnelly, *J. Opt. Soc. Am.* **70**:1126 (1980).

(Ki 76) Y. K. Kim and J. P. Desclaux, *Phys. Rev. Lett.* **36**:139 (1976).

(Kl 69) S. Klarsfeld, *Phys. Lett.* **30A**:382 (1969).

(Kn 73) E. J. Knystautas and R. Drouin, *Nucl. Instrum. Methods* **110**:95 (1973).

(Kn 76) E. J. Knystautas and R. Drouin, in *Beam-Foil Spectroscopy* (I. A. Sellin and D. J. Pegg, eds.) Plenum Press, New York (1976), p. 377.

(Kr 77) T. Krist, P. Kuske, A. Gaupp, W. Wittmann, and H. J. Andrä, *Phys. Lett.* **61A**:94 (1977).

(Ku 75) H. W. Kugel, M. Leventhal, D. E. Murnick, C. K. N. Patel, and O. R. Wood, *Phys. Rev. Lett.* **35**:647 (1975).

(Ku 77) H. W. Kugel and D. E. Murnick, *Rep. Progr. Phys.* **40**:297 (1977).

(La 47) W. L. Lamb and R. C. Retherford, *Phys. Rev.* **72**:241 (1947).

(La 72) G. P. Lawrence, C. Y. Fan, and S. Bashkin, *Phys. Rev. Lett.* **28**:1612 (1972).

(La 76) C. S. Lage and W. Whaling, *J. Quant. Spectrosc. Radiat. Transfer* **16**:537 (1976).

(Le 67) M. R. Lewis, T. Marshall, E. H. Carnevale, F. S. Zimnoch, and G. W. Wares, *Phys. Rev.* **164**:94 (1967).

(Le 69) M. R. Lewis, F. S. Zimnoch, and G. W. Wares, *Phys. Rev.* **178**:49 (1969).

(Le 72) W. N. Lennard, R. M. Sills, and W. Whaling, *Phys. Rev. A* **6**:884 (1972).

(Le 72a) M. Leventhal, D. E. Murnick, and H. W. Kugel, *Phys. Rev. Lett.* **28**:1609 (1972).

(Le 73) J. A. Leavitt, J. W. Robson, and J. O. Stoner, Jr., *Nucl. Instrum. Methods* **110**:423 (1973).

(Le 73a) M. Leventhal, *Nucl. Instrum. Methods* **110**:343 (1973).

(Le 73b) W. N. Lennard, W. Whaling, R. M. Sills, and W. A. Zajc, *Nucl. Instrum. Methods* **110**:385 (1973).

(Le 73c) W. N. Lennard and C. L. Cocke, *Nucl. Instrum. Methods* **110**:137 (1973).

(Le 74) M. Leventhal and P. E. Havey, *Phys. Rev. Lett.* **32**:808 (1974).

(Le 75) E. J. Lewis and J. D. Silver, *J. Phys. B* **8**:2697 (1975).

(Li 71) C. H. Liu, S. Bashkin, W. S. Bickel, and T. Hadeishi, *Phys. Rev. Lett.* **26**:222 (1971).

(Li 72) C. H. Liu, M. Druetta, and D. A. Church, *Phys. Lett.* **39A**:49 (1972).

(Li 74) C. H. Liu, S. Bashkin, and D. A. Church, *Phys. Rev. Lett.* **33**:993 (1974).

(Li 77) L. Lipsky, R. Aniana, and M. J. Conneely, *At. Data Nucl. Data Tables* **20**:127 (1977).

(Li 78) A. E. Livingston, H. G. Berry, and G. E. Thomas, *Nucl. Instrum. Methods* **148**:125 (1978).

(Li 78a) A. E. Livingston and H. G. Berry, *Phys. Rev. A* **17**:1966 (1978).

(Li 80) A. E. Livingston, S. J. Hinterlong, J. A. Poirier, R. DeSerio, and H. G. Berry, *J. Phys. B* **13**:L139 (1980).

(Li 80a) A. E. Livingston, L. J. Curtis, R. M. Schectman, and H. G. Berry, *Phys. Rev. A* **21**:771 (1980).

(Li 80b) A. E. Livingston and S. J. Hinterlong, *Phys. Lett.* **80A**:372 (1980).

(Li 81) A. E. Livingston, E. H. Pinnington, D. J. G. Irwin, J. A. Kernahan, and R. L. Brooks, *J. Opt. Soc. Am.* **71**:442 (1981).

(Li 81a) A. E. Livingston and S. J. Hinterlong, *Phys. Rev. A* **23**:758 (1981).

(Lo 75) M. Lombardi, *Phys. Rev. Lett.* **35**:1172 (1975).

(Lü 51) G. Lüders, *Ann. Phys. (Leipzig)* **6**:301 (1951).

(Lu 73) L. Lundin, B. Engman, J. Hilke, and I. Martinson, *Phys. Scr.* **8**:274 (1973).

(Lu 81) S. R. Lundeen and F. M. Pipkin, *Phys. Rev. Lett.* **46**:232 (1981).

(Ma 63) R. P. Madden and K. Codling, *Phys. Rev. Lett.* **10**:516 (1963).

(Ma 65) P. R. Malmberg, S. Bashkin, and S. G. Tilford, *Phys. Rev. Lett.* **15**:98 (1965).

(Ma 69) J. Macek, *Phys. Rev. Lett.* **23**:1 (1969).

(Ma 70) I. Martinson, J. Bromander, and H. G. Berry, eds. *Beam-Foil Spectroscopy, Nucl. Instrum. Methods* **90**:1–371 (1970).

(Ma 70a) I. Martinson, W. S. Bickel, and A. Ölme, *J. Opt. Soc. Am.* **60**:1213 (1970).

(Ma 70b) I. Martinson and W. S. Bickel, *Phys. Lett.* **31A**:25 (1970).

(Ma 70c) J. Macek, *Phys. Rev. A* **1**:618 (1970).

(Ma 72) R. Marrus and R. W. Schmieder, *Phys. Rev. A* **5**:1160 (1972).

(Ma 73) K. D. Masterson and J. O. Stoner Jr., *Nucl. Instrum. Methods* **110**:441 (1973).

(Ma 74) I. Martinson, in *Nuclear Spectroscopy and Reactions, Part C* (J. Cerny, ed.) Academic Press, New York (1974), p. 467.

(Ma 74a) I. Martinson and A. Gaupp, *Phys. Rep.* **15C**:113 (1974).

(Ma 76) I. Martinson, in *Beam-Foil Spectroscopy* (ed by S. Bashkin), Springer, Berlin (1976), p. 33.

(Ma 76a) R. Marrus, in *Beam-Foil Spectroscopy* (S. Bashkin, ed.), Springer, Berlin (1976), p. 209.

(Ma 77) I. Martinson, L. J. Curtis, P. L. Smith, and E. Biemont, *Phys. Scr.* **16**:35 (1977).

(Ma 78) R. Marrus and P. J. Mohr, *Adv. At. Mol. Phys.* **14**:181 (1978).

(Ma 80) S. Mannervik, G. Astner, and M. Kisielinski, *J. Phys. B* **13**:L441 (1980).

(Ma 81) S. Mannervik, *Phys. Scr.* **22**:575 (1981).

(Ma 81a) S. Mannervik, I. Martinson, and B. Jelenkovic, *J. Phys. B* **14**:L275 (1981).

(Mi 34) A. C. G. Mitchell and M. W. Zemansky, *Resonance Radiation and Excited Atoms*, Cambridge University Press (1934).

(Mi 74) R. Middleton and C. T. Adams, *Nucl. Instrum. Methods* **118**:329 (1974).

(Mo 75) J. R. Mowat, P. M. Griffin, H. H. Haselton, P. Laubert, D. J. Pegg, R. S. Peterson, I. A. Sellin, and R. S. Thoe, *Phys. Rev. A* **11**:2198 (1975).

(Mo 75a) P. J. Mohr, *Phys. Rev. Lett.* **34**:1050 (1975).

(Mo 76) P. J. Mohr, in *Beam-Foil Spectroscopy* (I. A. Sellin and D. J. Pegg, eds.), Plenum Press, New York (1976), p. 97.

(Mu 71) D. E. Murnick, M. Leventhal, and H. W. Kugel, *Phys. Rev. Lett.* **27**:1625 (1971).

(Mu 72) D. E. Murnick, unpublished work.

(Mu 79) D. E. Murnick, C. K. N. Patel, M. Leventhal, O. R. Wood, and H. W. Kugel, *J. Phys. (Paris)* **40**:C-1–34 (1979).

(No 70) L. C. Northcliffe and R. F. Schilling, *Nucl. Data Tables* **A7**:3–4 (1970).

(On 79) J. A. O'Neill, E. H. Pinnington, K. E. Donnelly, and R. L. Brooks, *Phys. Scr.* **20**:60 (1979).
(Oo 76) H. Oona and W. S. Bickel, *Rev. Sci. Instrum.* **47**:517 (1976).
(Oo 76a) H. Oona and W. S. Bickel, *J. Opt. Soc. Am.* **66**:278 (1976).
(Pa 76) H. P. Palenius, L. J. Curtis, and L. Lundin, *J. Phys. B* **9**:L473 (1976).
(Pe 73) D. J. Pegg, I. A. Sellin, R. Peterson, J. R. Mowat, W. W. Smith, M. D. Brown, and J. R. MacDonald, *Phys. Rev. A* **8**:1350 (1973).
(Pe 75) D. J. Pegg, H. Haselton, R. S. Thoe, P. M. Griffin, M. D. Brown, and I. A. Sellin, *Phys. Rev. A* **12**:1330 (1975).
(Pe 78) D. J. Pegg, P. M. Griffin, B. M. Johnson, K. W. Jonson, and T. H. Kruse, *Astrophys. J.* **224**:1056 (1978).
(Pe 79) D. J. Pegg, J. P. Forester, P. M. Griffin, G. D. Alton, S. B. Elston, B. M. Johnson, M. Suter, R. S. Thoe, and C. R. Vane, *J. Phys. (Paris)* **40**:C1–205 (1979).
(Pe 80) D. J. Pegg, in *Methods of Experimental Physics*, Vol. 17, Academic Press, New York (1980), p.529.
(Pi 70) E. H. Pinnington, *Nucl. Instrum. Methods* **90**:93 (1970).
(Pi 73) E. H. Pinnington, H. O. Lutz, and G. W. Carriveau, *Nucl. Instrum. Methods* **110**:55 (1973).
(Pi 73a) E. H. Pinnington, H. G. Berry, J. Desesquelles, and J. L. Subtil, *Nucl. Instrum. Methods* **110**:315 (1973).
(Pi 74) E. H. Pinnington and H. O. Lutz, *Can. J. Phys.* **52**:1253 (1974).
(Pi 79) E. H. Pinnington, R. N. Gosselin, J. A. O'Neill, J. A. Kernahan, K. E. Donnelly, and R. L. Brooks, *Phys. Scr.* **20**:151 (1979).
(Po 74) O. Poulsen and J. L. Subtil, *J. Phys. B* **7**:31 (1974).
(Po 75) O. Poulsen, P. S. Ramanujam, and D. B. Iversen, *J. Phys. B* **8**:L450 (1975).
(Po 75a) O. Poulsen, T. Andersen, and N. Skouboe, *J. Phys. B* **8**:1393 (1975).
(Pr 76) S. W. Provencher, *J. Chem. Phys.* **64**:2772 (1976).
(Ra 79) W. L. Randolph, B. I. Deutch and G. Weyer, *Hyperfine Interactions* **7**:279 (1979).
(Ri 73) P. Richard, *Phys. Lett.* **45A**:13 (1973).
(Rø 77) M. Rødbro, R. Bruch, P. Bisgaard, P. Dahl, and B. Fastrup, *J. Phys. B* **10**:L483 (1977).
(Rø 79) M. Rødbro, R. Bruch, and P. Bisgaard, *J. Phys. B* **12**:2413 (1979).
(Sa 67) J. A. R. Samson, *Techniques of Vacuum Ultraviolet Spectroscopy*, J, Wiley and Sons, New York (1967).
(Sa 79) U. Sander and H. H. Bukow, *Radiat. Effects* **40**:143 (1979).
(Sc 73) R. W. Schmieder and R. Marrus, *Nucl. Instrum. Methods* **110**:459 (1973).
(Sc 77) D. Schneider, R. Bruch, W. H. E. Schwarz, T. C. Chang, and C. F. Moore, *Phys. Rev. A* **15**:926 (1977).
(Sc 79) R. M. Schectman, L. J. Curtis, and H. G. Berry, *J. Phys. (Paris)* **40**:C1–275 (1979).
(Sc 80) R. M. Schectman, R. D. Hight, S. T. Chen, L. J. Curtis, H. G. Berry, T. J. Gay, and R. DeSerio, *Phys. Rev. A* **22**:1591 (1980).
(Se 68) I. A. Sellin, B. L. Donnally, and C. Y. Fan, *Phys. Rev. Lett.* **21**:717 (1968).
(Se 69) I. A. Sellin, C. D. Moak, P. M. Griffin, and J. A. Biggerstaff, *Phys. Rev.* **184**:56 (1969).
(Se 71) I. A. Sellin, D. J. Pegg, M. Brown, W. W. Smith, and B. Donnally, *Phys. Rev. Lett.* **27**:1108 (1971).
(Se 73) I. A. Sellin, J. R. Mowat, R. S. Peterson, P. M. Griffin, R. Laubert, and H. H. Haselton, *Phys. Rev. Lett.* **31**:1335 (1973).

(Se 76) I. A. Sellin and D. J. Pegg, eds. *Beam-Foil Spectroscopy*, Plenum Press, New York (1976).

(Se 76a) I. A. Sellin, in *Advances in Atomic and Molecular Physics*, Vol. 12, Academic Press, New York (1976), p. 215.

(Se 76b) I. A. Sellin, in *Beam-Foil Spectroscopy* (S. Bashkin, ed.), Springer Verlag, Berlin (1976), p. 265.

(Se 79) I. A. Sellin, L. Liljeby, S. Mannervik, and S. Hultberg, *Phys. Rev. Lett.* **42**:570 (1979).

(Sh 79) V. P. Shevelko and A. V. Vinogradov, *Phys. Scr.* **19**:275 (1979).

(Si 73) O. Sinanoglu, *Nucl. Instrum. Methods* **110**:193 (1973).

(Sø 76) G. Sørensen, in *Beam-Foil Spectroscopy* (I. A. Sellin and D. J. Pegg, eds.) Plenum Press, New York (1976), p. 165.

(St 71) J. O. Stoner, Jr. and J. A. Leavitt, *Appl. Phys. Lett.* **18**:477 (1971).

(St 73) J. O. Stoner, Jr. and J. A. Leavitt, *Opt. Acta* **20**:435 (1973).

(St 76) P. H. Stelson, in *Beam-Foil Spectroscopy* (I. A. Sellin and D. J. Pegg, eds.), Plenum Press, New York (1976), p. 401.

(St 76a) J. O. Stoner, Jr. and I. Martinson, in *Beam-Foil Spectroscopy* (I. A. Sellin and D. J. Pegg, eds.), Plenum Press, New York (1976), p. 259.

(Su 76) J. L. Subtil, P. Ceyzeriat, J. Desesquelles, and M. Druetta, in *Beam-Foil Spectroscopy* (I. A. Sellin and D. J. Pegg, eds.), Plenum Press, New York (1976), p. 791.

(Su 76a) J. L. Subtil, P. Ceyzeriat, A. Denis, and J. Desesquelles *J. Phys. (Paris)* **37**:1299 (1976).

(Ti 73) K. Tillmann, H. J. Andrä, and W. Wittmann, *Phys. Rev. Lett.* **30**:155 (1973).

(To 74) K. X. To, E. J. Knystautas, and R. Drouin, *Can. J. Spectrosc.* **19**:72 (1974).

(To 78) N. H. Tolk, J. C. Tully, J. S. Kraus, W. Heiland, and S. H. Neff, *Phys. Rev. Lett.* **41**:643 (1978).

(To 79) K. X. To, E. Knystautas, R. Drouin, and H. G. Berry, *J. Phys. (Paris)* **40**:C1–1 (1979).

(Tr 53) S. Triebwasser, E. S. Dayhoff, and W. E. Lamb, *Phys. Rev.* **89**:98 (1953).

(Tr 79) E. Träbert, I. A. Armour, S. Bashkin, N. A. Jelley, R. O'Brien, and J. D. Silver, *J. Phys. B* **12**:1665 (1979).

(Tr 79a) E. Träbert and B. C. Fawcett, *J. Phys. B* **12**:L441 (1979).

(Tr 80) E. Träbert and P. H. Heckmann, *Phys. Scr.* **22**:489 (1980).

(Va 76) S. L. Varghese, C. L. Cocke, B. Curnutte, and G. Seaman, *J. Phys. B* **9**:L387 (1976).

(Va 76a) S. L. Varghese, C. L. Cocke, and B. Curnutte, *Phys. Rev. A* **14**:1729 (1976).

(Ve 79) E. Veje, *J. Phys. (Paris)* **40**:C1–253 (1979).

(Vi 76) G. A. Victor, R. F. Stewart, and C. Laughlin, in *Beam-Foil Spectroscopy* (I. A. Sellin and D. J. Pegg, eds.) Plenum Press, New York (1976), p. 43.

(We 69) A. W. Weiss, *Phys. Rev.* **188**:119 (1969).

(We 70) A. W. Weiss, *Nucl. Instrum. Methods* **90**:121 (1970).

(We 73) A. W. Weiss, *Adv. At. Mol. Phys.* **9**:1 (1973).

(Wh 69) W. Whaling, R. B. King, and M. Martinez-Garcia, *Astrophys. J.* **158**:389 (1969).

(Wi 68) W. L. Wiese and A. W. Weiss, *Phys. Rev.* **175**:50 (1968).

(Wi 70) W. L. Wiese, *Nucl. Instrum. Methods* **90**:25 (1970).

(Wi 72) W. Wittmann, K. Tillmann, H. J. Andrä, and P. Dobberstein, *Z. Phys.* **257**:279 (1972).

(Wi 77) H. Winter and M. Gaillard, *Z. Phys.* **A281**:311 (1977).

(Wi 79) W. L. Wiese, in *Progress in Atomic Spectroscopy* (W. Hanle and H. Kleinpoppen, eds.) Plenum Press, New York (1979), p. 1101.

(Wi 79a) W. Wittmann, unpublished data (quoted by An 79).

(Wi 79b) H. Winter and H. J. Andrä, *Z. Phys. A* **291**:5 (1979).

(Wi 80) H. Winter and H. J. Andrä, *Phys. Rev. A* **21**:581 (1980).

(Wi 80a) H. Winter, P. H. Heckmann, and B. Raith, *Hyperfine Interactions* **8**:261 (1980).

(Yn 74) J. L. Yntema, *Nucl. Instrum. Methods* **122**:45 (1974).

(Yo 78) S. M. Younger and W. L. Wiese, *Phys. Rev. A* **17**:1944 (1978).

(Za 67) R. L. de Zafra and W. Kirk, *Am. J. Phys.* **35**:573 (1967).

Note Added in Proof (Chapter 1)

Since the completion of this review, a number of important advances has been made in the field of atomic physics in heavy-ion reactions at energies near the Coulomb barrier (cf. Section 7). An up to date presentation related to this subject can be found in (Me 84).

K-hole production in 7.5 MeV/u U + U collisions has been studied by Meyerhof and collaborators (St 84) as a function of the Q-value. According to the predictions of Figure 7.2b the vacancy rate should strongly decrease as a function of delay time. Since such an effect indeed has been observed experimentally, one may hope to obtain information on the magnitude of delay times and their relation to the energy loss in deep inelastic nuclear collisions. The data show indications that the delay time is underestimated by conventional reaction models.

Similarly, the energy spectra of electrons and positrons in deep inelastic nuclear collisions leading to nuclear fission have been studied (Ba 83). In general agreement with theoretical expectations, cf. Figure 7.6b, the slopes of the observed spectra are clearly steeper than those originating from elastic scattering events. The data are not yet, however, fully consistent with detailed coupled-channel calculations (Mü 84) based on friction models for the deep inelastic reaction.

The first observations of narrow line structures in the spectrum of positrons emitted in very heavy collision systems have been impressively corroborated and extended by two experimental groups working at the GSI accelerator (Darmstadt). Striking line structures have been observed in U + U, U + Th and U + U collisions by the EPOS group (Sch 83). The first two systems also were studied by the Munich group (C1 84) with similar results. Convincing evidence has been compiled by both groups which seems to rule out a trivial explanation of the effect in terms of conversion of γ-lines. The expected accompanying line structures in the electron or γ-ray spectrum are not observed. Furthermore the width of the positron line and its angular distribution strongly hint that the source of the positron emission moves with center-of-mass system. The most convincing interpretation of these facts assumes that in a small fraction of all collisions a combined nuclear system ("giant nucleus") is formed for up to $\approx 10^{-19}$ s (Re 81b, Mü 83).

Inspired by the experimental evidence attempts have been undertaken to go beyond the phenomenological analysis of the positron line in terms of a classical time delay picture. The fully quantal theory and its semiclassical limit have been discussed in (He 83a). To obtain a workable description one may start from quantum mechanical scattering theory and introduce two regions in the radial coordinate R. For large distances the nuclei do not interact and the semiclassical treatment is appropriate. In the interior region atomic excitations may be neglected while the nuclear scattering is described by the nuclear S-matrix. This leads to a factorization of the excitation amplitudes (To 84, He 84b). Averaging over the spread of the beam energy, it turns out that the electronic excitation is determined by the nuclear autocorrelation function (To 83).

Using an expression for the autocorrelation function motivated by the statistical model for nuclear reactions, Tomoda and Weidenmüller could qualitatively reproduce the line structure in the positron spectrum. This is not surprising, since it can be shown (Re 83) that such a description is completely equivalent to the semiclassical picture of a nuclear time delay employed earlier (Re 81b). The lifetime distribution function $\rho(T)$, cf. equation (7.9), is given by the Fourier transform of the nuclear autocorrelation function.

In an attempt to understand the mechanism behind reactions with long delay times, the heavy-ion internuclear potential has been examined. Seiwert *et al.* (Rh 83) predict that in system like U + Cm the potential under favourable conditions (head-on-head orientation of the deformed nuclei to minimize Coulomb repulsion) should exhibit a pocket. Based on this result Heinz *et al.* (He 83b), He 84a,b) constructed a simple reaction model assuming the presence of several rotational bands of resonance states in a potential pocket. The autocorrelation function and its Fourier transform, the distribution of reaction times, could be calculated analytically. The positron emission spectra resulting from delayed collisions exhibit a narrow positron line (He 83b). The effect has a narrow excitation function (He 84a) centered at the top of the barrier, since only here the required long-lived resonance levels can be excited with sufficient probability. To get a quantitative description of the process, the incorporation of inelastic mechanisms will be required. These may also lead to additional structure in the positron spectrum due to the conversion of nuclear excitations during the lifetime of the combined nuclear system (Re 81b, Mü 83).

At the time of writing no completely satisfactory understanding of the "positron line" has been achieved, in particular the weak dependence of its position on the charge of the colliding system is not understood. Nevertheless, there is much evidence that indeed the phenomenon is connected with the supercritical vacuum of QED discussed in Section 2. At the same time this means that remarkably long lived "giant" nuclear systems with $Z > 180$ have been observed. The study of these objects probably will reveal surprising facts about nuclear properties far off the domain of stable elements.

References to the Note Added in Proof

(Ba 83) H. Backe, P. Senger, W. Bonin, E. Kankeleit, M. Krämer, R. Krieg, V. Metag, N. Trautmann, J. B. Wilhelmy, *Phys. Rev. Lett.* **50**:1838 (1983).

(Cl 84) M. Clemente, E. Berdermann, P. Kienle, H. Tsertos, W. Wagner, C. Kozhuharov, F. Bosch, and W. Koenig, *Phys. Lett.* **137B**:41 (1984).

(He 83a) U. Heinz, B. Müller, and W. Greiner, *Ann. Phys.* (*N.Y.*) **151**:227 (1983).

(He 83b) U. Heinz, J. Reinhardt, B. Müller, W. Greiner, and U. Müller, *Z. Phys.* **A314**:125 (1983).

(He 84a) U. Heinz, J. Reinhardt, B. Müller, W. Greiner, and W. T. Pinkston, *Z. Phys.* **A316**:341 (1984).

(He 84b) U. Heinz, U. Müller, J. Reinhardt, B. Müller, and W. Greiner, *Ann. Phys.* (*N.Y.*) **158**:476 (1984).

(Me 84) W. E. Meyerhof and J. F. Chemin, in *Advances in Atomic and Molecular Physics, Vol. 20* (D. R. Bates and B. Bedersen, eds.), Academic Press, New York (1984).

(Mü 83) U. Müller, G. Soff, T. deReus, J. Reinhardt, B. Müller, and W. Greiner, *Z. Phys.* **A313**:263 (1983).

(Mü 84) U. Müller, G. Soff, J. Reinhardt, T. deReus, B. Müller, and W. Greiner, *Phys. Rev. C* **30**:1199 (1984).

(Re 81b) J. Reinhardt, U. Müller, B. Müller, and W. Greiner, *Z. Phys.* **A303**:173 (1981).

(Re 83) J. Reinhardt, B. Müller, W. Greiner, and U. Müller, *Phys. Rev. A* **28**:2558 (1983).

(Rh 83) M. J. Rhoades-Brown, V. E. Oberacker, M. Seiwert, and W. Greiner, *Z. Phys.* **A310**:287 (1983).

(Sch 83) J. Schweppe, A. Gruppe, K. Bethge, H. Bokemeyer, T. Cowan, H. Folger, J. S. Greenberg, H. Grein, S. Ito, R. Schule, D. Schwalm, K. E. Stiebing,N. Trautmann, P. Vincent, and M. Waldschmidt, *Phys. Rev. Lett.* **51**:2261 (1983).

(St 84) Ch. Stoller, M. Nessi, E. Morenzoni, W. Wölfli, W. E. Meyerhof, J. D. Molitoris, E. Grosse, and Ch. Michel, to be published.

(To 83) T. Tomoda and H. A. Weidenmüller, *Phys. Rev. C* **28**:739 (1983).

(To 84) T. Tomoda, *Phys. Rev. A* **29**:536 (1984).

Index